# Emerging Engineering Technologies and Industrial Applications

Younes El Kacimi
*Ibn Tofail University, Morocco*

Khaoula Alaoui
*Ibn Tofail University, Morocco*

IGI Global
PUBLISHER of TIMELY KNOWLEDGE

A volume in the Advances in Civil and Industrial Engineering (ACIE) Book Series

Published in the United States of America by
IGI Global
Engineering Science Reference (an imprint of IGI Global)
701 E. Chocolate Avenue
Hershey PA, USA 17033
Tel: 717-533-8845
Fax: 717-533-8661
E-mail: cust@igi-global.com
Web site: http://www.igi-global.com



Library of Congress Cataloging-in-Publication Data

Names: El Kacimi, Younes, 1980- editor. | Alaoui, Khaoula, 1992- editor.
Title: Emerging engineering technologies and industrial applications / edited by: Younes El Kacimi, Khaoula Alaoui.
Description: Hershey PA : Engineering Science Reference, [2024] | Includes bibliographical references. | Summary: "Emerging technologies have a rapidly transformative effect on the engineering sector and continue to play a crucial role in determining its trajectory as we move towards a future centered around the evolution of these disruptive forces. Adopting these technologies is critical for firms who want to stay ahead of the current technology trends and guarantee service deliverables competing with the industry's best offers"-- Provided by publisher.
Identifiers: LCCN 2023049103 (print) | LCCN 2023049104 (ebook) | ISBN 9798369313350 (hardcover) | ISBN 9798369313367 (ebook)
Subjects: LCSH: Engineering. | Technological innovations.
Classification: LCC TA145 .E45 2024 (print) | LCC TA145 (ebook) | DDC 658.5/14--dc23/eng/20240425
LC record available at https://lccn.loc.gov/2023049103
LC ebook record available at https://lccn.loc.gov/2023049104

This book is published in the IGI Global book series Advances in Civil and Industrial Engineering (ACIE) (ISSN: 2326-6139; eISSN: 2326-6155)

British Cataloguing in Publication Data
A Cataloguing in Publication record for this book is available from the British Library.

All work contributed to this book is new, previously-unpublished material. The views expressed in this book are those of the authors, but not necessarily of the publisher.

For electronic access to this publication, please contact: eresources@igi-global.com.

# Advances in Civil and Industrial Engineering (ACIE) Book Series

Ioan Constantin Dima
University Valahia of Târgovişte, Romania

ISSN:2326-6139
EISSN:2326-6155

## Mission

Private and public sector infrastructures begin to age, or require change in the face of developing technologies, the fields of civil and industrial engineering have become increasingly important as a method to mitigate and manage these changes. As governments and the public at large begin to grapple with climate change and growing populations, civil engineering has become more interdisciplinary and the need for publications that discuss the rapid changes and advancements in the field have become more in-demand. Additionally, private corporations and companies are facing similar changes and challenges, with the pressure for new and innovative methods being placed on those involved in industrial engineering.

The **Advances in Civil and Industrial Engineering (ACIE) Book Series** aims to present research and methodology that will provide solutions and discussions to meet such needs. The latest methodologies, applications, tools, and analysis will be published through the books included in **ACIE** in order to keep the available research in civil and industrial engineering as current and timely as possible.

## Coverage

- Hydraulic Engineering
- Engineering Economics
- Productivity
- Construction Engineering
- Materials Management
- Coastal Engineering
- Structural Engineering
- Operations Research
- Ergonomics
- Transportation Engineering

IGI Global is currently accepting manuscripts for publication within this series. To submit a proposal for a volume in this series, please contact our Acquisition Editors at Acquisitions@igi-global.com or visit: http://www.igi-global.com/publish/.

The Advances in Civil and Industrial Engineering (ACIE) Book Series (ISSN 2326-6139) is published by IGI Global, 701 E. Chocolate Avenue, Hershey, PA 17033-1240, USA, www.igi-global.com. This series is composed of titles available for purchase individually; each title is edited to be contextually exclusive from any other title within the series. For pricing and ordering information please visit http://www.igi-global.com/book-series/advances-civil-industrial-engineering/73673. Postmaster: Send all address changes to above address. 

## Titles in this Series

*For a list of additional titles in this series, please visit:* http://www.igi-global.com/book-series/advances-civil-industrial-engineering/73673.

*Emerging Materials, Technologies, and Solutions for Energy Harvesting*
Shilpa Mehta (Auckland University of Technology, New Zealand) Arij Naser Abougreen (University of Tripoli, Libya) and Shashi Kant Gupta (Eudoxia Research University, USA)
Engineering Science Reference • copyright 2024 • 368pp • H/C (ISBN: 9798369320037) • US $315.00 (our price)

*Innovations in Machine Learning and IoT for Water Management*
Abhishek Kumar (Chandigarh University, India) Arun Lal Srivastav (Chitkara University, India) Ashutosh Kumar Dubey (University of Castilla-La Mancha, Spain) Vishal Dutt (AVN Innovations Pvt. Ltd., India) and Narayan Vyas (AVN Innovations Pvt. Ltd., India)
Engineering Science Reference • copyright 2024 • 312pp • H/C (ISBN: 9798369311943) • US $300.00 (our price)

*Intelligent Engineering Applications and Applied Sciences for Sustainability*
Brojo Kishore Mishra (NIST Institute of Science and Technology (Autonomous), India & Saveetha College of Liberal Arts and Sciences, SIMATS, Chennai, India)
Engineering Science Reference • copyright 2023 • 542pp • H/C (ISBN: 9798369300442) • US $270.00 (our price)

*Global Science's Cooperation Opportunities, Challenges, and Good Practices*
Mohamed Moussaoui (National School of Applied Sciences of Tangier (ENSAT), Abdelmalek Essaadi University, Morocco)
Engineering Science Reference • copyright 2023 • 357pp • H/C (ISBN: 9781668478745) • US $245.00 (our price)

*Handbook of Research on Inclusive and Innovative Architecture and the Built Environment*
Ng Foong Peng (Taylor's University, Malaysia) and Ungku Norani Sonet (Taylor's University, Malaysia)
Engineering Science Reference • copyright 2023 • 526pp • H/C (ISBN: 9781668482537) • US $315.00 (our price)

*Artificial Intelligence and Machine Learning Techniques for Civil Engineering*
Vagelis Plevris (Qatar University, Qatar) Afaq Ahmad (University of Engineering and Technology, Taxila, Pakistan) and Nikos D. Lagaros (National Technical University of Athens, Greece)
Engineering Science Reference • copyright 2023 • 385pp • H/C (ISBN: 9781668456439) • US $250.00 (our price)

*Geoinformatics in Support of Urban Politics and the Development of Civil Engineering*
Sérgio António Neves Lousada (University of Madeira, Portugal)

IGI Global
PUBLISHER of TIMELY KNOWLEDGE

701 East Chocolate Avenue, Hershey, PA 17033, USA
Tel: 717-533-8845 x100 • Fax: 717-533-8661
E-Mail: cust@igi-global.com • www.igi-global.com

# Table of Contents

## Section 2

## Nanoscience, Nanotechnology, and Nanoengineering



## Section 3

## Medical and Biomedical Engineering Technologies

# Detailed Table of Contents

focusing on toxicity. Challenges in chemical health and safety research, such as primitive algorithms and skill requirements, limited its early use. Recent advancements in AI and computer science have highlighted the significance of ML and DL. Improved data, computing power, and algorithms have led to autonomous data analyses, transforming solutions in structural engineering (MPSE). In integrated computational materials engineering (ICME), ML and AI are assimilated into efforts like the materials genome initiative (MGI), accelerating autonomous experiments and improving data analytics.



Artificial intelligence and machine learning applications in image processing are examined in this chapter. It covers AI methods including supervised, unsupervised, reinforcement, and deep learning. Genetic algorithms, rule-based systems, expert systems, and fuzzy logic are AI methods. SVM, decision trees, random forests, K-means clustering, and PCA are machine learning methods. CNN, RNN, and GANs are utilised for object recognition, classification, and segmentation. The chapter discusses how artificial intelligence and machine learning affect accuracy, efficiency, and decision-making. The need to choose proper measurements and procedures for assessment and performance analysis is also stressed. Ethics like justice, privacy, transparency, and human-AI cooperation are covered in the chapter.



Fullerene molecules are composed entirely of carbon, in the form of a hollow sphere, ellipsoid, or tube. Fullerenes in the cylindrical form are called carbon nanotubes (CNTs) or buckytubes and fullerenes in the spherical form are referred to as buckyballs. Fullerenes have attracted considerable attention in different fields of science since. Their unique carbon cage structure coupled with immense scope for derivatization makes them a potential therapeutic agent. The fullerenes can be utilized in organic photovoltaic (OPV), portable power, medical purpose, antioxidants, and biopharmaceuticals and dentistry. Recently, fullerene-C60 contributed greatly to the field of biosensing and bio-nanotechnology. Fullerene allows its wide use for designing the highly sensitive chemical/biosensors. Fullerene is not dangerous to biological systems. Fullerene, as the first symmetric nanostructure in the carbon nanomaterials family, opened up new perspectives in nanomaterials field leading to discovery and research on other symmetric carbon nanomaterials.

Successful deployment of carbon nanocomposites in many applications, such as sensing, energy storage, and catalysis, relies on the selection, synthesis, and tailoring of the surface properties. Carbon nanotubes, due to their large surface areas, unique surface properties, and needle-like shape, can deliver a lot of therapeutic agents, including DNA, siRNAs and proteins to the target disease sites. Carbon nanotubes can be readily excreted through the renal route by means of degradation through myeloperoxidase enzyme. Nanotubes are categorized as single-walled carbon nanotubes and multiple walled carbon nanotubes. Theirs advances have been made in the delivery of anticancer, anti-inflammatory drugs, bioactive molecules, and proteins. Drugs and biomolecules can be loaded in carbon nanotubes, which can then be utilized as targeted molecules. Recently, carbon nanocomposites have attracted a lot of attention in the field of cancer diagnosis and therapy, with their ability to deliver therapeutic molecules and allow visualization of cells and tissues, which are necessary for the cure and treatment of diseased and damaged tissues.



To bridge the gap between patients' desire for active involvement in healthcare decisions and the availability of trustworthy resources, the chapter introduces SIKDA. This groundbreaking model aims to empower patients and caregivers by facilitating informed patient decision-making. SIKDA represents a transformative approach to consumer health information, providing patients with a single point of access to credible and up-to-date health information online and the knowledge and guidance of a librarian. By leveraging existing NHS online resources and collaborating with public and medical librarians, SIKDA ensures accessibility and inclusivity for diverse demographics, minimizing costs to public services. The study establishes the foundation for a novel paradigm, enhancing patient engagement, healthcare outcomes, information literacy, and collaborative efforts in the digital healthcare era.

Image processing has become more crucial in medical applications due to its ability to collect and evaluate data from medical images. This book chapter provides an overview of various image processing techniques used for medical applications, including deep learning algorithms, segmentation techniques, and a combination of both. Additionally, the authors discuss other studies that analysed X-rays and used image processing to identify cancer, brain tumours, and other disorders. The results of the study demonstrate how image processing techniques have the potential to significantly improve sickness detection speed and accuracy, facilitating early diagnosis and treatment. The planning of therapies and the accuracy of diagnoses can both be enhanced by the use of image processing tools. Healthcare workers' ability to recognise and manage a variety of medical conditions will undoubtedly increase.



Telemedicine is the use of immersive multimedia communication to transmit patient knowledge in order to conduct consultations, medical tests and operations, and medical professional partnerships from a distance. The fact that telemedicine is an "accessible and rapidly changing science" is often emphasized. The constant advancement of technology, which paves the way for the expansion of internet connections and the expansion of data processing capability, has increased the possibilities for the global health industry, especially telemedicine, to expand. Information exchange, data mining, the internet of things, wearables, digital computing, and robotics are all emerging as key drivers of creativity in the coming decade.



Radio frequency identification technology (RFID) and time series forecasts are used to create a dependable IoT-based traffic system. The system regulates city traffic. The suggested method estimates junction traffic volume over time using LSTM neural networks. RFID technology improves data collection accuracy and reliability. Data preparation includes outlier identification to remove anomalies. Training the LSTM model on preprocessed data reveals traffic volume trends. The trained model predicts traffic volume using historical data. Prediction performance is quantified by MAE, MAPE, and R2. The proposed approach is tested using four intersection traffic data. Results indicate that LSTM-based traffic volume estimation works. The optimal design is determined by evaluating system performance for 12-to-168-time steps. The experimental findings suggest that the proposed method can accurately anticipate traffic volume, helping traffic managers enhance flow. RFID and time series projections bolster traffic system reliability.

Block chain secures data. Blockchain imitates computerised trading markets. The 2008 cryptocurrency and transaction network Bitcoin leads blockchain technology. When buyers or sellers declare payments, blockchain transfers bitcoin. Blockchain can change finance like the Internet. Block chain technology may affect medicine. This fast-growing issue affects many healthcare providers. Open, irreversible, stable, and collaborative technology are preferred by cryptocurrency over regulated, hidden, proprietary, and changeable ones. Blockchain is used in medical records, security, database management, and biotechnology regulation. Blockchain helps biopharmaceuticals track illicit medications. It monitors medicines and counterfeiters. This book chapter presents a comprehensive block chain technology for health sector decision makers and explores difficulties and constraints using a suitable markup. Here are current and potential medical blockchain applications.



The book chapter reports on hybrid manufacturing (wire-arc additive manufacturing and metal forming) of 3D-printed bimetallic structure parts. After 3D printing, the metal flow behaviour of the printed material was studied using a Gleeble 3500 thermal-mechanical simulator. The test conditions were temperature range 850-1000°C and strain rate of 1s−1 and 10s−1 to strain of 0.6. The results show that the initial microstructure of the bimetallic structure had excellent bonding properties. The flow stress-strain curves increased with a decrease in deformation temperature and vice versa at a given strain rate. The study developed a constitutive equation that predicts the metal flow behaviour of the material studied. From the analysis, the material constants were a stress exponent of 4.76 and an activation energy of 213.7 kJmol−1. The study indicates that this method (hybrid manufacturing) is possible in the industrial production process.

# Foreword

Emerging technologies are changing how we attempt the built environment. The consumer gaming industry has made powerful and inexpensive 3D visualization tools available to designers, architects, and engineers. Fast data networks originally built for mobile phones are now connecting inexpensive sensors to power digital twins, offering a new level of asset management. Augmented reality and technology are improving the way our engineers plan buildings, making construction safer. Artificial intelligence and machine learning are powerful new analytical technologies that enhance our ability to design and manage the built environment in ways that meet human needs and deliver net zero goals. In fact, new technologies are appearing and evolving at incredible speed.

So, the book entitled *Emerging Engineering Technologies and Industrial Applications* underscores the urgency to adopt the new technologies, emphasizing that industries or companies which fail to integrate these technologies risk becoming relics of the past. Topics span a broad spectrum, including the pivotal overview of emerging technologies in engineering, Artificial Intelligence transformative applications, the agility and precision of Robotics, the interconnected world of IoT, the infinitesimal marvels of Nanotechnology, and the future trajectories of these technological juggernauts.

This book is written with a new approach and new philosophy, and it is hoped that it will fulfill their aspirations. This book will aim to provide relevant theoretical frameworks and the latest empirical research findings in the emerging engineering technologies. It will be written for provides a medium for the communication of ideas, best practices, developments in all aspects of this field and includes both fundamental and practical aspects of novel engineering technologies.

This book is one of indispensable resource for professionals and researchers working in the field of emerging engineering technologies and their industrial applications.

So, the Book «Emerging Engineering Technologies and Industrial Applications» contains four sections, which contains 18 chapters. In fact, the authors presented the recent advances, and future challenges for emerging engineering technologies and their industrial application, which provides a very good reference with the valuable information.

The first section of this book focuses on artificial intelligence, deep learning and machine learning and its applications in engineering, computational chemistry and computational visual analysis, and three specific chapters are presented.

Furthermore, the second section undertaken to completely assess the opportunities for research and development to enhance nanocomposites and nanobiotechnological industrial applications.

In section three, some Medical and Biomedical Engineering Technologies has carefully been highlighted.

Finaly, section four elucidates some engineering tools and technologies for futures challenges.

This book equips industrial and research engineers, alongside research scientists, with the tools needed to stay relevant and at the forefront of their industries. It offers valuable insights into harnessing these technologies to improve product design development processes and contribute significantly to their respective fields' growth.

This book is a comprehensive resource that presents new and up-to-date extensive description of the principles of Emerging Engineering Technologies and theirs Industrials Applications. The book also elucidates the recent advances, and future challenges for students, industrial and developers. I think that this book is a rich resource experiments and industrial applications of emerging engineering technologies solutions that are cost-effective and will be useful to academic, practicing, and industrial.

*Rachid Touir*
*Advanced Materials and Process Engineering Laboratory, Faculty of Science, Ibn Tofail University, Morocco & Regional Center of Education and Training Professions (CRMEF), Morocco*

# Preface

Since the environment has started paying back what humanity has done with the environment. People from all segments of life realized that living unsustainability is not an option for humans. During the last decade, the climate change crisis presents a multi-dimensional challenge to the development of the built environment. With finite global resources and increasingly unpredictable climate patterns, the need to improve our understanding of sustainable practices and materials for construction has never been more pressing. *The Handbook of Research on Corrosion Sciences and Engineering* covers corrosion engineering through an extensive description of the principles of corrosion, corrosion prevention strategies and design of corrosion protection systems. The Handbook of Research elucidate also the recent advances, and future challenges for students and practicing corrosion engineers.

## THE CHALLENGES

*The Handbook of Research on Corrosion Sciences and Engineering* is written with a new approach and new philosophy, and it is hoped that it will fulfill their aspirations. This book will aim to provide relevant theoretical frameworks and the latest empirical research findings in the corrosion engineering area. It will be written for provides a medium for the communication of ideas, best practices, developments in all aspects of this field and includes both fundamental and industrial aspects of corrosion sciences engineering. Occurrence of corrosion and its practical control is an area of study covering a wide range of scientific knowledge.

This Handbook of Research devoted to shed light on the recent developments in the usage of sustainable materials to protect metallic materials against corrosion and provides emerging research exploring the theoretical and practical aspects of corrosion engineering science and technology. Their objective is to provide scientists, engineers, technologists, industrial and students with useful and important information and data on corrosion investigation and control.

The target audience of *The Handbook of Research on Corrosion Sciences and Engineering* will be composed of scientists, engineers, technologists and industrials working in the field of corrosion and corrosion protection of materials in various disciplines.

## ORGANIZATION OF THE HANDBOOK OF RESEARCH

This Handbook of Research is organized into 21 chapters, and in order to cover all aspects of corrosion, we chose to devise this book into 5 sections. Each of these sections contains several chapters, which detail feedback from expert and specialist authors in their fields. The sections are:

Section 1: Advanced Engineering Tools and Technology for Corrosion Science
Section 2: Nanomaterials & Smart Coatings for Corrosion Protection
Section 3: Corrosion of Biomaterials and Biomimetic materials
Section 4: Green & Sustainable Corrosion Inhibition
Section 5: Applied Corrosion Science

## DESCRIPTION OF SECTIONS AND CHAPTERS

### Section One: Advanced Engineering Tools and Technology for Corrosion Science

Consideration of the blooming development of information technology can still provide surprising theoretical finding out of experimental observation and can often be described comprehensively and predictively. Moreover, we can also use Artificial Intelligence techniques to analyze and mine the data generating from the above research branches, to develop new materials with better corrosion resistance. In studying corrosion mechanisms researchers have a selection of methods at their requirement. Section 1 focuses on the modeling revolution in scientific research and developing tendency in corrosion study, and 4 specific chapters are presented.

The artificial intelligence and machine learning technology presented in chapter 1 is pointed out how the computer can pave the way of predicting corrosion degrees as well as designing their role in corrosion related research. While traditional corrosion scientists design new corrosion inhibitors through structural modification and testing hundreds of compounds in the laboratory, with long-term procedures and high costs by modifying the structures of new corrosion inhibitors or existing inhibitors. Chapter 2 introduces a computational protocol to differentiate the performance of organic corrosion inhibitors and design new anticorrosive inhibitors. Another aim of this chapter is to correct common mistakes in conducting computational calculations. DFT-examination of the reactivity of the isolated molecules before the adsorption process. Physical adsorption analysis of molecules on the metallic surface using Monte Carlo or Molecular Dynamics simulations. Chemical adsorption studies of molecules on the metallic surface using density functional-based tight binding "DFTB" or periodic density functional theory "DFT" methods. Finally, two a case study is presented and discussed in Chapters 3 and 4. The recommendations of this section offer a clear identification of research and development issues and create a roadmap for achieving them.

### Section Two: Nanomaterials and Smart Coatings for Corrosion Protection

Section 2 undertaken to completely assess the opportunities for research and development to enhance coating practices, coating materials and applications. This section consists of discussing the current status, challenges and prospects of nanomaterials for corrosion protection. Effectively, Carbon nanotubes have found great interest for application in industries, owing to their high mechanical strength, elasticity, and electrical conductivity. It has been shown that their application in the coatings can increase corrosion mitigation performance through an increase in coating integrity or hydrophobicity for barrier coatings and an increase in electrical conductivity for sacrificial coatings.

However, they usually need modification to have better compatibility with the coating matrix or to increase inhibition characteristics. To this end, several treatments were developed for the modification of carbon nanotubes.

A wide range of treatments based on oxidation, salinization, polymer grafting, surface decoration with inorganic materials, and corrosion inhibitor loading introduced for modification of carbon nanotubes were reviewed in Chapter 5. From the conjugation of Carbon nanotubes with inorganic hybrid it is expected to obtain nanocomposite coatings that combine high anti-corrosion efficiency with improved mechanical stability. Chapter 6 presents a concise review of microstructure and corrosion behaviour of different nanotube composite coatings. Science behind the corrosion and corrosion resistance of nanotube composite coatings, followed by a selection of current state and recent advances on promoted nanotube composite coating and based matrix composites coatings has been presented. Finally, discussion of the relevant topics, highlighting recent progress and unresolved questions was provided.

Chapter 7 focused on another aspect of coating is thus introduces the basic model of wettability, and then summarizes the advanced strategies for constructing superhydrophobic concrete, which is divided into preparing protective coatings on concrete surfaces and improving the overall hydrophobicity of concrete materials.

In addition, the influence of superhydrophobic modification on anticorrosion was summarized, and the theoretical basis of superhydrophobic technology applied to concrete anticorrosion was briefly discussed. Finally, the research and application of superhydrophobic concrete are summarized, the challenges and future application prospects of superhydrophobic concrete are pointed out.

Furthermore, Chapter 8 completes this section by studying the reinforcement of the steel protection against corrosion, using surface treatments by application of conversion organic coating. This chapter described and discussed the importance of optimization of adhesion of the polymer to the metal oxide. This chapter present and explains also that, the dispersive interactions and acid-base type interactions between the surface sites of the two materials in contact play an important role in adhesion phenomena. In order to understand and predict adhesion, it is necessary to quantify these interactions. Dispersive and Lewis acid-base interactions between the oxides present on the surface of galvanized sheets and model organic molecules by using the inverse gas chromatography at infinite dilution and the new models is presented. The obtained results proved that the dispersive surface energy of the Monogal surface is greater than that of zinc oxide surface but much lower than the zinc hydroxide surface. It was also showed that the Monogal exhibited the lower Lewis acid-base properties.

## Section Three: Corrosion of Biomaterials and Biomimetic Materials

Metals are often used in a wide range of biomedical applications since they have good mechanical characteristics, like higher strength, ductility and toughness. However, the primary disadvantage of metallic biomaterials is their rapid reactivity, which causes corrosion when exposed to physiological conditions like body fluids. When exposed to body fluids, the metallic biomaterial is subjected to wear and corrosion, hence, the mechanical properties are reduced. Corrosion resistance, which also has a significant impact on biocompatibility, affects the efficacy and longevity of an implant material. In the section 3, the body environment has been carefully examined and the potential impacts of corrosion on the biocompatibility of various biomaterials has been highlighted. The fundamentals of implant failure, recovery, and failure mechanisms will be mentioned, and the most common in-vitro and in-vivo corrosion processes has been discussed. Finally, the different methods for preventing biomaterial corrosion have been emphasized.

## Section Four: Green and Sustainable Corrosion Inhibition

Application of corrosion inhibitors is one of the most feasible ways for metal protection against corrosion reactions. The corrosion inhibitors can get adsorbed on metal surfaces causing inhibition of corrosion mainly through the blockage of many active sites. The tremendous drawbacks associated with the application of organic corrosion inhibitors, especially the significant toxicity to the environment, have significantly enhanced further examination of biodegradable and safe alternatives, such as the usage of green and sustainable inhibitors.

In view of these several forms of green corrosion inhibitors such as drugs or medicines, plant extracts, natural polysaccharides and synthetic inhibitors derived from multicomponent reactions are being employed.

Metabolites and phytochemicals that can be extracted from diverse plant species can be used for this specific regard. The naturally extracted phytochemicals have several advantages, namely being environmentally friendly, ecologically acceptable, low cost, abundantly available, easily renewable, and so on. Chapter 10 present an overview of the properties, applicability, and recent advancements of some natural products used as potential inhibitions in various corrosive systems.

Secondary, chapter 11 present the natural polysaccharides particularly chitin and chitosan extracted from Parapenaeus Longirostris shrimp shell waste gained substantial advancement. Chitin and Chitosan extract have been employed effective as corrosion inhibitors for Stainless steel in cleaning chemicals condition.

Chapter 12, present the chitosan oligosaccharide was modified with an 8-hydroxyquinoline derivative to develop a novel bio-based, green and sustainable mild steel corrosion inhibitor. The experimental results have been in good agreement with the theoretical study, which has shown that the developed material exhibits the excellent anti-corrosive activity for mild steel in HCl.

The section 4 discusses also some synthesized green corrosion inhibitors such as Oleic Sources-Based Green Inhibitors from Agro-Industrial Waste and cysteine. Among corrosion inhibitors, pharmaceutical drugs are most effective, water-soluble, eco-friendly, low-cost, good adsorbent, polar structural molecule, biodegradable or persistent, lipophilic, hydrophilic, non-volatile, recyclable and effective at low concentrations and high temperatures. In the chapter 15, the main inhibition characteristics of pharmaceutical drugs for various metal types in the acidic, saline and other corrosion medium were reviewed and discussed. Herein would be helpful as a reference for both industrial and academic researchers.

## Section Five: Applied Corrosion Science

Application of corrosion inhibitors is one of the most feasible ways for metal protection against corrosion reactions. Section 5 is dedicated to discussing fundamentals principles, recent advances, and future challenges in corrosion inhibition. Chapter 16 summarized the molecular structure and classification of organic sulfonate in the market. To disclosure corrosion inhibition properties of organic sulfonate, a series of test methodologies were established by mimicking different application environment from general and indoor storage to harsh exterior environments, including salt spray test, rust prevention test in the presence of water solution, seven day accelerated lamination test, water displacement test and single piece rust prevention test. The correlation between the molecular structure, like base number, cation type, micelle size, cation sulfur ration and the corrosion inhibition performance of organic sulfonate under various mimicked test conditions was systematically revealed, with the comprehensive analysis of their action mechanism based on the produce of hydrophobic adsorption film excluding water and air from reaching the metal surface. The further aspects with respect to the development, synergy effect and application of organic sulfonate was discussed as well.

Chapter 17, 18, 19 and 20 is focuses on the inhibition of corrosion of metals and alloys by organic compounds based on Imidazole derivative, Bolaamphiphile Surfactants, Benzimidazole derivatives and Benzoxazepine Derivatives in several medium such as Hydrochloride, Phosphoric acid solution and Mixed pickling baths. Corrosion inhibition abilities was studied using electrochemical techniques, scanning electron microscopy, Energy dispersive X-ray analysis and DFT method.

Chapter 21, Welding the austenitic stainless steel together with the carbon steel is a common practice in spot welding research but there is a lack of information about the effect of corrosion pertaining to the dynamic resistance and therefore some surface-corroded carbon steel plates are intentionally welded together with 304L austenitic stainless-steel plates. The welded samples underwent the tensile shear test, hardness test, metallurgical observation and electrical signal interpretation. Bonding strengths and hardness distributions are both significantly affected by corrosion presence, which reduces the weld strength in overall measure. Microscopic examinations show that the corrosion has randomly scattered around the vicinity of weld nuggets while the electrical signal distinguishes the corroded weld over good welds in terms of electrical resistances.

*Khaoula Alaoui*
*Ibn Tofail University, Morocco*

*Younes El Kacimi*
*Ibn Tofail University, Morocco*

# Acknowledgment

The editors would like to acknowledge the help of all the people involved in this project and, more specifically, to the authors and reviewers that took part in the review process. Without their support, this book would not have become a reality.

First, the editors would like to thank each one of the authors for their contributions. Our sincere gratitude goes to the chapter's authors who contributed their time and expertise to this book.

Second, the editor's wishes to acknowledge the valuable contributions of the reviewers regarding the improvement of quality, coherence, and content presentation of chapters. Several authors also served as referees; we highly appreciate their double task.

Finally, we would like to thank all the IGI publishing team for professional assistance and support.

Editorial Advisory Board

*Khaoula Alaoui*
*Ibn Tofail University, Morocco*

*Younes El Kacimi*
*Ibn Tofail University, Morocco*

Section 1

# Deep Learning, Machine Learning, and Artificial Intelligence

# Chapter 1
# Artificial Intelligence and Its Application in Engineering: A Comprehensive Review

**Tanima Sahoo**
*Brainwre University, India*

**Arijit Mondal**
*Brainwre University, India*

**Piyal Roy**
*Brainware University, India*

**Amitava Podder**
*Brainwre University, India*

## ABSTRACT

*Artificial Intelligence has transformative impact on precision medicine, structural health monitoring, and autonomous manufacturing systems, revolutionizing traditional approaches. This book chapter explores the challenges and ethical considerations inherent in AI implementation, addressing issues like data privacy, security, and human-AI collaboration. By scrutinizing real-world implementations and futuristic possibilities, this study paints a vivid picture of AI's profound influence on society. As Artificial Intelligence continues to evolve, its ethical integration into various sectors and the establishment of mutual trust between humans and machines emerge as paramount concerns. Amidst the rapid advancements, this research underscores the imperative of responsible AI development, ensuring a harmonious coexistence between artificial intelligence and humanity.*

DOI: 10.4018/979-8-3693-1335-0.ch001

*Figure 1. Application of AI in phenomics [21]*

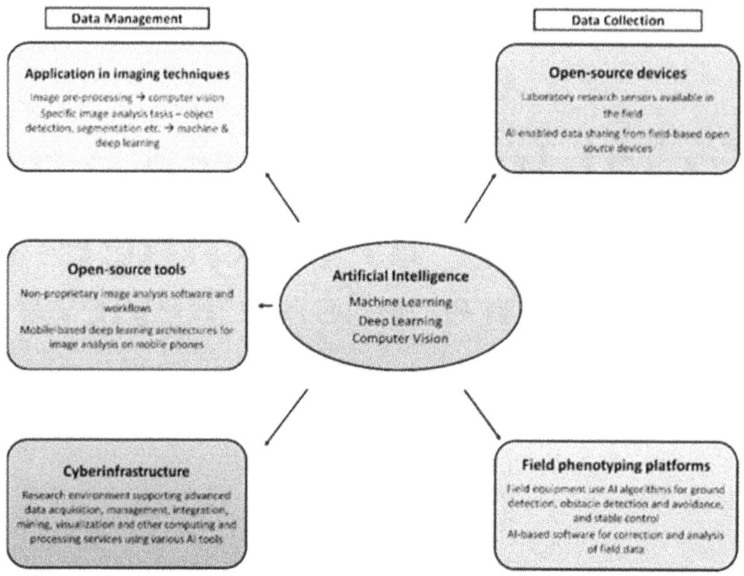


## 1. INTRODUCTION

Artificial intelligence methodologies have impacted not only individual human survival but also every project lifecycle. A decent quantity of consideration has been given to sectors such as engineering, product engineering, examination, surveillance, management, repairs and upkeep of infrastructure, and item evaluations and testing. Artificial intelligence can potentially be advantageous for power systems. Additionally, with the aid of artificial intelligence techniques, specialists' capacity to create, source, and oversee state-of-the-art products and technologies in the fields of health, insurance, power, petroleum, and natural resources gas, education, aviation, production, and automotive has significantly increased in recent years. Rationale computation techniques have greatly contributed to the creation of higher capacity, more fuel-efficient aircraft in the aviation industry, as well as reduced asset availability and repair costs and times, by promptly detecting, alerting, and fixing defects. The globe continues to be in the initial stages of an improved production revolt powered by AI, despite the broad usage of artificial intelligence in tackling various technical and production difficulties [3]. AI does, however, also provide significant difficulties and dangers for engineering, including moral, societal, legal, and security concerns. Engineers must therefore use AI properly and ethically by being aware of its advantages and drawbacks [2].

## 2. FUNDAMENTALS OF ARTIFICIAL INTELLIGENCE IN ENGINEERING

The topic "Fundamentals of Artificial Intelligence in Engineering" discusses the fundamental ideas and methods of Artificial Intelligence (AI) and how they relate to numerous engineering issues and fields. AI may be used to examine a lot of data and find patterns and insights.

*Figure 2. Components, types, and subfields of AI [24]*

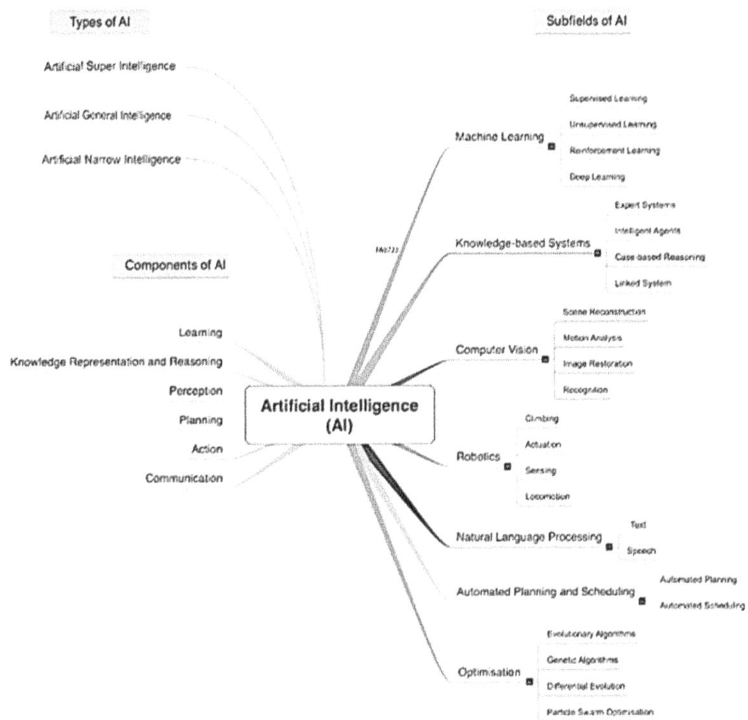


Machines in this application of AI naturally learn from experience rather than being explicitly taught to carry out certain tasks. Artificial neural networks are the foundation of the subset of machine learning known as "Deep Learning" that is used for predictive analysis. Unsupervised learning, supervised learning, and reinforcement learning are only a few examples of the many machine learning algorithms [2]. The algorithm in unsupervised learning does not use classified information to make decisions on its own without any direction. A function is derived from the training data in supervised learning, which consists of a set of input objects and the intended output. Machines utilize reinforcement learning to determine the best possibility that needs to be considered and to take appropriate actions to increase the reward [4]. Computers that have been programmed to process natural languages do so when they interact with human language. With the help of machine learning, it is possible to extract meaning from human languages through natural language processing. NLP involves having a machine record audio of a person speaking. The communication from audio to text then takes place, and the text is subsequently processed to turn the data into audio. In order to communicate with people, the machine then uses audio. Applications of NLP can be found in IVR (Interactive Voice Response) systems used in contact centres, in language translation systems like Google Translate, and in word processors like Microsoft Word that assess the precision of grammar in text. But because of the rules that are required for communicating information using natural language, which are difficult for computers to understand, natural language processing is challenging due to the nature of human languages [3]. Therefore, NLP uses algorithms to recognize and abstract the rules of the natural languages so that the unstructured data from the human languages can be translated into a format that is recognized by the computer.

*Figure 3. AI and ML applications in enhancing sustainable communities [23]*

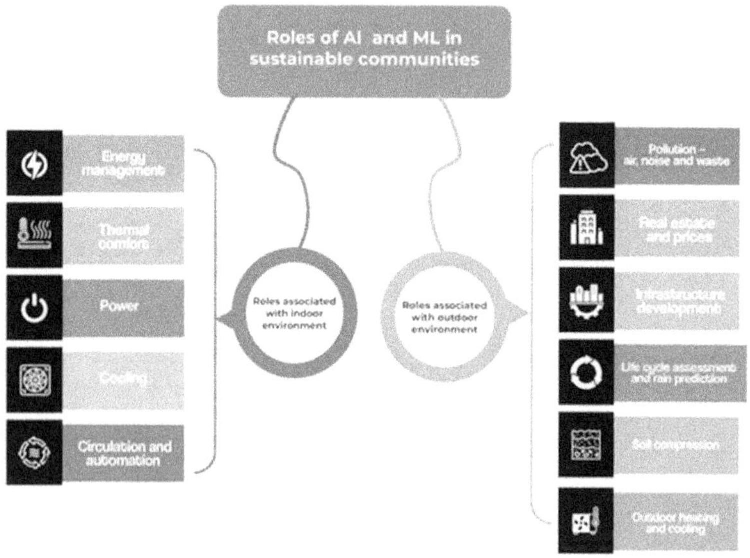


## 2.1 Keys Concepts of Artificial Intelligence

This study's keyword-based systematic mapping approach has a number of drawbacks. There were no recognized primary studies in the area of AI ethics because the focus was solely on the keywords. The exclusion method and the frequency of phrases in the articles were used to assess the relevancy of the papers. The definitions of the topics that the keywords stood for weren't examined. Notwithstanding these drawbacks, the keyword-based technique produced knowledge of pertinent keywords in cases where no previous listing was provided, covering a broad and ill-defined topic of AI ethics. The early work also provides pertinent keywords on AI ethics, which will aid future systematic mapping investigations [1][3].

## 2.2 Machine Learning and Deep Learning Basics

Artificial intelligence in the data science age aims to give computers human-level intelligence, and machine learning as well as deep learning are the tools used to do this. which facilitates the work of artificial intelligence. Device The area or subset of computational intelligence that focuses on learning teaches the robots to learn. Deep learning has limitations. It aids in increasing high standards in the educational setting. Artificial Intelligence and Deep Learning are essential for computer upgrades. systems to become knowledgeable entities capable of making decisions and making forecasts without the help of a person. The topic of artificial intelligence is whether computers are capable of thought and decision-making. Computer learning aids in the application of artificial intelligence to the framework, and using deep learning, machine learning objectives can be met on the system in a more methodical way. There are two sections to this work. The second portion will elucidate machine learning's methods, uses, and other aspects.

*Figure 4. Evolution of Artificial Intelligence [22]*

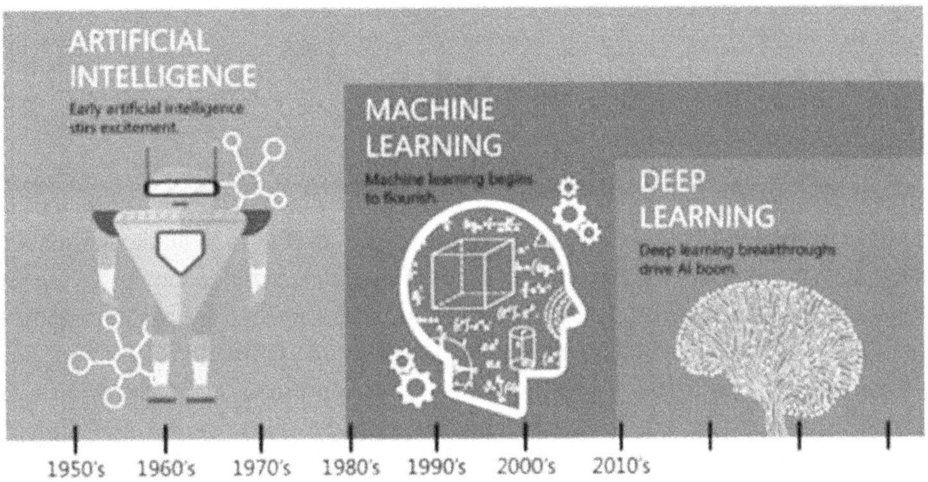


## 2.3 Natural Language Processing (NLP) Fundamentals

Human language, not computer language, is processed by computer software. Among the earliest and most well-known applications of NLP is spam detection, which examines the subject line of emails and text to decide if it's trash or not. The methods used in NLP today are based on machine learning. Among the NLP activities are text translation, an examination of emotions, and speech recognition. NLP involves having a machine record audio of a person speaking. The communication from audio to text then takes place, and the text is subsequently processed to turn the data into audio. In order to communicate with people, the machine then uses audio. Applications of NLP can be found in IVR systems used in contact centres, in language translation systems like Google Translate, and in word processors like Microsoft Word that assess the precision of grammar in text. But because of the rules that are required for communicating information using natural language, which are difficult for computers to understand, natural language processing is challenging due to the nature of human languages [3]. Therefore, NLP uses algorithms to recognize and abstract the rules of the natural languages so that the unstructured data from the human languages can be translated to a format that is recognized by the computer [5].

## 2.4 Computer Vision Fundamentals

Visual data can be captured and analysed by machines. In this instance, visual information is captured by cameras, image-to-digital data conversion occurs via analog-to-digital c conversion, and data processing is done via digital signal processing. A computer is then fed the resultant data. Sensitivity, or the machine's capacity to detect weak impulses, and resolution, or the range over which the device can discern between objects, The use of a device Pattern recognition, medical picture analysis, signature detection, and other tasks all involve eyesight. the understanding that letting machines see is okay. This method makes use of a digital signal, an analog-to-digital c converter, and camera. processing to record and examine visual data [6]. It is frequently compared to human vision because Machine vision, for instance, is not limited by being able to see past barriers and is a product of nature. It is employed in numerous applications, including acknowledgment of medical image research. System perception, which is concerned with computer-based picture processing, is usually connected to how machines look.

## 3. AI APPLICATIONS in CIVIL ENGINEERING

Circuits, robots, aircraft, bridges, and other complex engineering systems can all benefit from its use in the design and optimization process. The early diagnosis and avoidance of equipment breakdowns are made possible by AI-based condition monitoring systems' ability to evaluate sensor data and spot inconsistencies in the equipment's behaviour. This could facilitate maintenance planning and decrease costs [7]. saving time and money while idle. The design and operation of power systems can be improved with the help of AI algorithms. Leading to higher dependability and productivity. For instance, AI-based models can help with the prediction of electricity prices and optimize the production of electricity to save costs. Identification of electrical system breakdowns is achievable using AI algorithms. AI systems may analyse sensor data to find the mistakes' underlying causes and recommend remedies. Thus, failure rates might decrease. In order to reduce weight and material usage while preserving safety standards, AI can be utilized to optimize bridge and building designs. to determine other classes and an estimate of the amount of soil moisture content. Machine learning can be used in the field of structural engineering to identify the position and degree of defects by detecting them using sensory or visual data.

### 3.1 Structural Analysis and Design Optimisation

In reality, optimization is rarely utilized in the construction of buildings and other civil structures due to the combination of difficult issue formulation and lengthy simulation timeframes. Rapidly assessing design performance with surrogate models helps speed up design space exploration and cut down on optimization computation time. The primary focus for large-scale bridges is structural component monitoring, upkeep, and functioning. Nonetheless, keeping an eye on mechanical and electrical components is crucial for movable bridges as well. gathered from Sunrise Bridge's mechanical components in Florida. Accelerometers were installed to measure vibrations in the mechanical parts (e.g., rack and pinion) in order to check their condition. The suggested strategy calls for monitoring and training stages. The training stage concentrated on obtaining statistical features [8].

### 3.2 Construction Project Management

Ensuring worker safety in construction is a challenging endeavour due to the constantly changing and intricate work environments on construction sites. At the moment, computer vision methods for target recognition that track employees' whereabouts could take longer computational repeats. The information obtained by Ground-based Penetrating Radar (GPR) scanning of concrete bridge decks shows that, as is Picking rebars by hand is a tedious and time-consuming process. First, the pixels carrying the information are found using image processing techniques. possible peak for rebar. Next, the windowed images surrounding the possible pixel are taken from GPR imaging. Ultimately, a skilled CNN classifies those. While there are numerous ways to visually monitor construction sites, modern technology has not attained the level of precision necessary for reliable construction equipment identification in genuine pictures of construction sites [7]. In order to address this issue, a model is presented for detecting construction equipment based on a profound convolutional network, which is beneficial for managing building sites. This is a trained model. utilising a tiny quantity of data about construction equipment via transfer learning [8].

### 3.3 Traffic Management and Urban Planning

The primary cause of air pollution, particularly in crowded urban areas, is motor vehicle emissions. However, it has been discovered that the level of pollutants can be greatly lowered with proper traffic management, permitting the movement of city thoroughfares. The purpose of this study is to measure the amount of emissions generated by traffic and to determine how traffic management affects lowering emissions from vehicles and improving air quality. A model known as the adaptive fuzzy inference system for neurons was created. to calculate the amount of NO2 and PM10 emissions from traffic at specific crossings. Next, software for traffic control simulation was employed to create a scenario for improving traffic at these crossings. The next step was to measure the analysis of the increase in air quality brought about by changes in traffic management, utilizing the created the ANFIS model. The findings suggested that considerable reductions in delays at specific crossings would result in a decrease in NO2 and PM10 [9]. The hybrid model that was suggested raised the prediction precision and enhanced comprehension of the connection between features of traffic and emissions of pollutants. It also makes the work of the city easier. planners and aids in making decisions on the quality of urban air [9].

### 3.4 Risk Assessment and Disaster Management

Due to the high expense of human selection, artificial intelligence and machine learning are replacing human analysts in company operations. Built on machine learning, AI offers the maximum level of accuracy in calculations and data analysis by gradually learning new skills over time. AI can create procedures. Wherever needed, automation in the fields, astute analytics, and coherent ideas [10]. Chatbots driven by AI have demonstrated their potential as a method for customer service; therefore, they appear to be an uncompromised resource for businesses because they help them save money and time. Such creative technology would alter not just how businesses operate but also non-commercial endeavours. AI can be used in an organization to analyze many types of data, including data that describes the environment in which the business operates. various options to identify the best one, etc. Additionally, AI can be applied to knowledge mining to create and manage organizational knowledge strategies. Additionally, AI can be applied throughout the organization's departments for many types of operations carried out in these departments [10].

## 4. AI APPLICATIONS in MECHANICAL ENGINEERING

Automating processes like quality assurance, inspection, and assembly in the manufacturing industry uses AI. This could increase production and efficiency. Based on the basic composition and basic principles of the traditional expert system, a mechanical fault diagnosis expert system based on RBR and CBR reasoning is constructed. The system includes a manoeuvre case database, a fault diagnosis rule database, a fault diagnosis database, a fault reasoning machine, knowledge processing, a fault diagnosis process interpreter, a learning system, and an expert system with a man-machine interface.

## 4.1 Manufacturing Process Automation

Depending on the perspective, a system for manufacturing can mean several different things. As seen in the manufacturing systems used in this work, they include equipment, robots, conveyors, and related activities, including material handling and maintenance, that are set up to generate the required product. Numerous uncertainties and interdependencies make factory operations extremely nonlinear and stochastic [10][11]. The passage of materials through the job cells must be "optimally controlled" in order for such modern production systems to function well and remain globally competitive. Reviewing recent developments in AI techniques for manufacturing systems, such as systems modelling along with performance analysis, as well as optimal systemic-level control and decision-making, is the goal of this section.

## 4.2 Predictive Maintenance of Machinery

The goal of prognosis is to forecast how a machine's performance will deteriorate over time, from its current condition to its ultimate functional breakdown. Timely maintenance is facilitated by accurate RUL estimation. Generally speaking, AI-based prognosis is a component of the data-driven approach that depends on building a model of the evolution of machine performance in order to forecast future device performance based on past and present conditions. Iteratively performing one-step-ahead prediction until the projected value crosses a failure threshold is how RUL is estimated. When assessing machine performance directly proves to be challenging, sensor data is frequently utilized to generate an artificially generated Health Index (HI) that serves as a proxy for the device's performance. The RUL estimate and the HI-based prognosis are demonstrated. This section presents related efforts for AI-based machine prognosis.

## 4.3 Robotic Systems and Automation

The use of AI in relation to HRC is particularly examined in this section. The subject of HRC is one that distinctively crosses both the system level and machine/process layers, which is consistent with the hierarchical approach this study uses to organize and structure the assessment of AI in industrial systems. A purposefully built robot technology and an operator who work on concurrent activities within an interconnected workspace, or where the system of robotics and a human can accomplish tasks continuously or even jointly, is what is referred to as a Human-Robot Collaboration (HRC) [11]. Examining the effect of artificial intelligence on HRC is especially crucial in light of the growing number and variety of robotic uses throughout manufacturing. The goal of using robotics in industry has historically been to harness the benefits robots have over humans in terms of strength, endurance, repeatability, and the capacity to function in dangerous conditions. The industry has long known that by using robots to carry out these kinds of jobs, human operators may more effectively concentrate on their innate natural advantages, which are related to flexibility, ambiguity, adaptability, and intellect. However, the corresponding distribution of skills, efficiency, and types of abilities continually changes and transforms into less clear as these technologies' sophistication, computational power, and application domain increase with the introduction of increasingly potent and capable AI/ML tools.

## 4.4 CAD and CAE Integration With AI

It is well established that the existing mechanical computer-aided design framework has the ability to express design for elements, assemblies, and drafting and drawing. or properly set up. Both feature-based and parametric design modelling are widespread technologies present in CAD systems in 3D. There are two typical methods of 3D design: an original design or a redesign of an already-existing layout. Regarding fresh design, the artist will need to create a new model of the component, beginning either by following a 2D sketch, extruding, rotating, lofting, or sweeping; or by utilising 3D primitive and boolean operations to create a 3D section. Other features come after them. procedures like chamfer, fillet, and hole. Utilizing After that, assembly restrictions were created to create a 3D assembly form. diverse components combined. While adjustments to the current layout, measurements, and further characteristics are altered or modified to create a new assembly and part [12]. The. The technology that underpins parametric design is dimension-driven geometry, in which geometric modifications are produced when the values of the dimensions change. limit and relationship/equation in a parametric computer algebra system, the geometric components of the models are linked to specifications. Therefore, choosing the appropriate parameters that fulfil the design specifications is a crucial part of design. an engineer. The background, education, and skill set of a CAD designer who works with product geometry and properties play a critical role in creating a first-rate design. cutting the cycle of product development short and wasting time. With the use of CAE systems, the designed structure can be examined from a variety of angles, such as loads and power conditions (using the MES method, for example). fluxes (such as injection-moulded die filling), kinematic and dynamic studies (such as aerodynamic evaluation or combustion engine simulation of an aeroplane wing), or the functioning and application of the intended product (such as a car's ergonomic cockpit functionality). and utilizing the Solid to create a cutting-edge work schedule. The suggestion was made that the L-shaped. The ideal body type is one that requires a positive relationship between body mass and strength; it is required to include extra ribs against the rear wall. The purpose of the machine's tool is to machine parts. 500 kg using drilling and milling techniques (holes up to 1600 mm in length). When creating, try to keep the size of the body and the body's weight in mind when making the decision. standard. The outcomes of simulations and computations examining the changes in the body caused by the action of loads determined by the machined elements' weight [12].

# 5. AI APPLICATIONS IN ELECTRICAL AND ELECTRONIC ENGINEERING

Artificial intelligence-based electrical system automation could be considered a wise improvement over existing industrial processes. Developments in artificial technology for intelligence, which integrates intelligence, are essential to its utilization for a variety of reasons [1]. Mechanical automation control technology is becoming more and more of a technologically advanced system, and the application of artificial intelligence technology opens up new options for the industry. The application of AI in electrical automation might improve control effectiveness and expand the system's operational range even further. It is because of the union of cloud computing and big data technology, electrical automation, and the financial The maximum potential of pertinent data can be stimulated. Enabling cloud computing to solve the problem of information data while also permitting precise and impartial data analysis should be carried out. The first strategy is indicative of a focus on the investigation of human conduct and is typically an empirical branch of science. The subsequent method is a blend of technology, mathematics, and computing science. Regarding system development for the electrical industry, building on the second is more practical. methodology and design mechanisms that function on the foundation of specific laws rather than copying ideas and procedures in individuals' jobs [13].

## 5.1 Smart Grid and Energy Management

In addition to robots and voice assistants, the Internet of Things (IoT) can include smart objects (SOs) or gadgets that simplify tasks. AI-enabled SOs employ object recognition, while humans employ speech recognition, facial recognition, and applications such as speech and expression recognition, deep neural networks, computer vision, learning transfer, etc. By June, Smart Oven promises to always prepare food. altogether It features food and an HD camera. thermometer that aids in tracking the food's baking automatically in the oven and can change the cooking mode if needed. The oven can be controlled by Alexa, which, through user preference analysis, will recommend and adjust an automatic cooking program. The vehicle AI from Effective is an in-cabin sensor system. It is suitable for self-driving taxis and extremely automated automobiles. It recognizes the sentiment and uses in-cabin cameras to record the faces and voices of the vehicle's occupants, as well as audio devices. Programs intended to incentivize end users to temporarily reduce their energy use in reaction to a price indication from the daily electricity market or a trigger from the electricity grid controller are referred to as Demand Response (DR). Energy consumers' power consumption patterns are altered by DR to better balance supply and demand. By altering their patterns of electricity usage at times of high consumption and disasters to adapt to an hourly price structure, it gives users the chance to participate in grid operations. Financial incentives are also extended to astute shoppers. Under incentive-based systems, customers that cut back on their electricity use during peak hours are given set or time-varying financial advantages. In recent years, a number of alternative strategies for DR applications have been actively studied. They suggest a privacy-preserving plan that would allow demand response providers to calculate private demand restrictions and demand response awards in the smart grid while protecting the privacy of their customers. The genuine identity and the extremely fine metering data were kept apart by the scheme to protect customer privacy. This means that the DR can only access one of the two at a time and cannot connect the two. The DR has entered the field of electronic commerce, where intelligent procedures are integrated into the pool, in line with the advancements in ICT. This improves the intelligence of communication between the utility and the Electricity Management System (EMS) [13].

## 5.2 Circuit Design and Optimization

Although systematic approaches for designing processes in multiphase extraction of minerals are actively being developed, they have not yet been implemented in the industry, as far as we are aware. The quest to boost output, cut expenses, lessen waste's detrimental effects on the environment, and create simpler, more cost-effective procedures has driven the growth of these techniques [58]. Heuristic-based approaches, hybrid approaches, and rigorous methods are the three general approaches used in process design. The heuristic-based approach assists in identifying process alternatives by using general guidelines. Hybrid approaches create a workable process design by fusing the knowledge of the designer with basic principles. Strict approaches employ an algorithm for optimization to get the best answers while using a model of mathematics to represent a group of options. When they travel, The mathematical difficulty of the issue increases, the likelihood of receiving better solutions increases, the value of the designer's experience reduces, and the creation process moves from being more closely aligned with art to being more closely aligned with science when heuristics and rigorous methods are used. The majority of the research that has been published in peer-reviewed journals deals with flotation circuit design. Published works utilizing heuristic design techniques are scarce. Unit by unit, the procedure design is added, and it is stopped when adding more does not result in a higher profit. Flotation circuits were the application of the technology. The scientific community has lost interest in heuristic design methods as a means of solving design problems since they cannot ensure the discovery of optimal solutions. The primary hybrid technique for linear circuit analysis is the most significant hybrid approach used in the design of mineral processing facilities. This method offers a basic understanding of how unit operations behave and interact in multiple-stage processing circuit configurations. By describing the circuit's global recovery as an outcome of each process unit's transfer function, the mass balance technique can be applied to the circuit [14] [13].

## 5.3 Signal Processing and Image Recognition

Numerous ISPs that make use of machine learning have been proposed in recent years. They can be divided roughly into two groups: techniques for fine-tuning the traditional ISP functionalities and techniques that use a DNN-based encoder-decoder model in place of them. In the first category, machine learning techniques are used to fine-tune the current ISP parameters. Depending on the context, it specifies perception loss or recognition of image loss. aim to improve the accuracy of recognition or perceptual quality. Sometimes, algorithms that evolve over time or alternative search techniques are used since the majority of ISPs in use today are black-box solutions with many local minima in the search space. Still, these techniques have limited performance improvement and only optimize ISP characteristics for a specified downstream portion. In part to solve the issue, a study suggests a combined training regimen in which downstream CNN and ISP are optimized in turn while the other remains constant. Other techniques use gradient-based backpropagation to optimize the parameters. In addition to optimizing the hyperparameters, Reconfig ISP uses a neural architecture searching approach to investigate functions. It's crucial to remember that the computational expenses of their ISPs for inference time in this category are low. comprise traditional functions. We contend that the issue When compared to other categories, these methods have less expressive capacity. In the alternative category, techniques establish an ISP with DNN on an encoder-decoder basis. Lately, amazing model architectures are suggested as a replacement for ISP functions, like deblurring and denoiser color consistency. Certain works suggest general models that can take the place of different operations using the same design, but a few others swap out the entire ISP pipeline. using DNN. Although ISPs based on DNNs and ISP functions provide cutting-edge capabilities, computational cost is still a problem, particularly when high-resolution outputs are required. At the moment, smartphones are among the main applications for ISPs, so ISPs with low computing costs are much sought after. We suggest a dynamic ISP control strategy that doesn't fall into any of these two categories based on the information provided. Through dynamic parameter management, classical ISPs have modest computing costs and are powerful. Using our approach, optimization from ISP to picture is conceivable end-to-end recognition, surpassing customary individual, both the internet service provider and the identification model were optimized.

## 5.4 IoT Enabled System

IoT has a wide range of applications across numerous industries, in addition to smart home use. These remedies execute a financial and statistical examination of the entire company and provide forecasts utilizing some ML and AI formulas. One alluvium product called Primer provides commercial remedies. Primer generates a synopsis of the real-time stability score derived from the gathered data, the sensors, and the features of the gadget. This seeks to recognise any issues far in advance, inform operators of any anomalies, and make necessary modifications to the complete facility, from anything as basic as a sensor. Another IoT-based industry solution is Pluto Shift. It enables business enterprises to continuously monitor their asset performance, calculate the impact on finances, and provide assistance in making well-informed decisions. IoT is a general term that encompasses an excessive number of actuators, data storage, data processing, and Internet-connected devices. Consequently, every IoT-enabled device may sense its surroundings, send, store, and evaluate the information gathered, as well as take action. correspondingly. The last action to take in accordance with this relies totally on the processing phase. True to an IoT service Strong-mindedness is measured by processing level in intelligence. and/or performing it. An IoT device that isn't smart will possess limited abilities and be unable to change the information. However, a more intelligent IoT framework will have AI and could help achieve the real objective of automation and adjustment. Given this, there aren't many instances of current We talk about IoT services here, with the functioning of AI underlying them. [1][2].

## 6. AI APPLICATIONS IN CHEMICAL AND PROCESS ENGINEERING

Creating AI-powered models is critical to understand. The state of machine learning today is a collection of impromptu techniques, somewhat akin to alchemy. Additionally, as over time, alchemy became more formal and severe and changed into chemical engineering and chemistry, ML must change. When feasible, the application of first-principles knowledge could counteract this restriction, which can impose some discipline and rigor on models that are solely data-driven. As mentioned, there are numerous programs that are prepared to produce rapid success in this new phase of AI data research. But still, the truly fascinating and thought-provoking issues are found in Creating conceptual frameworks like hybrid models requires analytical models of emergence, domain-specific knowledge discovery engines, and mechanism-based causal explanations. Despite all the enthusiasm about machine learning right now, these advances will involve moving beyond merely datacentric ML and using more techniques for reasoning and representing knowledge from the initial stages of AI. They would necessitate a suitable integration of data-driven processing with symbolic reasoning. AI can help construct data-driven models of chemical processes, systems, and phenomena, such as reaction kinetics, thermodynamics, fluid dynamics, heat and mass transfer, etc. Chemical engineering problems that require sophisticated solutions, such as process design, synthesis, scheduling, planning, etc., can be helped by AI. Artificial neural networks, fuzzy logic, expert systems, and other techniques can be used by AI to monitor and regulate chemical processes [15]. Through the use of methods like pattern recognition, anomaly detection, classification, and others, AI can assist in the detection and diagnosis of errors in chemical processes. Chemical engineering applications of AI can make use of methods like data mining, machine learning, and natural language.

## 6.1 Process Control and Optimization

Neural networks have been widely utilized in chemical engineering for many applications such as fault detection, nonlinear process identification, and sensor data processing due to their strong modelling capabilities. However, its application in process control has just become widely recognized in the last few years due to the growth in nonlinear control research. This work aims to present a comprehensive overview of the numerous applications—both in simulated and online implementation—that use neural networks as algorithms for chemical process management. The article has been divided into three main control schemes: adaptive oversight, inverse-model-based control, and predictive control. We provide an overview of the main uses, goals, and outcomes of the work in every one of these categories. The article highlights the enormous potential of adopting neural networks for managing processes. It also demonstrates the lack of current successful online applications and the popularity of the hierarchical neural network for such control of process applications [16].

## 6.2 Quality Control and Assurance

The creation, along with the incorporation of process evaluation techniques, offers an alternative to traditional methods of quality assessment. surveillance systems. These devices recognize flaws in the welding process and the adverse effects of unclear intervention. Factors can be successfully decreased, and defective components are separable. In terms of laser welding, procedures Three phases comprise the monitoring process. Within the initial phase of the preparatory procedure, the weld seam is followed. Further on, weld flaws, splatters, melt pools, and so forth are observed throughout the procedure. Finally, the weld's shape and obvious flaws are analized after processing. using weld characteristics, AI-based techniques, flaws, and the weld's condition might be anticipated and modified. Also, remedial actions are suggested by means of comments to, if necessary, system control, and the complete Over time, the process grows more efficient as additional data is gathered. There are various techniques for process tracking and evaluation of quality [13].

## 6.3 Safety Monitoring and Hazard Analysis

Modern manufacturing facilities are becoming more complex due to the interconnectedness of process units brought about by plantwide management and optimisation. In such a plant, several sensors and actuators are connected to control systems, which in turn regulate plant operations. While the actuators physically modify the system's variables to regulate the process, the sensors help monitor the circumstances of the process. Processes may wander beyond their safe range of operation in spite of these control and surveillance mechanisms if there are actuator, detector, or system malfunctions. These flaws could result in a system breakdown, which would then trigger a plant accident. Process systems employ multiple levels of safety to avert potential hazards. Such layer failures raise the likelihood of accidents and their effects. Identification of hazards, evaluation of likelihood, and the fallout from dangerous events, taking into account the layers of protection within a system, give insight into the safety state of the system. These papers, however, do not define the interrelationships between the various process safety aspects or the general Advanced Semiconductor Manufacturing (ASM) process [17]. To get a comprehensive understanding of the risk management framework for process risk management, it is necessary to study the models and methodologies utilized in tandem. For that reason, the methodologies and models of process safety elements are reviewed and analyzed in this article, with an emphasis on how they relate to each other.

### 6.4 Sustainable Process Development

This study looks at creative approaches to assisting AI applications in order to produce intelligent buildings in a sustainable manner. AI in buildings refers to the intelligence displayed by software-driven electronics and systems that observe their surroundings in facilities and take appropriate measures to maximize performance under specific limitations or in a given setting. A dynamic and intelligent building is one that has an adaptable architecture that offers each occupant productive, economical, and environmental conditions that are deemed acceptable by an ongoing interplay between its four fundamental elements: people, processes (automation, control, and systems), and places (fabric, structure, and facilities) (users, services); and management (upkeep, effectiveness); and the relationship between these. The ideas presented in this work that can facilitate the use of AI in buildings are lean construction, building information modelling, and nanotechnology. This project is significant in that it looks at artificial intelligence support systems in buildings as part of smart cities and offers a case study of the ongoing Lagos, Nigeria-based Eko Atlantic project. The study suggests creative green architecture and integrated project delivery for the best application of these ideas to promote the long-term growth of artificial intelligence in structures.

## 7. REAL WORLD IMPLEMENTATIONS

Precision medicine is one of the most significant applications of AI in healthcare. The goal of precision medicine is to streamline the processes of prognosis, therapeutic intervention, and diagnosis. It makes use of enormous multidimensional biological data sets that include genetic variation in individuals as well as other risk variables like age, gender, and race. It also considers potential therapeutic targets, including immunological profile, metabolism, and environmental sensitivity. This enables clinicians to customize early interventions, whether they be preventative or therapeutic, for individual patients [7]. Numerous programs are available for precision medicine. These fall into three categories: genomic-based testing, sophisticated algorithms, and digital health applications. 23 people with a high risk of cardiovascular disease could be found with an accuracy of 85%. This cognitive assistant includes medical knowledge and reasoning [18].

### 7.1 AI Driven Structural Health Monitoring

Most traditional infrastructure inspection methods rely on visual evaluations (i.e., existence, position, and width of cracks) that depend on the judgement and experience of specialists, which may not always be accurate. Furthermore, these methods are expensive and time-consuming. Vision-based checks can be carried out using autonomous image processing programs that need computers and machine learning algorithms, enhancing photos, applying simple picture analyzing adjustments to magnify and monitor edges to speed up the inspection, or human inspection of raw images without post-processing. It has been continually enhanced to solve complicated picture patterning problems, like the detection of vehicles, pedestrians, and faces. The majority of computer vision research is focused on creating end-to-end learning algorithms with artificial intelligence, especially with the use of Deep Convolutional

Neural Networks (DCNNs), which have been shown to be able to classify images with more than 95% accuracy. Additionally, by utilizing asymmetric filters and feature maps, CNNs have been applied to pixel-level labelling inside an image to recognize and localize various items of interest. These technologies are becoming more well-known in SHM due to recent advancements in computer vision. They are among the most useful instruments for non-contact evaluation of flexion, corrosion, concrete spalling, fractures in pavement and concrete, fatigue identification, and superficial and subsurface damages [19].

## 7.2 Autonomous Manufacturing System

The market has gotten more competitive due to growing client demands and economic globalization. Manufacturing businesses need to increase their production in dispersed settings and set up numerous factories in different distant geographic regions. At the moment, dispersed manufacturing systems are widely used in many different kinds of manufacturing businesses, including steel, automotive, and chemical and food processing. The planning and modelling of systems for distributed manufacturing have drawn interest. substantial attention due to their important effects on raising the effectiveness of operations. What a crucial role scheduling plays in industrial systems in lowering production costs and enhancing customer contentment [19]. A few decades ago, numerous studies on scheduling issues in systems for production and services were conducted. These issues fall under the category of single-machine scheduling involving parallel machines, the job-shop, flow-shop scheduling, the schedule, and variations of them. Recent times have seen researchers suggest a novel scheduling technique, namely distributed scheduling, which seeks to arrange dispersed manufacturing techniques [17] [18].

## 7.3 AI Assisted Healthcare Equipment Design

In this study, we have discussed smart healthcare frameworks, emphasizing topics like machine learning-based disease diagnosis, wearable and smartphone-based health monitoring systems, IoMT, and social robotics for AAL, including software integration architectures to create these types of helpful frameworks. An overview of the systems for monitoring health based on wearable technology and mobile phones is depicted. It is clear that nearly all of the systems are capable of monitoring patients' body temperatures and heartbeats, making it clear how important the essentials are. Furthermore, a portion of the developed frame. Additionally, take your blood pressure in addition to the body's temperature and heartbeat. Most of the time, the Node MCU is the most commonly utilized edge device; however, one of the plans employed two gadgets as edge computer apparatus. Nearly all of the systems made use of WiFi for data transmission; only the setups implemented in them used Bluetooth to send and receive data. Items Peak and online apps are widely used for data visualization which helps doctors keep an eye on their patients. Moreover, in data, Adafruit and instruments are utilized. illustration. A prevalent unfavourable feature among the polled research indicates that the structures are not sufficiently manufactured for use in clinical settings. It's been discovered that most of the examined systems kept an eye on a single indicator for the patients aside from the defined frameworks. Most of the time, the front and back camera modules are utilized. as a sensing component; only the presented schemes made use of an accelerometer and microphone for data impression. For the experiment, various smartphones from the Samsung and iPhone brands are used, though the prototype demonstrated did not cite any model of smartphone. The maximum quantity of participants is 205 individuals, and the minimum quantity is 5 persons. The maximum and minimum lengths of the recorded videos were 15 minutes and 10 seconds, roughly. Almost always, a 30 Hz data sample rate is used. Most smartphone-based apps address security issues without any systems for monitoring health.

## 8. CHALLENGES AND ETHICAL CONSIDERATION

Organizations must choose their ethical pillars before they can create an ethical framework. Essential elements for such an ethical reference frame include what standards should be followed and how the company should be run. This framework must be adaptable but yet sufficiently defined to support research and enable managers to assess and scrutinize intricate organizational choices. For example, managers should be able to lead the company as a whole (e.g., cross-functional activities like human resources) in addition to evaluating novel offerings and services under this framework. Using the positioning matrix that was previously presented, many AI-related projects can be analyzed and prioritized in terms of their possible effects on people and society, as well as the need for ethical consideration. AI ought to adhere to ethical norms as well, especially if it has a high degree of self-learning and a significant potential impact on people and communities. Corporate offerings must not only satisfy client needs but also adhere to additional standards, such as ethical considerations, in order to maintain a company's competitiveness. We suggest that three factors should interact in order for us to be successfully able to regulate AI ethics. Firstly, while making administrative decisions connected to AI, ethical considerations must be incorporated if the attempt is morally questionable (projects in two or three). Secondly, in order to integrate ethical elements, managers must possess an ethical framework that allows them to align various possible choices with moral implications. Third, many aspects of the organizational environment, such as stakeholder groups, among others, must be taken into account. Future real-time artificial intelligence would employ almost every one of the currently existing fields of AI study. Nonetheless, there are a number of AI domains that may have a big influence on how real-time AI systems are developed because they are connected to these systems basic operating limitations [20].

### 8.1 Data Privacy and Security

Software that enables computer network-based social interaction, connection, gaming, and collaboration is known as social networking. This term encompasses the common social media platforms, such as wikis, podcasts, blogs, tags, and search engines in more recent times. More still Lately, the usage of online forums to create and create relationships amongst people who have similar interests has assimilate into the business sector as well. Since social networks for professionals like LinkedIn and Blue Chip Expert are still expanding, associations for professionals are increasing in popularity on hitherto private websites like Facebook, as well as risk management and enterprise security for MySpace. Professionals need to acknowledge the truth that these websites are becoming an avenue for the unapproved release of private company information. Although there are many advantages for social networking among the network solutions, the dangers associated with social media ought to be addressed. Like every new technology, social media is progressing quickly, so security experts must stay informed about the dangers connected to it. There is an age group joining the workforce with the expectation that this technology will not only be accessible for them to utilize, but is also necessary for how they interact with co-workers and business associates. While there are numerous advantages that arise from utilizing social media platforms both internally and externally, the architecture and policies to protect against The hazards need to be handled early on and shouldn't be considered a joke [17].

### 8.2 Human AI Collaboration and Trust

The continuous tradition of studying trust as an important influence on human-technology interaction across disciplines has produced a variety of viewpoints and definitions. The literature on psychological and human-technology interaction suggests that confidence in people and the willingness to depend on and be susceptible to AI in human-human interactions are two gatherings. In these domains, definitions frequently draw from organizational studies, arguing that a teammate's behaviour has an effect on interpersonal trust. Integrity is defined as how people view a team member's dependability and credibility. Kindness is reflected in perspectives regarding a team member's consideration for the group's welfare. Recent research on AI agent trust further demonstrates that dependable, independent, and consistent artificial intelligence agents are thought to be more reliable. Depending on how reliable each member of the human intelligence group is, team members will modify their conduct and cooperation (e.g., choosing not to disclose information in the event that it is not viewed as trustworthy). Furthermore, both human and artificial intelligence traits and states can serve as antecedents of trust among colleagues in addition to their common space. As an illustration, the impact of a system's Variations in AI system types may have an impact on trustworthiness. As opposed to Team and interpersonal trust are described as the belief that every team member shares that allows for the unrestricted exchange of ideas and information and is one of the most important characteristics. of a successful team. Effective teamwork requires human beings, and AI colleagues need to think well of one another and believe that their opinions are being trusted. Research from several fields confirms that in order for HAIT to work together efficiently, there must be suitable degrees of reciprocal trust, as demonstrated by both the AI agent and the human. The engineering literature suggests design strategies aimed at achieving this objective by establishing suitable reciprocal trust (i.e., trust engineering) between humans and AI, for example, by taking care of issues like explanation, security, or training [20] [18].

## 9. CONCLUSION AND FUTURE SCOPE

Through the application of concepts, artificial intelligence enables robots to think analytically. Outstanding support for the different regions has been produced using approaches to artificial intelligence. in the previous 20 years. Artificial intelligence keeps becoming more and more significant in different fields. The idea of this piece of writing is based on artificial intelligence, its various applications, and The methods of artificial intelligence applied in the field of Power System Stabilizers (PSS) are used to maintain system stability, oscillation dampening, and high-calibre performance in the detection of network intrusions. To prevent network intrusions in the medical field When it comes to the classification of medical images in the databases for accounting, I explained how these AI methods are applied in video games to address the typical issues and to add features to existing games to enjoy myself. A promising future lies in the examination of Additionally, there will unquestionably be network intrusion detection in the field of energy systems and stabilizers in the future. As we determine that more study in this field is possible, there are incredibly encouraging and successful outcomes that are obtained by these methods. Even though scientists have not yet fully grasped the capacity and promise of artificial intelligence, the use of this technology will probably have a significant impact on human lives in the years to come. Given its advantages and broad range of applications, artificial intelligence seems like the best option. Given the advancement of AI, does this mean that the world of the future is getting more artificial? The old, established paradigm regarding biological understanding is fixed, whereas the emerging paradigm of non-biological computing and intellect is expanding rapidly. The estimated number of binary digits in the human brain is ten thousand million. However, memorising is probably how most stuff is used—visual perceptions, among other relatively inefficient methods. Thus, given the limitations of natural intelligence and the volatile nature of the world, we may now rely on computers to function properly. AI is a genuinely ground-breaking. achievement in computer science, soon to be fundamental during the ensuing years and decades, and a part of every piece of contemporary software. That offers a chance as well as a threat. Artificial intelligence will be used to support both offensive and defensive cyber operations. Furthermore, new methods of cyberattack will be developed to exploit the unique flaws in AI technology. Last but not least, AI's demand for massive quantities of training data will increase the value of data and redefine how we must manage it. Consider data security. Globally prudent governance will be necessary to guarantee that this historic technology will lead to wealth and safety for all.

## REFERENCE

Abioye, S. O., Oyedele, L. O., Akanbi, L., Ajayi, A., Delgado, J. M. D., Bilal, M., & Ahmed, A. (2021). Artificial intelligence in the construction industry: A review of present status, opportunities and future challenges. *Journal of Building Engineering*, *44*, 103299. doi:10.1016/j.jobe.2021.103299

Chahal, A., & Gulia, P. (2019). Machine learning and deep learning. *International Journal of Innovative Technology and Exploring Engineering*, *8*(12), 4910–4914. doi:10.35940/ijitee.L3550.1081219

Chowdhary, K. (2020). Natural language processing. *Fundamentals of artificial intelligence*, 603-649.

Dibike, Y. B., Velickov, S., & Solomatine, D. (2000, March). Support vector machines: Review and applications in civil engineering. In *Proceedings of the 2nd Joint Workshop on Application of AI in Civil Engineering* (pp. 215-218). Research Gate.

Diliman, Q. C. (2005). Traffic impact assessment for sustainable traffic management and transportation planning in urban areas. *Proceedings of the Eastern Asia Society for Transportation Studies*, *5*, 2342–2351.

Dokas, I. M., Feehan, J., & Imran, S. (2013). EWaSAP: An early warning sign identification approach based on a systemic hazard analysis. *Safety Science*, *58*, 11–26. doi:10.1016/j.ssci.2013.03.013

Dominey-Howes, D. (2002). Documentary and geological records of tsunamis in the Aegean Sea region of Greece and their potential value to risk assessment and disaster management. *Natural Hazards*, *25*(3), 195–224. doi:10.1023/A:1014808804611

Jiang, J. L., & Wong, W. K. (2018). Fundamentals of common computer vision techniques for textile quality control. In *Applications of computer vision in fashion and textiles* (pp. 3–15). Woodhead Publishing. doi:10.1016/B978-0-08-101217-8.00001-4

Katam, R., Kalapatapu, P., & Pasupuleti, V. D. K. (2022, June). A review on technological advancements in the field of data driven structural health monitoring. In *European workshop on structural health monitoring* (pp. 371-380). Cham: Springer International Publishing.

Kazeem, K. O., Olawumi, T. O., & Osunsanmi, T. (2023). Roles of Artificial Intelligence and Machine Learning in Enhancing Construction Processes and Sustainable Communities. *Buildings*, *13*(8), 2061. doi:10.3390/buildings13082061

Kimera, R., Kaggwa, F., Mwavu, R., Mugonza, R., Tumuhimbise, W., Munguci, G., & Kamuganga, F. (2020). *An Overview of Data Science Innovations.* Challenges and Limitations Towards Real-World Implementations in Global Health.

Koch, F. A., & Oldham, W. K. (1985). Oxidation-reduction potential–a tool for monitoring, control and optimization of biological nutrient removal systems. *Water Science and Technology*, *17*(11-12), 259–281. doi:10.2166/wst.1985.0237

Lee, D., & Yoon, S. N. (2021). Application of artificial intelligence-based technologies in the healthcare industry: Opportunities and challenges. *International Journal of Environmental Research and Public Health*, *18*(1), 271. doi:10.3390/ijerph18010271 PMID:33401373

Long, D. D. (2018). Practice-informed research: Contemporary challenges and ethical decision-making. *The Right for the Elderly to Commit Suicide*, *15*(2), 15.

Lu, P., Chen, S., & Zheng, Y. (2012). Artificial intelligence in civil engineering. *Mathematical Problems in Engineering*, 2012.

Nabwire, S., Suh, H. K., Kim, M. S., Baek, I., & Cho, B. K. (2021). Application of artificial intelligence in phenomics. *Sensors (Basel)*, *21*(13), 4363. doi:10.3390/s21134363 PMID:34202291

Obaid, A. J., & Sharma, S. (2020). Recent trends and development of heuristic artificial intelligence approach in mechanical system and engineering product design. *Saudi Journal of Engineering and Technology*, *5*(2), 86–93. doi:10.36348/sjet.2020.v05i02.008

Pannu, A. (2015). Artificial intelligence and its application in different areas. *Artificial Intelligence*, *4*(10), 79–84.

Patel, A. R., Ramaiya, K. K., Bhatia, C. V., Shah, H. N., & Bhavsar, S. N. (2021). Artificial intelligence: Prospect in mechanical engineering field—a review. Data Science and Intelligent Applications. *Proceedings of ICDSIA*, *2020*, 267–282.

Strong, A. I. (2016). Applications of artificial intelligence & associated technologies. *Science [ETEBMS-2016]*, *5*(6).

Stulov, A., Tikhonov, A., & Snitko, I. (2020, September). Fundamentals of artificial intelligence in power transformers smart design. In *2020 International Ural Conference on Electrical Power Engineering (UralCon)* (pp. 34-38). IEEE. 10.1109/UralCon49858.2020.9216245

Venkatasubramanian, V. (2019). The promise of artificial intelligence in chemical engineering: Is it here, finally? *AIChE Journal. American Institute of Chemical Engineers*, *65*(2), 466–478. doi:10.1002/aic.16489

Zaabab, A. H., Zhang, Q. J., & Nakhla, M. (1995). A neural network modeling approach to circuit optimization and statistical design. *IEEE Transactions on Microwave Theory and Techniques*, *43*(6), 1349–1358. doi:10.1109/22.390193

Zhao, S., Blaabjerg, F., & Wang, H. (2020). An overview of artificial intelligence applications for power electronics. *IEEE Transactions on Power Electronics*, *36*(4), 4633–4658. doi:10.1109/TPEL.2020.3024914

## ABBREVIATIONS

**AAL:** Ambient Assisted Living
**AI:** Artificial Intelligence
**ASM:** Advanced Semiconductor Manufacturing
**DCNNs:** Deep Convolutional Neural Networks
**DCNNs:** Deep Convolutional Neural Networks
**DR:** Demand Response
**EMS:** Electricity Management System
**GPR:** Ground-based Penetrating Radar
**HAIT:** Human-Artificial Intelligence Teaming
**HI:** Health Index
**HRC:** Human-Robot Collaboration
**ISP:** Internet Service Provider
**IVR:** Interactive Voice Response
**IoT:** Internet of Things
**ML:** Machine Learning
**NLP:** Natural Language Processing
**PSS:** Power System Stabilisers
**SHM:** Structural Health Monitoring
**SOs:** smart objects

# Chapter 2
# Deep Learning and Machine Learning and Their Application on Computational Chemistry

**Hüseyin Fatih Çetinkaya**
*Faculty of Engineering, Turkey*

**Ali Fazıl Yenidünya**
*Sivas Cumhuriyet Universitesi, Turkey*

**Serap Çetinkaya**
*Sivas Cumhuriyet Universitesi, Turkey*

**Burak Tüzün**
*Sivas Cumhuriyet University, Turkey*

## ABSTRACT

*Machine learning (ML), coined by Arthur Samuel in 1959, is now a versatile tool in various fields like data mining, computer vision, and medical diagnosis. It builds models from training data for predictions without explicit programming. Deep learning (DL), a subset of machine learning, utilizes neural networks like DNN, CNN, and RNN. In health and safety research, ML applications started in the mid-1990s, focusing on toxicity. Challenges in chemical health and safety research, such as primitive algorithms and skill requirements, limited its early use. Recent advancements in AI and computer science have highlighted the significance of ML and DL. Improved data, computing power, and algorithms have led to autonomous data analyses, transforming solutions in structural engineering (MPSE). In integrated computational materials engineering (ICME), ML and AI are assimilated into efforts like the materials genome initiative (MGI), accelerating autonomous experiments and improving data analytics.*

DOI: 10.4018/979-8-3693-1335-0.ch002

## 1. INTRODUCTION

Artificial intelligence (AI) has started with Bayes' statistics and Laplace's least squares, and Markov chain derivatives. Machine learning (ML) was initiated by Arthur Samuel in 1959 (Samuel, 1959). It has since improved the existing data mining strategies, computer science, natural language processing and developed the means of biometric recognition, medical diagnosis, credit-card fraud detection, stock market analysis, speech and handwriting recognition, strategy games, and robotics (Pierson and Gashler, 2017; Jiao et al., 2020a).

Machine learning works on mathematical models, as it uses many mathematical tools, including probability theory, statistics, approximation theory, convex analysis, algorithm complexity theory, and many others (Jiao et al., 2020a).

Deep learning (DL) is a multi-layer algorithm that exploits data using artificial neural networks and creates an abstract feature. It is a newer area of ML and is structured by several frameworks: deep neural network (DNN), convolutional neural network (CNN), and recurrent neural network (RNN) (Jiao et al., 2020a; LeCun et al., 2015).

Practical ML applications have started quite recently, in 1990s, and in health and safety (Lee et al., 1995). Related research aimed at classifying toxic materials and future predictions. Swift developments inartificial intelligence and computer science allowed the use of ML and DL algorithms in cumbersome statistical calculations. By the turn of the century, these algorithms have been used in automated/autonomous data analysis (Pan et al., 2008; Pan et al., 2009).

Today, machine learning is deeply involved in the process and structural engineering (MPSE) application fields, in making machine-based autonomous decisions. This involves simultaneous and sequential operation steps: (i) perception of a knowledge space (ii) anticipation of perceived space, and (iii) unsupervised action of a robotic system (Dimiduk et al., 2018).

In existing MPSE operations, "Big Data" often refers to the data itself and the data sources. Data collection, and the accuracy of the collected data constitute very important issues. The "semi-structured" MPSE data is then used by industrial developers and designers. Here, ML/AI tools can again lead to transformations in the improvement of this data structure problem, as the application of these tools to the collected data, in turn, help their evolution and improve the performance of the users (Dimiduk et al., 2018).

The scale of depth depends on the number of independent observations on the states of a system. Independent observations essentially involve the creation of experimental data. This step is rather costly and all of created data must be utilised. Thus, developing ML/AI technologies could also be used in data mining in the near future. Such developments are needed in the creation of models in order to reduce the experimental workload and its cost. Here two types of data are created, by measurements and by synthesis. MPSE users might thus need to develop data production methods to fulfil the needs ML/AI data structuring efforts (Dimiduk et al., 2018).

Although ML/AI is promptly integrated into ICME, integrated computational materials engineering, and in the MGI, Materials Genome Initiative, efforts, the application of ML/AI in ICME and MGI still seems to be in progress. While material data has been a focus of interest in MPSE for some time, ML/AI has not been stated in available ICME reports (National Science and Technology Council Committee on Technology Subcommittee on the Materials Genome Initiative, 2011; National Science and Technology Council Committee on Technology Subcommittee on the Materials Genome Initiative, 2014).

Machine learning and artificial intelligence already have their places in ordinary life facilities. Yet, practitioners and scientists involved, are expected now to be interested in Machine Learning and Artificial Intelligence technologies for MPSE, to clarify their relatedness to physics, to evaluate the usability of evolving AI-based systems in material selection and improvement, and manufacturing processes in light of the prediction that machine learning will have the essential transformative power in the industrial realms throughout 21st century (Dimiduk et al., 2018; National Science and Technology Council Committee on Technology Subcommittee on the Materials Genome Initiative, 2014).

## 2. ML/DL ALGORITHMS AND DATA COLLECTION

Machine learning (ML) is the central subsection of artificial intelligence (AI). Its algorithms fashion mathematical models by drilling data which are then used to make predictions or take decisions. Deep learning (DL) is, on the other hand, a fairly new domain of ML. It creates and operates artificial neural networks. Its architecture is a multilayer perceptron with manifold veiled strata. DL combines low-level features to create or highlight abstract high-level data. As mentioned above, the use of ML algorithms in health and safety researches has been started mid-1990s. The applications of ML/DL algorithms in other domains awaited late 2000s (Dimiduk et al., 2018; Baduge et al., 2022).

This section provides a brief introduction to the ML/DL process, commonly used ML/DL algorithms, and the latest sensors and imaging technologies that can be used to collect data to be input into these algorithms (Baduge et al., 2022).

### 2.1. Machine Learning

Machine learning is a subdomain of artificial intelligence that involves computer observations on a particular deposit of data and built models on input data to solve specified problems. Unlike traditional computing, ML builds predictive models by which unseen data can be employed. Machine learning techniques can be classified in innumerable ways. One conspicuous classification is based on the amount of supervision during the model training process. Here two core supervised errands are classification and regression. In classification, a separate class "label" is anticipated. Supervised learning dataset comprises both predictors and outcomes, titled labels. In regression, a continuous value is estimated. Some of the recurrent supervised learning algorithms include k-Nearest Neighbours, Support Vector Machines (SVMs), Logistic Regression, Linear Regression, and Neural Networks. ML models can be mainly categorized as supervised learning, unsupervised learning, or reinforcement learning (Fig. 1). At the outset, the supervised model is trained using the labelled dataset (Step 1) in Figure 1, and the unseen data is then formed by the model's inferences (Baduge et al., 2022).

Unsupervised learning, on the other hand, uses an unlabelled dataset to envision internal data structure. This approach is widely taken in the objectives of clustering, anomaly detection and novelty detection, visualization and dimensionality reduction (Baduge et al., 2022).

Reinforcement learning is one of the ML routines. It observes the environment and then trains a particular ML model by paying desired behaviours and punishing undesirable behaviours. The aim here is to find the best strategy, called a 'policy', by achieving maximum rewards over time. The action to be taken in a particular situation is determined by policy (Baduge et al., 2022).

*Figure 1. Classification of machine learning*

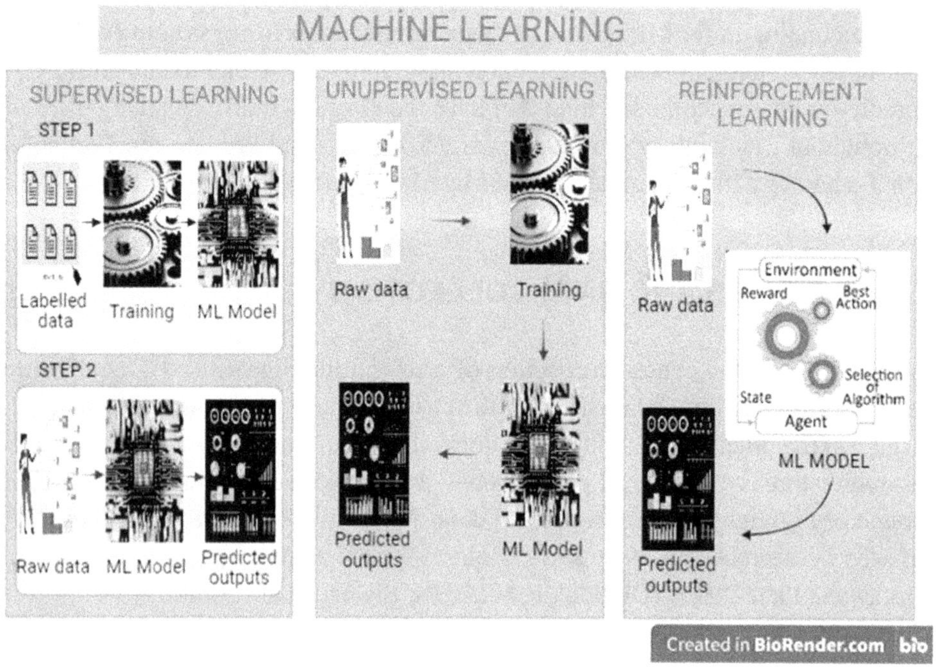


## 2.2. Deep Learning

DL is a subfield of machine learning in Figure 2, and it is to do with the study of artificial neural networks and other related machine learning algorithms having manifold hidden layers. Therefore, a deep learning algorithm should have multilayered input and output. This architecture, in comparison with ML, enables DL algorithms to be generally more useful for higher dimensional data such as images, video, and audio, than ML (Baduge et al., 2022).

### 2.2.1. Feedforward Neural Networks

Feed-forward neural networks (FNN), in other words "Multilayer Perceptrons" (MLPs), are a widely used for the information flow in DL algorithms in the frontward direction (Figure 3, circles denote neurons). FNNs are made up of interconnected neuron multilayers. Input data is served to the input layer and the data flows through the hidden layers. Output layers produce the end result in Figure 3 (Baduge et al., 2022).

Feed-forward neural networks can be trained using a particular dataset. In this training process, the loss of data is worked out by matching the output with the anticipated factual values. Loss computing and summing may involve on of the following methods: mean squared error, mean absolute error, and binary cross-entropy. The aim here is to curtail the cost function by backpropagation, which calculates the slope between error and weights. Optimization algorithms can be Adam, NAdam, Adadelta, or Gradient descent. The same dataset should be reiteratively treated to fine-tune the weights in order to minimize the error rate of the trained model (Baduge et al., 2022).

*Figure 2. AI, ML, DL fields and commonly used algorithms*

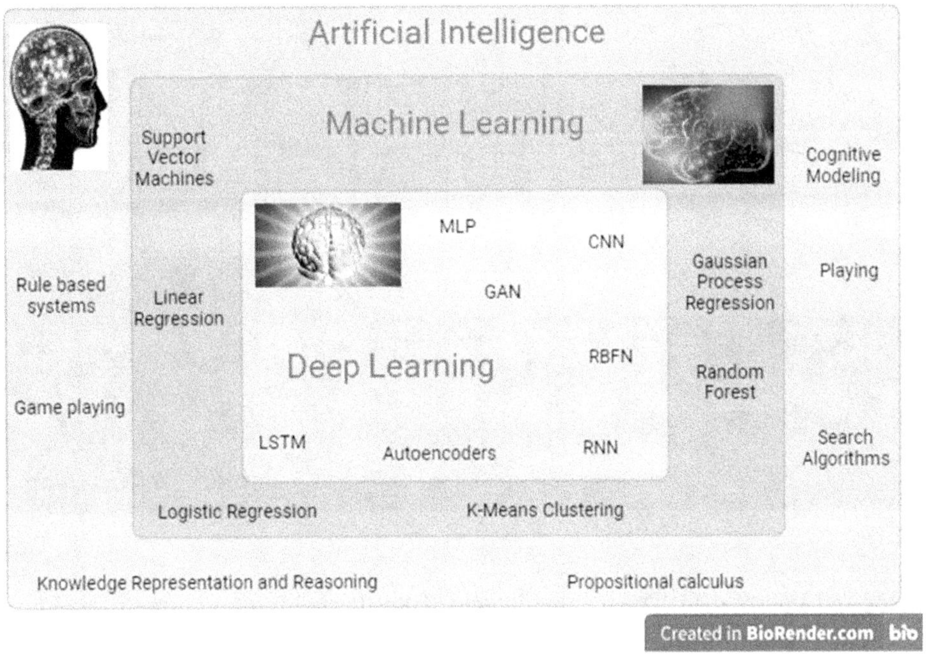


## 2.2.2. Convolutional Neural Networks

Convolutional Neural Network (CNN), by itself, forms a distinct category in Artificial Neural Network (ANN) and processes data in the grid-like topology, made up of the convolution grid, pooling grid, and fully connected (FC) grid, respectively (Figure 4). CNNs use images to produce visions in the computer. The convolution grid is the main elementary unit of CNN and includes the components of filter/kernel/

*Figure 3. Diagram of a feedforward neural network and a single neuron*

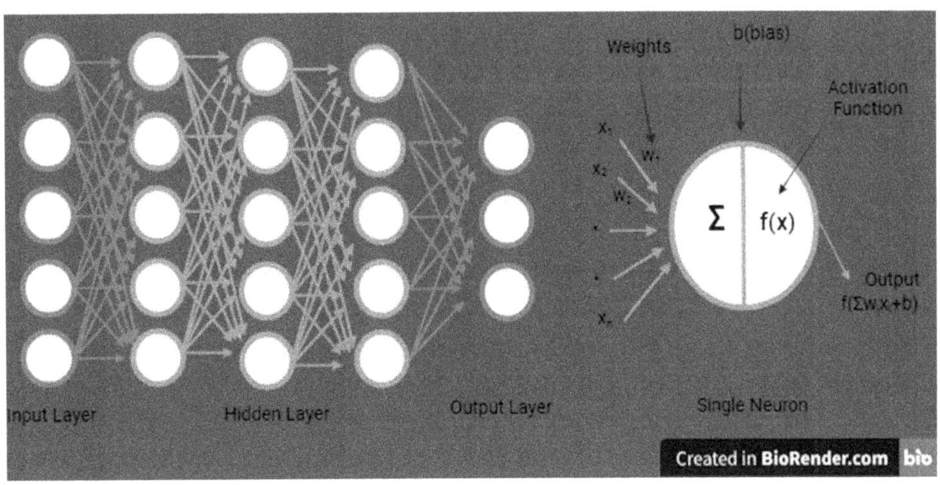

*Figure 4. Typical architecture of a CNN*

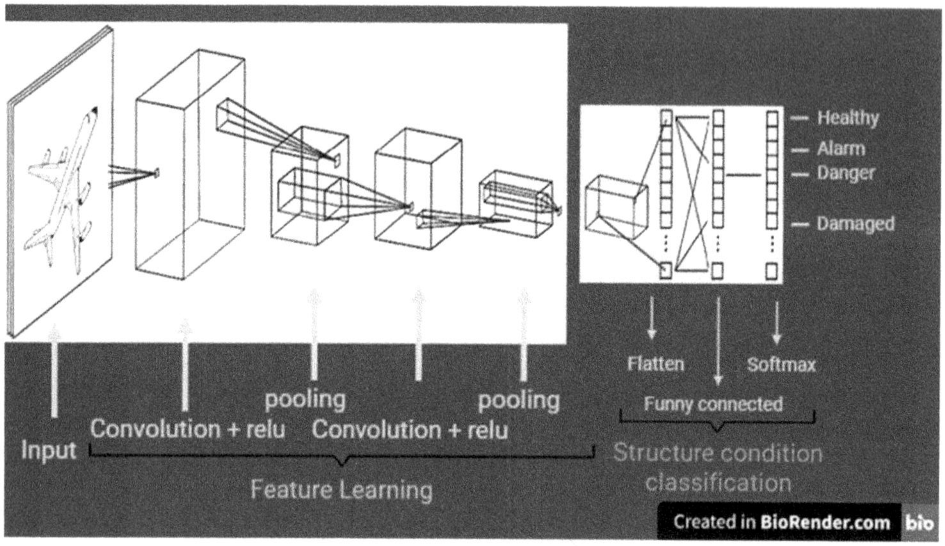


feature detector and feature map. The feature detector/filter/kernel is a two-dimensional arrangement of weights. Its size is smaller than that of the image. A dot product is calculated between the pixel value of the image and the weights of the filters, and the result is fed to an output array. This is the essence of convolution process. The feature detector is progressed through the intact image to accomplish this calculation and detect features. In this scheme, unlike in NN, neurons of different layers are not interconnected (Figure 3). Merely the filter neurons are connected to those of the following layer (Figure 4). The output of CNN is called feature map or activation map or convolved feature. This technology has transformed DL by reducing the number of the interconnections, resulted in lower memory needs and the use of relatedly larger data inputs in the form of images, videos, or sound. The number of filters in the convolution determines the depth of the output. A mediating activation appears to be necessary for the process to proceed beyond the convolution layer. The pooling layer creates a much smaller subsample by editing the size of the authentic image. The fully connected layer forms the terminal portion of the network in Figure 4. Filter size, number of filters, padding, and steps by and large determine the architectural design of DL. Some of the outstanding CNN types are AlexNet, VGGNet, and ResNet (Baduge et al., 2022).

### 2.2.3. Generative Adversarial Networks

Generative Adversarial Network (GAN) forms another deep learning algorithm that can also create images, and audio-visual products. It uses similar training datasets that are explained so far and it is structured by two kinds of deep neural networks: a generator and a discriminator in Figure 5. The former creates new topographies, resembling the trained dataset. The latter part is responsible for identifying the real data from that produced by the generator, and providing it with feedback on the quality of the newly produced images with respect to that of the real one. Early training efforts can result in the production of some fake images which the generator can detect. Later along the process, the generator may also continue to produce such images. However, in a successful training process, the discriminator can dis-

*Figure 5. Architecture of GAN and cGAN, a and b, respectively*

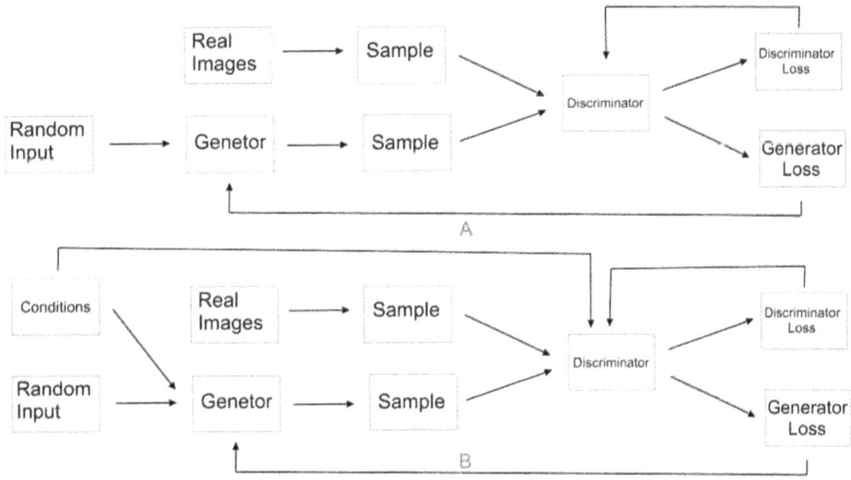


criminate the fake images The available types of GANs include Progressive GANs, Conditional GANs, Loop GANs, Text-to-image GANs, Super-Resolution GANs, InfoGANs, DCGANs, and Wasserstein GANs (Baduge et al., 2022).

### 2.2.4. Variational Autoencoders

Variational Autoencoder (VAE, another deep generative algorithm, is an autoencoder that organize coding distribution during training. It has an encoder, a decoder neural network, and a latent (encoded) field in Figure 6. The encoder transforms old features into a new feature representation in the latent space, and the decoder reverses this process during reconstruction (Baduge et al., 2022).

An autoencoder is trained using data and searches for the best encoding-decoding scheme through an iterative optimization process. Variational autoencoders reorganise training to warrant that the latent space is well organized to facilitate the productive process. When the latent space is well organized, new data can be generated via the decoder by decoding the points sampled from the latent space (Baduge et al., 2022).

### 2.2.5. Recurrent Neural Networks

Recurrent Neural Network (RNN) generates predictions for successive data. Similar to ANNs and CNNs, RNNs work on a dataset to train the algorithm. However, unlike ANNs and CNNs, RNNs use previous inputs to impact the existing input and output in the network in Figure 7. Its well-known variants are Long Short Term Memory (LSTM), Bidirectional Recurrent Neural Networks (BRNN), and Gated Recurrent Units (GRUs) (Baduge et al., 2022).

*Figure 6. Schematic of an autoencoder*

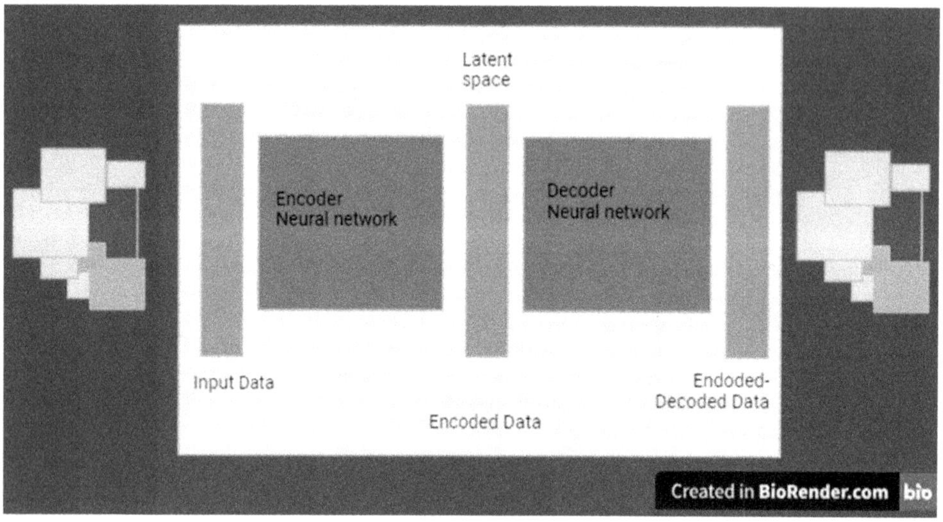


## 3. APPLICATIONS

### 3.1. QSPR/QSAR Model Development

Quantitative structure-activity/property relationship analysis (QSAR/QSPR) embraces regression and classification models and is often applied in biological and pharmaceutical realms (Tabian et al., 2019; Verma et al., 2010; Quintero et al., 2012). Common machine learning tools are sufficient for the construction of QSAR/QSPR models, since QSAR/QSPR has a well-built data pipeline which can impressively expedite the advancement of machine learning processes. Its prediction capacity appears to be highly accurate and reliable in chemical health and safety investigations (Wang et al., 2017; Wang et al., 2019; Zhou et al., 2017).

*Figure 7. RNN diagram*

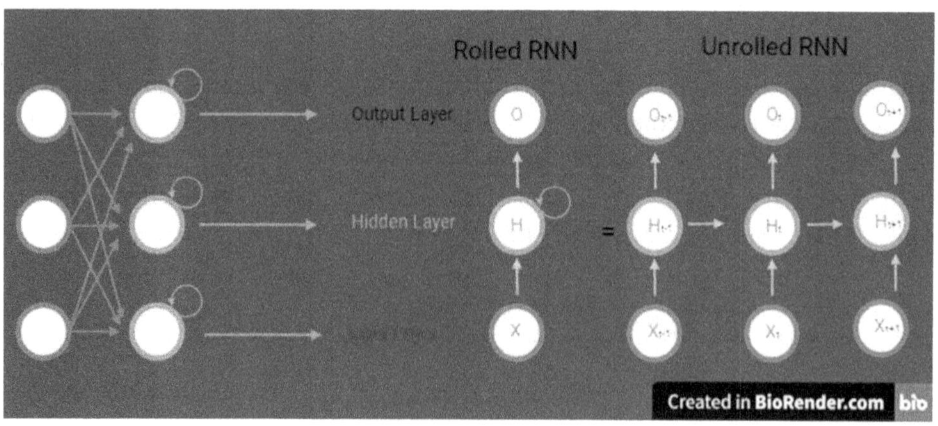

*Figure 8. QSAR/QSPR flowchart*

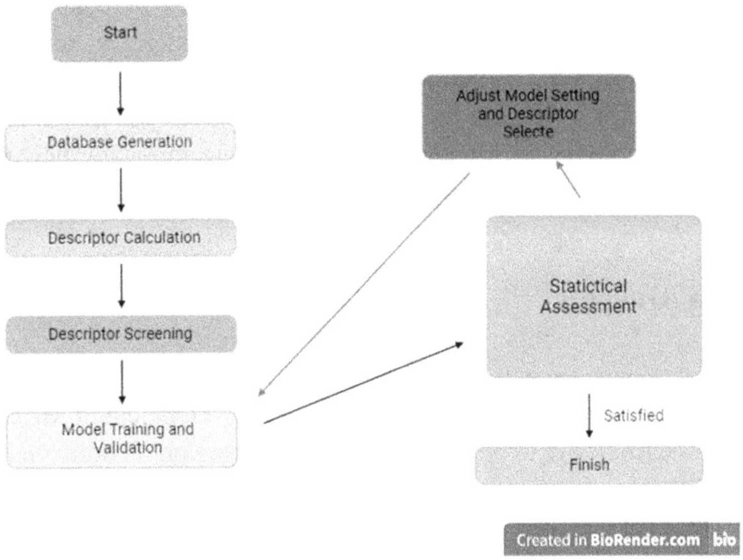


In the development of a QSAR/QSPR model, the initial phase is to create the dataset in Figure 8. The dataset should have minimum 50 data points. During experimental data collection the same experimental conditions should be maintained and the property/toxicity measurements should be judiciously inspected (Jiao et al., 2020a).

In a QSAR/QSPR study, determinants/features can be physicochemical characteristics or theoretical molecular descriptors. After descriptor calculation, they can be categorized between 0-D and 4-D descriptors. A QSAR/QSPR model can only have a limited number of features/predictors which are mostly limited between 4 and 8, depending on the size of the dataset. Training and validation constitute the second step (Jiao et al., 2020a).

The foremost target properties for property predictions are lower flammability limit (LFL), upper flammability limit (UFL), auto-ignition temperature (AIT), and flash point (FP). The other properties are the ignition energy (MIE) and self-accelerating decomposition temperature (SADT) (Jiao et al., 2020a).

Multiple linear regression (MLR) was the standard methodology to produce a QSPR model. The relationship between descriptors and identified features is considered non-linear and their mechanism of interaction is unidentified. A major limitation occurs during the application of a linear regression to a dataset. This is mostly for the reason that MATLAB has a pre-installed machine learning toolbox and there is no degree of freedom when using proprietary software. To overcome this obstacle, substantial amount of work has been performed using both particle swarm optimization (PSO) and genetic function approximation (GFA) (Jiao et al., 2020b; Bagheri et al., 2012; Khajeh and Modarress,2010). For the future development of hazardous trait QSPR analysis, more advanced machine learning algorithms are needed for the descriptive screening - and regression processes. These improve the prediction accuracy of the QSPR model (Jiao et al., 2020a).

In the analysis of compromised traits by QSPR, k-fold or leave-one-out cross-validation are needed to increase the reliability. Furthermore, on account of the difficulty of creating a large enough database, a more comprehensive database will also be necessary (Jiao et al., 2020a).

QSAR has been established before the QSPR. With QSAR both acute and chronic toxicities can be predicted. Acute toxicity mainly refers to $LD_{50}$ (lethal dose 50) and $LC_{50}$ (lethal concentration 50). $LD_{50}$ represents the quantity of a chemical that gives rise to 50% (half) mortality in a group of test animals if eaten, inoculated, or applied to the animal skin. $LC_{50}$ is the airborne or waterborne concentration of a substance that eradicates 50% of test animals in a single exposure. Chronic toxicity involves hazardous effects on reproduction, mutagenicity, and carcinogenicity. Some other QSAR studies have been involved in skin- and respiratory sensitization (Wu and Wang, 2018; Hao et al., 2019; Tan et al., 2009; Luechtefeld et al., 2018).

### 3.2. Consequence Analysis

The chief contemporary scheme for output analysis is computational fluid dynamics (CFD) modelling (Shen et al., 2020; Jiao et al., 2019; Mi et al., 2020). With the development of ML algorithms, artificial neural networks (ANN) have begun to be widely used in outcome predictions such as gas distribution and resource terms prediction. It can also be combined with other distribution models such as PHAST or FLACS to overwhelm the limitations caused by absent information source in emergencies (Jiao et al., 2020a; Ma et al., 2020).

A backpropagation neural network was developed to relate the gas concentration sensed by gas detectors (Wang et al., 2015). Therefore, perilous gas distributions can be predicted independent of the information source. Uneven input parameters between neural network training processes can aggravate the conditions of model training and interfere with its prediction accuracy. To overwhelm this strain, the original tracking parameters for both backpropagation neural networks (BPNNs) and SVM models have been replaced with integrated Gaussian parameters (Wang et al., 2018) and a series of machine learning algorithm models were combined with the classical Gaussian model (Ma and Zhang, 2016).

During the last two decades, deep learning has become prevalent due to its superiority with its meticulousness when applied to large datasets. To eliminate the discordances between accuracy and efficiency, deep belief networks (DBNs) and convolutional neural networks (CNNs) have been exploited (Ni et al., 2019). A specifically designed LSTM model for direct gas distribution in real environment was proposed (Qian et al., 2019). The dropout approach has been introduced to overcome overfitting and ameliorate the simplification ability of a model. ANN has been demonstrated to be handy both in explosion risk analysis and flammable material distribution (Shi et al., 2019). Trial applications of QSAR/QSPR can also been seen in outcome analysis modelling studies (Bjerrum and Threlfall, 2017; Segler et al., 2018).

## 4. ARTIFICIAL INTELLIGENCE IN COMPUTATIONAL CHEMISTRY

The importance of ML/DL/AI has been appreciated by computational chemists [36-39] and by those in synthetic chemistry (Wei et al., 2016; Goh et al., 2017; segler et al., 2017). In these fields ML has been employed both in the mining of historical experiment results and in the complex nonlinear variable relationships (Liu et al., 2016; wu et al., 2017). ML has also been utilized in computational quantum chemistry to envisage especially the stability and reactivity of molecules. Both quantum chemistry and density functional theory (DFT) essentially use limited computational scaling ($O(N^6)$ and $O(N^3)$ respectively) (Carter, 2008). In recent studies employing ML/AI algorithms has enabled the manipulations of multi-body wave functions and force fields in computational chemistry (Carleo and Troyer, 2017; Hush, 2017; Faber et al., 2017; Schütt et al., 2017; Tüzün et al., 2017; Tüzün and Sarıpınar, 2020)

## 5. CONCLUSION

The ground of artificial intelligence has expanded significantly in recent years and is creating practical impressions on social strata in general. The capability of computer software to execute complex linguistic and visual endeavours, central issues that have motivated the people in this field since 1950s, has evolved expressively. The present gains in AI still falls short in terms of fashioning human-like intelligence in mechanics, although a continuous scientific effort uses its total means. Successful applications reside mostly in healthcare related industrial disciplines.

Future projections clearly demonstrate the essentiality of governmental back-up is critical in the multi-dimensional progress and implementation of AI products in the globe and beyond. By and large bureaucracy is expected to keep pace with the evolution rate attained in the AI front. Necessary adjustments in educational curriculums respecting the needs of AI, for example, unquestionably require administrative participation. Social sciences should also be driven in the AI field in order to curtail negative impressions and create positive awareness.

## REFERENCES

Baduge, S. K., Thilakarathna, S., Perera, J. S., Arashpour, M., Sharafi, P., Teodosio, B., Shringi, A., & Mendis, P. (2022). Artificial intelligence and smart vision for building and construction 4.0: Machine and deep learning methods and applications. *Automation in Construction*, *141*, 104440. doi:10.1016/j.autcon.2022.104440

Bagheri, M., Borhani, T. N. G., & Zahedi, G. (2012). Estimation of flash point and autoignition temperature of organic sulfur chemicals. *Energy Conversion and Management*, *58*, 185–196. doi:10.1016/j.enconman.2012.01.014

Bjerrum, E. J., & Threlfall, R. (2017). *Molecular generation with recurrent neural networks (RNNs).* arXiv preprint arXiv:1705.04612.

Carleo, G., & Troyer, M. (2017). Solving the quantum many-body problem with artificial neural networks. *Science*, *355*(6325), 602–606. doi:10.1126/science.aag2302 PMID:28183973

Carter, E. A. (2008). Challenges in modeling materials properties without experimental input. *Science*, *321*(5890), 800–803. doi:10.1126/science.1158009 PMID:18687955

Dimiduk, D. M., Holm, E. A., & Niezgoda, S. R. (2018). Perspectives on the impact of machine learning, deep learning, and artificial intelligence on materials, processes, and structures engineering. *Integrating Materials and Manufacturing Innovation*, *7*(3), 157–172. doi:10.1007/s40192-018-0117-8

Faber, F. A., Hutchison, L., Huang, B., Gilmer, J., Schoenholz, S. S., Dahl, G. E., & von Lilienfeld, O. A. (2017). *Fast machine learning models of electronic and energetic properties consistently reach approximation errors better than DFT accuracy.* arXiv preprint arXiv:1702.05532.

Goh, G. B., Hodas, N. O., & Vishnu, A. (2017). Deep learning for computational chemistry. *Journal of Computational Chemistry, 38*(16), 1291–1307. doi:10.1002/jcc.24764 PMID:28272810

Hao, Y., Sun, G., Fan, T., Sun, X., Liu, Y., Zhang, N., Zhao, L., Zhong, R., & Peng, Y. (2019). Prediction on the mutagenicity of nitroaromatic compounds using quantum chemistry descriptors based QSAR and machine learning derived classification methods. *Ecotoxicology and Environmental Safety, 186*, 109822. doi:10.1016/j.ecoenv.2019.109822 PMID:31634658

Hush, M. R. (2017). Machine learning for quantum physics. *Science, 355*(6325), 580–580. doi:10.1126/science.aam6564 PMID:28183936

Jiao, Z., Hu, P., Xu, H., & Wang, Q. (2020a). Machine learning and deep learning in chemical health and safety: A systematic review of techniques and applications. *Journal of Chemical Health and Safety, 27*(6), 316–334. doi:10.1021/acs.chas.0c00075

Jiao, Z., Ji, C., Yuan, S., Zhang, Z., & Wang, Q. (2020b). Development of machine learning based prediction models for hazardous properties of chemical mixtures. *Journal of Loss Prevention in the Process Industries, 67*, 104226. doi:10.1016/j.jlp.2020.104226

Jiao, Z., Yuan, S., Ji, C., Mannan, M. S., & Wang, Q. (2019). Optimization of dilution ventilation layout design in confined environments using Computational Fluid Dynamics (CFD). *Journal of Loss Prevention in the Process Industries, 60*, 195–202. doi:10.1016/j.jlp.2019.05.002

Khajeh, A., & Modarress, H. (2010). QSPR prediction of flash point of esters by means of GFA and ANFIS. *Journal of Hazardous Materials, 179*(1-3), 715–720. doi:10.1016/j.jhazmat.2010.03.060 PMID:20381958

LeCun, Y., Bengio, Y., & Hinton, G. (2015). *Deep learning. nature, 521*(7553), 436-444.

Lee, Y., Buchanan, B. G., Mattison, D. M., Klopman, G., & Rosenkranz, H. S. (1995). Learning rules to predict rodent carcinogenicity of non-genotoxic chemicals. *Mutation Research. Fundamental and Molecular Mechanisms of Mutagenesis, 328*(2), 127–149. doi:10.1016/0027-5107(94)00202-G PMID:7739598

Liu, R., Ward, L., Wolverton, C., Agrawal, A., Liao, W. K., & Choudhary, A. (2016, August). Deep learning for chemical compound stability prediction. In *Proceedings of ACM SIGKDD workshop on large-scale deep learning for data mining (DL-KDD)* (pp. 1-7). ACM.

Luechtefeld, T., Rowlands, C., & Hartung, T. (2018). Big-data and machine learning to revamp computational toxicology and its use in risk assessment. *Toxicology Research, 7*(5), 732–744. doi:10.1039/c8tx00051d PMID:30310652

Ma, D., Gao, J., Zhang, Z., Zhao, H., & Wang, Q. (2020). Locating the gas leakage source in the atmosphere using the dispersion wave method. *Journal of Loss Prevention in the Process Industries, 63*, 104031. doi:10.1016/j.jlp.2019.104031

Ma, D., & Zhang, Z. (2016). Contaminant dispersion prediction and source estimation with integrated Gaussian-machine learning network model for point source emission in atmosphere. *Journal of Hazardous Materials, 311*, 237–245. doi:10.1016/j.jhazmat.2016.03.022 PMID:27035273

Mi, H., Liu, Y., Jiao, Z., Wang, W., & Wang, Q. (2020). A numerical study on the optimization of ventilation mode during emergency of cable fire in utility tunnel. *Tunnelling and Underground Space Technology, 100*, 103403. doi:10.1016/j.tust.2020.103403

National Science and Technology Council Committee on Technology Subcommittee on the Materials Genome Initiative. (2011). *Materials Genome Initiative for global competitiveness*. National Science and Technology Council.

National Science and Technology Council Committee on Technology Subcommittee on the Materials Genome Initiative. (2014). *Materials Genome Initiative strategic plan*. National Science and Technology Council.

Ni, J., Yang, H., Yao, J., Li, Z., & Qin, P. (2019). Toxic gas dispersion prediction for point source emission using deep learning method. *Human and Ecological Risk Assessment*.

Pan, Y., Jiang, J., Wang, R., & Cao, H. (2008). Advantages of support vector machine in QSPR studies for predicting auto-ignition temperatures of organic compounds. *Chemometrics and Intelligent Laboratory Systems, 92*(2), 169–178. doi:10.1016/j.chemolab.2008.03.002

Pan, Y., Jiang, J., Wang, R., Cao, H., & Cui, Y. (2009). A novel QSPR model for prediction of lower flammability limits of organic compounds based on support vector machine. *Journal of Hazardous Materials, 168*(2-3), 962–969. doi:10.1016/j.jhazmat.2009.02.122 PMID:19329246

Pierson, H. A., & Gashler, M. S. (2017). Deep learning in robotics: A review of recent research. *Advanced Robotics, 31*(16), 821–835. doi:10.1080/01691864.2017.1365009

Qian, F., Chen, L., Li, J., Ding, C., Chen, X., & Wang, J. (2019). Direct prediction of the toxic gas diffusion rule in a real environment based on LSTM. *International Journal of Environmental Research and Public Health, 16*(12), 2133. doi:10.3390/ijerph16122133 PMID:31212880

Quintero, F. A., Patel, S. J., Munoz, F., & Sam Mannan, M. (2012). Review of existing QSAR/QSPR models developed for properties used in hazardous chemicals classification system. *Industrial & Engineering Chemistry Research, 51*(49), 16101–16115. doi:10.1021/ie301079r

Samuel, A. L. (1959). Some studies in machine learning using the game of checkers. *IBM Journal of Research and Development, 3*(3), 210–229. doi:10.1147/rd.33.0210

Schütt, K., Kindermans, P. J., Sauceda Felix, H. E., Chmiela, S., Tkatchenko, A., & Müller, K. R. (2017). Schnet: A continuous-filter convolutional neural network for modeling quantum interactions. *Advances in Neural Information Processing Systems*, •••, 30.

Segler, M. H., Kogej, T., Tyrchan, C., & Waller, M. P. (2018). Generating focused molecule libraries for drug discovery with recurrent neural networks. *ACS Central Science, 4*(1), 120–131. doi:10.1021/acscentsci.7b00512 PMID:29392184

Segler, M. H., Preuss, M., & Waller, M. P. (2017). Learning to plan chemical syntheses. arXiv preprint arXiv:1708.04202.

Shen, R., Jiao, Z., Parker, T., Sun, Y., & Wang, Q. (2020). Recent application of Computational Fluid Dynamics (CFD) in process safety and loss prevention: A review. *Journal of Loss Prevention in the Process Industries*, *67*, 104252. doi:10.1016/j.jlp.2020.104252

Shi, J., Li, X., Khan, F., Chang, Y., Zhu, Y., & Chen, G. (2019). Artificial bee colony Based Bayesian Regularization Artificial Neural Network approach to model transient flammable cloud dispersion in congested area. *Process Safety and Environmental Protection*, *128*, 121–127. doi:10.1016/j.psep.2019.05.046

Tabian, I., Fu, H., & Sharif Khodaei, Z. (2019). A convolutional neural network for impact detection and characterization of complex composite structures. *Sensors (Basel)*, *19*(22), 4933. doi:10.3390/s19224933 PMID:31726762

Tan, N. X., Rao, H. B., Li, Z. R., & Li, X. Y. (2009). Prediction of chemical carcinogenicity by machine learning approaches. *SAR and QSAR in Environmental Research*, *20*(1-2), 27–75. doi:10.1080/10629360902724085 PMID:19343583

Tüzün, B., Çağlar Yavuz, S., & Sarıpınar, E. (2017). 4D-QSAR analysis and pharmacophore modeling: Propoxy methylphenyl oxasiazole derivatives by electron conformatitional-genetic algorithm method. *Journal of Physical & Theoretical Chemistry*, *14*(2), 149–164.

Tüzün, B., & Saripinar, E. (2020). Molecular docking and 4D-QSAR model of methanone derivatives by electron conformational-genetic algorithm method. *Journal of the Indian Chemical Society*, *17*, 985–1000.

Verma, J., Khedkar, V. M., & Coutinho, E. C. (2010). 3D-QSAR in drug design-a review. *Current Topics in Medicinal Chemistry*, *10*(1), 95–115. doi:10.2174/156802610790232260 PMID:19929826

Wang, B., Chen, B., & Zhao, J. (2015). The real-time estimation of hazardous gas dispersion by the integration of gas detectors, neural network and gas dispersion models. *Journal of Hazardous Materials*, *300*, 433–442. doi:10.1016/j.jhazmat.2015.07.028 PMID:26223017

Wang, B., Xu, K., & Wang, Q. (2019). Prediction of upper flammability limits for fuel mixtures using quantitative structure–property relationship models. *Chemical Engineering Communications*, *206*(2), 247–253. doi:10.1080/00986445.2018.1483350

Wang, B., Yi, H., Xu, K., & Wang, Q. (2017). Prediction of the self-accelerating decomposition temperature of organic peroxides using QSPR models. *Journal of Thermal Analysis and Calorimetry*, *128*(1), 399–406. doi:10.1007/s10973-016-5922-8

Wang, R., Chen, B., Qiu, S., Zhu, Z., Wang, Y., Wang, Y., & Qiu, X. (2018). Comparison of machine learning models for hazardous gas dispersion prediction in field cases. *International Journal of Environmental Research and Public Health*, *15*(7), 1450. doi:10.3390/ijerph15071450 PMID:29996467

Wei, J. N., Duvenaud, D., & Aspuru-Guzik, A. (2016). Neural networks for the prediction of organic chemistry reactions. *ACS Central Science*, *2*(10), 725–732. doi:10.1021/acscentsci.6b00219 PMID:27800555

Wu, H., Lorenson, A., Anderson, B., Witteman, L., Wu, H., Meredig, B., & Morgan, D. (2017). Robust FCC solute diffusion predictions from ab-initio machine learning methods. *Computational Materials Science*, *134*, 160–165. doi:10.1016/j.commatsci.2017.03.052

Wu, Y., & Wang, G. (2018). Machine learning based toxicity prediction: From chemical structural description to transcriptome analysis. *International Journal of Molecular Sciences, 19*(8), 2358. doi:10.3390/ijms19082358 PMID:30103448

Zhou, L., Wang, B., Jiang, J., Pan, Y., & Wang, Q. (2017). Quantitative structure-property relationship (QSPR) study for predicting gas-liquid critical temperatures of organic compounds. *Thermochimica Acta, 655*, 112–116. doi:10.1016/j.tca.2017.06.021

# Chapter 3
# Artificial Intelligence and Machine Learning and Its Application in the Field of Computational Visual Analysis

**Digvijay Pandey**
https://orcid.org/0000-0003-0353-174X
*Department of Technical Education, Government of Uttar Pradesh,, India*

**Vinay Kumar Nassa**
https://orcid.org/0000-0002-9606-7570
*Department of Information Communication Technology, Tecnia Institute of Advanced Studies, India*

**Binay Kumar Pandey**
https://orcid.org/0000-0002-4041-1213
*Department of Information Technology, College of Technology, Govind Ballabh Pant University of Agriculture and Technology, Pantnagar, India*

**Blessy Thankachan**
*School of Computer Applications, JECRC University, India*

**Pankaj Dadheech**
https://orcid.org/0000-0001-5783-1989
*Swami Keshvanand Institute of Technology, Management, and Gramothan, India*

**Darshan A Mahajan**
https://orcid.org/0000-0002-1239-6343
*NICMAR University, India*

**A. Shaji George**
https://orcid.org/0000-0002-8677-3682
*Business System Department, Almarai Company, TSM, Riyadh, Saudi Arabia*

## ABSTRACT

*Artificial intelligence and machine learning applications in image processing are examined in this chapter. It covers AI methods including supervised, unsupervised, reinforcement, and deep learning. Genetic algorithms, rule-based systems, expert systems, and fuzzy logic are AI methods. SVM, decision trees, random forests, K-means clustering, and PCA are machine learning methods. CNN, RNN, and GANs are utilised for object recognition, classification, and segmentation. The chapter discusses how artificial intelligence and machine learning affect accuracy, efficiency, and decision-making. The need to choose proper measurements and procedures for assessment and performance analysis is also stressed. Ethics like justice, privacy, transparency, and human-AI cooperation are covered in the chapter.*

DOI: 10.4018/979-8-3693-1335-0.ch003

## 1. INTRODUCTION

Images are vital sources of information in the digital age, and they are utilized in a broad variety of disciplines, including business, science, law, education, politics, the media, the military, medical imaging and diagnostics, art, digital forensics, and sports, to name a few. Recently, fraudulent photographs have caused problems for various applications. Digital photos can be edited using image-editing tools such as digital cameras, personal computers, and image-processing software. This allows images to be altered in a variety of ways. These tools come with functionality for the user interface and can be scaled as needed. The authenticity and reliability of an image are two concerns that are targeted by attempts made at forgery detection. To make images that are unrecognizable and appear so real that the authenticity is lost, techniques such as image splicing, cloning, and editing are used to create fake photographs. These techniques are used to create photos that have been falsified. Thanks to developments in graphics technology and advanced hardware and software tools, the process of creating forged photographs has become straightforward, and digital images may be instantly manipulated. This is possible due to the fact that digital images are stored in a digital format. images can now be altered by anyone, regardless of their level of experience, and the methods for detecting forged images are still in the experimental stages of development. picture tampering and picture forgery have the same fundamental objective, which is to change some of the image's data in order to either hide traces of tampering or alter specific aspects of the image in order to communicate false information.

In recent years, the contribution of artificial intelligence has expanded significantly across all industries. Image processing techniques and artificial intelligence are intertwined. Both of them have collaborated on the development of a number of technologies that are now used in our everyday lives to make even the most difficult tasks easier to accomplish. Image processing and computer vision are often considered to be two of the most important components when constructing an AI model. Image processing can be illustrated as the task of categorization of images into one of various classes, which provides the backbone for other computer vision functions such as detection, segmentation, and localization. AI in the health-care system has been facing strenuous challenges to design and develop intelligent and autonomous health-care solutions. Image processing can be illustrated as the task of categorization of images into one of various classes(Pandey, B. K.,et al.,2011).

According to Rawat et al. (2017), deep learning models have an effect on multiple layers of nonlinear acquired knowledge processing, including the extraction and transformation of functions, pattern classification, and analysis for issue solving. Processing images with the use of a computer has come to be seen as absolutely necessary for finding solutions to problems in the field of medical imaging. Deep learning approaches led to a main breakthrough in the field of image interpretation by assisting, recognizing, classifying, and quantifying patterns in medical imaging. As a result, deep learning looks to be a new building block for obtaining increased efficiency in numerous medical applications (Shen et al,2017b).

According to Wut et al. (2017), the use of AI in medical imaging plays a crucial role in achieving higher levels of diagnostic accuracy and therapeutic efficacy in clinical settings. Utilizing deep learning can lead to significant improvements in the performance of other AI methodologies. For instance, the capability of deep learning in medical images acquired with computed tomography (CT), magnetic resonance imaging (MRI), and positron emission tomography (PET) for better results (Shen et al,2017b). Although the applications of image processing are numerous, the strategies and principles that are utilized in each of these applications are identical. In the field of artificial intelligence recognition technology, there is still some doubt regarding the effectiveness and precision of the technology. As a result, the researchers had a hard time locating the flaws and maintaining control over the product's functionality.

The primary use of image processing is one of the areas of agriculture that is experiencing the most rapid expansion. In almost all facets of agriculture, image processing techniques are utilized as an instrument for doing data analysis. Agricultural plant diseases like, extraction of sick leaves, stems, fruit quantify, the affected area by disease and determines the illness's color, shape, are identified with the assistance of artificial intelligence and image processing (Hemamalini et al,2022,Raghuvanshi et al,2022). Image Segmentation and Feature extraction is accomplished utilizing the K-Means, principle component analysis. RBF-SVM, SVM, random forest, and ID3 are the additional techniques that are utilized in the process of image classification.

The subfield of Artificial Intelligence known as "computer vision" is what enables a computer to derive meaningful information and make recommendations based on what it sees based on what it is fed. Convolutional neural networks are used in computer vision for the processing of visual data at the pixel level, and recurrent neural networks are used in deep learning for the purpose of determining how one pixel is related to another pixel (Daoud, M. K.,et al.,2024). The term "Edge Intelligence" refers to the deployment of artificial intelligence to edge devices, which enables the implementation of computer vision that is scalable, efficient, robust, safe, and private. When combined with artificial intelligence techniques, computer vision algorithms can lead to a wide variety of real-world applications (Xi Zhang,2022). Some of these applications include automatic face identification and recognition.

## 2. PHOTOS IN DIGITAL FORMAT

Because digital images can more easily convey messages than words can, they are utilized for the purpose of achieving successful communication. An image can be thought of as a two-dimensional function of the coordinates x and y, with f representing the amplitude of the grey level or the intensity of the x and y coordinates and f(x,y) representing the picture's spatial coordinates (Daoud, M. K.,et al.,2023). Each individual component that makes up an image is referred to as a "pixel," and this term describes the smallest unit that can be found in digital images. A digital image is represented as a matrix with a fixed number of rows and columns, and the values in each row and column correspond to the intensity value of a pixel located at a particular position in the image.

### 2.1 Different Kinds of Digital Pictures

- **Grayscale Image:** A grayscale image is comprised of 8-bit pixels with intensities that vary from 0 to 255, with 0 representing complete darkness and 255 representing complete lightness.
- **Binary picture:** An picture that only contains the numbers 0 and 1 for each pixel, a binary image is a type of image that is used in processing applications.
- **Colour image**: In order to record information about colours, colour images make use of raster maps. These maps measure the chrominance and intensity of the light.

## 2.2 Image Features

Image characteristics are essential components of any algorithm designed to detect photo forgery or manipulation. The process of feature extraction contributes to the generation of the feature set that is required for the detection of picture forgery. The feature vector is a vector that stores the extended set of features. The matching of local interest points needs the usage of picture attributes and descriptors.

- Natural features are expressions of an object's real nature that are visible to the naked eye.
- Artificial features are features that are derived from modified photographs.
- The ability of a feature to hold constant when shifted is known as shift invariance.
- The ability of a characteristic to maintain its original state despite being rotated is referred to as its rotation invariance.

The ability of a feature to keep its characteristics the same even when its size changes is referred to as size invariance.

- The invariances of mirror, shear, and affine are preserved even when operations are performed on the data.
- The appearance of a feature is not altered when it is obscured by something else; it continues to be the same.
- There shouldn't be any overlapping features, and each object should have its own distinctive characteristics.
- It is important to ensure the dependability of shared values between objects.
- Features are said to be independent when they do not have a statistically significant correlation between them. This is because changes to one value should not have an effect on the other values.
- Resistance to noise and artefacts: An excellent characteristic should be impervious to both types of interference.
- If you want to portray something in a concise manner, you should use fewer features.
- There needs to be sufficient material in the features so that they can differentiate between the many fields of interest.
- The extracted feature needs to be robust enough to withstand shifts in geometric shape, changes in lighting conditions and intensity, and distortions in the image.

# 3. TECHNIQUES USED TO FAKE DIGITAL IMAGES

Image forgery or tampering is a common form of visual content falsification that is used for the purpose of keeping confidential or otherwise important information hidden. The photographs are either tampered with or falsified in a variety of ways, and this is done so that it is more difficult to spot the changes.

## 3.1 Various Forms

- **Copy-move or Cloning**: Copy-move image forgery is a form of image forgery in which a forged image is created by copying and pasting or moving small or distinct portions of an image within the same image. To deal with this form of tampering, algorithms are developed. These algorithms vary in terms of the properties that are extracted and the matching technique.

• **Image Splicing**: A sort of image manipulation called "photomontage" mixes two or more genuine pictures to make a distinct fake image. By fusing two or more images into one, false images are formed.

• **Image Retouching**: Image fraud or manipulation can be discovered via the procedure known as image resampling (Mirroring, Stretching, Rotation, and Resize). It may be necessary to resize, rotate, stretch, or mirror an item in order to bring its height into proportion with its relative proportions. It can be placed in either the Active or the Passive category depending on the context.

• **Active approach-based techniques**, such as steganography and watermarking, require already embedded image data in order to function properly. From the original message, which is often an encrypted version of some kind of hash value, an external authentication code known as a "digital signature" is generated and used. A digital signature, on the other hand, will only work if an authentication message is also transmitted alongside the media. Once the message has been successfully received and decoded, the digital signature will no longer prevent unauthorized copies from being made. Complex cryptographic algorithms make it difficult to break into the system but developing them is an expensive endeavor. It is necessary to have specialized imaging hardware in order to implement the active method of photo authentication, which is based on digital signatures or watermarking. Passive photo authentication is an alternate option when a digital signature or watermark cannot offer active side information. It is dependent on the markings that various image-forgery processing procedures leave on the image and is detected using various image-forgery detection methods.

## 3.2 The Use of Algorithms

There have been five different kinds of algorithms suggested for the purpose of image tampering, forgery, and attacks.

• **Algorithms that are based on Blocks Images** are broken up into the spatial domain and the transform domain by using block-based algorithms. While the Transform Domain employs a variety of image transformation methods to identify instances of image forgery, the Spatial Domain examines blocks in relation to the values of their constituent pixels. Post-transformation methods that are supported by AI and statistical processing are also used.

• **Key point-based Algorithms**: Key point-based methods reduce the computational complexity of feature matching and post-processing by identifying and choosing high-entropy picture regions to extract a feature vector for each key point. This process is known as key point-based method reduction. SIFT, MIFT, and SURF are three examples of the many different forms that key point-based feature vectors can take.

• **Brute Force Algorithms**: The detection of brute force can be accomplished by comparing matching segments with versions that have been circularly shifted using exhaustive search and autocorrelation.

• **Algorithms that are based on Hybrid Methods**: Hybrid algorithms use both block strategies and key point strategies in their computations.

• **Algorithms that are based on Deep Learning**: Deep neural networks are able to operate effectively on enormous datasets and can learn abstract features directly from photographs with very little to no preparation required. They are able to identify structural changes that occurred post-forgery and can automatically learn relevant and discriminative properties.

## 4. MACHINE LEARNING (ML)

Machine learning is a technique for analyzing large amounts of data and deriving information that is meaningful from that analysis. Machine learning employs a set of rules and a logical model structure. This area of study draws from a wide range of disciplines, including artificial intelligence, probability, statistics, information theory, as well as philosophy, psychology, and neurobiology. The development of algorithms that are based on previous relationships and data trends requires the utilization of machine learning, which is an essential component of fake consciousness. It has applications in many different fields, such as bioinformatics, the detection of viruses, image deconvolution, information retrieval, game play, and marketing. It is broken up into four primary categories according on the type of learning signal or feedback that is accessible(George, A. H.,et al.,2021).

A classifier is taught using recognized classifications, and it is then watched while it learns the data in the training dataset. This type of learning is referred to as "supervised learning." This process is known as "supervised learning." Classification and regression are the two subcategories of supervised learning problems. Regression problems are continuous, whereas classification tasks are discrete. Examples include estimating the selling price of a certain product based on the cost of similar products from other brands.

**Unsupervised Learning**: Unsupervised learning is a method for extracting labels from data for which there is no established classification. It is employed to explore data and discover patterns between items, and it performs well on transactional data. Self-organizing maps, nearest-neighbor mapping, k-means clustering, and singular value decomposition are examples of well-liked unsupervised approaches.

**Semi-supervised Learning**: Unlabeled and labelled data are both used in semi-supervised learning, a type of learning. When labelling is too expensive to support a completely labeled training procedure, it is helpful. It can be applied in conjunction with techniques like classification, regression, and prediction.

**Reinforcement Learning**: In contrast to supervised learning, reinforcement learning focuses on online performance, foraying into new terrain, and making use of existing information. Usually, it is described as a Markov decision process (MDP). Through the multi-armed bandit issue and finite MDPs, the exploration vs. exploitation trade-off in reinforcement learning has been investigated. Traditional methods require knowledge of the MDP, but reinforcement learning algorithms focus on large MDPs, which are too complex for accurate methods to be practical.

### 4.1 The Learning Method Known as Machine Learning

The development of machine learning algorithms allows for the implementation of a variety of models and iterative processes.

- Using different methods, such as OLSR, linear, logistic, stepwise, multivariate adaptive regression spline, and locally estimated scatterplot smoothing, regression is the process of predicting the value of a continuously varying variable.
- **A decision problem** is modeled using instances or examples of training data that the model considers crucial or necessary. These techniques usually create a database of training data and use a similarity measure to compare test data to the database in order to determine the best match and generate a prediction. Lazy learning essentially retains the training data and waits to learn the lesson after receiving the test data. k-Nearest Neighbor (kNN), Learning Vector Quantization (LVQ), Self-Organizing Map (SOM), and Locally Weighted Learning (LWL) are among the numerous types of algorithms.

- **A predictive model** is used in decision tree learning to link observations about an item to judgments about its intended value. Different types of algorithms include Classification and Regression Trees (CART), Iterative Dichotomizer 3 (ID3), C4.5 and C5.0 (differing iterations of a powerful method), Decision Stump, M5, and Conditional Decision Trees.
- Naive Bayes, Gaussian, Multinomial, AODE, BBN, and BBN are used in Bayesian algorithms—which are based on probability theory—to describe uncertainty.
- **Clustering** is a technique for categorizing objects into distinct groups while utilizing unsupervised learning. With k-means, k-medians, expectation maximization (EM), and hierarchical clustering techniques, it is classified as both hierarchical and partitional clustering.
- **Apriori and Eclat algorithms** are examples of association rule learning algorithms that extract rules to find significant and meaningful associations in large multidimensional datasets.
- **ANNs** are supervised learning models that are derived from biological neural networks. They use artificial neurons with highly weighted interconnections, and they learn by adjusting the weights of the connections to carry out distributed processing in parallel. Perceptron, back-propagation, Hopfield Network, and Radial Basis Function Network are among the several categories of algorithms.
- **Deep learning techniques** are a contemporary update to ANN that make use of plentiful computing power to construct bigger and more intricate neural networks. Convolutional neural networks (CNN), stacked auto-encoders, deep belief networks (DBN), and deep Boltzmann machines (DBM) are a few examples.
- **Dimensionality reduction**, which lowers the number of dimensions and eliminates redundant and irrelevant data to lower the computational cost, is an effective remedy for the dimensionality curse. The following algorithms are used: Sammon Mapping, Multidimensional Scaling (MDS), Projection Pursuit, Linear Discriminant Analysis (LDA), Mixture Discriminant Analysis (MDA), Quadratic Discriminant Algorithms (QDA), and Flexible Discriminant Applications (FDA).
- **Unsupervised learning models** called ensemble algorithms integrate subsets of training data to produce a valid hypothesis. They are based on weak learner models. Boosting, bagging, bootstrapping aggregates, AdaBoost, stacked generalization, gradient boosting machines (GBM), gradient boosted regression trees (GBRT), and random forest are examples of popular ensemble algorithms.

### 4.2 Machine Learning Applications

Applications of machine learning are broken down into supervised, unsupervised, semi-supervised, and reinforcement learning categories. Applications for semi-supervised learning include web page categorization, audio analysis, protein sequence classification, and text document classification.

## 5. MACHINE LEARNING ALGORITHMS: IMAGE FORGERY DETECTION

Deep learning is a method of data analysis that divides incoming data into many classes using computational methods. Image fraud is recognized using discriminant analysis (DA), k-nearest neighbor (kNN), and random forest (RF).

**Naive Bayes:** Assuming that characteristics and class labels are independent of one another, the Naive Bayes method, which is based on the Bayes theorem, determines the probability of each feature. It is simple to construct, effective for big datasets, and used for binary and multiclass classification issues.

**Support Vector Machine (SVM):** Modern classification methods like SVM are used to categories and examine data using learning algorithms. By locating the hyperplane with the greatest gap between two classes, it classifies an instance into two or more classes.

**Decision Tree:** A decision tree-based classifier employs a hierarchical rule-based method to divide inputs into consistent classes. It is a straightforward and often used classification technique. It employs a non-parametric methodology, with leaf nodes serving as class labels and non-leaf nodes serving as attributes.

**Discriminant Analysis:** A decision tree-based classifier is a straightforward and often used classification method that divides inputs into consistent classes using a hierarchical rule-based method. With non-leaf nodes representing attributes and leaf nodes representing class labels, it employs a non-parametric method.

**k-Nearest Neighbor (kNN):**When classifying images based on the nearby training examples in the feature space, kNN is a reliable and adaptable supervised learning technique that is applied. Its two steps are training and testing/classification. Without engaging in actual learning and model creation, training data with labelled classes or known instances is held during this phase. During testing, the K nearest neighbors of unknown instances are calculated, and a class of query instances is decided by voting among those neighbors.

**Random Forest:** RFC is a tree-based algorithm that was created by Tin Kam and improved by Leo Breiman. It uses a variety of models to learn and produce more accurate predictions. It is a meta estimator that reduces overfitting by fitting the number of trials of various subsamples of data to produce the complete forest.

## 6. DEEP LEARNINGANDIMAGE FORGERY DETECTION

Image forgery detection uses developments in machine learning (ML) and deep learning (DL). For decision-making, machine learning algorithms assist in locating the underlying linkages of data. For image processing tasks such as segmentation, classification, detection, image context, and retrieval, convolutional neural networks (CNNs) are frequently used. CNNs have paved the path for answers to more difficult computer vision (CV) challenges. In this thesis, the lightweight, residual, and spatial CNN bases are investigated.

**Alex net:** With only a few floating-point operations, CNN trained on the ImageNet dataset can categories images into up to 1000 categories with 61.0 million parameters. The supplied image size is 227 by 227.

**VGG16:** it is a CNN trained on the ImageNet dataset with 16 layers, 144 million parameters, and 224 224 image input.

**VCG-19:**Using 144 million parameters and an image input size of 224 × 224, the VGG-19 CNN trained on the ImageNet dataset can categorize images into up to 1000 separate categories.

**Google Net:**With 7.0 million parameters and a 224 × 224 picture input size, CNN can classify images into up to 1000 different categories.

**Squeeze Net:**With only a few floating-point operations, CNN trained on the ImageNet dataset can classify photos into up to 1000 categories with 1.24 million attributes.

**Shuffle Net:**50 layers deep CNN trained on the ImageNet dataset can classify images into up to 1000 different categories.

*Figure 1. Classification of artificial intelligence techniques*

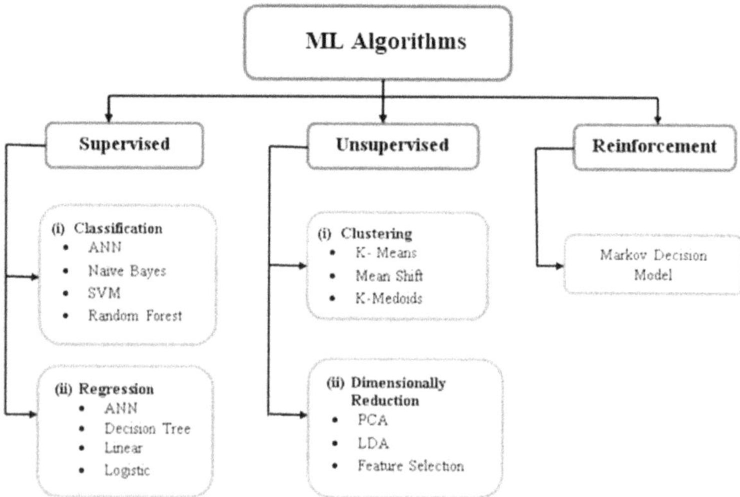


**MobileNetV2**:With the use of rich representations, CNN trained on the ImageNet dataset can classify images into up to 1000 different categories.

**ResNet-18**:With 11.7 million parameters and a 224-by-224 picture input size, CNN trained on the ImageNet dataset can recognize up to 1000 categories.

**ResNet-50**:With 25.6 million parameters and a 224-by-224 picture input size, CNN trained on the ImageNet dataset can categorize images into up to 1000 separate groups.

**ResNet-101**:Using 44.6 million parameters and an image input size of 224 × 224, CNN trained on the ImageNet dataset can categorize images into up to 1000 different categories.

## 7. CLASSIFICATION OF AI TECHNIQUES FOR IMAGE PROCESSING

Machine learning is the subfield of artificial intelligence that is responsible for providing information to computers and has the capacity to learn statistical data rather than program code. Figure 1 illustrates that machine learning may be broken down into three distinct subfields: supervised learning, unsupervised learning, and reinforcement learning. In supervised learning, a single supervisor is utilized to train the data via training examples. This type of learning is referred to as "supervised learning." Labelling datasets for the purpose of accurate algorithm training, data categorization, and data prediction is supported by this feature. Artificial Neural Networks, Bayesian Networks, Decision Trees, Support Vector Machines (SVM), ID3, k-Nearest Neighbor, Markov Model, and other similar systems are all examples of Supervised Learning. Unsupervised learning is when a program is given access to an infinite amount of data in order to identify patterns and determine how those patterns are related to one another (Babu, Sritha Zith Dey, et al. (2022)). Discovering previously hidden patterns in the datasets is possible through the use of unsupervised learning. k-Means Clustering, k-Nearest Neighbor, and other similar methods are examples. One of the approaches to supervised and unsupervised learning known as reinforcement learning is used to train machine learning models to make a series of decisions through the process of reinforcement learning. Robots that are powered by reinforcement learning find widespread use

*Figure 2. Structure of artificial neural network*

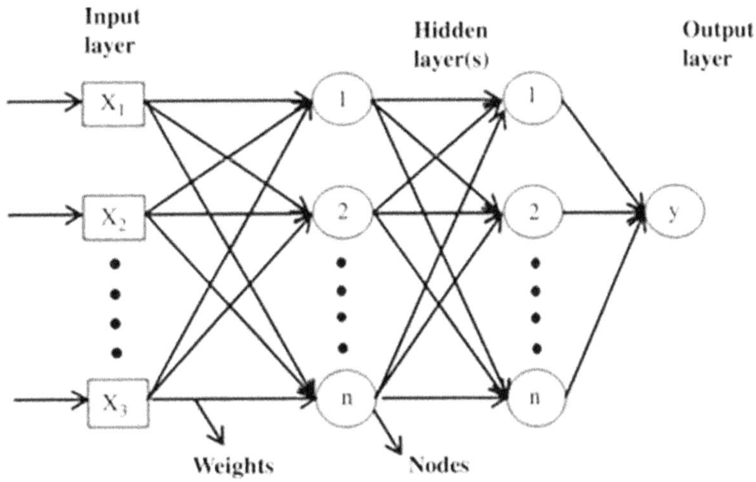


in industry, where they can be found carrying out a variety of tasks. Image enhancement, restoration, segmentation, object detection, image compression, manipulation, and image generation are all areas in which artificial intelligence has proven to be useful in the field of image processing. This is due to the enormous amount of data that has become available.

## 7.1 Neural Networks Created Artificially

(ANN) stands for "artificial neural network," which refers to a type of computational model that is based on the structure and functions of biological neural networks. The variation in input and output has an effect on the structure of the artificial neural network (ANN), which is affected by the information that travels through the network. Considering ANN as a nonlinear statistical tool allows for the modelling of intricate relationships that exist between inputs and outputs (Xi Zhang, 2022). The ability of an ANN to learn from the observation of data sets is put to use in the preprocessing of input images and the approximation of the output of random functions. These techniques are often regarded as the most effective ways to derive novel solutions while simultaneously defining computing functions or distributions in an economical manner.

## 7.2 The Architecture of the ANN

The Artificial Neural Network (ANN) has three layers: the input layer, the output layer, and the hidden layer. Through the use of nodes, the input layer is linked to the hidden layer, and the output layer may be reached directly from the nodes that make up the hidden layer. The input data is obtained from the network.

## 7.3 Application of ANN in Image Processing

Imaging techniques in medicine allow for the diagnosis of a wide variety of illnesses, including kidney stones, glaucoma, brain tumors, cancer of the liver, and so on. The difficulties were fixed by utilizing an

*Figure 3. Structure of the decision tree*

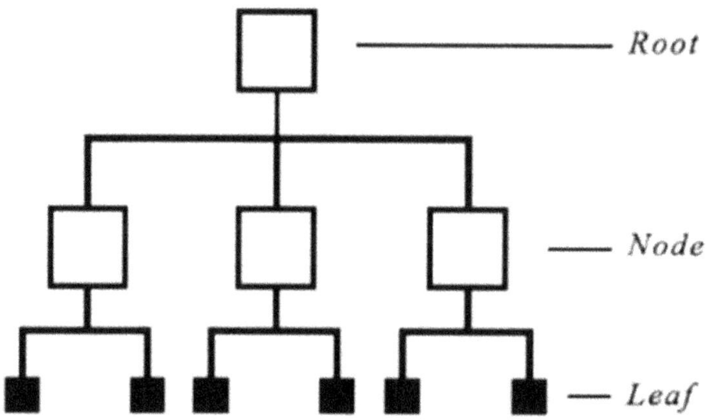


ANN method for picture segmentation in real-time diagnosis with wavelets filters (Figure 2) (Pandey, B. K.,et al.,2023). This allowed for the problems to be resolved. The supervised learning approach of an artificial neural network is utilized for Sign Language Recognition (SLR), which stands for Sign Language Recognition. In the field of remote sensing, the Artificial Neural Network has emerged as a useful tool for the gathering and evaluation of data obtained through remote sensing (Parthiban, K.,et al.,2021).In the field of civil engineering, the detection of flaws on the wall surface of high-rise structures, such as cracks and spalls, is an essential part of the process of using image processing and machine learning algorithms to maintain the buildings and reach a high level of classification accuracy.

## 7.4 Decision Tree

The Decision Tree algorithm has found widespread use in the process of picture classification, particularly for usage with aerial and information-based images (Figure 3). According to Sharma et al. (2013), this classification method is beneficial for remote sensing applications since it is efficient, straightforward, adaptable, and advantageous. Image classification, in general, seeks to obtain thespectral, spatial, accuracy, and other qualities to make the classification trustworthy and to build thematic maps of digital pictures (Pooja et al,2011). This is stated as the primary objective of image classification. Classification methods may be broken down into two distinct categories: supervised and unsupervised classification methods. The decision tree is a form of supervised classification approach, which is one of the inductive learning algorithms. This type of technique falls under the category of inductive learning. It is composed of a hierarchical structure that is subdivided into several layers, beginning with the root node and progressing upward to the sub-nodes and leaves.

According to Sharma et al. (2013), developing a decision tree calls for intensive training in addition to the use of a training data set that is made up of two distinct categories of variables: response variables and explanatory factors. The Shannon entropy theory and the gain information theory are the two fundamental algorithms that are used in the decision tree. These two theories combined to establish the mathematical foundation upon which the decision trees were constructed. The decision tree construction algorithms ID3 and C4.5 have their origins in Shannon's ideas, which serve as their foundation.

## 7.5 Support Vector Machine

The purpose of the support vector machine (SVM) is to categorize data in a binary form, which produces a linear separation hyperplane. Conventional support vector machine (SVM) classification greatly enhanced the dimension of an initial feature by utilizing a kernel approach. In most cases, it is broken down into the categories of classification, regression, and clustering (Zaidi, A., et al., (2023)). SVM techniques that are often used include support vector regression (Smola, 1999), least squares support vector machine (Suykens et al, 2002), and successive projection algorithm-support vector machine (Galvao et al, 2008). An technique offered by Support Vector Machine in Image Processing is used for recognizing and categorizing pictures in a variety of fields, including agriculture, healthcare, remote sensing, and others.

## 7.6 Random Forests Algorithm

The approach known as random forests is utilized for supervised learning applications such as classification and regression. The collection of learning strategies known as a random forest is made up of a number of different decision trees. The accuracy of the forecast is improved by using a random forest to average the results of numerous decision trees. There are several other models that may be used to make predictions about data categorization. The Random Forest Classification Algorithm yields a high level of accuracy for image data analysis while also reducing the amount of processing time required.

## 7.7 K- Means Clustering

(Shaikh et al., 2017) utilized the illness in order to do picture segmentation. The dataset consists of the collection of both healthy and ill pictures. It is possible to recognize a diseased image by first preprocessing the data and then segmenting it into three subimages that each have distinct characteristics (A. Y. A. B. Ahmad, et al.(2019)). As a direct result of this, the detection method in question does not provide a high level of precision.

## 7.8 Markov Chain Model

One form of reinforcement learning that is used in machine learning techniques is called the Markov Chain Model. It is a collection of a random variable process that contains the analytical model in a probabilistic manner in which output depends on the prior results. This model was constructed in this way because the output depends on the previous results. The Markov chain technique is used in (Matis et al.,1989) for the forecasting of image using multiple linear and rank regression models. In order to make accurate projections of the picture, the second-order Markov Chain Model is utilized, in which the data are utilized concurrently. The conclusion that can be drawn from the findings of the photographs is that the effectiveness of the second-order Markov Chain Model is superior to that of the first-order model.

## 8. PARTICLE SWARM OPTIMIZATION (PSO)

According to Meneses et al. (2009), the particle swarm optimization (PSO) algorithm is a metaheuristic algorithm that is capable of solving hard mathematical problems (Fraihat, B. A. M.,et al.,2023). This method is built on the notion of swarm intelligence. Compared to other optimization techniques PSO provides greater advantages with lower parameters (Pandey, B. K., et al., (2021)).

PSO algorithm (or) symbiotic cooperative algorithm is a population-based search algorithm, which depends on the modeling of the social behavior of birds inside a flock. The goal of particle is to graphically mimic the elegant and unexpected dance of a bird flock. The Objective of uncovering patterns that regulate the capacity of birds to fly synchronously and promptly changes the direction by regrouping an ideal formation (A. Y. A. B. Ahmad, et al. (2023). This principle is formalized into a simple and efficient optimization technique. According to Kumar Gaurav et al. (2013), while using PSO, the particles, which are referred to as people, are "flown" over hyper dimensional search space. The positions of the particles inside the search space are modified based on the social psychological propensity (Pandey, D., et al., (2021))..

According to Tajbakhsh et al. (2016) and Shen et al. (2017), Deep Learning is a spectacular state-of-the-art accomplishment for a variety of medical image analysis tasks. These tasks include image denoising, augmentation, segmentation, and anomaly detection (Dutta, S.,et al.,2023). The majority of picture improvement issues are solved by using evolutionary optimization algorithms including genetic algorithms, bacterial forging, artificial bee colonies, ant colonies, and particle swarm optimization, etc. (Li et al., 2008; Marinakis et al., 2013). Since real photographs cannot be relied upon to provide reliable results in the medical industry, any critical features, such as aberrant cell development or dilated blood arteries, will be undetectable after image augmentation. In order to circumvent the issues described above, an image improvement strategy based on the Type 2 Fuzzy Set Theory and the PSO system has been used to low contrast color pictures. In order to keep the image's quality at its highest possible level, one might consider utilizing PSO in a type-2 fuzzy domain for the fitness function (B. Muniandi et al.,2019).

## 9. TYPES OF PARTICLE SWARMOPTIMIZATION

Particle Swarm Optimization algorithms are classified as Gradient PSO, Hybrid PSO and Newtons method (Figure 4).

Deterministic techniques and heuristic methods are the two primary categories that make up optimization strategies in general. According to Bruno et al.'s 2019 research, the purpose of the deterministic technique is to determine whether or not an iterative process that involves a gradient would, after a predetermined number of iterations, converge to the minimal value of the objective function (Pandey, B. K., et al., (2022)).

The iterative procedure is written as:

$X^{k+1} = X^k + \alpha^k d^k$

Where,

x - variable vector

$\alpha$ - step size

d - descent direction

k - iteration number

*Figure 4. Classification of PSO algorithm*

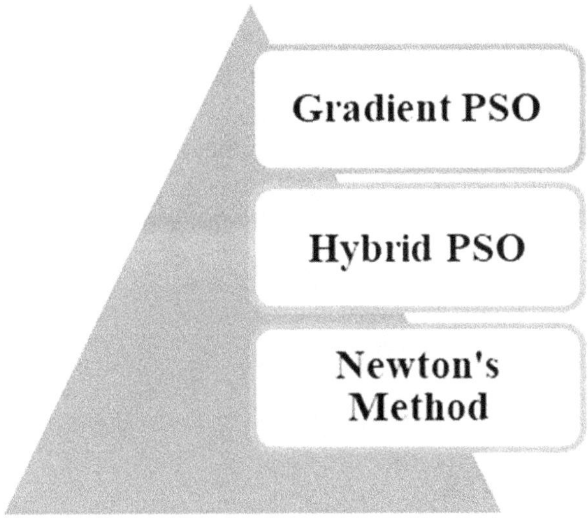


In contrast to deterministic approaches, heuristic methods do not make use of the objective function gradient in the descending direction (Ahmad, A. Y. B.,et al.,2024). In order to determine whether the target function is maximal or minimal, it is necessary to choose, in a tasteful and well-organized fashion, the points at which such a function will be computed in order to realize the goal of imitating nature (Pandey, J. K.,et al., 2022).

The mix of deterministic and heuristic approaches is what's known as a hybrid technique. A heuristic approach is often used by hybrid approaches in order to identify the most global minimal area (Ahmad, A. Y. B.,et al.,2023). Following the identification of this area, the algorithm for the hybrid formulation shifts to a deterministic approach in order to arrive at the minimal point. The heuristic technique is utilized as the general approach for this formulation (Sengupta, R.,et al.,2020). The heuristic method is used to provide excellent candidates for an optimum solution, and the best point identified is used as the starting point for the deterministic procedures, which converge to local minimums.

Further, three deterministic methods in which the PSO was coupled:

**1. Conjugate Gradient Method:** This method enhances the convergence rate of the steepest descent method by selecting downward directions which is a linear combination of the gradient direction with the descending directions of previous iterations.

$X^{k+1}= x^k+\alpha^k d^k$

$d^k= -\nabla(x^k)+\gamma^k d^{k-1}$

where,

$\gamma$ - Conjugation Coefficient

**2. Newton's Method:** This method uses second derivative information for accelerating the convergence of the iterative process. First derivative information is employed the steepest descent and conjugate gradient methods. This method is written as:

$X^{k+1}= x^k+\alpha^k d^k$

$d^k=-[H(x)]^{-1}\nabla U(x^k)$

Where,

H(x) - Hessian of the function

Usually, this method requires little iteration to converge. But, it requires a matrix that grows with the size of the problem. If the estimate is distant from the minimum point, the Hessian matrix is weakly conditioned. Also, it involves inserting a matrix, which formulates the method computationally more expensive.

**3.Quasi Newton method:**By making use of the gradient information provided by the function, it aims to approximate the Hessian inverse. Second derivatives are not a part of this approximation because of its nature. As a result, despite being computationally faster than Newton's methods, this one has a slower rate of convergence.

$X^{k+1} = x^k + \alpha^k d^k$

$d^k = -H^k \nabla U(x^k)$

$H^k = H^{k-1} + M^{k-1} + N^{k-1}$

## 10. FUTURE DEVELOPMENTS

• Improved Methods for Detecting photos That Have Been modified or Forged The next step in the evolution of image forgery will be the creation of more complex methods that can detect and identify photos that have been modified or forged (Ni, L., Ahmad,et al.,2023). In order to do this, it may be necessary to investigate deep learning techniques such as CNNs and GANs in order to increase the precision of forgery detection algorithms (Sharma, M.,et al.,2023).

• **Real-Time Image Scam Detection**: Real-time applications require efficient and quick image scam detection methods, and future development will involve the design of AI and ML algorithms that can process images in real time to prevent scams, fake profiles, and fraudulent activities (Anand, R.,et al.,2023) .

• **Developments in Deep Learning:** Deep learning has been an essential component of image processing, and next research and development will concentrate on further refining and enhancing it to do more difficult jobs (Madhumathy, P., & Pandey, D. (2022)). This might entail the development of novel designs, the optimisation of network topologies, or the investigation of new training methodologies (Kumar, M. S.et al.,2021).

• **Integration of many Modalities of Data:** Future development will require the integration of many modes of data in order to improve image processing skills. Artificial intelligence algorithms will be built to integrate and analyse data from a variety of sources, which will enable a more thorough comprehension of pictures and their interpretation (Nishitha, P., & Pandey, D. (2021)).

• **Explainable AI in Image Processing**: Explainable AI is an approach to artificial intelligence that strives to offer outcomes that are both visible and interpretable. This gives users the ability to comprehend the decision-making process that AI models utilise. Both predict that the future of image processing will centre on the development of explainable artificial intelligence methods. These approaches will be used to increase trust and dependability in essential applications such as medical imaging and autonomous cars(Deepa, R.,et al.,2022).

• **Privacy and Security**: Future research will entail incorporating privacy-preserving techniques into AI and ML algorithms to safeguard sensitive information in pictures. Robust security measures will also be implemented to avoid adversarial attacks or unauthorised access. Both of these aspects will be addressed in the context of protecting sensitive information.

• **Applications in Multiple Fields**: The AI and ML approaches used in image processing have the potential to be used in a variety of fields, including but not limited to healthcare, agriculture, robotics, and environmental monitoring (Sennan, S.,et al.,2022). To prepare for future development, current algorithms will need to be adapted, and domain-specific models will need to be developed (George, A. S.,et al.,2022) so that they can meet particular issues and needs (M. K. Daoud,et al.,2023).

## 11. CONCLUSION

The use of cheap technology and software makes digital photo fraud harder to detect. It must be detected using an authorized picture based on image alteration traces. Image fraud detection methods may reveal copying, shifting, splicing, retouching, and resampling. An genuine signature or watermark is needed for active picture authentication, which limits the usage of specialized imaging equipment. Passive image authentication may replace active authentication without active side information.

Image counterfeiting is increasingly detected using passive techniques as image-editing software gets simpler to use to create convincing forgeries. Detecting forged digital photographs requires dependable, precise, and effective methods. No algorithms exist to automatically verify digital photos. Most photo forgeries leave behind unnoticed visual distortions like fingerprints, making it hard to establish a uniform approach for recognising all forgeries. To identify various digital content alteration signatures, many forgery detection algorithms must be constructed.

Researchers have been exploring methods to spot fraudulent or manipulated photographs since the Internet and digital photography became popular. There are several ways to detect changed or fabricated photos, however the failure rate influences their effectiveness. Some techniques are resilient to additive noise and lossy compression and have high temporal complexity, but they cannot recognize rotational or scaling changes. Post region duplication, noise invasion, blurring, compression, and block-matching search space affect picture counterfeit detection.

## REFERENCES

Ahmad, A., Abusaimeh, H., Rababah, A., Alqsass, M., Al-Olima, N., & Hamdan, M. (2024). Assessment of effects in advances of accounting technologies on quality financial reports in Jordanian public sector. *Uncertain Supply Chain Management, 12*(1), 133–142. doi:10.5267/j.uscm.2023.10.011

Ahmad, A. Y. A. B. (2019). Empirical Analysis on Accounting Information System Usage in Banking Sector in Jordan. *Academy of Accounting and Financial Studies Journal, 23*(5), 1–9.

Ahmad, A. Y. A. B., Gongada, T. N., Shrivastava, G., Gabbi, R. S., Islam, S., & Nagaraju, K. (2023). E-Commerce Trend Analysis and Management for Industry 5.0 using User Data Analysis. *International Journal of Intelligent Systems and Applications in Engineering, 11*(11s), 135–150.

Ahmad, A. Y. A. B., & Kumari, S. S. (2023). Blockchain Implementation in Financial Sector and Cyber Security System. *2023 International Conference on Artificial Intelligence and Smart Communication (AISC)*, Greater Noida, India. 10.1109/AISC56616.2023.10085045

Ahmad, A. Y. B. (2024). Ethical implications of artificial intelligence in accounting: A framework for responsible ai adoption in multinational corporations in Jordan. *International Journal of Data and Network Science*, *8*(1), 401–414. doi:10.5267/j.ijdns.2023.9.014

Ahmad, A. Y. B. (2024). Firm Determinants that Influences Implementation of Accounting Technologies in Business Organizations. *WSEAS Transactions on Business and Economics*, *21*, 1–11. doi:10.37394/23207.2024.21.1

Ahmad, A. Y. B., Gongada, T. N., Shrivastava, G., Gabbi, R. S., Islam, S., & Nagaraju, K. (2023). E-Commerce Trend Analysis and Management for Industry 5.0 using User Data Analysis. *International Journal of Intelligent Systems and Applications in Engineering*, *11*(11s), 135–150.

Ahmad, A. Y. B., Hannoon, A., Al-Daoud, K. I., Abu-Alsondos, I. A., & Al-Qaisieh, M. S. (2023). Assessment of Cloud Based Accounting Technology Adoption and Business Performance. *Kurdish Studies, 11*(3).

Ahmad, A. Y. B., Kumari, D. K., Shukla, A., Deepak, A., Chandnani, M., Pundir, S., & Shrivastava, A. (2024). Framework for Cloud Based Document Management System with Institutional Schema of Database. *International Journal of Intelligent Systems and Applications in Engineering*, *12*(3s), 672–678.

Ahmad, A. Y. B., Kumari, D. K., Shukla, A., Deepak, A., Chandnani, M., Pundir, S., & Shrivastava, A. (2024). Framework for Cloud Based Document Management System with Institutional Schema of Database. *International Journal of Intelligent Systems and Applications in Engineering*, *12*(3s), 672–678.

Ahmad, A. Y. B., & Kumari, S. S. (2023). Blockchain Implementation in Financial Sector and Cyber Security System. *2023 International Conference on Artificial Intelligence and Smart Communication (AISC)*, Greater Noida, India. 10.1109/AISC56616.2023.10085045

Ahmad, A. Y. B., Tiwari, A., Nayeem, M. A., Biswal, B. K., Satapathy, D. P., Kulshreshtha, K., & Bordoloi, D. (2024). Artificial Intelligence Perspective Framework of the Smart Finance and Accounting Management Model. *International Journal of Intelligent Systems and Applications in Engineering*, *12*(4s), 586–594.

Ahmad, A. Y. B., William, P., Uike, D., Murgai, A., Bajaj, K. K., Deepak, A., & Shrivastava, A. (2024). Framework for Sustainable Energy Management using Smart Grid Panels Integrated with Machine Learning and IOT based Approach. *International Journal of Intelligent Systems and Applications in Engineering*, *12*(2s), 581–590.

Anand, R., Lakshmi, S. V., Pandey, D., & Pandey, B. K. (2023). An enhanced ResNet-50 deep learning model for arrhythmia detection using electrocardiogram biomedical indicators. *Evolving Systems*, 1–15.

Babu, S. Z. D., & Pandey, D. (2022). Analysation of Big Data in Smart Healthcare. In *Artificial Intelligence on Medical Data: Proceedings of International Symposium, ISCMM 2021*, (pp. 243-251). Singapore: Springer Nature Singapore.

Colaço, M. J., Orlande, H. R. B., & Dulikravich, G. S. (2006). Inverse and optimization problems in heat transfer. *Journal of the Brazilian Society of Mechanical Sciences and Engineering*, *28*(1), 1–24. doi:10.1590/S1678-58782006000100001

Daoud, M. K., Al-Qeed, M., Ahmad, A. Y. B., & Al-Gasawneh, J. A. (2023). Mobile Marketing: Exploring the Efficacy of User-Centric Strategies for Enhanced Consumer Engagement and Conversion Rates. *International Journal of Membrane Science and Technology, 10*(2), 1252–1262. doi:10.15379/ijmst.vi.1425

Daoud, M. K., Al-Qeed, M., Ahmad, A. Y. B., & Al-Gasawneh, J. A. (2023). Mobile Marketing: Exploring the Efficacy of User-Centric Strategies for Enhanced Consumer Engagement and Conversion Rates. *International Journal of Membrane Science and Technology, 10*(2), 1252–1262. doi:10.15379/ijmst.vi.1425

Daoud, M. K., Taha, S., Al-Qeed, M., Alsafadi, Y., Ahmad, A., & Allahham, M. (2024). EcoConnect: Guiding environmental awareness via digital marketing approaches. *International Journal of Data and Network Science, 8*(1), 235-242.

Deepa, R., Anand, R., Pandey, D., Pandey, B. K., & Karki, B. (2022). Comprehensive performance analysis of classifiers in diagnosis of epilepsy. *Mathematical Problems in Engineering, 2022*, 2022. doi:10.1155/2022/1559312

Dutta, S., Mukherjee, U., & Pandey, D. (2023). Recent Trends of Addressing COVID-19 Disease by AI/ML. In Handbook of Research on Thrust Technologies' Effect on Image Processing (pp. 395-419). IGI Global.

Fraihat, B. A. M., Ahmad, A. Y. B., Alaa, A. A., Alhawamdeh, A. M., Soumadi, M. M., Aln'emi, E. A. S., & Alkhawaldeh, B. Y. S. (2023). Evaluating Technology Improvement in Sustainable Development Goals by Analysing Financial Development and Energy Consumption in Jordan. *International Journal of Energy Economics and Policy, 13*(4), 348–355. doi:10.32479/ijeep.14438

Galvão, R. K. H., Araújo, M. C. U., Fragoso, W. D., Silva, E. C., José, G. E., Soares, S. F. C., & Paiva, H. M. (2008). A variableelimination method to improve the parsimony of MLR models using the successive projections algorithm. *Chemometrics and Intelligent Laboratory Systems, 92*(1), 83–91. doi:10.1016/j.chemolab.2007.12.004

Gaurav, K., & Bansal, H. (2013), Particle Swarm Optimization (PSO) Technique for Image Enhancement. *IJECT, 4.*

George, A. H., Fernando, M., George, A. S., Baskar, T., & Pandey, D. (2021). Metaverse: The next stage of human culture and the internet. [IJARTET]. *International Journal of Advanced Research Trends in Engineering and Technology, 8*(12), 1–10.

George, A. S., Baskar, T., George, A. H., Pandey, D., & Martin, A. G. (2022). A Review of 6G: Towards The Future. *Partners Universal International Research Journal, 1*(4), 1–12.

Hemamalini, V., Rajarajeswari, S., Nachiyappan, S., Sambath, M., Devi, T., Singh, B. K., & Raghuvanshi, A. (2022). Food quality inspection and grading using efficient image segmentation and machine learning-based system. *Journal of Food Quality, 2022*, 1–6. doi:10.1155/2022/5262294

Kumar, M. S., Sankar, S., Nassa, V. K., Pandey, D., Pandey, B. K., & Enbeyle, W. (2021). Innovation and creativity for data mining using computational statistics. In *Methodologies and Applications of Computational Statistics for Machine Intelligence* (pp. 223–240). IGI Global. doi:10.4018/978-1-7998-7701-1.ch012

Li, L., & Li, D. (2008). Image texture classification based on immune particle swarm optimization. *Acta Geodaetica et Cartographica Sinica.*, *37*(2), 185–189.

Madhumathy, P., & Pandey, D. (2022). Deep learning based photo acoustic imaging for non-invasive imaging. *Multimedia Tools and Applications*, *81*(5), 7501–7518. doi:10.1007/s11042-022-11903-6

Marinakis, Y., Iordanidou, G. R., & Marinaki, M. (2013). Particle swarm optimization for the vehicle routing problem with stochastic demands. *Applied Soft Computing*, *13*(4), 1693–1704. doi:10.1016/j.asoc.2013.01.007

Matis, J. H., Birkett, T., & Boudreaux, D. (1989). An Application of the Markov Chain Approach to Forecasting Cotton Yields from Surveys. *Agricultural Systems*, *29*(4), 357–370. doi:10.1016/0308-521X(89)90097-8

Meneses, A. A. M., Machado, M. D., & Schirru, R. (2008). Particle swarm optimization applied to the nuclear reload problem of a pressurized water reactor. *Progress in Nuclear Energy*, *51*(2), 319–326. doi:10.1016/j.pnucene.2008.07.002

Muniandi, B., Huang, C., Kuo, C., Yang, T., Chen, K., Lin, Y., Lin, S., & Tsai, T. (2019). A 97% maximum efficiency fully automated control turbo boost topology for battery chargers. *IEEE Transactions on Circuits and Systems. I, Regular Papers*, *66*(11), 4516–4527. doi:10.1109/TCSI.2019.2925374

Muniandi, B., Huang, C.-J., Kuo, C.-C., Yang, T.-F., Chen, K.-H., Lin, Y.-H., Lin, S.-R., & Tsai, T.-Y. (2019, November). A 97% Maximum Efficiency Fully Automated Control Turbo Boost Topology for Battery Chargers. *IEEE Transactions on Circuits and Systems. I, Regular Papers*, *66*(11), 4516–4527. doi:10.1109/TCSI.2019.2925374

Ni, L., Ahmad, S. F., Alshammari, T. O., Liang, H., Alsanie, G., Irshad, M., Alyafi-AlZahri, R., Bin-Saeed, R. H., Al-Abyadh, M. H. A., Abu Bakir, S. M. M., & Ayassrah, A. Y. B. A. (2023). The role of environmental regulation and green human capital towards sustainable development: The mediating role of green innovation and industry upgradation. *Journal of Cleaner Production*, *421*, 138497. doi:10.1016/j.jclepro.2023.138497

Nishitha, P., & Pandey, D. (2021). A study on student perception towards online education during covid-19 crisis. *Augmented Human Research*, *6*(1), 16. doi:10.1007/s41133-021-00055-1

Pandey, B. K., Pandey, D., Alkhafaji, M. A., Güneşer, M. T., & Şeker, C. (2023). A Reliable Transmission and Extraction of Textual Information Using Keyless Encryption, Steganography, and Deep Algorithm with Cuckoo Optimization. In *Micro-Electronics and Telecommunication Engineering: Proceedings of 6th ICMETE 2022* (pp. 629–636). Springer Nature Singapore. doi:10.1007/978-981-19-9512-5_57

Pandey, B. K., Pandey, D., Wairya, S., & Agarwal, G. (2021). An advanced morphological component analysis, steganography, and deep learning-based system to transmit secure textual data. [IJDAI]. *International Journal of Distributed Artificial Intelligence*, *13*(2), 40–62. doi:10.4018/IJDAI.2021070104

Pandey, B. K., Pandey, D., Wairya, S., Agarwal, G., Dadeech, P., Dogiwal, S. R., & Pramanik, S. (2022). Application of integrated steganography and image compressing techniques for confidential information transmission. *Cyber Security and Network Security,* 169-191.

Pandey, B. K., Pandey, D., Wariya, S., & Agarwal, G. (2021). A deep neural network-based approach for extracting textual images from deteriorate images. *EAI Endorsed Transactions on Industrial Networks and Intelligent Systems, 8*(28), e3–e3. doi:10.4108/eai.17-9-2021.170961

Pandey, B. K., Pandey, S. K., & Pandey, D. (2011). A survey of bioinformatics applications on parallel architectures. *International Journal of Computer Applications, 23*(4), 21–25. doi:10.5120/2877-3744

Pandey, D., Ogunmola, G. A., Enbeyle, W., Abdullahi, M., Pandey, B. K., & Pramanik, S. (2021). COVID-19: A framework for effective delivering of online classes during lockdown. *Human Arenas,* 1-15.

Pandey, D., Pandey, B. K., & Wairya, S. (2021). Hybrid deep neural network with adaptive galactic swarm optimization for text extraction from scene images. *Soft Computing, 25*(2), 1563–1580. doi:10.1007/s00500-020-05245-4

Pandey, D., Wairya, S., Al Mahdawi, R. S., Najim, S. A. D. M., Khalaf, H. A., Al Barzinji, S. M., & Obaid, A. J. (2021). Secret data transmission using advanced steganography and image compression. *International Journal of Nonlinear Analysis and Applications, 12*(Special Issue), 1243–1257.

Pandey, J. K., Jain, R., Dilip, R., Kumbhkar, M., Jaiswal, S., Pandey, B. K., & Pandey, D. (2022). Investigating Role of IoT in the Development of Smart Application for Security Enhancement. In *IoT Based Smart Applications* (pp. 219–243). Springer International Publishing.

Parthiban, K., Pandey, D., & Pandey, B. K. (2021). Impact of SARS-CoV-2 in online education, predicting and contrasting mental stress of young students: A machine learning approach. *Augmented Human Research, 6*(1), 10. doi:10.1007/s41133-021-00048-0

Pooja, A.P., Jayanth, J. and Koliwad, S.(2011). Classification of RS data using Decision Tree Approach. *International Journal of Computer Applications, 23*(3).

Raghuvanshi, A., Singh, U. K., & Joshi, C. (2022). A review of various security and privacy innovations for IoT applications in healthcare. *Advanced Healthcare Systems, 4*, 43–58. doi:10.1002/9781119769293.ch4

Rawat, W., & Wang, Z. (2017). Deep Convolutional Neural Networks for Image Classification: A Comprehensive Review. *Neural Computation, 29*(9), 1–98. doi:10.1162/neco_a_00990 PMID:28599112

Seixas Gomes de Almeida, B., & Coppo Leite, V. (2019, December 4). Victor CoppoLeite (2019),Particle Swarm Optimization: A Powerful Technique for Solving Engineering Problems. *Swarm Intelligence.* Advance online publication. doi:10.5772/intechopen.89633

Sengupta, R., Sengupta, D., Kamra, A. K., & Pandey, D. (2020). Artificial Intelligence and Quantum Computing for a Smarter Wireless Network. *Artificial Intelligence, 7*(19), 2020.

Sennan, S., Pandey, D., Alotaibi, Y., & Alghamdi, S. (2022). A Novel Convolutional Neural Networks Based Spinach Classification and Recognition System. *Computers, Materials & Continua, 73*(1), 343–361. doi:10.32604/cmc.2022.028334

Shaikh, T. N., & Mukane, S. M. (2017). Segmentation of Pomegranate Leaf for Detection of Disease Using Image Processing. *International Journal of Innovative Research in Computer and Communication Engineering*, *5*(4), 378–384.

Sharma, M., Saripalli, S. R., Gupta, A. K., Palta, P., & Pandey, D. (2023). Image Processing-Based Method of Evaluation of Stress from Grain Structures of Through Silicon Via (TSV). *International Journal of Image and Graphics*, 2550008. doi:10.1142/S0219467825500081

Sharma, M., Sharma, B., Gupta, A. K., & Pandey, D. (2023). Recent developments of image processing to improve explosive detection methodologies and spectroscopic imaging techniques for explosive and drug detection. *Multimedia Tools and Applications*, *82*(5), 6849–6865. doi:10.1007/s11042-022-13578-5

Sharma, R., Ghosh, A., & Joshi, P. K. (2013). Decision tree approach for classification of remotely sensed satellite data using open source support. Indian Academy of Sciences. *Journal of Earth System Science*, *122*(5), 1237–1247. doi:10.1007/s12040-013-0339-2

Shen, D., Wu, G., & Suk, H.-I. (2017b). Deep Learning in Medical Image Analysis. *Annual Review of Biomedical Engineering*, *19*(1), 221–248. doi:10.1146/annurev-bioeng-071516-044442 PMID:28301734

ShenL.YueZ.FengF. (2017). MSR-net: Low-light Image Enhancement Using Deep Convolutional Network. arXiv:1711.02488v1

Smola, A. (1999). *Regression Estimation with Support Vector Learning Machines* [Master's Thesis, The Technical University of Munich].

Suykens, J. A. K., Van Gestel, T., De Brabanter, J., De Moor, B., & Vandewalle, J. (2002). *Least Squares Support VectorMachines*. World Scientific. doi:10.1142/5089

Tajbakhsh, N., Jae, Y., & Shin, Y. (2016). Convolutional Neural Networks for Medical Image Analysis: Full Training or Fine Tuning? *IEEE Transactions on Medical Imaging*, *35*(5), 1299–1312. doi:10.1109/TMI.2016.2535302 PMID:26978662

Wu, G., Kim, M., Wang, Q., Munsell, B. C., & Shen, D. (2016). Scalable High Performance Image Registration Framework by Unsupervised Deep Feature Representations Learning. *IEEE Transactions on Biomedical Engineering*, *63*(7), 1505–1516. doi:10.1109/TBME.2015.2496253 PMID:26552069

Zaidi, A. (2023). Predicting wildfires in Algerian forests using machine learning models. *Heliyon*, *9*(7), e18064. doi:10.1016/j.heliyon.2023.e18064 PMID:37519679

Zaidi, A. (2023). Smart Implementation of Industrial Internet of Things using Embedded Mechatronic System. *IEEE Embedded Systems Letters*. doi:10.1109/LES.2023.3309985

Zaidi, A., Ajibade, S.-S. M., Musa, M., & Bekun, F. V. (2023). New Insights into the Research Landscape on the Application of Artificial Intelligence in Sustainable Smart Cities: A Bibliometric Mapping and Network Analysis Approach. *International Journal of Energy Economics and Policy.*, *13*(4), 287–299. doi:10.32479/ijeep.14683

Zaidi, A., & Al-Harbi, O. (2023). Statistical Analysis of a Linear Multi-Step Numerical Treatment. *Journal of Statistics Applications & Probability*, *12*(1), 26. doi:10.18576/jsap/120126

Zaidi, A., & Al Luhayb, A. (2023). Two Statistical Approaches to Justify the Use of the Logistic Function in Binary Logistic Regression. *Mathematical Problems in Engineering, 2023*, 5525675. doi:10.1155/2023/5525675

Zhang, X. (2022). *Application of Artificial Intelligence Recognition Technology in Digital Image Processing*. Wireless Communications and Mobile Computing- Hindawi., doi:10.1155/2022/7442639

Section 2

# Nanoscience, Nanotechnology, and Nanoengineering

# Chapter 4
# Fullerene-Based Modifier:
## Biosensing and Nanobiotechnological Industrial Applications

**Loutfy H. Madkour**
https://orcid.org/0000-0002-3101-8356
*Tanta University, Egypt*

## ABSTRACT

*Fullerene molecules are composed entirely of carbon, in the form of a hollow sphere, ellipsoid, or tube. Fullerenes in the cylindrical form are called carbon nanotubes (CNTs) or buckytubes and fullerenes in the spherical form are referred to as buckyballs. Fullerenes have attracted considerable attention in different fields of science since. Their unique carbon cage structure coupled with immense scope for derivatization makes them a potential therapeutic agent. The fullerenes can be utilized in organic photovoltaic (OPV), portable power, medical purpose, antioxidants, and biopharmaceuticals and dentistry. Recently, fullerene-C60 contributed greatly to the field of biosensing and bio-nanotechnology. Fullerene allows its wide use for designing the highly sensitive chemical/biosensors. Fullerene is not dangerous to biological systems. Fullerene, as the first symmetric nanostructure in the carbon nanomaterials family, opened up new perspectives in nanomaterials field leading to discovery and research on other symmetric carbon nanomaterials.*

## 1. INTRODUCTION

Fullerene was discovered as carbon allotrope by Kroto et al. in 1985, who was awarded with Nobel prize of chemistry in 1996 for their amazing discovery (Kroto et al.1985). Since then, fullerene has been widely explored in cancer theranostics because of its special structures and distinct physio-chemical characteristics. Fullerene is viewed as a form of graphene sheet that rolled into a hollow sphere, ellipsoid, or tube shape. It entirely consists of 60 carbon atoms linked by 30 carbon double bonds with abundant π-π conjugation in cage-type structure (Augustine et al.2017) . Interestingly, the unique geometry and molecular topology of fullerene endow it with prominent properties such as large surface area, small particle size, and

DOI: 10.4018/979-8-3693-1335-0.ch004

high reactivity. Furthermore, fullerene also exhibits electronic, optical, thermal, mechanical properties, and other versatile physicochemical features (Jariwala et al.2013) . The large specific area and unique 3D scaffold structure of fullerene enable the loading of multiple drugs in the fullerene cage through covalent bonding (Augustine et al.2017). Given the excellent characteristics of penetration into solid tumors and high chemical reactivity, fullerene can also be designed as stimuli-responsive systems based on the special conditions of the tumor sites to achieve targeted delivery of therapeutic molecules (Ding et al.2016) . In addition, due to its inherent optical and thermodynamic properties, fullerene has been proved to act as a good photosensitizing agent, making it a promising nanomaterial in cancer imaging, hyperthermia, PDT, and photoacoustic/thermoacoustic assisted theranostics (Shi et al.2019). Notably, owing to its anticancer activity and sensitization effect on cancer cells, fullerene can serve as a strong antineoplastic agent (Ye et al.2021). Besides, fullerene can be employed as an antioxidant due to its free radical scavenging capacity (Ye et al.2021). Interestingly, metal atoms can also be incorporated into fullerene to generate metallofullerene, which inherits the combined properties from carbon cages and internal metal, showing promising prospects for application as contrasts in magnetic resonance imaging (MRI) and X-ray, radiotracers and anticancer agents (Wang et al.2019).

Recently, fullerene-$C_{60}$ contributed greatly to the field of biosensing and bio-nanotechnology. The unique electrochemical and physicochemical properties, together with biocompatibility characteristics of fullerene, allow its wide use for designing the highly sensitive chemical/biosensors.

Concerning all-carbon fullerene polymers, these are fascinating materials which continue to be the subject of intensive research. One-, two- and three-dimensional structures have been synthesized, and many more potential all-carbon polymers are possible. Therefore, electrochemical biosensors based on fullerene-$C_{60}$ with their cost-effectiveness and suitability for micro-fabrication can be expected to become increasingly popular in the near future.

## 1.1. Nanotechnology for Carbon Nanomaterials

In 1985 Kroto, et al., the serendipitous discovery of fullerene triggered the research of carbon structures into the world of symmetric nanomaterials. Consequently, Robert F. Curl, Harold W. Kroto and Richard E. Smalley were awarded the Noble prize in chemistry for their discovery of the buckminsterfullerene ($C_{60}$ with a cage-like fused-ring structure).Since its first discovery, fullerene-$C_{60}$ has been the object of extensive research. Their unique and favorable characteristic of easy chemical modification, conductivity, and electrochemical properties has led to its tremendous use in (bio) sensor applications. Nanotechnology is becoming increasingly important in the field of (bio) sensors. Bio-nanotechnology is a new emerging field of nanotechnology and combines knowledge from engineering, physics, and molecular engineering with biology, chemistry, and biotechnology aimed at the development of novel devices, such as biosensors, nanomedicines, and bio-photonics (Baghchi, et.al.2013).

It is the aim of this chapter to present the most recent and relevant contributions in the development of biosensors based on fullerene-$C_{60}$, graphine and different biological components. A brief introduction and history of fullerene-$C_{60}$ is first presented. Available methods for synthesis and functionalization of fullerene-$C_{60}$ are mentioned also.

The important parameters in the exact diagnosis are sensitivity and accuracy. These two terms can describe the technological power of the diagnostic systems. In the development of sensitivity and accurate measurement, nanomaterials have an important place in today's technologies (Lim et.al.2015).

## 1.2. Carbon

Carbon is one of the most abundant elements in nature and definitely holds one of the first places in nanomaterials research field. Entering the carbon-based nanomaterials world was triggered by the discovery of buckminsterfullerene (shortened to fullerene or buckyball), the third carbon allotrope, after graphite and diamond, in 1985 by Robert et al. have various allotrope like a diamond, graphite, etc., .The third allotropic form of carbon is fullerene.

## 1.3. Carbon Nanotubes

Carbon nanotube are exactly as they sound. They are Fullerenes that have formed a tube or cylindrical shape from a sheet of carbon molecules in a hexagonal shape. The nanotube can be of various lengths, and although most are capped at the end, some are open. As they are Fullerenes, they are all internally hollow.

## 1.4. What Are Buckminsterfullerene ($C_{60}$)?

Fullerenes exist in various number forms in both natural and human-made sources and amongst all Buckminsterfullerene, (C60) is the most common and abundant form. Functionalized Fullerenes are mostly used in diagnostic, therapeutic applications. Fullerene is a very common compound known as an allotrope of carbon that consists of carbon atoms that are attached through either single or double bonds. A fullerene is a molecule with a hollow center and is made up of carbon. Many shapes can represent Fullerenes, such as spheres, cubes, or tubes. Fullerenes are the allotrope form of carbon and closed convex polyhedron, composed of an even number of carbon atoms. The name fullerene originated from that of Richard Buck-minster Fuller (1895-1983).

Fullerenes were first discovered by three scientists in 1985. Sir Harold W. Koto, Dr Richard E. Smalley, and Robert F. Curl discovered the first fullerene after subjecting carbon cores to the atmosphere of helium gas and using a laser to vaporize the rods. Eleven years later, in 1996, they were awarded the Nobel Prize for their groundbreaking discovery.

A fullerene is the three-dimensional allotrope of carbon shaped like a sphere, tube or ellipsoid with the carbon molecules bound to each other through single and double bonds forming pentagonal and hexagonal fused rings. Due to their π-electrons, Fullerenes exhibit interesting phenomena that are easily manipulated by chemical methods.

The first fullerene molecule was discovered and prepared in 1985 by Richard Smalley, Robert Curl, James Heath, Sean O'Brien, and Harold Kroto at Rice University. The name comes from Buckminster Fuller, whose geodesic domes it resembles.

Fullerene (C60) Just like CNTs and graphene-based materials, fullerene C60 has also been widely used for PTT/PDT. Having a soccer-ball-like fully-conjugated carbon structure consisting of 12 pentagons and 20 hexagons facing symmetrically, fullerene (C60) can strongly absorb visible light with interesting photo-excitation properties upon light exposure (Li et al. 2018). Upon irradiation, C60 is excited to a triplet state, which can transfer its energy to surrounding oxygen to form singlet oxygen ($^1O_2$) (Hu, Z. Zhao, F. Wang, Y. et al.., 2014). However, the poor NIR absorption and insolubility in the biological solution of unmodified C60 have strongly limited C60 for PDT (Li, Hong, et.al.2017).

UV–vis excitation of C60 has been applied for anti-bacteria under visible light and fullerenes could also be used in PTT, particularly modified C60 is already used for photo-dynamic cancer therapy (Krishna, et al.., 2010). Since 1997, C60 has been reported as the PDT agent for $^1O_2$ formation under visible light for virus inactivation in biological fluids (Semliki Forest virus (SFV, Togaviridae) and vesicular stomatitis virus (VSV, Rhabdoviridae)) (Käsermann and Kempf, 1997). In 2003, Yamakoshi et al. reported C60 for ROS generation in aqueous media under visible light and demonstrated a DNA cleavage capacity in DNA extracts (Yamakoshi, et al. (2003). To overcome the insolubility limitation of C60 mentioned above, serval modification strategies by introducing hydrophilic moieties, including water-soluble tri-malonic acid functional groups like N-methylpyrrolidinium (BB4) (Mroz, et,al. 2011), PEG (Liu, et, al.2007), and block copolymers micelles, have been reported to acquire hydrophilic properties (Tegos, et al. 2005). It was found that cationic C60 showed rapid and broad killing effects on more than 99.99% of bacterial and fungal cells and showed better selectivity than that widely used antimicrobial PDT agents (e.g., TBO, toluidine blue O) toward mouse fibro blast cells (L929) (Tegos, et al. 2005). Later in 2011, Hamblin and coworkers used the same cationic Fullerenes for intraperitoneal (IP) carcinomatosis PDT in the mice model (colon adenocarcinoma cell line (CT26) bearing BALB/c mice, IP injection) under visible light (400–700 nm) (Mroz, et,al. 2011). Krishna et al.2010 reported that up to 72% tumour size reduction under 785 nm laser irradiation of a water-soluble polyhydroxy fullerene (HPF) for 10 min (500 mW/cm$^2$). Besides, HPF could also be used to image tumours with photoacoustic tomography (PAT) in BT474-bearing mice after intratumoural injection. To achieve *in vivo* targeted area delivery of C60, PEG conjugated C60 has been studied for PDT of cancer *in vivo* under visible light (Tabata, et, al.1997). PEG-modified C60 could be cooperated with other particles to achieve multifunctional systems. For example, in 2007, Liu et al. chelated $Gd^{3+}$ to C60-PEG to enhance PDT efficacy with MRI under visible light (400–700 nm, 89.2 mW/cm$^2$, 10 min), devising a therapeutic and diagnostic hybrid system in tumour-bearing mice (AR1/CDF1 mice) (Liu, et, al.2007). After intravenously injecting C60-PEG-Gd into mice, a sufficiently strong MRI signal from the tumour tissue was obtained for clinical MRI diagnosis.

Since C60-based PDT shows no specificity to tumour cells, tumour targeting agents, including peptides (Shi, et al. 2013), aptamers (Liu, et al. 2013), antibodies, and specific receptor-related molecules, were applied to further improve selectivity for selective inhibition of the tumour growth. In conjugation with tumour-targeting ligands (e.g., NGR that could cognize the CD13 isoform and tumour vascular), C60 showed a specific PDT inhibition of MCF-7 cells (Shi, et al. 2013). In 2010, Liu & Tabata functionalized C60 with pullulan (a polysaccharide with high affinity for asialoglycoprotein receptors; highly expressed at hepatocytes surface) for PDT of a HepG2 hepatoma cell line with high selectivity, leading to a 60% death of HepG2 cancer cells 6 h (Liu and Tabata, 2010) . C60 could also be doped with other anti-cancer drugs (DOX (Shi, J. Liu, Y. Wang, L. et al.., 2014), metals (Shi, Yu, Wang, et al.., 2013), other CNMs or PAs (e.g., Ce6) (Guan, Ge, Wu, et al.., 2016) for CT-PDT and PTT-PDT synergic therapies. Indeed, Shi et al. 2013) conjugated C60 with PEG, iron oxides (IONPs), and PDT anti-cancer drug HMME for PDT. By combining with C60, HOMME showed a 23-fold higher uptake of the tumour over HMME with significantly stronger killing effects. The resultant C60 hybrid was found in the liver, spleen, kidney, and tumour 3 h post-injection and showed excellent *in vitro* and *in vivo* PDT efficacies under 532 nm laser irradiation (300 mW/cm$^2$, 5 min) (Shi et al. 2013).

Overall, FCNVs showed no obvious damage to major organs and could be excreted by the liver and kidney with long-time blood circulation Guan, Ge, Wu, et al.., 2016). PEG crosslinked C60 was reported to be able to photoactively ablate various malignant cells after multimeric modification. In this context, Lee et al. reported (Lee, Ahn, et al. 2012) that Ce6 and folate conjugated PEG-C60 (multimeric C60) showed promising PTT and PDT therapeutic effects toward KB cancer, but with arthritis, after in intravenous injection. The surface temperature was found to reach 44 °C with tremendous singlet oxygen generation under 670 nm laser (300 mW/cm$^2$, 10 min), resulting in arthritic progress inhibition in the arthritis-induced DBA/1 J mice mode and significant tumour volume regression in KB tumour-bearing mice (10 mg/kg).

Fullerene, with its unique structure consisting of sp$^2$ carbons, presents a high symmetric cage with different sizes ($C_{60}$, $C_{76}$ and so on). The most abundant fullerene in the as-synthesized composition is $C_{60}$, the molecular structure of which, together with the one of $C_{70}$, can be observed. $C_{60}$ consists of 60 carbon atoms with $C_5$–$C_5$ single bonds which form 12 pentagons and $C_5$–$C_6$ double bonds which from 20 hexagons (Kroto, et al. 1985) . Indeed, each fullerene with 2n + 20 carbon atoms contains 'n' hexagon (Georgakilas, V. et al. 2015). As shape, fullerene $C_{60}$ is like a soccer balls, and Yadaf et al. have stated that the ratio of fullerene molecule to soccer ball is the same as the ratio of soccer ball to the earth, where the diameter of the earth, soccer ball and fullerene molecule are $12.75 \times 10^6$ m, $2.2 \times 10^{-1}$ m and $7.0 \times 10^{-10}$ m, respectively.

## 2. BASICS AND HISTORY OF FULLERENE ($C_{60}$)

Fullerenes are the allotropic form of carbon and closed convex polyhedral, composed of an even number of carbon atoms. The name fullerene originated from that of Richard Buckminster Fuller (1895-1983), an American engineer, architect, inventor, and writer, known for having discovered the structural principle of tensegrity, it is the property of a structure that stand by the forces of tension and compression. Fullerene is built up of fused pentagons and hexagons forming a curved structure. The smallest stable, and the most abundant, fullerene obtained by the usual preparation method is the Ih-symmetrical buckminsterfullerene $C_{60}$. The next stable homologue is $C_{70}$ followed by higher fullerenes $C_{74}$, $C_{76}$, $C_{78}$, $C_{80}$, $C_{82}$, $C_{84}$, and so on.

The newly found molecule was named by these scientists after Richard Buckminster Fuller, in 1967; he made a dome of shape like the carbon cluster, whereas, in 1970, the possibility of fullerenes existence was predicted. Later, it's observed that fullerene occur in interstellar dust and also in Earth's geological formations, only in the ppm-range. Fullerenes, also known as buckyballs, are carbon's new allotrope and their appearance also resembles soccer balls. Fullerenes consist out of 20 hexagons and 12 pentagons with a carbon atom in each edge of this lattice structure. Fullerenes also exist in different stones, naturally. Earlier, on the cave walls, they were deposited as soot.

Each carbon in a fullerene-$C_{60}$ atom is bonded to three others and is sp$^2$ hybridized. The $C_{60}$ molecule has two bond lengths, the 6:6 ring bonds can be considered as double bonds and are shorter than the 6:5 bonds. $C_{60}$ is not "super aromatic" as it tends to avoid double bonds in the pentagonal rings, resulting in poor electron delocalization. The estimated values of electron affinity (EA) (2.7 eV) and ionization potential (IP) (7.8 eV) of C60 indicate that it can easily contribute to the electron transfer reaction and reveal very rich electrochemical which makes them attractive candidates for electroanalytical applications (Jariwala, Sangwan,et al. 2013).

*Figure 1. Fullerenes in high symmetry*

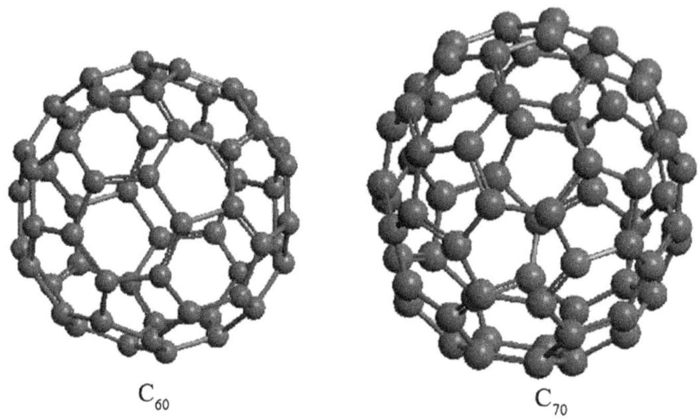


## 2.1. FULLERENE STRUCTURE

Fullerenes can serve as stages for the transportation of drugs and imaging agents. Fullerenes, when surface Functionalized, can have direct antioxidant activity.

The fullerenes are constituted by a network formed by 12 pentagons and 20 hexagons, closing the sphere. Each carbon of a fullerene is sp hybridized in the form of a sigma bond (simple bond) with three other carbon atoms, subtracting one electron from each carbon. The most important property of Fullerenes is its high symmetry (Figure 1). In this one, there are 120 symmetry operations, such as axis rotations or reflections in the plane. This makes the C60 molecule the most symmetrical molecule since it has the largest number of symmetry operations; there are three types of rotation axes C2, C3 and C5.

Production and applications of carbon-based nanomaterials have gained speed in recent years. Carbon nanomaterials that can be found in different conformations with $Sp^2$ hybridization can be extremely useful (Figure 2). These nanomaterials include nanotubes, graphene, carbon nanoparticles, carbon fibers, and Fullerenes (Sotiropoulou, Gavalas et al. 2003). Among these materials, fullerenes are nanomaterials that have gained speed in recent years due to their structure and unique properties.

Fullerenes are spherical carbon nanomaterial derivatives. This structure, unlike other carbon structures, consists of a closed form of pentagonal and hexagonal carbon structures fused together.

*Figure 2. $Sp^2$ hybridization*

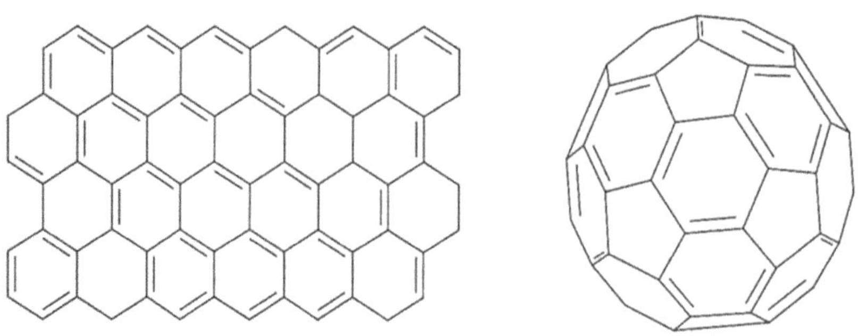

The spherical form of the fullerene nanomaterial gives this nanomaterial a large surface area (Chen, Guan, Yang et al. 2019). This feature is a sought after feature for biosensors and sensor systems. The most advantageous feature of the double bonds formed by the carbon structure is that they can be modified as they can easily respond to chemical reactions.

## 3. DIFFERENT FORMATIONS OF FULLERENES

Fullerenes are presented with the empirical formula of Cn to adopt closed mesh morphology with the most famous member called buckminsterfullerene (sometimes Buckyball) with the molecular formula of $C_{60}$ which is named after the Americana architect and system theorist, Buckminster Fuller. The term Buckyball is used for $C_{60}$ because of its similarity to a soccer ball. Fullerenes resemble the structure of graphite, which is constructed of sheets of carbon molecules in hexagonal shapes. Unlike graphite, Fullerenes are typically pentagonal in shape and therefore do not lie in flat sheets. The most common Fullerenes are constructed of sixty carbon atoms ($C_{60}$) and are bound by single and double bonds. These are the most common types of Fullerenes and are referred to as buck-minster **Fullerenes**, or simply, buckyballs. We also must remember that all forms of Fullerenes have a hollow core similar to footballs and soccer balls.

Since geodesic domes designed by Buck-minster Fuller in 1960s provided clues to realize fullerene $C_{60}$'s structure, this molecule was named after this American architecture, buckminsterfullerene. After this discovery, in 1990, Wolfgang Kra¨tschmer of the Max Planck Institute for Nuclear Physics, and Donald Huffman of the University of Arizona, with their students, prospered in producing fullerene in bulky quantities. Finally, in 1996 Kroto, Smalley and Curl were awarded the Noble Prize in Chemistry for fullerene discovery.

Although the emergence of 2D carbon nanomaterials, as grapheme reported by Geim and Novoselov in 2004, who were awarded the Noble Prize in 2010, marginalized fullerenes and moved the majority of the studies towards other carbon nanostructures (Schwerdtfeger, Wirz, Avery, 2013), the incredible properties of fullerene could not be underestimated.

As stated previously, Buckminster fullerenes also go by "buckyballs". Buckyballs were named after the famed architect and designer R. Buckminster Fuller. They are the most common of all the fullerenes consisting of 60 carbons ($C_{60}$). Buckminster fullerenes form what resembles a soccer ball and have 32 sides, of which 12 sides are pentagonal, and 20 sides are hexagons. Buckyballs are extremely strong, making them valuable in the construction of different types of materials.

The following Figures (3-7) (Bhakta, Poulomi; Barthunia, Bhavna, 2020) will detail the different formations that Fullerenes can take. Other Spherical Fullerenes can consist of as few as 20 carbons and range to being over 100 carbons.

In relation to crystallographic properties of fullerene molecule, the presence of symmetric elements including 30 twofold axes, 20 threefold axes and 12 fivefold axes has made fullerene the most symmetrical molecule (as already mentioned) that is controlled by Golden Mean rule (Yadav, Kumar, 2008). $C_{60}$, with face centered cubic lattices (FCC) in solid phase, has a stable structure to the extent that the destruction of fullerene cage happens at temperatures higher than 1000°C.

As mentioned above, different strategies have been exploited in order to promote fullerenes' behavior in solvents, especially aqueous dispersion, which play a crucial role in biological applications (Astefanei, et al. 2015). One of them is involved with transferring fullerenes from an organic solution

*Figure 3. Different formations that Fullerenes can take*

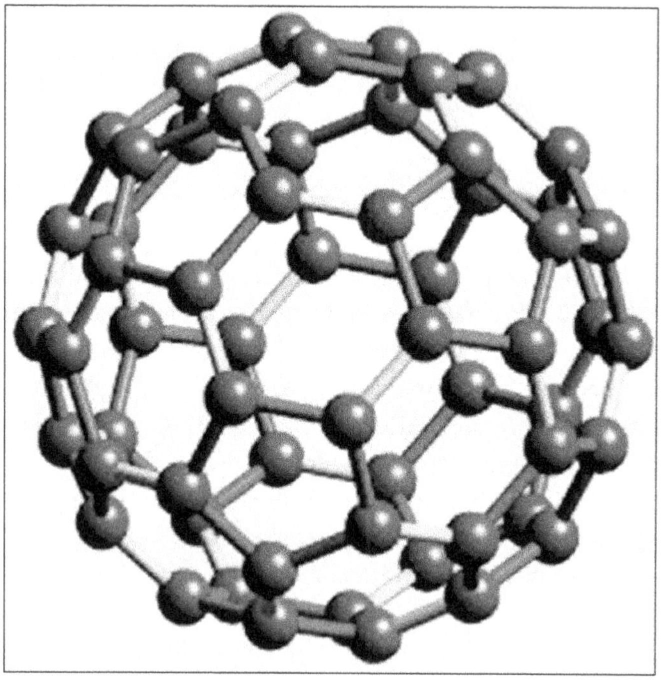

such as toluene or benzene to an aqueous phase with ultrasound treatment with following removal of the organic solvent. The mechanism of fullerene dispersal in aqueous solutions with the aforementioned

*Figure 4. Different formations that Fullerenes can take*

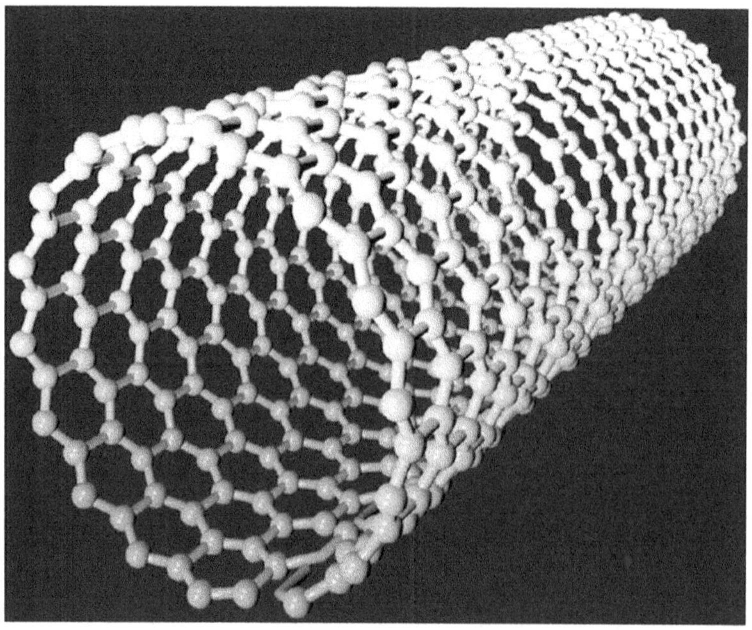

*Figure 5. Different formations that Fullerenes can take*

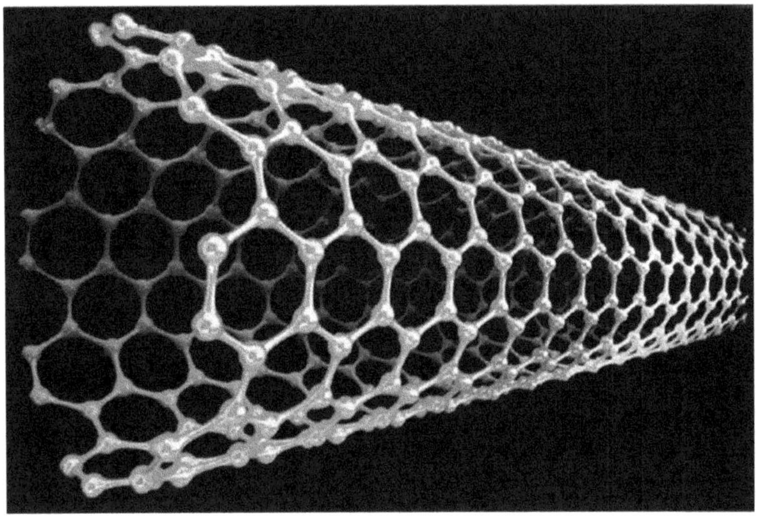

method could be explained in two ways: (i) the formation of water shell via H bonding and charge transfer between $C_{60}$ and water molecules stabilizes fullerene in water and (ii) ultrasound treatment creates a covalent bond between hydroxides and carbons in fullerene cage, which culminates in fullerenol moieties and consequent easy fullerene dissolution. Other attempts include encapsulation in special carriers like cyclodextrins, calixarenes, polyvinylpyrrolidone, micelles and liposomes, which give advantages for bio-pharmaceutical applications like drug delivery (Youle, et al. 2005). In addition to the

*Figure 6. Different formations that Fullerenes can take*

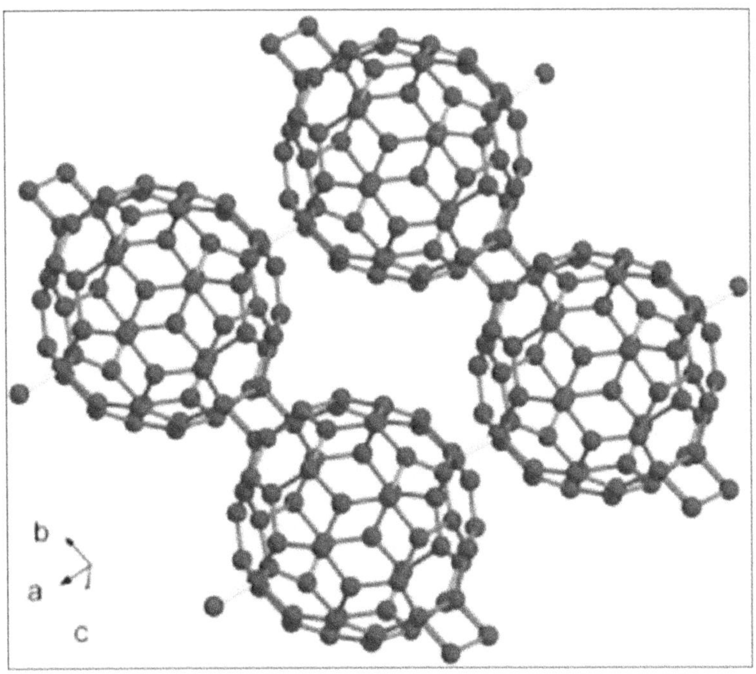

*Figure 7. Different formations that Fullerenes can take*

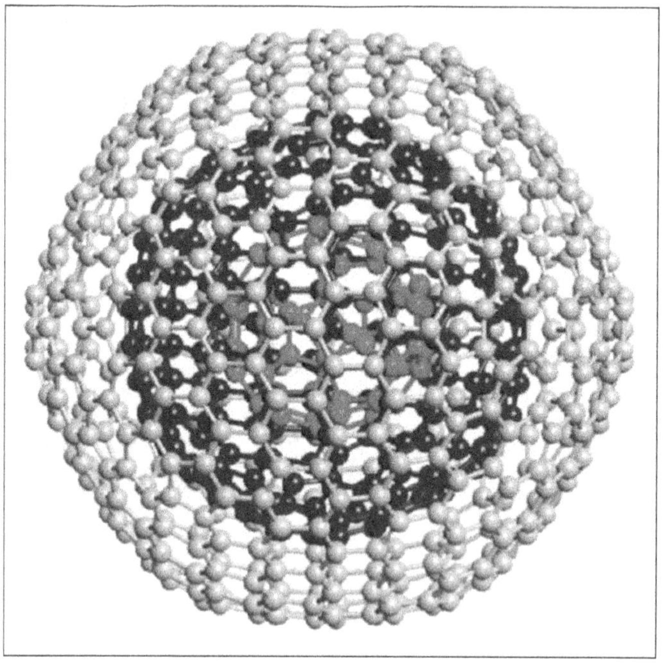

mentioned methods, chemical modification of fullerene with hydrophilic substance such as amino acids, carboxylic acids, polyhydroxyl groups, or amphiphilic polymers enhances hydrophilicity of Fullerenes in biological systems (Biju, 2014). Once rendered soluble in biological media, the excellent properties of buckminsterfullerene make it useful for biomedical applications including antiviral activity, antioxidant activity as a radical scavenger, encapsulating medically materials like magnetic metals for theranostics applications and as a drug carrier in scaffolds.

**Closed mesh topology:** Fullerenes are capable of exhibiting a closed mesh topology and are depicted by an empirical formula which is Cn in which n is known to be the number of carbon atoms that are present. Although, it is possible to have more than one isomers as the values can vary in different ways.

**Naming:** The naming system for each compound and its family is different and so is the case with fullerene as well. The family of fullerene is named after a compound known as buckminsterfullerene (C60) which is the most popular member of this family. This one is named after Buck-minster Fuller. These are the closed Fullerenes and can also be called buckyballs as they have a lot of similarities with the game of soccer. Likewise, the nested closed fullerenes are known as buck onions whereas the cylindrical Fullerenes are known as nanotube or buckytubes. There is a term called fullerite which is the solid mixture of either pure or mixed fullerene.

**Re-diction:** Fullerenes were initially just predicted and no discovery of them was yet made until 1985 when they were found out in nature as well as outer space.

**Source of attraction:** Ever since the Fullerenes have been discovered in 1985, they have gained a lot of attention and attraction in various fields and specifically in the science field. Various properties of fullerene in physical, chemical, and biological aspects have also been identified. Specifically, in the medical field, they are known to have a definite size, hydrophobic properties, and possible electronic configurations. They consist of mainly a carbon cage structure and due to the potentially strong properties; it enables them to be a therapeutic agent. Various studies have been led in the biological applications, all of which have made it evident that fullerenes are an excellent compound being utilized in the field of medicine.

## 4. CHARACTERISTICS OF FULLERENES

There are various characteristics of fullerene which are the basis of this compound and are the key functionalizing aspect that enhances the growth and production of fullerene. The entire family of fullerene and most specifically C60 has a very appealing nature and is thus referred to as the best compounds to exhibit their physical and electrochemical properties. As discussed earlier, fullerene has a hydrophobic nature and this enables it to fit itself in HIV protease to stop the movement of substrates towards the site of the enzyme.

## 5. PROPERTIES AND PRODUCTION OF FULLERENES

Fullerene has a lot of properties (Gokhale M.M, Somani,2015) as follows:

1- Pure fullerenes have a good close packing than impure Fullerenes.

2- C60 derivatives show level of solubility (Taylor, Hare, Abdul-Sada, Kroto, 1990).

3- Fluorinated derivatives are more soluble and some Bromo derivatives are much less soluble.

4- Fullerenes have low density (1.65 g/cc) relative to diamond (3.51 g/cc).

5- Fullerenes are stable up to a temperature of 1,000°C.

6- Fullerenes can react with nucleophiles.

7- Fullerenes can undergo various reactions such as reduction, oxidation, hydrogenation, halogenation, nucleophilic reactions, radical reaction, transition metal complex reaction, regioselective reaction, and so on (Taylor,1999).

8- Bind with single and double covalent bonds.

9- Nanotubes make up the strongest/toughest material known

10- Nanotubes contain strong graphene bonds

11- Due to the negative space, nanotube s are highly compressible, and that compression is easily reversed.

12- Buckyballs react differently under different types of temperature and pressure.

13- The hardness of fullerene is harder than diamond

14- The ductility of fullerene is 100 times stronger than steel

15- Fullerene conducts electricity. Fullerene has stronger conductivity than copper,

16- Fullerene's weighs only one 6 of copper material.

17- Fullerene made of carbon, so fullerene can be extracted from waste

They have a great future in the field of medicine, pharmacy, structural materials, and electronics.

**Topology:** Topology is a mathematical study of certain products whose properties are dependent upon the structural variations that come along within a product. The prominent term is Schlegel diagrams which help in the identification of 3D structures of Fullerenes having closed shells because the 2D technology is not much applicable to it.

**Closed Fullerene:** The most common type of fullerene is a closed fullerene which consists of a sphere-like shell having cycles of both, either pentagons or heptagons. When the shell's faces have 5 to 6 sides then a formula is followed by them which is known as Euler's polyhedron formula.

**Bonding:** Every carbon atom consists of either a single or a double covalent bond due to the connectivity level that it shares with the neighboring atoms. There is always a mixture of these bonds present. However, a technique known as Raman spectroscopy is the most common one through which the bonding state can be identified and then worked upon. Others include IR spectroscopy and X-ray spectroscopy.

**Encapsulation:** Endohedral Fullerenes are one of the fullerene derivatives having small molecules or ions incorporated inside the cage atoms.

**Aromaticity:** Buckminsterfullerene is not capable of possessing the feature of super-aromaticity which means the localization of hexagonal rings over the molecules does not take place. Several types of research have also been conducted in this regard to enhance the reactivity of fullerenes.

A $C_{60}$ molecule has a diameter of 7 Å. $C_{60}$ molecules form a solid of weakly bound molecules together. Fullerites are fullerenes' crystalline state. It looks like a yellow powder, but after dissolving in toluene, it turns to pink. The molecule is quite stable, chemically. Other than $C_{60}$ and $C_{78}$, $C_{70}$, $C_{86}$, $C_{76}$ $C_{84}$, have been separated and discussed in detail. Carbon disulphide, toluene, and xylene are good solvents.

**Spherical fullerene:** Spherical fullerene is another different shaped type of fullerene in which n acts as the number of carbon atoms that are free to perform any delocalization. They are allowed to try delocalizing over the entire molecule. A certain series of the number of carbon atoms is provided for this arrangement where the stable shell must be equivalent to n = 2, 8, 18, 32, 50, 72, 98, 128, etc.

## 6. SYNTHESIS METHODS OF $C_{60}$ FULLERENE

For large quantities, however, water laser along with arc heating of graphite are applied (Nimibofa, Newton, et al. 2018). The first bulky production of fullerene, which has been done by Kratschmer et al. in 1990, was based upon evaporation and recondensation of graphite, its schematic is shown in Fig. 8a. In 1987, fullerene ions in flames were found and in 1991. Howard et al. (1992) succeeded in extracting significant yields of fullerene from laminar sooting flame of premixed mixture of benzene and oxygen at low pressure, the schematic of the premixed chamber is shown in Figure 8b Howard et al. (1992). In this method, flame conditions like pressure, C/O ratio, residence time and gas velocity in combustion of benzene influence fullerene amount and composition (Youle, et al. 2005). Among them, pressure is an influential factor and formation of fullerene could be done at low pressure like the astrophysical atmosphere at temperatures lower than 1700 K (Mansurov, 2012). Furthermore, combustions methods including laser pyrolysis of vapor hydrocarbons and laser irradiation of polycyclic aromatic hydrocarbons (PAHs) have been used to produce fullerene.

## 7. TOXICITY AND BIO-COMPATIBILITY EVALUATION OF FULLERENES

Since pristine and derivatized $C_{60}$ can access to intracellular space and accumulate into cell membrane and cytoplasm, they might be a threat for cell functioning and integrity (Zhang, 2011).

Recently, an investigation has been accomplished in order to study the cytotoxicity of two type fullerene solutions on Chinesehamster V-79 cells (Kyzyma, et al. 2015). Two fullerene solutions, one

*Figure 8. Schematic of the premixed chamber*

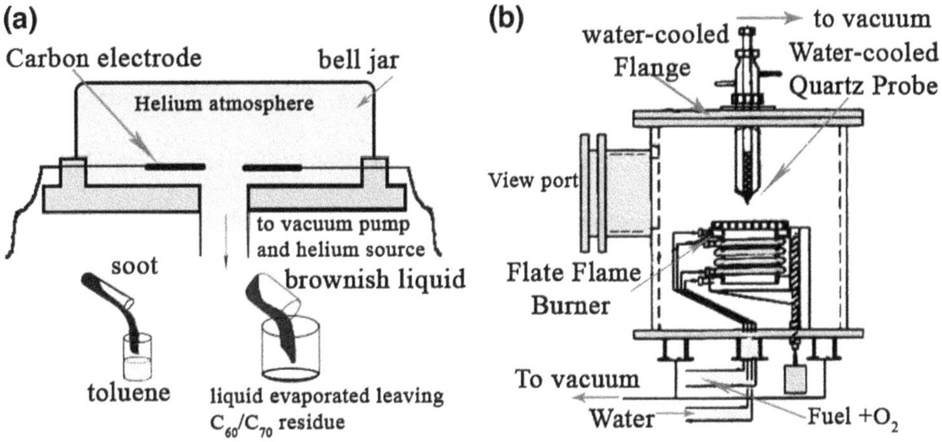


prepared via solvent exchange method and the other prepared via N-methylpyrrolidone (NMP), in which cluster size is smaller, were prepared and tested, and in both cases no toxicity was reported in *in vitro* tests, demonstrating that N-methylpyrrolidone can be exploited as solvent in preparing aqueous fullerene solutions (Kyzyma, et al. 2015). The first *in vivo* study was accomplished in 1996 by injecting an aqueous suspension of micronized $C_{60}$ to rodents. In these experiments, $C_{60}$ demonstrated to be no deadly, and did not show acute or sub-acute toxicity to this animal species (Moussa, et al. 1997).

### 7.1. Disadvantages of Fullerene

1- Fullerenes are prone to degradation or decomposition in the presence of light and oxygen (Gokhale M.M, Somani, 2015 and Arbogast, Darmanyan et al. 1991).

2- The intersystem crossing of singlet excited state to energetically lower triplet excited state results in the decomposition of C60.

## 8. TYPES OF FULLERENES AND THEIR SPECIFIC USES ($C_{60}$, $C_{70}$, FULLERENOLS)

Fullerene has many areas for using such as:

1. Pharmaceutical: Diagnostic reagents, super drugs, cosmetics, nuclear magnetic resonance (NMR) with the developer. DNA affinity

2. Energy: Solar battery, fuel cell, secondary battery.

3.Industry: Wear resistant material, flame retardant materials, lubricants, polymer additives, high-performance membrane, catalyst, artificial diamond, hard alloy, electric viscous fluid, ink filters, high-performance coatings, fire retardant coatings, manufacturing bioactive materials, memory materials, embedded molecular and other characteristics, composite materials etc.

4. Information industry: Semiconductor record medium, magnetic materials, printing ink, toner, ink, paper special purposes.

5. Electronic parts: Superconducting semiconductor, diodes, transistors, inductor.

6. Optical materials, electronic camera, fluorescence display tube, nonlinear optical materials.

## 9. GENERAL APPLICATIONS OF FULLERENES: (Yadav, Kumar, 2008)

There are various nanobiotechnological and industrial applications of fullerene which are discussed as below:

- As a material for semiconductor technology
- Uses of Fullerenes in solar cells
- As organic photovoltaic (OPV) applications
- Incorporation of Fullerenes in the polymers as polymer additives
- As catalysts
- As water purification and bio-hazard protection catalysts
- In portable power devices
- As vehicles compounds
- Fullerenes in fire retardant paint
- Growth potential of fullerenes

## 10. NANOBIOTECHNOLOGICAL AND BIOLOGICAL APPLICATIONS OF FULLERENES

Along with their efficient distribution in blood and local organs, these Fullerenes are extensively used in angiography, organ-specific imaging, oxidative stress reduction and gene/drug delivery. Fullerenes Functionalized with iodine is a potential X-ray contrast agent with prolonged retention in the blood. It also provides excellent *in vivo* biocompatibility compared to other clinically available X-ray agents. Metal lo-fullerenes are advantageous compared to other metal nanoparticles in terms of toxicity as they release fewer metals/ions due to the stable caging effect provided by C60 (Isobe et al.2001).

### 10.1 Fullerene-Based Nanosystems

Recently, fullerene and its derivatives are used as effective tool for cancer imaging because of their optical features such as excellent fluorescence properties. Kwag et al. synthesized a kind of hydrogenated Fullerenes with strong NIR fluorescence intensity, which enabled the *in vivo* florescent imaging of tumor sites with high-resolution (Kwag et al.2013). The generated hydrogenated Fullerenes showed great water solubility, photosensitivity, specificity tumor targeting, and robust fluorescent signal without conjugating with other fluoroscopes or isotopes, which achieved the non-invasive photo-luminescent imaging of KB and HCT-116 tumor-bearing mice. Shi et al. reported a PAI-guided strategy for cancer diagnosis based on fullerene (Shi et al.2019). They conjugated D-A antenna onto the fullerene to red-shift their main absorption into the NIR window, followed by DSPE-mPEG coating to enhance the biocompatibility of the whole nanocomplexes. *In vivo* results obtained from HeLa cell-bearing mice showed that the constructed nanocomplexes could produce a strong PA signal for biological imaging, which guided the synergistic PDT and PTT for efficient cancer elimination.

## 10.2 Fullerenes in Drug and Gene Delivery

The delivery of drugs/genes directly into the cells is a major challenge due to the three membrane barriers. Fullerenes are inorganic nanoparticles with a hydrophobic core. The functional groups attached to the core provide complexity to the behavior of the fullerene molecule. Attaching hydrophilic moieties makes them water-soluble and capable of carrying drugs and genes for delivery (Rouse et al.2007). DNA Functionalized Fullerenes show better efficiency than commercially available lipid-based vectors in COS-1 cells, increasing the lifetime of DNA in endosomes by supporting their chromosomal incorporation (Mori et al.2006). The peptide-fullerene conjugate was evaluated for its ability to penetrate through flexed and unflexed skin. The mechanical flexion that alters the structural organization of skin increases penetration by conceding the permeability barrier of the epidermis, which suggests that the fullerenes can also penetrate through intact skin (Thrash et al.1999). The genotoxicity of the Fullerenes was determined in bacterial reverse mutation in hamster lung cells followed by oral administration of fullerenes to the rats. The Fullerenes did not induce toxicity *in vitro* or *in vivo* (Yamakoshi et al.2003) .

## 10.3 Fullerenes in the Diagnostic Applications (Gokhale M.M, Somani, 2015)

Taken together, fullerene and its derivatives can be tailored to meet specific demands of multifarious modalities in cancer treatment and diagnosis (Figure 9)(Tang et al.2022) . Nevertheless, there are still several issues that limit the application of fullerene, such as poor water solubility and intrinsic toxicity. To cope with these limitations, fullerene can be covalently modified with water-soluble functional biopolymers to enhance its water solubility and tumor homing capability. Regarding its intrinsic toxicity, fullerene can be modified with bio-compatible molecules to enhance their bio-safety. For instance, many polar functional groups such as hydroxide, carboxy and amino groups can be attached on its surface to enhance its biocompatibility (Ye et al.2021). Therefore, many fullerene derivatives like hydroxyfullerenes, carboxyfullerenes and glycofullerenes have emerged, which displays outstanding performance against cancer in multiple angles including angiogenesis inhibition, immune system activation, antioxidation and reversal of multidrug resistance (Ye et al.2021) .

### 10.3.1 Fullerene for Intracellular Target-Based Anticancer Strategies

It is well known that intracellular signaling pathways and metabolic processes are vital players in oncogenesis, tumor progression, and neoplastic dissemination. Therefore, based on the unique advantages of fullerene and its derivatives in cancer treatment, they have been broadly investigated as potential drug delivery platforms for intracellular targeting over the years (Zhang et al.2015) . Serda et al. fabricated a novel targeted anticancer agent by constructing a kind of highly water-soluble hexakis-glucosamine fullerene derivative for treating pancreatic cancer, which was termed as Sweet-C60 and could majorly accumulate in pancreatic stellate cells (PSCs) nucleus (Serda et al.2018). Because glucose metabolism-associated energy was necessary for the proliferation and tumorigenesis, the glycoconjugation of fullerene could enhance the cancer-targeting properties to some extent. Confocal microscopy experiments towards PSCs confirmed the outstanding targeting and permeation ability of this synthesized fullerene derivative. In addition, *in vitro* results of phototoxicity studies indicated that Sweet-C60 was nontoxic within 1 mg/ml concentrations, and exhibited strong photoactive and photodynamic cytotoxic effect on PSCs under blue and green light, illustrating the great potential of Sweet-C60 in PDT.

*Figure 9. Fullerene and its derivatives*

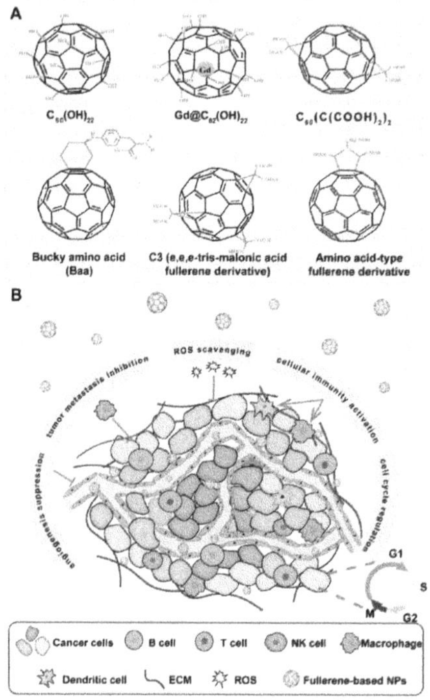


RNA interference (RNAi) has shown great therapeutic potential in various diseases including cancers due to its ability in silencing certain oncogenes, however, the efficiency of RNAi can be greatly affected due to the lysosomal-mediated degradation (Wang et al.2012). To address this issue, Wang et al. constructed a kind of amphiphilic fullerene derivative (C60-Dex-$NH_2$) to deliver siRNA into cancer cells, which not only promoted the lysosomal entrapment, but also cause lysosomal membrane destruction through triggering controllable ROS under visible light irradiation (Wang et al.2017). The MDA-MB-231 cells were used to assess the cellular uptake of the constructed nanocomplexes, which showed that their lysosomes became swollen and most nanocomplexes could be observed scattered in their cytoplasm after irradiating with visible light, indicating the successful lysosomal escape of these Nanocomposite. Moreover, a remarkable gene silencing efficiency of this fullerene-based siRNA carrier was verified both *in vitro* and *in vivo*, presenting an innovative approach for siRNA delivery.

Tumorigenesis is often accompanied with metastasis, which is the primary reason of mortality. A variety of proteins that exist in the cytoplasm are associated with tumor metastasis, therefore, these cell metastasis-associated proteins are becoming specific anti-neoplastic targets in precise cancer treatment (Soung et al.2016). Myosin heavy chain 9 (MYH9), a cytoplasm-located protein overexpressed in many cancer cells, involves in the metastasis-related cell motility and epithelial-mesenchymal transition (EMT) regulation process, which is also responsible for undesirable prognosis (Wang et al.2019). As functionalized fullerene derivatives display excellent abilities in inhibiting cancer metastasis and survival, among them, aminated fullerene shows much higher cellular uptake in comparison with other fullerene derivatives, therefore, Zhou et al. prepared a kind of C70 fullerene derivatives (C70-EDA) modified with multiple ethylenediamine (EDA) moieties, which was able to inhibit cancer cell migration, modulate intracellular MYH9 distribution, and hinder EMT through binding on MYH9 protein (Zhou et al.2020).

After treating A549 cells with C70-EDA, an inhibitory effect on cancer cell migration and EMT process reversion was observed *in vitro*. Therefore, this work by Zhou's group unveiled a novel therapeutic target based on MYH9 and opened a new door to use C70-EDA as a promising agent against cancer metastasis.

## 10.3.2 Fullerene for TME-Based Anticancer Strategies

ECM can not only provide structural and mechanical support for TME components, but also determines cell behavior and phenotype. Dysregulated interactions between ECM and other cell types can lead to serious pathological consequences, such as the occurrence of hepatocellular carcinoma (HCC) (Wu et al.2018). Therefore, ECM remodeling is an approach to prevent oncogenesis. A fullerene nanofilm with an artificial ECM-like structure was fabricated by Sosnowska et al. via the arc discharge method, which had a high adhesive capacity with cancer cells to change their behavior and reduce their proliferation (Sosnowska et al.2019). *In vitro* cell viability and proliferation assessment using HepG2 and C3A liver cancer cells demonstrated that this nanofilm could suppress the proliferation of HCC cells with good biocompatibility. Moreover, other *in vitro* results revealed that this fabricated nanofilm could contribute to the termination of cell cycle of HCC through mechanotransduction, implying that the application of fullerene nanostructures as artificial ECM-like structures could effectively attenuate cancer cell malignancy to improve anticancer therapeutic efficacy. Reprogramming the immunosuppressive TME (ITM) is very important to achieve successful immunotherapy effect. Li et al. proposed a strategy by modifying gadofullerene (Gd@C82) with β-alanines (GF-Ala) to rebuild ITM, which could induce macrophages to transform from tumor-supportive M2 type to M1 type, ultimately triggering robust antitumor immunity (Li et al.2020). *In vitro* results based on RAW264.7 cells, 4T1 and A549 cancer cells confirmed that both the M1-related cytokines and protein expression were significantly upregulated, and the proliferation of tumor cells showed a remarkable reduction up to 57.9% after GF-Ala treatment, indicating that GF-Ala was able to polarize macrophages to exert an inhibitory tumor growth.

## 10.3.3. Fullerenes' Antibacterial and Antiviral Action

Anionic type of Fullerenes possesses antioxidant activity and cationic type shows antiproliferative and antibacterial properties (Mashino, Shimotohno et al. 2005). Fullerenes can also engage in reactions with superoxide compounds in the case of an illness and recognize the free radicals in within mitochondria to sustain their membranes integrity with the subsequent cell protection from apoptosis and toxins (Hsu, & Luo, 2019).

## 10.3.4. Antioxidant and Neuroprotective Activity

Fullerenes have neuroprotective activity because fullerenes have capability to can react with oxygen species such as $O_2$ (Superoxide) and •OH (Hydroxyl) radicals which can attack on lipids, proteins, DNA and other macromolecules with no consumption. Fullerenes are termed as radical sponges because they are considered as the world's most efficient radical scavenging agent (Krusic, Wasserman, et al. (1991) and Tokuyama, Yamago et al. (1993). Fullerenes also act as a medical antioxidant. In disease condition cellular (reactive oxygen species) ROS production can cause apoptosis. So fullerene derivatives can defend apoptosis by neutralization of ROS (Yang, Chen, Shi. (2019)).

### 10.3.5. Free Radical Scavenging and Antioxidant Activity

Fullerene compounds have a unique cage structure, combines in the core with numerous conjugated double bonds, interact with the biomolecules, and have avid reactivity with free radicals.

### 10.3.6. Anti-HIV Activity of Fullerenes

Several reports are made on fullerenes' ability to inhibit the activity of HIV-1 protease. Fullerene enters into the pocket, and then through van der Waals interactions attaches to the active site causing antiviral activity. Some fullerene derivatives inhibit the HIV-1 viruses that have resistance against multiple drugs. Fullerene derivatives act as anti-HIV agents by fitting smartly in the active site of the viral protease. The fullerene derivatives of cationic and anionic type inhibit the replication of HIV-RT and hepatitis C virus.

### 10.3.7. Fullerenes in Drug and Gene Therapy

Fullerenes belong to a class of inorganic nanoparticles having a small size (~ 1 nm). The fullerene core being very hydrophobic, Fullerenes can become water-soluble and be capable of carrying drugs and genes for the cellular delivery by attaching hydrophobic moieties. The derivatives of fullerenes are able to can cross the cell membrane to bind to mitochondria (Azzam, Domb (2004))..

### 10.3.8. Tumor Research and Anti-tumor Drugs

In cancer research, HeLa cells were used in recent experiments. This includes the production of new Photosensitizers which get absorbed by cancer cells and still trigger the death of the cell. To prevent damage to the cell, the body shouldn't contain a new Photosensitizers in it for long. When cancer cells absorb them, and they get exposed to light radiation, creating reactive oxygen that damages the lipids, proteins, and DNA that makes up the cancer cell.

### 10.3.9. Crossing the blood-brain barrier

A blood-brain barrier is made of the tight endothelial junctions that inhibit the para-cellular permeability, thus creating it a big task to deliver drugs into the central nervous system. Fullerene and its water-soluble derivatives have a high potential to complete that task.

### 10.3.10. As an X-ray contrast agent

The fullerene derivatives can be used as X-ray contrast agents. C60 fullerenes have been proposed as contrast agents for MRI method. Especially promising for *in vitro* and *in vivo* NMR imaging are their complexes with gadolinium (containing $Gd^{3+}$ ions entrapped inside the fullerene cage) known also as gadofullerene. The toxic heavy radio metal cannot exit from fullerenes cage once it has been placed. This property suggests the application of Fullerenes as radio-tracers *in vivo* (Shinohara, (2000).

*Figure 10. Results of the distribution*

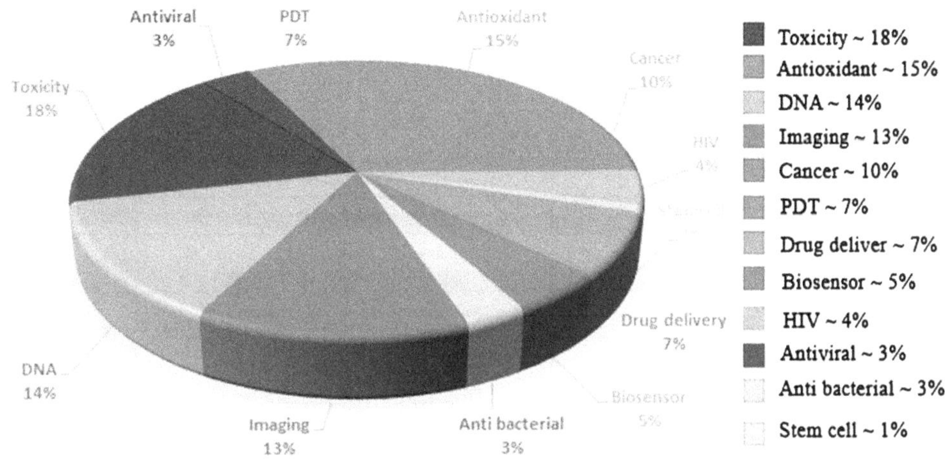


### 10.3.11. Antimicrobial Activity

The Fullerenes found to have potential antimicrobial activity due to its intercalation into biological membranes. The different strains of fungi and bacteria such as Candida albicans, Bacillus subtilis, Escherichia coli, and Mycobacterium avium showed positive results (Mashino, Okuda et al., 1999)).

In order to investigate divergence of publications regarding the biomedical applications of fullerene, we searched the publications within the scientific community using Scopus database (date of search: 21 November 2016). We used keywords including cancer, photodynamic therapy, antiviral, anti-bacterial, HIV, DNA, RNA, toxicity, biosensor, antioxidant, imaging, drug deliver, and stem cell. The papers that have these keywords in their title, abstract and keywords list have been counted and the results of the distribution of them in various biomedical applications are demonstrated in Figure 10. The number of papers with word 'toxicity' is more than other keywords and it demonstrates the great importance that was placed to the toxicity of the Fullerenes which is a prerequisite for the biomedical applications of Fullerenes.

Now, in the 30th anniversary of $C_{60}$ discovery, we consider that is time to review the progress of the fullerene-based research from its birth until now. In this chapter, we will review 30 years of investigations on Fullerenes, started in 1985 with a serendipitous discovery.

The various applications of fullerene derivatives in bio-medicine are summarized in Table 1.

## 11. THERAPEUTIC APPLICATIONS

### 11.1. Carrier for Drug Delivery

Carbon nanotubes (CNTs) with various functionalities are described for targeting of Amphotericin B to Cells.The intracellular penetration was enhanced when doxorubicin with nanotube have given. For their nanosize carbon nanotube finds an application in tablet manufacturing as lubricants or glidant (Sebastien, Giorgia et al. (2005)).

*Table 1. Fullerene derivatives and their various applications in biomedical engineering*

| Fullerene derivatives | Applications | Refs |
|---|---|---|
| Fullerenol or fullerenol, $C_{60}(OH)_{12}$ | Antioxidant and reducing apoptosis in cortical neurons cultures | Dugan, et al. (1996) |
| Carboxyfullerene, $C_{63}((COOH)_2)_3$ | Inhibiting neurodegeneration involved in ALS | Dugan, et al. (1997) |
| $C_{60}(ONO_2)_{7\pm2}$ | With antioxidant activity can prevent oxidant-induced pulmonary diseases | Lai, et al. (2003) |
| Amino acid derivatives of fullerene | Antiviral activity | Mashino, et al. (2005) |
| Hexa(sulfobutyl)fullerene, $[C_{60}(CH_2\text{-}CH_2\text{-}CH_2\text{-}CH_2SO_3Na)_6;(FC_4S)]$ | Antiproliferative agent in the arteriosclerosis and inhibition of plasma lipid peroxidation | Hsu, et al. (2000) |
| Fullerenol | Non-invasive image-guided cancer therapy | Krishna, et al. (2010) |
| Fullerenol | Tumor inhibitory activity | Zhu, et al. (2008) |
| bis-Fulleropyrrolidines | Inhibitory activity against HIV | Marchesan, et al. (2005) |
| Fullerenol | Anti-metastatic activity, cancer treatment | Jiao, et al. (2010) |
| $C_{60}$(Nd) nanoparticles | Inducing autophagy in cancer cells and enhancing chemotherapeutic killing of cancer cells | Wei, et al. (2010) |
| $C_{60}[>M(C_3N_6{+}C_3)_2][>M(C_3N_6C_3)_2]\text{-}(^{-}I)10$ | Antimicrobial photodynamic inactivation | Yin, et al. (2015) |
| tris-Malonic acid fullerenes | Radical scavenging | Chi, et al. (1998) |
| Dendrofullerene | Antioxidant activity | Witte, et al. (2007) |
| C60-PEI-FA | Drug delivery system | Shi, et al. (2013) |
| Fulleropyrrolidines bearing 3,5-di-tert-butyl-4-hydroxyphenyl | Antioxidant | Enes, et al. (2006) |
| Porphyrin-$C_{60}$ | PDT cell inactivation | Alvarez, et al. (2006) |
| C60-PEG-Gd | Anti-tumor activity | Liu, et al. (2007) |
| Pd@Cys-$C_{60}$ nanoparticles | Glucose biosensor | Zhong, et al. (2012) |
| Cucurbit[8]uril–fullerene | Photodynamic therapy | Jiang, et al. (2006) |
| $Gd@C_{60}[C(COOH)_2]_{10}$ | MRI contrasts agent | Bolskar, et al. (2003) |
| 6-Amino-g-cyclodextrin (ACD)/C60 | Photodynamic therapy | Nobusawa, et al. (2012) |
| Hexasulfobutyl fullerenes | Treatment of neurodegenerative diseases via radical scavenging | Chi, et al. (1998) |
| Fullerene-cryptand-22 | Urea biosensor | Wei,et al. (2001) |
| Carboxyfullerene | Antimicrobial and antibacterial activity | Tsao, et al. (2002) |
| $C_{60}(OH)_{16}$AMBP | Osteoporosis treatment | Gonzalez, et al. (2002) |
| $C60[C(COOH)_2]_2$/glucose oxidase (GOx)/TMB | Glucose colorimetric biosensor | Li, et al. (2013) |
| Carboxyfullerene | Targeted drug delivery | [266] Xu, et al. (2006) |
| Carboxyfullerene | Urea biosensor | Saeedfar, et al. (2013) |
| Hoechst 33258–conjugated hyaluronated fullerene | Photodynamic tumor therapy | Kim, et al. (2015) |
| Fullerene-phenylalanine-poly lactic acid | Drug carrier – photodynamic therapy | Li, et al. (2015) |
| Cystine $C_{60}$ derivative | Inhibiting apoptosis | Hu, et al. (2007) |
| Amphiliphilic fullerene bisadducts | Inhibiting HIV | Bosi, et al. (2003) |
| DNA-fullerene nanoparticles | Carrying drug | Isobe, et al. (2006) |
| Porphyrin adducts of cyclohexyl fullerene-$C_{60}$ | Targeted delivery of paramagnetic Mg $^{2+}$ isotope | Rezayat, et al. (2009) |
| Fullerenol | Stimulate osteogenesis in bone marrow | Yang, et al. (2014) |
| Fullerene containing free amino group and N-Fmoc-L-glutamic acid a-tert-butyl ester | Antibacterial activity | Pellarini, et al. (2001) |
| Malonodiserinolamide derivatized fullerene | Drug delivery system | Raoof, et al. (2012) |
| Carboxyfullerene | Antioxidant activity | Lin, et al. (1999) |

## 11.2. Fullerenes for Topical Drug Delivery and Controlled Release of Drugs

Fullerenes are used in cosmetic applications and trans-dermal delivery because of its antioxidant activity and its interactions with epidermal keratinocytes which provides many spectacular abilities to them. Due to these properties, different forms are used often, like anti-inflammatory substances, moisturizers, cytoprotective, barrier-repair agents, UVB inhibitory, anti-melanoma-genesis agents, and hair growth stimulators.

## 11.3. Cancer Treatment and Photo-Dynamic Therapy

Cancer Nanotechnology, using nanoscale materials or techniques to overcome the fatal hurdles in cancer treatments or detection, is a newly emerging area.

## 11.4. Genetic Engineering

CNTs and CNH (Carbon nanohorns) are used to manipulate genes and atoms in the development of bioimaging genomes, proteomics, and in tissue engineering. Nanotube and nanohorns can adhere to various antigens on their surface which can a source of antigens in vaccines (Pai, Nair et al. 2006).

## 11.5. Nucleic Acid Delivery and DNA Photo-Cleavage

$C_{60}$ derivatives cytotoxicity was mediated by its DNA cleaving capability. So, other usages of Fullerenes are related to its easy photo-excitation, that's why Photodynamic compounds, which are fullerene-based, are being established for cancer treatment.

## 11.6. Preservative

They are used as a preservative in anti-aging cosmetics and with zinc oxide in formulations to prevent oxidation (Hirlekar, Yamagar et al. 2009).

## 11.7. Implant Prosthesis

The body exhibits a rejection reaction for implants associated with postoperative pain. This can be overcome by the use of miniature sized nanotube and nanohorns get incorporated with other proteins and amino acids. Due to their high tensile strength, CNTs filled with calcium and arranged or grouped in the bone structure can act as a bone substitute or in the form of artificial joints (Ding, Lu, et al., 2001).

## 11.8. Diagnostic Tool

Because of their ability of fluorescence with specific biomolecules, protein-encapsulated protein filled nanotubes have been used as implantable biosensors. Nanosize robots and motors with nanotube can be used for the study of cells and biological systems (Ding, Lu et al. 2001 and Lacerda, Bianco et al. 2006).

## 12. DENTAL APPLICATIONS

### 12.1. Diagnosis and Treatment of Oral Cancer

The exosome is a membrane bound secretory vesicle containing a proteomic and genomic marker whose level is elevated in malignancy. This marker has been studied by using atomic force microscopy, which uses nanoparticles. Undergoing trials are nanoparticle-coated, radioactive sources placed close to or within the tumor to destroy it (Shetty, Swati, et al. 2013). Gadolinium-containing Endohedral Fullerenes are MRI contrast agents for the diagnosis of several diseases.

### 12.2. Tissue Engineering

Potential applications of tissue engineering and stem cell research in dentistry include the treatment of orofacial fractures, bone augmentation, cartilage regeneration of the temporomandibular joint, pulp repair, periodontal ligament regeneration, and implant osseointegration. Bone grafts can be better developed with the use of nanocrystalline hydroxyapatite because it stimulates the cell proliferation required for remodeling of periodontal tissue (Kamboj, Arora, and Gupta, 2016).

### 12.3. Local Anesthesia

The gingiva of the patients is instilled with a colloidal suspension containing millions of active, analgesic, micron-sized dental robots that respond to input supplied by the dentist. After completion of treatment, they restore sensation providing the patient with anxiety free and needless comfort. Anesthesia is fast acting, and reversible, with no side effects or complications associated with its use (Freitas et al., 2000).

### 12.4. Hypersensitivity Cures

Changes in the pressure transmitted hydrodynamically to the pulp may cause hypersensitivity. The dentinal tubules of a hypersensitive tooth have twice the diameter and eight times the surface density of those in non sensitive teeth. Dental nanorobots could selectively and precisely occlude selected tubules in minutes using native logical materials, offering patients a quick and permanent cure (Saravana, Vijayalakshmi, 2006).

### 12.5. Orthodontic Treatment

Using excessive orthodontic force might cause loss of anchorage and root resorption. Reduction in the frictional force produced by an orthodontic movement by coating the orthodontic wire with inorganic fullerene-like tungsten disulfide Nanoparticles (IF-WS2) known for their excellent dry lubrication properties. Orthodontic nanorobots could directly manipulate the periodontal tissues, allowing rapid and painless tooth straightening, rotating and vertical repositioning and rapid tissue repair within minutes to hours (Saravana, Vijayalakshmi, 2006).

## 12.6. Nanocomposite

The latest advancement in the manufacturing process of dental composite resins is the utilization of Nanoparticles technology. Inorganic fillers in nano dimensions are diffused homogeneously without any accumulation in the matrix in the artificial teeth produced from Nanocomposite. Studies have shown that Nanocomposite artificial teeth are more durable than acrylic teeth and microfill composite teeth and have a higher resistance to abrasion (Ghazal, Albashaireh, Kern, 2008).

## 12.7. Prevention of Dental Caries

Using toothpaste containing nanosized calcium carbonate enabled remineralization of early enamel lesions (Nakashima, Yoshie, et al., 2009).

# 13. FUNCTIONALIZED FULLERENES AS DRUG DELIVERY NANOPARTICLES

PEBs which are known as paclitaxel embedded buckysomes are a type of amphiphilic fullerene while consisting of nanostructures that are approximately 100 to 200 nm. They have hydrophobic pockets inside them which contain the anti-cancer drug paclitaxel. The US FDA has also approved of them as these are the drugs that protect a human body from some life-threatening conditions, for example, breast cancer. This water-soluble fullerene and its derivatives protect the person from any side effects and discomforts that come along.

## 13.1. Advantageous Approach

This is a thoroughly advantageous approach because it helps a person in fighting the disease and certainly improves the circulation of blood, as well as the anti-cancer drugs, cause various kinds of side effects one of which is poor blood circulation. However, the size of these PEBs is very small equivalent to 200 nm so that RES uptake can be avoided at all costs.

## 13.2. Production of the Lipophilic Slow-Release System

As discussed earlier that fullerene is capable of producing an ideal lipophilic system so it helps in creating several drugs which can be used as a single dose naming them as the drug cocktails. This was discovered by Zakharian and colleagues. This is the most common practice for the therapy of lung cancer.

## 13.3. Transfection Vectors

Other than giving out drug molecules, Fullerenes also work as the Transfection vectors for the deliverance of exogenous DNA so that they can be worked up to check for mediate gene transfer. This is a highly beneficial technique in the process of gene therapy.

### 13.4. Transfection Efficiency

However, Transfection efficiency is brought along only with the help of fullerene and its derivatives. It is capable of possessing the properties and characteristics which are essential in carrying out this process.

## 14. REACTIVE OXYGEN SPECIES (ROS) QUENCHING BY FUNCTIONALIZED FULLERENES

Krusic and colleagues are known to figure out all the potentials of fullerene for scavenging reactive oxygen species. This has initiated and increased the interest of marketers to start using fullerene as an antioxidant.

### 14.1. Antioxidant Compounds

It was suggested in early phases that C60 and its derivatives are capable of having antioxidant compounds which are highly beneficial in the biological systems. Carboxyfullerenes is the most important one among all as it occupies excellent antioxidant compounds.

### 14.2. Influence of Size

Along with that, the size of fullerene is also very important as it influences the processes that are going on. An experiment was led out which depicted that fullerenes are capable of reducing cytotoxicity, damage to the mitochondria, and formation of free radicals as human lung carcinoma cell line and capillary endothelial cell line of a rat's brain was used in this experiment.

### 14.3. Iron-induced Lipid Peroxidation

Another study was presented by Lin and colleagues that carboxy fullerenes consist of anti-oxidative properties so these are also capable of repressing iron-induced lipid peroxidation.

### 14.4. Process of Apoptosis

Fullerenes are highly beneficial for the inhibition of apoptosis which was discovered by Monti and colleagues as carboxy Fullerenes is a type of fullerene which helps protect blood cells from going through apoptosis along with the involvement of mitochondrial membrane. This is known to be carried at the earliest stage of apoptosis.

## 15. FULLERENES AS PHOTOSENSITIZERS

Fullerenes captured with metal ions have potential in the diagnostic application and are known as endofullerenes or metallofullerenes. These metallofullerenes are applied as magnetic resonance imaging contrast agent (MRI), X-ray imaging agents and radio-pharmaceuticals. Holmium a metallofullerol molecule, could accumulate in the liver and can be detected in the bone. This shows that these fullerenes are targeted to tissues that are rich in macrophages. It could also be useful as a chemotherapeutic agent for the treatment of leukemia and bone cancer (Rancan et al.2002). Another study revealed that when fullerene derivatives were used for carrying serum protein profiling, it helped identify proteins for pathologies and biomarker discovery using the material-enhanced laser desorption (MALDI) technique. Direct laser irradiation of the adsorbed protein was applied over the range m/z 2000–3000 for screening the proteins present in serum.

Photo-excitation of Fullerenes is another important medical application. These can be excited from ground state to 1 C60 by photo-irradiation. Whereas in the presence of molecular oxygen, its excitation can result in highly toxic species forming super-oxide anion radical $O_2^-$ (Ji ZQ, Sun H et al.2006). In the presence of biological agents like guanosine, these exited fullerenes can be reduced, making them potential Photosensitizers for their use in photodynamic therapy (PDT). C60 Fullerenes were conjugated with oligonucleotide or acridine for their interaction with nucleic acids for increasing cytotoxicity in cancer cells. Dendrite monoadduct C60 was added to Jurkat cells, and on exposure to UV light, the cell count has decreased approximately by 19% within two weeks (Liu et al.2007). They have also been studied for biodistribution in tumour cells by radio-tracer 125I-labeled C60 (OH)x (Mroz P, Pawlak A et al.2007). The polyethylene glycol-conjugated fullerene containing $Gd^{3+}$ ions was used for PDT combined with MRI and the data signifies that Photosensitizers significantly promoted the tumour target ability and MRI activity (Semeniuk et al. 2019). Fullerenes conjugated with hydrophilic cationic groups were used for PDT against mouse cancer cell lines, and they were highly effective photosensitizers for killing cancer cells by induction of apoptosis after illumination (Wang et al.2015).

One of the cancer treatment strategies using fullerenes is lightbased therapies such as photo-dynamic therapy and photo-thermal therapy (PTT). Photo-dynamic therapy is a non-invasive and nonsurgical treatment for some type of tumors and some of non-malignant diseases. Photo-dynamic is a photochemical method that uses photo-sensitizing agents (Photosensitizers [PS]) and light irradiation to produce reactive oxygen species in tumor tissue, to induce damage and apoptosis. The schematic of this process is shown in Figure 11 (Chen, et al. (2012).

## 16. USE OF POLYETHYLENE GLYCOL

PEG known as polyethylene glycol is a conjugated type of fullerene which was initially procured by $Gd^{+3}$ ions for a process called photo-dynamic therapy while combining with the magnetic resonance imaging commonly known as MRI.

## 17. FULLERENES FOR MEDICAL THERAPEUTICS AND DIAGNOSTICS

Being the best compound to be used in the medical field, it sure brings a lot of benefits to medical therapeutics and diagnostics. Fullerenes have been found out to be great antioxidants along with being radical scavengers.

*Figure 11. Schematic of this process*

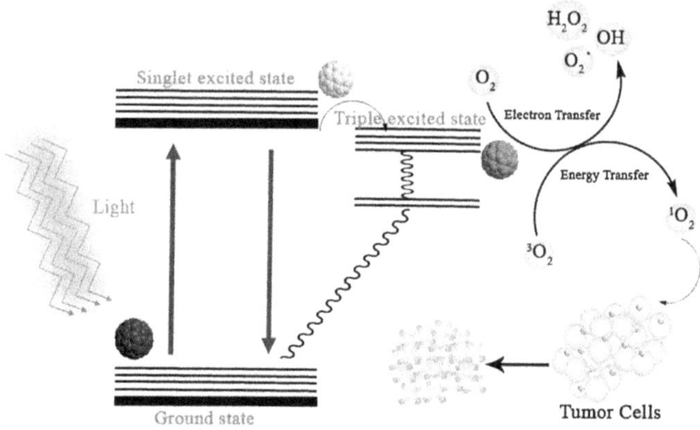


## 17.1. Fullerenes for Allergies

Mast cells are the ones that can be identified in body tissues of a human body and these are the ones that initiate allergic responses in a human body. The allergies are triggered when one comes in contact with any sort of allergens and in that case the cells reveal mediators which are responsible for these triggers.

## 17.2. Fullerenes for Asthma

Asthma is one of the most common diseases nowadays and the greatest percentage of it lies on mast cells. The main trigger of asthma is different types of allergens and allergy-causing organisms. Fullerenes are being used and studied to identify ways in which they can be found helpful to protect from allergic reactions being generated in the human lungs.

## 17.3. Fullerenes for Arthritis

The building blocks of arthritis are mast cells and in that case, Fullerenes are being discovered and utilized to present how the growth of mast cells can be inhibited for the prevention of arthritis.

## 17.4. Medical Diagnostics

Another one of the major applications of fullerene is the field of medical diagnostics. A lot of procedures go by every day to diagnose diseases that are either known to man or not known. To keep this method free from any toxicity, advancements and adaptations are being applied at every level.

## 17.5. Fullerenes in Cosmetics

Personal care and skin care products intricately intertwined with chemical industries. The emergence of nanotechnology in cosmetics triggered about 40 years ago with using liposomes in moisturizing creams (Bhat, et al., 2013). Fullerenes, thanks to their antioxidant activity, are used in moisturizers (Bhat, et al., 2013).

## 18. FULLERENES AS MRI CONTRAST AGENT FOR DIAGNOSIS

MRIs are physiological tests performed on people who are suspected of any disease or are supposed to go through some procedure. This can be done through making an increase in the relaxivity by the use of Trimetaspheres as this enhances the performance ability of MRI.

### 18.1. Coronary Artery Disease

Coronary artery disease is one such disease that has been thoroughly studied due to Trimetasphere derivatives. This compound has a high potential to be dealt with potentially strong and stable components which are capable of dealing with the plaque that builds up in the blood vessels of humans due to this disease.

### 18.2. Applications of Fullerenes in Biosensors

Fullerenes can be used as mediator between recognition site and biosensor electrode, amplifying the electron transfer rate created by biochemical reactions of analyte and biological component in recognition site (Afreen, et al., 2015).

## 19. FUTURE DIRECTIONS

Nanotechnology has the potential to significantly alter the way cancer is treated. Nanomaterials have unique properties that make them highly effective for targeted drug delivery and cancer therapy (Lin YX, Wang Y et al.2020). The potential to improve cancer detection and treatment through nanotechnology is tantalizing, but the journey from the laboratory to the clinic is a road laden with obstacles. Nevertheless, with perseverance and sustained investment in research, we can unlock the full potential of nanotechnology in the fight against cancer. The future holds the promise of more effective, targeted, and less invasive treatments that will significantly improve the lives of cancer patients, and it is our collective responsibility to ensure that this promise becomes a reality. However, there are still many challenges that need to be addressed to improve the clinical translation of nanomaterials. These include reducing toxicity, improving targeting specificity, and understanding the interactions between nanomaterials and the human body (Hu X, Yu S, et al.2022). The unique applications of Fullerenes are due to the extraordinary structure of fullerene. Because of its poor solubility in common organic solvents, spectroscopic analysis of products, purification, separation, and assessment of purity will become difficult. Today, the conjugation of fullerene-based biomolecules has become a primary strategy in overcoming the intrinsic hydrophobic obstacles encountered by Fullerenes to use such molecules for biological purposes, Research toward the application of Fullerenes in different physicochemical and pharmacological fields is still very active, and breakthroughs and applications are constantly being discovered. Most of the studies performed on the toxicity of C60-Fullerenes showed that these molecules are not toxicity to human and animal cells *in vitro* and their acute toxicity against animal tissues *in vivo* is low (Fiorito, Serafino et al. 2006).

## 20. CONCLUSION

Fullerene is an allotrope of carbon is known as one of the best compounds that are being used in various industries but specifically in the biomedical industry. There are several industries in which the applications of Fullerenes can be expressed but the main and most important ones can be found in the biomedical industry. Fullerene is working as an excellent compound for enhancing the credibility of the biomedical industry. Fullerene has the advantage of offering high sensitivity to various biological and biomedical analysts. Moreover, Fullerenes those are not dangerous to biological systems and materials. They can also easily exchange electrons to species surrounding the bio-molecules. Furthermore, they serve as an ideal substrate to receive and release electrons to a transducer. Consequently, it possesses a high potential for biosensor applications.

Fullerenes have become important molecules in different health related subjects during the last three decades. The biomedical applications presented in this manuscript represent potential uses of this class of carbon family; due to its extraordinary properties (see Table 1). As more applications of these nanomaterials become available, the demand for different forms of Fullerenes will increase. For future works, Functionalized Fullerenes will considerably change the face of biological applications, and the credits go to the collaborative efforts of scientists who focus on both chemistry and medicine. These collaborative efforts are critical for introducing the most effective modifications on the carbon cage for successful diagnosis and therapeutic outcomes in different biomedical applications. Here, we have provided an overview of the ever-expanding research on Fullerenes in biomedical engineering applications ranging from drug delivery to noninvasive imaging in the human body. These studies are mostly limited to *in vitro* and *in vivo* researches and have not reached the final stages of clinical trials yet. It is expected that, with the rapid technological advancement in this field, this trend could change in the coming years so that the unique features of fullerenes could be exploited in clinical use. Although there are still major issues to overcome, the use of fullerenes for biomedical engineering will certainly have a bright future.

## REFERENCES

Alvarez, M. G., Prucca, C., Milanesio, M. E., Durantini, E. N., & Rivarola, V. (2006). Photodynamic activity of a new sensitizer derived from porphyrin-C60 dyad and its biological consequences in a human carcinoma cell line. *The International Journal of Biochemistry & Cell Biology*, *38*(12), 2092–2101. doi:10.1016/j.biocel.2006.05.019 PMID:16899389

Arbogast, J.W, Darmanyan, A.P, Foote, C.S, Rubin, Y, Diederich, F.N, & Alvarez, M.M. (1991). Photophysical properties of C60. *Journal of Physical Chemistry*, *95*(1), 11–12. doi:10.1021/j100154a006

Astefanei, A. (2015). Nu'n˜ez, O. Galceran, M.T. *Analytica Chimica Acta*, *882*, 1–21. doi:10.1016/j.aca.2015.03.025 PMID:26043086

Augustine, S., Singh, J., Srivastava, M., Sharma, M., Das, A., & Malhotra, B. D. (2017). Recent advances in carbon based nanosystems for cancer theranostics. *Biomaterials Science*, *5*(5), 901–952. doi:10.1039/C7BM00008A PMID:28401206

Azzam, T., & Domb, A. J. (2004). Current developments in gene transfection agents. *Current Drug Delivery*, *1*(2), 165–193. doi:10.2174/1567201043479902 PMID:16305382

Baghchi, M., Moriyama, H., & Shahidi, F. (2013). *Bio-Nanotechnology: A Revolution in Food, Biomedical and Health Sciences*. Wiley-Blackwell. doi:10.1002/9781118451915

Bhakta, P., & Barthunia, B. (2020, April–June). Fullerene and its applications: A review. *Journal of Indian Academy of Oral Medicine and Radiology*, *32*(2), 159–163. doi:10.4103/jiaomr.jiaomr_191_19

Bhat, P. (2013). Mulgund, S. *Indo Am. J. Pharm. Res.*, *3*(8), 6549–6554.

Biju, V. (2014). Chemical modifications and bioconjugate reactions of nanomaterials for sensing, imaging, drug delivery and therapy. *Chemical Society Reviews*, *43*(3), 744–764. doi:10.1039/C3CS60273G PMID:24220322

Bolskar, R. D., Benedetto, A. F., Husebo, L. O., Price, R. E., Jackson, E. F., Wallace, S., Wilson, L. J., & Alford, J. M. (2003). First Soluble M@C $_{60}$ Derivatives Provide Enhanced Access to Metallofullerenes and Permit in Vivo Evaluation of Gd@C $_{60}$ [C(COOH) $_{2}$ ] $_{10}$ as a MRI Contrast Agent. *Journal of the American Chemical Society*, *125*(18), 5471–5478. doi:10.1021/ja0340984 PMID:12720461

Bosi, S., Da Ros, T., Spalluto, G., Balzarini, J., & Prato, M. (2003). Synthesis and Anti-HIV properties of new water-soluble bis-functionalized[60]fullerene derivatives. *Bioorganic & Medicinal Chemistry Letters*, *13*(24), 4437–4440. doi:10.1016/j.bmcl.2003.09.016 PMID:14643341

Chen, M., Guan, R., & Yang, S. (2019). Hybrids of Fullerenes and 2D Nanomaterials. doi:10.1002/advs.201800941

Chen, Z., Ma, L., Liu, Y., & Chen, C. (2012). Applications of Functionalized Fullerenes in Tumor Theranostics. *Theranostics*, *2*(3), 238–250. doi:10.7150/thno.3509 PMID:22509193

Chi, Y., Bhonsle, J. B., Canteenwala, T., Huang, J.-P., Shiea, J., Chen, B.-J., & Chiang, L. Y. (1998). Novel Water-soluble Hexa(sulfobutyl)fullerenes as Potent Free Radical Scavengers. *Chemistry Letters*, *27*(5), 465–466. doi:10.1246/cl.1998.465

Ding, C., Tong, L., Feng, J., & Fu, J. (2016). Recent Advances in Stimuli-Responsive Release Function Drug Delivery Systems for Tumor Treatment. *Molecules (Basel, Switzerland)*, *21*(12), 1715. doi:10.3390/molecules21121715 PMID:27999414

Ding, R., Lu, G., Yan, Z., & Wilson, M. (2001). Recent advances in the preparation and utilization of carbon nanotubes for hydrogen storage. *Journal of Nanoscience and Nanotechnology*, *l*(1), 7–29. doi:10.1166/jnn.2001.012 PMID:12914026

Dugan, L. L., Gabrielsen, J. K., Yu, S. P., Lin, T.-S., & Choi, D. W. (1996). Buckminsterfullerenol Free Radical Scavengers Reduce Excitotoxic and Apoptotic Death of Cultured Cortical Neurons. *Neurobiology of Disease*, *3*(2), 129–135. doi:10.1006/nbdi.1996.0013 PMID:9173920

Dugan, L. L., Turetsky, D. M., Du, C., Lobner, D., Wheeler, M., Almli, C. R., Shen, C. K.-F., Luh, T.-Y., Choi, D. W., & Lin, T.-S. (1997). Carboxyfullerenes as neuroprotective agents. *Proceedings of the National Academy of Sciences of the United States of America*, *94*(17), 9434–9439. doi:10.1073/pnas.94.17.9434 PMID:9256500

Enes, R. F., Tomé, A. C., Cavaleiro, J. A. S., Amorati, R., Fumo, M. G., Pedulli, G. F., & Valgimigli, L. (2006). Synthesis and Antioxidant Activity of [60]Fullerene–BHT Conjugates. *Chemistry (Weinheim an der Bergstrasse, Germany)*, *12*(17), 4646–4653. doi:10.1002/chem.200501495 PMID:16534828

Fiorito, S., Serafino, A., Andreola, F., Togna, A., & Togna, G. (2006). Toxicity and biocompatibility of carbon nanoparticles. *Journal of Nanoscience and Nanotechnology*, *6*(3), 591–599. doi:10.1166/jnn.2006.125 PMID:16573109

Freitas, R. A. Jr. (2000). Nanodentistry. *The Journal of the American Dental Association*, *131*(11), 1559–1565. doi:10.14219/jada.archive.2000.0084 PMID:11103574

Georgakilas, V., Perman, J. A., Tucek, J., & Zboril, R. (2015). Broad Family of Carbon Nanoallotropes: Classification, Chemistry, and Applications of Fullerenes, Carbon Dots, Nanotubes, Graphene, Nanodiamonds, and Combined Superstructures. *Chemical Reviews*, *115*(11), 4744–4822. doi:10.1021/cr500304f PMID:26012488

Ghazal, M., Albashaireh, Z. S., & Kern, M. (2008). Wear resistance of nanofilled composite resin and feldspathic ceramic artificial teeth. *The Journal of Prosthetic Dentistry*, *100*(6), 441–448. doi:10.1016/S0022-3913(08)60262-0 PMID:19033028

Gokhale, M. M., & Somani, R. R. (2015). Fullerenes: Chemistry and its applications. *Mini-Reviews in Organic Chemistry*, *12*(4), 355–366. doi:10.2174/1570193X12666150930224428

Gonzalez, K. A. (2002). Synthesis and In vitro characterization of a tissue-Selective fullerene: Vectoring C60(OH)16AMBP to mineralized bone. *Bioorganic & Medicinal Chemistry*, *10*(6), 1991–1997. doi:10.1016/S0968-0896(02)00049-4 PMID:11937359

Guan, M., Ge, J., Wu, J., Zhang, G., Chen, D., Zhang, W., Zhang, Y., Zou, T., Zhen, M., Wang, C., Chu, T., Hao, X., & Shu, C. (2016). Fullerene/photosensitizer nanovesicles as highly efficient and clearable phototheranostics with enhanced tumour accumulation for cancer therapy. *Biomaterials*, *103*, 75–85. doi:10.1016/j.biomaterials.2016.06.023 PMID:27376559

Hirlekar, R., Yamagar, M., Garse, H., Vij, M., & Kadam, V. (2009). Carbon nanotubes and its applications: A review. *Asian Journal of Pharmaceutical and Clinical Research*, *2*, 17–27.

Howard, J. B. (1992). *Symposium (International) on Combustion*. Elsevier.

Hsu, H.-C., Chiang, L.-Y., Chen, W.-J., & Lee, Y.-T. (2000). Water-Soluble Hexasulfobutyl [60] Fullerene Inhibits Plasma Lipid Peroxidation by Direct Association with Lipoproteins. *Journal of Cardiovascular Pharmacology*, *36*(4), 423–427. doi:10.1097/00005344-200010000-00002 PMID:11026641

Hsu, S., & Luo, P. (2019). *From Nanoarchitectonics to Tissue Architectonics: Nanomaterials for Tissue Engineering. Advanced Supramolecular Nanoarchitectonics*. Elsevier Inc., doi:10.1016/B978-0-12-813341-5.00012-7

Hu, X., Yu, S., Yang, G., Long, W., Guo, T., Tian, J., Liu, M., Li, X., Zhang, X., & Wei, Y. (2022). Facile synthesis of inorganic–organic hybrid fluorescent nanoparticles with AIE feature using hexachlorocyclotriphosphazene as the bridge. *Journal of Molecular Liquids*, *345*, 117693. doi:10.1016/j.molliq.2021.117693

Hu, Z., Guan, W., Wang, W., Huang, L., Xing, H., & Zhu, Z. (2007). Protective effect of a novel cystine C60 derivative on hydrogen peroxide-induced apoptosis in rat pheochromocytoma PC12 cells. *Chemico-Biological Interactions*, *167*(2), 135–144. doi:10.1016/j.cbi.2007.02.009 PMID:17353010

Hu, Z., Zhao, F., Wang, Y., Huang, Y., Chen, L., Li, N., Li, J., Li, Z., & Yi, G. (2014). Facile fabrication of a C60-polydopamine-graphene nanohybrid for single light induced photothermal and photodynamic therapy. *Chemical Communications (Cambridge)*, *50*(74), 10815–10818. doi:10.1039/C4CC04416A PMID:25089303

Isobe, H., Nakanishi, W., Tomita, N., Jinno, S., Okayama, H., & Nakamura, E. (2006). Nonviral Gene Delivery by Tetraamino Fullerene. *Molecular Pharmaceutics*, *3*(2), 124–134. doi:10.1021/mp050068r PMID:16579641

Isobe, H., Tomita, N., Jinno, S., Okayama, H., & Nakamura, E. (2001). Synthesis and transfection capability of multi-functionalized fullerene polyamine. *Chemistry Letters*, *12*(12), 1214–1215. doi:10.1246/cl.2001.1214

Jariwala, D., Sangwan, V. K., Lauhon, L. J., Marks, T. J., & Hersam, M. C. (2013). Carbon nanomaterials for electronics, optoelectronics, photovoltaics, and sensing. *Chemical Society Reviews*, *42*(7), 2824–2860. doi:10.1039/C2CS35335K PMID:23124307

Ji, Z. Q., Sun, H., Wang, H., Xie, Q., Liu, Y., & Wang, Z. (2006). Biodistribution and tumour uptake of C60(OH)x in mice. *Journal of Nanoparticle Research*, *8*(1), 53–63. doi:10.1007/s11051-005-9001-5

Jiang, G. (2006). Li, G. J. *Photochem. Photobiol. B: Biol.*, *85*(3), 223–227. doi:10.1016/j.jphotobiol.2006.08.001

Jiao, F., Liu, Y., Qu, Y., Li, W., Zhou, G., Ge, C., Li, Y., Sun, B., & Chen, C. (2010). Studies on anti-tumor and antimetastatic activities of fullerenol in a mouse breast cancer model. *Carbon*, *48*(8), 2231–2243. doi:10.1016/j.carbon.2010.02.032

Kamboj, M., Arora, R., & Gupta, H. (2016). Comparative evaluation of the efficacy of synthetic nanocrystalline hydroxyapatite bone graft (Ostim ®) and synthetic microcrystalline hydroxyapatite bone graft (Osteogen ®) in the treatment of human periodontal intrabony defects: A clinical and denta scan study. *Journal of Indian Society of Periodontology*, *20*(4), 423–428. doi:10.4103/0972-124X.184036 PMID:28298825

Käsermann, F., & Kempf, C. (1997). Photodynamic inactivation of enveloped viruses by buckminsterfullerene. *Antiviral Research*, *34*(1), 65–70. doi:10.1016/S0166-3542(96)01207-7 PMID:9107386

Kim, S., Park, J., Youn, Y. S., Oh, K. T., Bae, J. H., & Lee, E. S. (2015). Hoechst 33258–conjugated hyaluronated fullerene for efficient photodynamic tumor therapy and necrotic tumor targeting. *Journal of Bioactive and Compatible Polymers*, *30*(3), 275–288. doi:10.1177/0883911515574136

Kratschmer, W., Lamb, L. D., Fostiropoulos, K., & Huffman, D. R. (1990). Solid C60: A new form of carbon. *Nature*, *347*(6291), 27. doi:10.1038/347354a0

Krishna, V., Singh, A. P., Sharma, P., Iwakuma, N., Wang, Q., Zhang, Q., Knapik, J., Jiang, H., Grobmyer, S. R., Koopman, B., & Moudgil, B. (2010). Polyhydroxy fullerenes for non-invasive cancer imaging and therapy. *Small*, *6*(20), 2236–2241. doi:10.1002/smll.201000847 PMID:20818623

Kroto, H. W., Heath, J. R., O'Brien, S. C., Curl, R. F., & Smalley, R. E. (1985). C60: Buckminsterfullerene. *Nature*, *318*(6042), 162–163. doi:10.1038/318162a0

Krusic P.J, Wasserman E, Keizer P.N, Morton J.R, Preston K.F. Radical reactions of C60 Science. 1991;254:1183–5

Kwag, D. S., Park, K., Oh, K. T., & Lee, E. S. (2013). Hyaluronated fullerenes with photoluminescent and antitumoral activity. *Chemical Communications (Cambridge)*, *49*(3), 282–284. doi:10.1039/C2CC36596K PMID:23174913

Kyzyma, E. (2015). J. Surf. Investig. *X-ray Synchrotron Neutron Tech.*, *9*(1), 1–5. doi:10.1134/S1027451015010127

Lacerda, L., Bianco, A., Prato, M., & Kostarelos, K. (2006). Carbon nanotubes as nanomedicines: From toxicology to pharmacology. *Advanced Drug Delivery Reviews*, *58*(14), 1460–1470. doi:10.1016/j.addr.2006.09.015 PMID:17113677

Lai, Y. L. (2003). Murugan, P. Hwang, K. *Life Sciences*, *72*(11), 1271–1278. doi:10.1016/S0024-3205(02)02374-3 PMID:12570927

Lee, D. J., Ahn, Y. S., Youn, Y. S., & Lee, E. S. (2012). Poly(ethylene glycol)-crosslinked fullerenes for high efficient phototherapy. *Polymers for Advanced Technologies*, *24*(2), 220–227. doi:10.1002/pat.3074

Li, L., Zhen, M., Wang, H., Sun, Z., Jia, W., Zhao, Z., Zhou, C., Liu, S., Wang, C., & Bai, C. (2020). Functional Gadofullerene Nanoparticles Trigger Robust Cancer Immunotherapy Based on Rebuilding an Immunosuppressive Tumor Microenvironment. *Nano Letters*, *20*(6), 4487–4496. doi:10.1021/acs.nanolett.0c01287 PMID:32407113

Li, Q., Hong, L., Li, H., & Liu, C. (2017). Graphene oxide-fullerene C60 (GO-C60) hybrid for photodynamic and photothermal therapy triggered by near-infrared light. *Biosensors & Bioelectronics*, *89*, 477–482. doi:10.1016/j.bios.2016.03.072 PMID:27055602

Li, Q., Huang, C., Liu, L., Hu, R., & Qu, J. (2018). Enhancing type I photochemistry in photodynamic therapy under near infrared light by using antennae–fullerene complexes. *Cytometry. Part A*, *93*(10), 997–1003. doi:10.1002/cyto.a.23596 PMID:30230226

Li, R., Zhen, M., Guan, M., Chen, D., Zhang, G., Ge, J., Gong, P., Wang, C., & Shu, C. (2013). A novel glucose colorimetric sensor based on intrinsic peroxidase-like activity of C60-carboxyfullerenes. *Biosensors & Bioelectronics*, *47*, 502–507. doi:10.1016/j.bios.2013.03.057 PMID:23628844

Li, Z., Zhang, F., Pan, L., Zhu, X., & Zhang, Z. (2015). Preparation and characterization of injectable Mitoxantrone poly (lactic acid)/fullerene implants for in vivo chemo-photodynamic therapy. *Journal of Photochemistry and Photobiology. B, Biology*, *149*, 51–57. doi:10.1016/j.jphotobiol.2015.05.018 PMID:26046749

Lim, E.-K., Kim, T., Paik, S., Haam, S., Huh, Y.-M., & Lee, K. (2015). Nanomaterials for theranostics: Recent advances and future challenges. *Chemical Reviews*, *115*(1), 327–394. doi:10.1021/cr300213b PMID:25423180

Lin, A. M., Chyi, B. Y., Wang, S. D., Yu, H.-H., Kanakamma, P. P., Luh, T.-Y., Chou, C. K., & Ho, L. T. (1999). Carboxyfullerene Prevents Iron-Induced Oxidative Stress in Rat Brain. *Journal of Neurochemistry*, *72*(4), 1634–1640. doi:10.1046/j.1471-4159.1999.721634.x PMID:10098871

Lin, Y. X., Wang, Y., Blake, S., Yu, M., Mei, L., Wang, H., & Shi, J. (2020). RNA nanotechnology-mediated cancer immunotherapy. *Theranostics*, *10*(1), 281–299. doi:10.7150/thno.35568 PMID:31903120

Liu, J., Ohta, S., Sonoda, A., Yamada, M., Yamamoto, M., Nitta, N., Murata, K., & Tabata, Y. (2007). Preparation of PEG-conjugated fullerene containing $Gd^{3+}$ ions for photodynamic therapy. *Journal of Controlled Release*, *117*(1), 104–110. doi:10.1016/j.jconrel.2006.10.008 PMID:17156882

Liu, J., & Tabata, Y. (2010). Photodynamic therapy of fullerene modified with pullulan on hepatoma cells. *Journal of Drug Targeting*, *18*(8), 602–610. doi:10.3109/10611861003599479 PMID:20180750

Liu, Q., Xu, L., Zhang, X., Li, N., Zheng, J., Guan, M., Fang, X., Wang, C., & Shu, C. (2013). Enhanced photodynamic efficiency of an aptamer-guided fullerene photosensitizer toward tumour cells. *Chemistry, an Asian Journal*, *8*(10), 2370–2376. doi:10.1002/asia.201300039 PMID:23907978

Mansurov, Z. (2012). Producing nanomaterials in combustion. *Combustion, Explosion, and Shock Waves*, *48*(5), 561–569. doi:10.1134/S0010508212050073

Marchesan, S., Da Ros, T., Spalluto, G., Balzarini, J., & Prato, M. (2005). Anti-HIV properties of cationic fullerene derivatives. *Bioorganic & Medicinal Chemistry Letters*, *15*(15), 3615–3618. doi:10.1016/j.bmcl.2005.05.069 PMID:15978810

Mashino, T., Okuda, K., Hirota, T., Hirobe, M., Nagano, T., & Mochizuki, M. (1999). Inhibition of E-coli growth by fullerene derivatives and inhibition mechanism. *Bioorganic & Medicinal Chemistry Letters*, *9*(20), 2959–2962. doi:10.1016/S0960-894X(99)00515-6 PMID:10571155

Mashino, T., Shimotohno, K., Ikegami, N., Nishikawa, D., Okuda, K., Takahashi, K., Nakamura, S., & Mochizuki, M. (2005). Humanimmunodeficiency virus-reverse transcriptase inhibition and hepatitis C virus RNA-dependent RNA polymerase inhibition activities of fullerene derivatives. *Bioorganic & Medicinal Chemistry Letters*, *15*(4), 1107–1109. doi:10.1016/j.bmcl.2004.12.030 PMID:15686922

Mori, T., Takada, H., Ito, S., Matsubayashi, K., Miwa, N., & Sawaguchi, T. (2006). Preclinical studies on the safety of fullerene upon acute oral administration and evaluation for no mutagenesis. *Toxicology*, *225*(1), 48–54. doi:10.1016/j.tox.2006.05.001 PMID:16782258

Moussa, F. (1997). *Fullerenes*, *97*(42), 332–336.

Mroz, P., Pawlak, A., Satti, M., Lee, H., Wharton, T., Gali, H., Sarna, T., & Hamblin, M. R. (2007). Functionalized fullerenes mediate photodynamic killing of cancer cells: Type I versus Type II photochemical mechanism. *Free Radical Biology & Medicine*, *43*(5), 711–719. doi:10.1016/j.freeradbiomed.2007.05.005 PMID:17664135

Mroz, P., Xia, Y., Asanuma, D., Konopko, A., Zhiyentayev, T., Huang, Y.-Y., Sharma, S. K., Dai, T., Khan, U. J., Wharton, T., & Hamblin, M. R. (2011). Intraperitoneal photodynamic therapy mediated by a fullerene in a mouse model of abdominal dissemination of colon adenocarcinoma. *Nanomedicine (London)*, *7*(6), 965–974. doi:10.1016/j.nano.2011.04.007 PMID:21645643

Nakashima, S., Yoshie, M., Sano, H., & Bahar, A. (2009). Effect of a test dentifrice containing nano-sized calcium carbonate on remineralization of enamel lesions in vitro. *Journal of Oral Science*, *51*(1), 69–77. doi:10.2334/josnusd.51.69 PMID:19325202

Nimibofa, A., Newton, E. A., Cyprain, A. Y., & Donbebe, W. (2018). Fullerenes:Synthesis and Applications. *J. Mater. Sci. Res.*, *7*, 22.

Nobusawa, K., Akiyama, M., Ikeda, A., & Naito, M. (2012). pH responsive smart carrier of [60] fullerene with 6-amino-cyclodextrin inclusion complex for photodynamic therapy. *Journal of Materials Chemistry*, *22*(42), 22610–22613. doi:10.1039/c2jm34791a

Pai, P., Nair, K., Jamade, S., Shah, R., Ekshinge, V., & Jadhav, N. (2006). Pharmaceutical applications of carbon tubes and nanohorns Curr. *Pharmaceutical Research*, *1*, 11–15.

Pellarini, F., Pantarotto, D., Da Ros, T., Giangaspero, A., Tossi, A., & Prato, M. (2001). A Novel [60] Fullerene Amino Acid for Use in Solid-Phase Peptide Synthesis. *Organic Letters*, *3*(12), 1845–1848. doi:10.1021/ol015934m PMID:11405726

Rancan, F., Rosan, S., Boehm, F., Cantrell, A., Brellreich, M., Schoenberger, H., Hirsch, A., & Moussa, F. (2002). Cytotoxicity and photocytotoxicity of a dendritic C60 mono-adduct and a malonic acid C60 tris-adduct on Jurkat cells. *Journal of Photochemistry and Photobiology. B, Biology*, *67*(3), 157–162. doi:10.1016/S1011-1344(02)00320-2 PMID:12167314

Raoof, M., Mackeyev, Y., Cheney, M. A., Wilson, L. J., & Curley, S. A. (2012). Internalization of C60 fullerenes into cancer cells with accumulation in the nucleus via the nuclear pore complex. *Biomaterials*, *33*(10), 2952–2960. doi:10.1016/j.biomaterials.2011.12.043 PMID:22245558

Rezayat, S., Boushehri, S. V. S., Salmanian, B., Omidvari, A. H., Tarighat, S., Esmaeili, S., Sarkar, S., Amirshahi, N., Alyautdin, R. N., & Orlova, M. A. (2009). The porphyrin–fullerene nanoparticles to promote the ATP overproduction in myocardium: 25Mg2+-magnetic isotope effect. *European Journal of Medicinal Chemistry*, *44*(4), 1554–1569. doi:10.1016/j.ejmech.2008.07.030 PMID:18782645

Rouse, J. G., Yang, J., Ryman-Rasmussen, J. P., Barron, A. R., & Monteiro-Riviere, N. A. (2007). Effects of mechanical flexion on the penetration of fullerene amino acid-derivatized peptide nanoparticles through the skin. *Nano Letters*, *7*(1), 155–160. doi:10.1021/nl062464m PMID:17212456

Saeedfar, K., Heng, L., Ling, T., & Rezayi, M. (2013). Potentiometric Urea Biosensor Based on an Immobilised Fullerene-Urease Bio-Conjugate. *Sensors (Basel)*, *13*(12), 16851–16866. doi:10.3390/s131216851 PMID:24322561

Saravana, K. R., & Vijayalakshmi, R. (2006). Nanotechnology in dentistry. *Indian Journal of Dental Research*, *17*(2), 62–65. doi:10.4103/0970-9290.29890 PMID:17051869

Schwerdtfeger, P., Wirz, L., & Avery, J. (2013). Program fullerene: A software package for constructing and analyzing structures of regular fullerenes. *Journal of Computational Chemistry*, *34*(17), 1508–1526. doi:10.1002/jcc.23278 PMID:23559399

Sebastien, W., Giorgia, W., Monica, P., Cedric, B., Jean-Paul, K., & Renato, B. (2005). Targeted delivery of amphotericin B to cells by using functionalized carbon nanotubcs. *Angewandte Chemie*, *117*(39), 6516–6520. doi:10.1002/ange.200501613

Semeniuk, M., Yi, Z., Poursorkhabi, V., Tjong, J., Jaffer, S., Lu, Z. H., & Sain, M. (2019). Future perspectives and review on organic carbon dots in electronic applications. *ACS Nano*, *13*(6), 6224–6255. doi:10.1021/acsnano.9b00688 PMID:31145587

Serda, M., Ware, M. J., Newton, J. M., Sachdeva, S., Krzykawska-Serda, M., Nguyen, L., Law, J., Anderson, A. O., Curley, S. A., Wilson, L. J., & Corr, S. J. (2018). Development of photoactive Sweet-C60 for pancreatic cancer stellate cell therapy. *Nanomedicine (London)*, *13*(23), 2981–2993. doi:10.2217/nnm-2018-0239 PMID:30501557

Shetty, N. J., Swati, P., & David, K. (2013). Nanorobots: Future in dentistry. *The Saudi Dental Journal*, *25*(2), 49–52. doi:10.1016/j.sdentj.2012.12.002 PMID:23960556

Shi, H., Gu, R., Xu, W., Huang, H., Xue, L., Wang, W., Zhang, Y., Si, W., & Dong, X. (2019). Near-Infrared Light-Harvesting Fullerene-Based Nanoparticles for Promoted Synergetic Tumor Phototheranostics. *ACS Applied Materials & Interfaces*, *11*(48), 44970–44977. doi:10.1021/acsami.9b17716 PMID:31702130

Shi, J., Liu, Y., Wang, L., Gao, J., Zhang, J., Yu, X., Ma, R., Liu, R., & Zhang, Z. (2014). A tumoural acidic pH-responsive drug delivery system based on a novel photosensitizer (fullerene) for in vitro and in vivo chemo-photodynamic therapy. *Acta Biomaterialia*, *10*(3), 1280–1291. doi:10.1016/j.actbio.2013.10.037 PMID:24211343

Shi, J., Wang, Z., & Wang, L. (2013). Photodynamic therapy of a 2-methoxyestradiol tumour-targeting drug delivery system mediated by Asn-Gly-Arg in breast cancer. *International Journal of Nanomedicine*, *8*, 1551. doi:10.2147/IJN.S40011 PMID:23637528

Shi, J., Zhang, H., Wang, L., Li, L., Wang, H., Wang, Z., Li, Z., Chen, C., Hou, L., Zhang, C., & Zhang, Z. (2013). PEI-derivatized fullerene drug delivery using folate as a homing device targeting to tumor. *Biomaterials*, *34*(1), 251–261. doi:10.1016/j.biomaterials.2012.09.039 PMID:23069706

Shi, J. Yu., & Wang, X. (2013). *PEGylated fullerene/iron oxide nanocomposites for photodynamic therapy, targeted drug delivery and MR imaging* (Vol. 34). Biomaterials. doi:10.1016/j.biomaterials.2013.08.049

Shinohara, H. (2000). Endohedral metallofullerenes. *Reports on Progress in Physics*, *63*(6), 843–892. doi:10.1088/0034-4885/63/6/201

Sosnowska, M., Kutwin, M., Jaworski, S., Strojny, B., Wierzbicki, M., Szczepaniak, J., Łojkowski, M., Święszkowski, W., Bałaban, J., Chwalibog, A., & Sawosz, E. (2019). Mechano-signalling, induced by fullerene C60 nanofilms, arrests the cell cycle in the G2/M phase and decreases proliferation of liver cancer cells. *International Journal of Nanomedicine*, *14*, 6197–6215. doi:10.2147/IJN.S206934 PMID:31496681

Sotiropoulou, S., Gavalas, V., Vamvakaki, V., & Chaniotakis, N. (2003). Novel carbon materials in biosensor systems. *Biosensors & Bioelectronics*, *18*(2-3), 211–215. doi:10.1016/S0956-5663(02)00183-5 PMID:12485767

Soung, Y. H., Nguyen, T., Cao, H., Lee, J., & Chung, J. (2016). Emerging roles of exosomes in cancer invasion and metastasis. *BMB Reports*, *49*(1), 18–25. doi:10.5483/BMBRep.2016.49.1.239 PMID:26592936

Tabata, Y. (1997). Photodynamic effect of polyethylene glycol–modified fullerene on tumour. *Japanese Journal of Cancer Research*, *88*, 1108. doi:10.1111/j.1349-7006.1997.tb00336.x PMID:9439687

Tang, L., Li, J., Pan, T., Yin, Y., Mei, Y., Xiao, Q., Wang, R., Yan, Z., & Wang, W. (2022). Versatile carbon nanoplatforms for cancer treatment and diagnosis: Strategies, applications and future perspectives. *Theranostics*, *12*(5), 2290–2321. doi:10.7150/thno.69628 PMID:35265211

Taylor, R., Hare, J. P., Abdul-Sada, A. K., & Kroto, H. W. (1990). Isolation, separation and characterization of the fullerenes C60 and C70: The third form of carbon. *Journal of the Chemical Society. Chemical Communications*, (20), 1423–1425. doi:10.1039/c39900001423

Tegos, G. P., Demidova, T. N., Arcila-Lopez, D., Lee, H., Wharton, T., Gali, H., & Hamblin, M. R. (2005). Cationic fullerenes are effective and selective antimicrobial photosensitizers. *Chemistry & Biology*, *12*(10), 1127–1135. doi:10.1016/j.chembiol.2005.08.014 PMID:16242655

Thrash, T. P., Cagle, D. W., Alford, J. M., Wright, K., Ehrhardt, G. J., Mirzadeh, S., & Wilson, L. J. (1999). Toward fullerene-based radiopharmaceuticals: High-yield neutron activation of endohedral 165Ho metallofullerenes. *Chemical Physics Letters*, *308*(3-4), 329–336. doi:10.1016/S0009-2614(99)00581-3

Tokuyama, H., Yamago, S., Nakamura, E., Shiraki, T., & Sugiura, Y. (1993). Photoinduced biochemical activity of fullerene carboxylic acid. *Journal of the American Chemical Society*, *115*(17), 7918–7919. doi:10.1021/ja00070a064

Tsao, N., Luh, T.-Y., Chou, C.-K., Chang, T.-Y., Wu, J.-J., Liu, C.-C., & Lei, H.-Y. (2002). In vitro action of carboxyfullerene. *The Journal of Antimicrobial Chemotherapy*, *49*(4), 641–649. doi:10.1093/jac/49.4.641 PMID:11909838

Wang, B., Qi, X., Liu, J., Zhou, R., Lin, C., Shangguan, J., Zhang, Z., Zhao, L., & Li, G. (2019). MYH9 Promotes Growth and Metastasis via Activation of MAPK/AKT Signaling in Colorectal Cancer. *Journal of Cancer*, *10*(4), 874–884. doi:10.7150/jca.27635 PMID:30854093

Wang, J., Xie, L., Wang, T., Wu, F., Meng, J., Liu, J., & Xu, H. (2017). Visible light-switched cytosol release of siRNA by amphiphilic fullerene derivative to enhance RNAi efficacy in vitro and in vivo. *Acta Biomaterialia*, *59*, 158–169. doi:10.1016/j.actbio.2017.05.031 PMID:28511875

Wang, T., Upponi, J. R., & Torchilin, V. P. (2012). Design of multifunctional non-viral gene vectors to overcome physiological barriers: Dilemmas and strategies. *International Journal of Pharmaceutics*, *427*(1), 3–20. doi:10.1016/j.ijpharm.2011.07.013 PMID:21798324

Wang, T., & Wang, C. (2019). Functional Metallofullerene Materials and Their Applications in Nanomedicine, Magnetics, and Electronics. *Small*, *15*(48), e1901522. doi:10.1002/smll.201901522 PMID:31131986

Wang, Y., Meng, Y., Wang, S., Li, C., Shi, W., Chen, J., Wang, J., & Huang, R. (2015). Direct solvent-derived polymer-coated nitrogen-doped carbon nanodots with high water solubility for targeted fluorescence imaging of glioma. *Small, 11*(29), 3575–3581. doi:10.1002/smll.201403718 PMID:25808813

Wei, L.-F. (2001). Shih, J.-S. *Analytica Chimica Acta, 437*(1), 77–85. doi:10.1016/S0003-2670(01)00941-2

Wei, P., Zhang, L., Lu, Y., Man, N., & Wen, L. (2010). C60(Nd) nanoparticles enhance chemotherapeutic susceptibility of cancer cells by modulation of autophagy. *Nanotechnology, 21*(49), 495101. doi:10.1088/0957-4484/21/49/495101 PMID:21071824

Witte, P., Beuerle, F., Hartnagel, U., Lebovitz, R., Savouchkina, A., Sali, S., Guldi, D., Chronakis, N., & Hirsch, A. (2007). Water solubility, antioxidant activity and cytochrome C binding of four families of exohedral adducts of C60 and C70. *Organic & Biomolecular Chemistry, 5*(22), 3599–3613. doi:10.1039/b711912g PMID:17971989

Wu, S., Zheng, Q., Xing, X., Dong, Y., Wang, Y., You, Y., Chen, R., Hu, C., Chen, J., Gao, D., Zhao, Y., Wang, Z., Xue, T., Ren, Z., & Cui, J. (2018). Matrix stiffness-upregulated LOXL2 promotes fibronectin production, MMP9 and CXCL12 expression and BMDCs recruitment to assist pre-metastatic niche formation. *Journal of Experimental & Clinical Cancer Research, 37*(1), 99. doi:10.1186/s13046-018-0761-z PMID:29728125

Xu, Z. P., Zeng, Q. H., Lu, G. Q., & Yu, A. B. (2006). Inorganic nanoparticles as carriers for efficient cellular delivery. *Chemical Engineering Science, 61*(3), 1027–1040. doi:10.1016/j.ces.2005.06.019

Yadav, B. C., & Kumar, R. (2008). Structure, properties and applications of fullerenes. *International Journal of Nanotechnology and Applications, 2*(1), 15–24.

Yamakoshi, Y., Umezawa, N., Ryu, A., Arakane, K., Miyata, N., Goda, Y., Masumizu, T., & Nagano, T. (2003). Active oxygen species generated from photoexcited fullerene (C 60) as potential medicines: O 2-• versus 1 O 2. *Journal of the American Chemical Society, 125*(42), 12803–12809. doi:10.1021/ja0355574 PMID:14558828

Yang, B., Chen, Y., & Shi, J. (2019). Reactive oxygen species (ROS) based Nanomedicine. *Chemical Reviews, 119*(8), 4881–4985. doi:10.1021/acs.chemrev.8b00626 PMID:30973011

Yang, X., Qin, L., Wang, Y., Zhang, B., Liu, Z., Ma, H., Lu, J., Huang, X., Shi, D., & Hu, Z. (2014). Detection of Mycobacterium tuberculosis based on H37Rv binding peptides using surface functionalized magnetic microspheres coupled with quantum dots – a nano detection method for Mycobacterium tuberculosis. *International Journal of Nanomedicine, 9*, 77. doi:10.2147/IJN.S71700 PMID:24379667

Ye, L., Kollie, L., Liu, X., Guo, W., Ying, X., Zhu, J., Yang, S., & Yu, M. (2021). Antitumor Activity and Potential Mechanism of Novel Fullerene Derivative Nanoparticles. *Molecules (Basel, Switzerland), 26*(11), 3252. doi:10.3390/molecules26113252 PMID:34071369

Yin, R., Wang, M., Huang, Y.-Y., Landi, G., Vecchio, D., Chiang, L. Y., & Hamblin, M. R. (2015). Antimicrobial photodynamic inactivation with decacationic functionalized fullerenes: Oxygen-independent photokilling in presence of azide and new mechanistic insights. *Free Radical Biology & Medicine, 79*, 14–27. doi:10.1016/j.freeradbiomed.2014.10.514 PMID:25451642

Youle, R. J. (2005). Karbowski, M. *Nature Reviews. Molecular Cell Biology*, *6*(8), 657–663. doi:10.1038/nrm1697 PMID:16025099

Zhang, H., Hou, L., Jiao, X., Ji, Y., Zhu, X., & Zhang, Z. (2015). Transferrin-mediated fullerenes nanoparticles as Fe(2+)-dependent drug vehicles for synergistic anti-tumor efficacy. *Biomaterials*, *37*, 353–366. doi:10.1016/j.biomaterials.2014.10.031 PMID:25453964

Zhang, M., Li, J., Xing, G., He, R., Li, W., Song, Y., & Guo, H. (2011). Variation in the internalization of differently sized nanoparticles induces different DNA-damaging effects on a macrophage cell line. *Archives of Toxicology*, *85*(12), 1575–1588. doi:10.1007/s00204-011-0725-y PMID:21881955

Zhong, X. (2012). Yuan, X. Chai, Y. *Chemical Communications (Cambridge)*, *48*(4), 597–599. doi:10.1039/C1CC16081H

Zhou, W., Huo, J., Yang, Y., Zhang, X., Li, S., Zhao, C., Ma, H., Liu, Y., Liu, J., Li, J., Zhen, M. M., Li, J., Fang, X., & Wang, C. (2020). Aminated Fullerene Abrogates Cancer Cell Migration by Directly Targeting Myosin Heavy Chain 9. *ACS Applied Materials & Interfaces*, *12*(51), 56862–56873. doi:10.1021/acsami.0c18785 PMID:33305958

Zhu, J., Ji, Z., Wang, J., Sun, R., Zhang, X., Gao, Y., Sun, H., Liu, Y., Wang, Z., Li, A., Ma, J., Wang, T., Jia, G., & Gu, Y. (2008). Tumor-Inhibitory Effect and Immunomodulatory Activity of Fullerol $C_{60}(OH)_x$. *Small*, *4*(8), 1168–1175. doi:10.1002/smll.200701219 PMID:18574800

# Chapter 5
# Carbon Nanocomposites:
## Emerging Engineering Technologies and Industrial Applications

**Loutfy H. Madkour**
https://orcid.org/0000-0002-3101-8356
*Tanta University, Egypt*

## ABSTRACT

*Successful deployment of carbon nanocomposites in many applications, such as sensing, energy storage, and catalysis, relies on the selection, synthesis, and tailoring of the surface properties. Carbon nanotubes, due to their large surface areas, unique surface properties, and needle-like shape, can deliver a lot of therapeutic agents, including DNA, siRNAs and proteins to the target disease sites. Carbon nanotubes can be readily excreted through the renal route by means of degradation through myeloperoxidase enzyme. Nanotubes are categorized as single-walled carbon nanotubes and multiple walled carbon nanotubes. Theirs advances have been made in the delivery of anticancer, anti-inflammatory drugs, bioactive molecules, and proteins. Drugs and biomolecules can be loaded in carbon nanotubes, which can then be utilized as targeted molecules. Recently, carbon nanocomposites have attracted a lot of attention in the field of cancer diagnosis and therapy, with their ability to deliver therapeutic molecules and allow visualization of cells and tissues, which are necessary for the cure and treatment of diseased and damaged tissues.*

## 1. ADVANCEMENT, AND APPLICATIONS OF CARBON IN COMPOSITES

Carbon nanotubes (CNTs) are a novel class of materials with a diverse range of applications (Madkour, 2019,2020,2022;2023) due to their unique physical properties.The physical, optical, and electronic properties of the CNTs further provide means for the precise control over their transport to specific target organs or tissues, and their controlled release according to the desired dosing regimen. Recently, several research groups have attempted to measure the electrical properties of carbon nanotubes experimentally.

DOI: 10.4018/979-8-3693-1335-0.ch005

The development of scanning tunneling microscopy (STM) and spectroscopy-based techniques has been particularly helpful in these experimental developments.

Carbon nanotubes is a fullerene molecule, it was first discovered in 1991 by Japanese scientist Iijima (Iijima, 1991) as tube-shaped of graphitic carbon, can be obtained either single or multi-walled nanotube, having a diameter measuring on the nanometer scale, and generally known as buckytubes. Carbon nanotubes (CNTs) can be defined as allotropes of carbons in tubular shape, which looks like graphene sheet rolls. The diameter of CNTs varies from 1 to 100 nm, while the length of CNTs may be up to millimeter scale. CNTs can be classified based on the number of layers: single-walled carbon nanotubes (SWCNTs) and multiple-walled carbon nanotubes (MWCNTs). In addition to this classification, SWCNTs can be categorized as armchair carbon nanotubes, zigzag carbon nanotubes, and chiral carbon nanotubes by their crystallographic configurations. Each type of CNTs has advantages, disadvantages, and unique properties due to their distinctive characteristics.

CNTs have large specific surface areas on which a wide range of molecular moieties can be attached. Also CNTs can be functionalized to enhance their physicochemical properties and permeation abilities through cell membranes.

CNTs are a very promising technique in sensing technology (Marquez et al.2023,Qureshi et al.2023) since they offer structural competence and measurable response under applied stresses and strains. They have functional capabilities such as actuation, sensing, and power harvesting even when operating at very low voltage. Integrating CNTs into polymers results in a whole new range of smart structure applications, advanced sensors, and actuators . For example, CNTs-based laminate composites had been used as strain sensors with wireless transmission systems and also as a sensing skin for damage detection. It had been also used to develop hybrid composites with self-sensing properties . Conductive thread created by twisting the CNT forests into a wire had also been used as a sensor to monitor deformation including delamination in composites . There is a wide range of in situ strain sensors based on multi-walled CNTs which are insensitive to temperature variation . Studies have shown that the delivery of drugs and gènes to brain can be achieved with functionalized CNTs. Thus, CNTs provide a new drug delivery approach to CNS diseases. However, problems with the alignment of CNTs with fibers in an epoxy matrix and their dispersion is often difficult so alternative approaches such as radial in-situ growth of CNTs on fiber surface are considered. In addition, the electrical response of the CNTs-based sensing film depends on their concentration in the matrix because more concentration leads to more nanotube-to-nanotube junctions thus increasing the conductance but, it is not favorable to increase the concentration beyond the percolation threshold. So, it is vital to evaluate the concentration of CNTs in the sensing film to optimize their sensing performance. Some research studies had also been conducted to develop CNTs-based wireless embedded sensors for composite civil structures . CNTs dispersed in cement matrix not only improve their mechanical properties and develop a smart material for real-time damage detection but also result in an efficient way of crack bridging during initial crack propagation . Recently, the University of Cincinnati developed an artificial neural system consisting of a long film of CNTs as a grid/sensor network attached to the surface of the structure. Furthermore, some researchers had also studied the strain sensing behavior of CNTs-based nanoscale sensors using Raman spectroscopy by indirect measurement of the resistance of a nanocomposite but it was very huge to be used in a sensor mechanism. In addition, MEMS of these nanoparticles were also developed using lithography and aligned SWNTs which can detect small cracks and measure small strains. But, this system required large signal processing because of sensors array to cover large areas for real-time SHM. Recently, a new real-time SHM technique termed nano-engineered thermal (NET) sensing had been developed using

CNTs and other conductive nanofillers. In this technique, composites having aligned CNTs were heated ohmically through electrical contact and a crack was visualized by thermal imaging. Any discontinuity present could affect both thermal and electrical resistance in these structures thus enabling tomographic full-field damage evaluation for in-situ monitoring in structures such as aircraft, automobiles, and wind turbines, among others.

Carbon nanotubes, due to their specific electrical properties, potentially use in plant genetic engineering (Kumari et al.2021) for delivering DNA into plant cells. Carbon nanotubes can be applied to deliver both linear and plasmid DNA, siRNA, co-delivery of protein and RNA, as reported in various plant species; Nicotiana benthamiana, Eruca sativa, Triticum aestivum, Gossypium hirsutum, Nicotiana tobacum, Nasturtium officinale, Spinacia oleracea, Gossypium hirsutum (Liu et al., 2009; Zhang et al., 2019a; Kwak et al., 2019). In recent study, Demirer et al., 2019 worked on applicability of polyethylenimine coated carbon nanotube for genetic transformation in plant. They reported that high electrostatics strength of carbon nanotube coated with positively charged polyethylenimine will provide co-advantage and shown strong affinity of negatively charged DNA to get stick to the nanomaterials.

Here we will focus on (CNTs), also called buckytubes and first evidenced in 1991 by the Japanese electron microscopist Sumio Iijima. Since his first report on multi-walled carbon nanotubes (MWNTs), followed by their single-walled counterparts (SWNTs) (Iijima and Ichihashi 1993),

## 2. STRUCTURE OF CARBON NANOCOMPOSITES

CNTs are graphene films tied up in a cylinder form with diameters of 1 nm (Chatterjee & Deopura, 2002). The high surface area, and fast kinetics, it has high advantages toward wastewater treatment. It has a high adsorption capacity for a wide variety of pollutants. Dichlorobenzene, ethylbenzene, metal ions ($Zn^{2+}$, $Pb^{2+}$, $Cu^{2+}$, $Cd^{2+}$), dyes, and aromatics are effectively removed by CNTs (Khin et al., 2012). The single-walled carbon nanotubes (SWCNTs) are composed of single-layer graphene films tied up into cylindrical and multiwalled carbon nanotubes (MWCNTs) are composed of multiple layers of graphene films with the spacing of 0.34 nm between the graphene layers (Zhao & Stoddart, 2009). Zhan et al. (2019) synthesized a modified multiwalled carbon nanotube with polydopamine (MWCNT-PDA). The resulting hybrid CNTs have high efficiency toward the metals ions due to their numerous active sites, highly porous structure, and large surface area. As per the study, the adsorption capacity by MWCNT-PDA reported was 318.47 mg/g for $Cu^{2+}$ and 350.87 mg/g for $Pb^{2+}$ (Zhan et al., 2019).

Carbon nanotubes (CNTs), a hollow cylindrical form of carbon allotrope, were first reported by Iijima in 1991. CNTs have a one-dimensional structure of rolled-up graphene layers with diameters ranging from one to a few tens of nanometers. The cylindrical carbon structure was composed of carbon atoms covalently linked to each other with a hexagonal structure. CNTs have unique optical properties, high conductivity and surface area, and less weight features which can also be tuned by the functionalization process. Generally, they have strong hydrophobicity, pore structure, and multiple surface groups. Based on the number of layers, CNTs are usually classified as single-walled and multiwalled carbon nanotubes. Single-walled carbon nanotubes (SWCNTs) have a single layer of graphene, while multiwalled carbon nanotubes (MWCNTs) have many concentrically arranged walls of graphene sheets with a distance of 0.34 nm. Apart from the number of layers, some other parameters also substantially differentiate SWCNTs from MWCNTs. In the case of SWCNT, large-scale synthesis with high purity is difficult to achieve. Also, the catalyst is essential for its synthesis. However, high defects can be created within SWCNTs by functionalization, while highly pure MWCNTs can be prepared on a large scale even without using catalysts. However, defects will be less for MWCNTs in comparison with SWCNTs.

An understanding of SWNT structure requires familiarity with the concept of nanotube chirality, since the chirality of a SWNT dictates many of its properties. The chirality map, illustrated in Scheme 1, has been developed as a tool for understanding chirality and its implications.

## 3. SCHEME ONE

A SWNT can be envisioned as a sheet of graphene one atom thick rolled into a tube (inset in Scheme 1). Most single-walled nanotubes (SWNT) have a diameter of close to 1 nanometer, with a tube length that can be many millions of times longer (Kalaiselvan Shanmugam, et al.2020).

CNTs are large molecules of pure carbon that are long, thin, tubular and cylindrical shape and having a size range between 2-3 nm (Shikha Singh, et al. 2021). CNTs are also defined as tubular fullerene or cylindrical graphene having $sp^2$ hybridization carbon atoms. CNTs are allotropes, which is having a tubular figure and prepared from graphite. They are classified into three categories (Rajwant, et al. 2018).

Carbon nanotubes have emerged worldwide applications: such as medical applications, coatings and films, microelectronics, energy storage, environment and biotechnology (Sidi Salah,et al.2020), and CNT cotton super fiber materials (Ismail, et al.2019); CNTs are also used as composite membrane (Han, Dong,et al.2020), transparent electrodes, and catalysis and can act as a transistor (Hills, Lau, Wright et al.2019).

### 3.1 Forms and Nature of CNT Composites

Carbon-based nanomaterials either nanotubes, nanofibers, Fullerenes, and graphene, are some famous structures because of their various types, structures, and dimensions; these such materials are known as nanomaterials (Mathur et al., 2017; Zaytseva and Neumann, 2016). Such materials have properties like semiconductor behavior, thermal conductivity, high structural properties, and light absorption properties (Dolez, 2015). Therefore, they are used in solar cells, structural materials, electronics, and semiconductors due to the properties and applications they have. Classification of Carbon nanotubes based on the number of walls present single-wall, double-wall, or multi-wall and these walls decide the properties of Carbon nanotubes they are responsible for their structural and conductive characteristics (Pérez-Sánchez et al., 2013; Sudha et al., 2018). It can also be used for herbicide production.

CNT strip (length = 30 cm and diameter =3 mm) with mesopores and the hydrophobic surface was used for removing o-cresol from water. Compared to pure CNT powder, CNT strip has less surface area and lower o-cresol adsorption capacity. However, strips can be recovered more easily than powder form and utilize 10 successive processes. Further, the strength of the strips is also high under compression.

### 3.2 Functionalized Carbon Nanotubes Composites

Like activated carbon, functionalized CNTs also performed well in the adsorption process. To remove uranyl ions ($UO_2^{2+}$) from water (pH = 4; time = 120 min), poly(amidoxime) functionalized CNTs were used as adsorbent. Amidoxime functional group is a ligand that shows a strong affinity with uranyl ions. Functionalized CNTs showed excellent uranyl ion adsorption (adsorption capacity = ~ 247 mg g−1) due to its abundant amidoxime group and porosity (surface area = 58.169 $m^2$ $g^{-1}$). Based on the Fourier transform infrared and X-ray photo-electron spectroscopic studies, two ways of coordination between amidoxime groups and uranyl ions were proposed (Scheme 2A). In another investigation, poly(acrylic acid) and polyacrylamide functionalized CNTs were fabricated by oxidative polymerization method and applied as sorbent for phenol removal in water. This adsorbent exhibited great performance for four cycles which is due to π-π interactions, hydrogen bonding, and electrostatic interactions between phenol rings and functionalized CNTs (Scheme 2B).

MWCNTs functionalized with tetra-n-butyl-ammonium bromide showed high adsorption performance toward Ni(II) ions. After 68 min of contact time, the highest Ni(II) adsorption capacity of 115.8 mg $g^{-1}$ (removal rate = 93%) was obtained. Coulombic force of attraction is the key reason for the adsorption of Ni(II) ions on functionalized MWCNTs.

## 4. SCHEME TWO

### 4.1 CNT—Based Composites

Numerous CNT-based binary and ternary composites were reported as adsorbents. For example, SWCNT/$Fe_3O_4$ composite adsorbent was fabricated for the elimination of xylene ($C_8H_{10}$) from aqueous solutions. Under the optimized conditions (pH = 8, dosage = 2000 mg $L^{-1}$, time period = 20 min), 99.2% (qe = 50 mg g−1) xylene (initial concentration = 100 mg $L^{-1}$) was captured by SWCNT/$Fe_3O_4$ composite adsorbent. In this process, contact time showed maximum effect among other influencing factors. Recently in 2019, hybrid carbon fiber-carbon nanotubes reinforced polymer composites has been studied by Zakaria et al. As a result of its impressive 3D network structure form, a great deal of interest has thus been generated among the science community on the hybridization of carbon fiber (CF) and carbon nanotube (CNT). Although there had been a large amount of studies conducted on this novel hybrid material, there are however, limited reviews being published on their fabrication methods and performances in polymer composites. For this reason, apart from collectively introducing and discussing the advantages and limitations of the hybrid CF-CNT based on various fabrication methods such as those of chemical vapour deposition, electrophoretic deposition, electrospray deposition and chemical functionalization, this study (Zakaria et al.2019) had also examined and compared the detailed performances of each of the hybrid CF-CNT reinforced polymer composite based on their fabrication methods through its mechanical, electrical and thermal properties.Zhang et al.2016,synthesized 3D and ternary composite hydrogels containing reduced graphene oxide, multi-walled carbon nanotubes and $Fe_3O_4$ nanoparticles (rGO/MCNTs/$Fe_3O_4$) via hydrothermal process. The fibrous MCNTs are involved in 3D porous network rGO structure, and the 100–200 nm $Fe_3O_4$ nanoparticles are uniformly dispersed onto the rGO and MCNTs surface. The rGO/MWCNTs/$Fe_3O_4$ composite hydrogels exhibit excellent microwave absorbability. The composite with a coating layer thickness of only 2.0 mm exhibits a maximum absorption value of −36 dB at 13.44 GHz. And the composite hydrogel shows a bandwidth of 11.4 GHz (from frequency of 6.5–17.9 GHz) corresponding to the reflection loss at −10 dB (90% absorption) with a thickness range of 2.0–4.0 mm. Thus, the as-prepared rGO/MCNTs/Fe3O4 composite hydrogels serve as a new member of promising lightweight and high-performance EM wave absorbing materials. The great reason for their excellent microwave absorption properties is also ascribed to the better impedance match.

## 5. HISTORY OF CARBON NANOTUBE COMPOSITES

CNTs can be assumed to have a history of approximately 30 years. At long last during the year 1991 after all the exploration work Japanese researcher and specialist Iilima has found carbon nanotube by circular segment release strategy at NEC (Nippon electric organization) (Iijima and Ichihashi 1993, Rajwant, et al. 2018 and Beatriz, R.C.M. Karla, F. R. et.al. (2019).

## 6. CHARACTERIZATION OF CNTS COMPOSITES

### 6.1. Transmission Electron Microscopy (TEM)

The internal microstructure and crystal structure of samples which are thin enough to transmit electrons can be analysed with TEM. (Anamika Sahu, et al. 2017).

### 6.2. Scanning Electron Microscopy (SEM)

SEM images the sample morphology by scanning the surface with a high-energy beam of electrons. (Anamika Sahu, et al. 2017).

### 6.3. Raman Spectroscopy

This technique gives information with details about configuration of CNTs. A number of walls, the presence of crystalline and amorphous carbon and diameter of SWNTs can be determined with the Raman spectroscopy.

### 6.4. Proton Nuclear Magnetic Resonance (Proton NMR)

H1-NMR of functionalized CNTs is portrayed by wide peaks for protons close to the CNT, fitting sharper with separation.

### 6.5. Absorption Spectroscopy (UV, Vis, and IR)

The possibilities of the use of CNTs for catalytic removal of NOx (NO, $NO_2$, $NO_3$) were also investigated by FT-IR (Anamika Sahu, et al. 2017). For a correct characterization of CNTs, all these techniques described herein cannot be used individually but must be used in complementary ways.

### 6.6. Thermal Analysis (TGA/DTG)

Thermal gravimetric analysis (TGA) and derivative thermal gravimetric (DTG) analysis methods controlled oxidation process that gives quantitative data on the weight fractions of carbon and metal catalyst in the sample, and the temperatures of bulk oxidation events (Anamika Sahu, et al. 2017).

## 6.7. Atomic Force Microscopy (AFM)

This method also provides us with information on the nanotube's length and an estimated assessment on the diameter of bundles.

## 6.8. Elemental Analysis

The purification of raw MWCNTs allowed the elimination of some catalyst impurities in oxidised form, and also raw MWCNTs could be functionalized with some oxygen-containing groups' like–COOH groups' and–OH groups that would be later confirmed by the elemental analysis (Anamika Sahu, et al. 2017).

## 6.9. X-Ray Diffraction (XRD)

This technique is used to obtain some information on the interlayer spacing, the structural strain, the impurities, SWNT bundles, several layers of MWNTs and allows morphological information for an unlimited number of crystals.

# 7. CLASSIFICATION OF CATEGORIES CARBON NANOTUBES TYPES AND THEIR PROPERTIES

Elemental carbon displays a complex allotropy depending on the nature of hybridization; diamond ($sp^3$ hybridized), graphite, graphene, Fullerenes, and carbon nanotubes ($sp^2$ hybridized), as depicted in Figure 1.

CNTs are hexagonally arranged, honey-combed lattice of carbon atoms formed by the rolling of graphene into seamless cylindrical structures (see Figure 2a). The representation of a zigzag (m = 0), armchair (n = m), and chiral (n ≠ m) nanotube is depicted in Figure 2(b-d).

These are Single-walled carbon nanotubes (SWNTs) and Multi-walled carbon nanotubes (MWNTs). Classifications of CNTs are described in Figure 3 (Hasnain and Nayak, 2019).

Various features of single-walled carbon nanotubes (SWNTs) and multi-walled carbon nanotubes (MWNTs) are described in Figure 4.

Based on the number of layers, structures of CNTs are classified into two types: single-walled carbon nanotubes (SWCNTs) and multiwalled carbon nanotubes (MWCNTs) (Figure 5).

The differences between the two classes are highlighted in Figure 6. Zigzag and armchair structures are obtained when $(n,0)$ or $m=0$ and $n=m$, respectively, while chiral nature exists when $n>m>0$ (Onyancha, Aigbe, et al. 2021).

## 7.1. Unique Properties of Single-Walled Carbon Nanotube (SWCNTs)

The structural features of SWNTs are generally illustrated by means of a (n, m) vector that describes the chirality and diameter (Anamika Sahu, et al. 2017). There are three SWNTs: (a) Armchair (10, 10), (b) Chiral (13, 6), and (c) Zigzag (14, 0) (Figure 7).

*Figure 1. Diamond (sp³ hybridized), graphite, graphene, Fullerenes, and carbon nanotubes (sp² hybridized)*

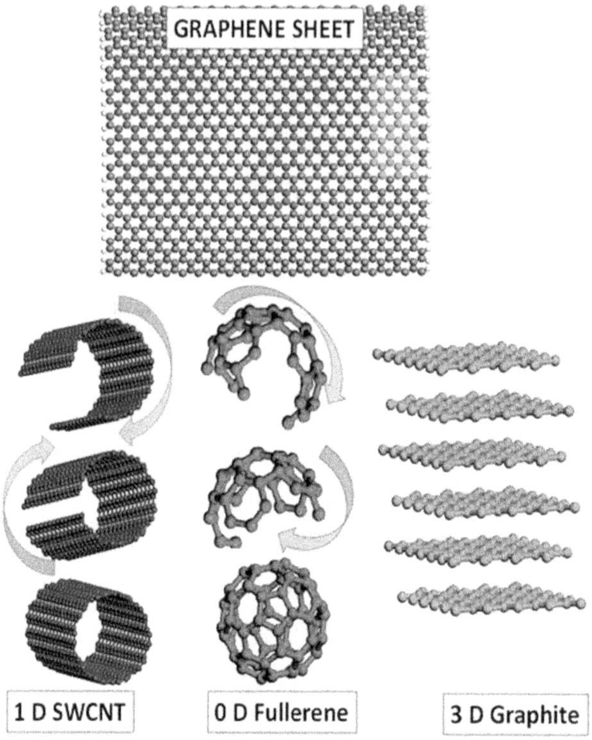


## 7.2. Double-Walled CNTs (DWCNTs)

Since the name is given, this CNT is comprised of two concentric CNTs, where an external layer of the carbon chamber is completely encased inside the internal chamber (outside cylinder is encased in inside

*Figure 2. CNTs*

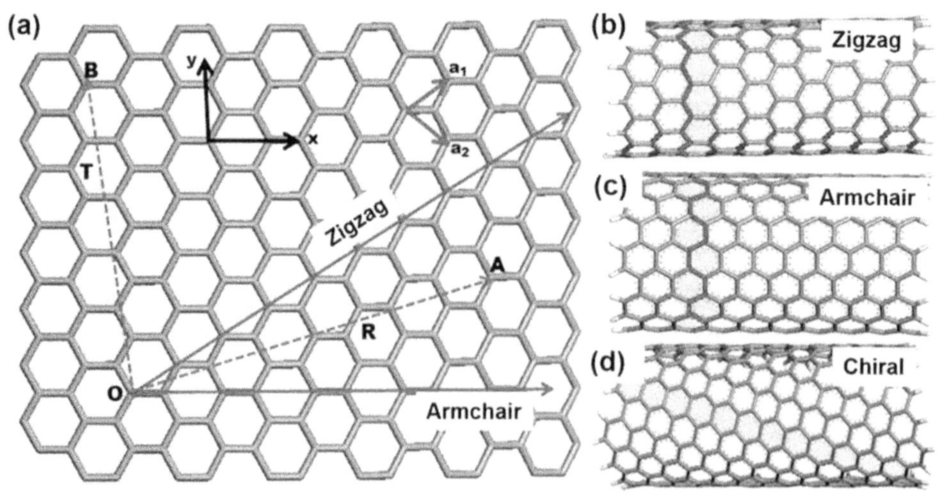

*Figure 3. Classifications of CNTs*

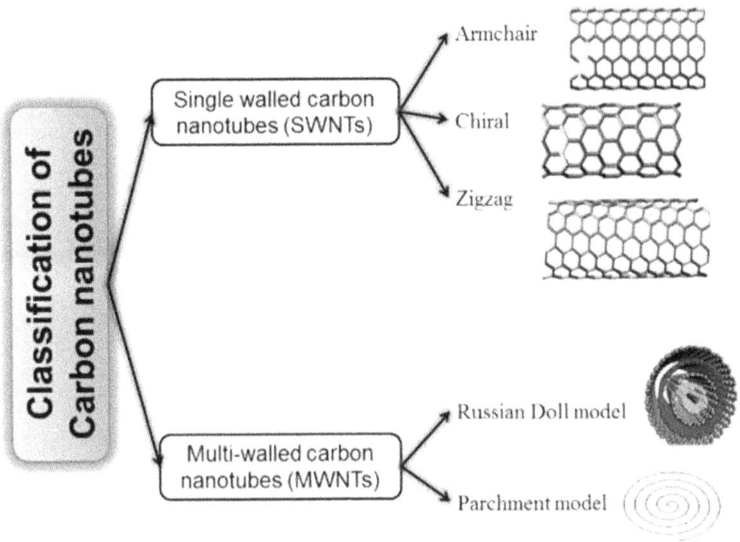


cylinder) (Gul, Zainab, Asad, Salma, (2019).

## 7.3. Multi-Walled CNTs (MWCNTs)

Multi-walled carbon nanotubes (MWCNTs) are a special form of carbon nanotubes in which multiple single-walled carbon nanotubes are nested inside one another.

*Figure 4. Features of single-walled carbon nanotubes (SWNTs) and multi-walled carbon nanotubes (MWNTs)*

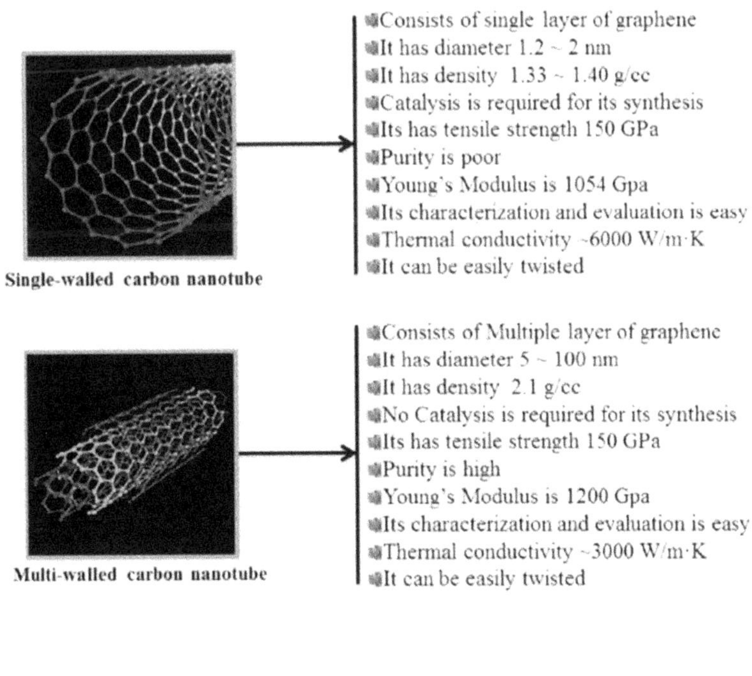

*Figure 5. Structures of CNTs*

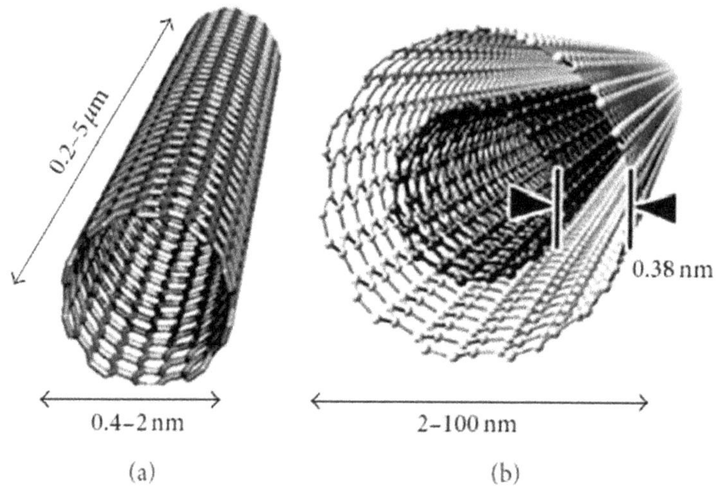


*Figure 6. Differences between the two classes*

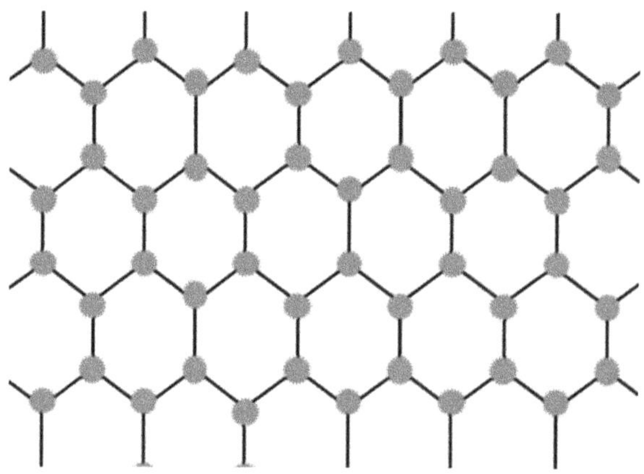

*Figure 7. Three SWNTs*

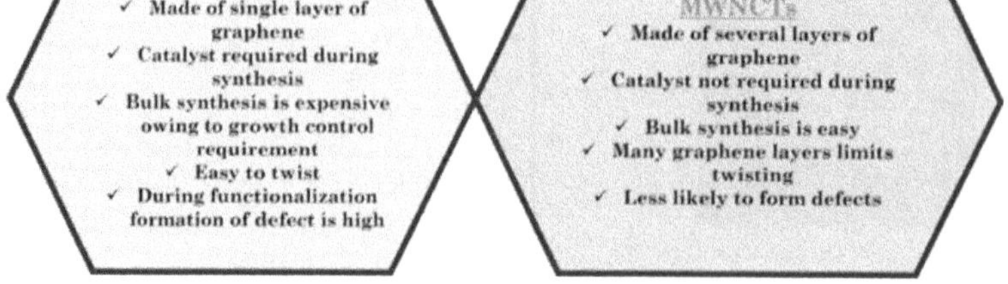

*Table 1. Comparative characteristics of SWCNT and MWCNT (Saifuddin, Raziah, et al.2013)*

| SWCNT | MWCNT |
|---|---|
| Single layer of graphene. | Multiple layer of graphene |
| Catalyst is required for synthesis. | Can be produced without catalyst. |
| Bulk synthesis is difficult as it requires proper control over growth and atmospheric condition. | Bulk synthesis is easy. |
| Not fully dispersed, and form bundled structures. | Homogeneously dispersed with no apparent bundled formation. |
| Resistivity usually in the range of $10^{-4}$–$10^{-3}$ Ω·m. | Resistivity usually in the range of $1.8 \times 10^{-5}$–$6.1 \times 10^{-5}$ Ω·m |
| Purity is poor. Typical SWCNT content in as-prepared samples by chemical vapour deposition (CVD) method is about 30–50 wt%. However high purity up to 80% has been reported by using arc discharge synthesis method. | Purity is high. Typical MWCNT content in as-prepared samples by CVD method is about 35–90 wt%. |
| A chance of defect is more during functionalization. | A chance of defect is less especially when synthesized by arc-discharged method. |
| Characterization and evaluation is easy. | It has very complex structure. |
| It can be easily twisted and are more pliable. | It cannot be easily twisted. |
| Less accumulation in body | More accumulation in body |

## 7.4. Differences in SWNTs and MWNTs

Carbon nanotubes are mainly classified in th two types: SWCNTs—Single-walled carbon nanotubes and MWCNTs—Multiple-walled carbon nanotubes. Comparison between SWCNT and MWCNT is as presented in Table 1 (Saifuddin, Raziah, et al.2013).

## 7.5. Types of CNTs justified on the basis of chirality

In determining the electrical characteristics of CNTs, chirality is a very significant variable. There is around three varieties of CNTs based on chirality, i.e. armchair, zigzag and also chiral. All the above mentioned three forms are illustrated in Figure 8.

### 7.5.1. Torus

Nanotorus is a carbon nanotube bent into a torus (doughnut) shape.

### 7.5.2. Nanobud

Nanobuds are the newly synthesized combination of carbon nanotubes and fullerenes that possess useful properties of both.

*Figure 8. Three forms*

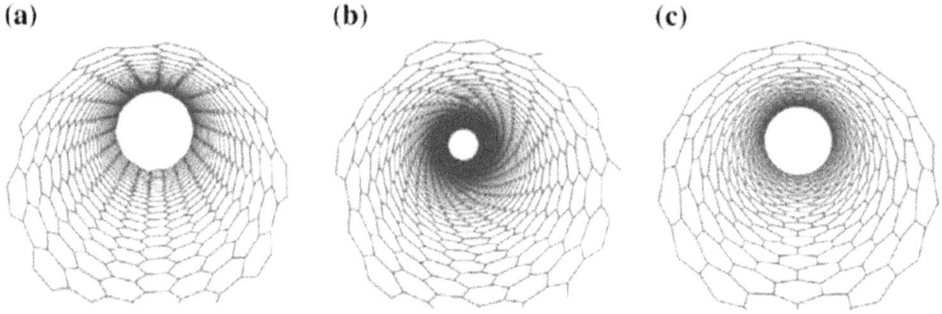


### 7.5.3. Cup stacked carbon nanotubes (CSCNTs)

CSCNTs exhibit semiconducting behaviors due to the stacking microstructure of graphene layers and thus differ from other quasi-1D carbon structures that behave as a metallic conductor of electrons.

### 7.5.4. Extreme carbon nanotubes

Characterization of carbon nanotube type was done by a combination of highresolution transmission electron microscopy (HRTEM), However, they were found inside of doublewalled carbon nanotubes (Guan, Suenaga, Iijima. 2008) (Figure 9).

## 8. DIFFERENT ALLOTROPES OF CARBON RELATED TO DEVICES

Moreover, demonstration of graphene as the basal structure for all the allotropes can be understood from Figure 9.

### 8.1. Zero-Dimensional Allotropes

### 8.1.1. Fullerene

It is made up of truncated icosahedral symmetry with 32 faces, 60 corners, and 90 edges composed of 20 hexagons and 12 pentagons where the 6:6 ring bonds (between two hexagons).

### 8.1.2. Graphene Quantum Dots (GQDs)

Carbon quantum dots (CQDs), widely regarded as carbon dots (CDs), are a 0-D allotrope composed of carbon having size <10 nm. GQDs have significance inherited from both graphene and CDs, which could be idealized as smithereens of graphene. The modification of 2-D graphene sheets into 0-D GQDs with a radius smaller than exciton Bohr's radius (Tian, Tang, et al. (2018)) generates modification in electron distribution

*Figure 9. Demonstration of graphene as the basal structure for all the allotropes*

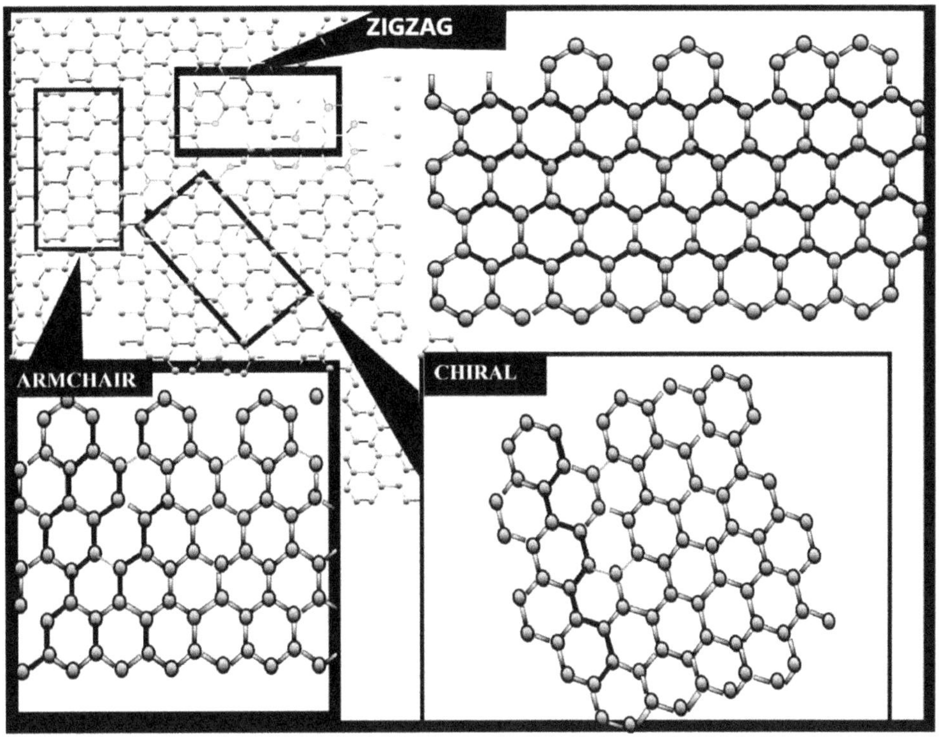


## 8.2. One-Dimensional Allotropes

### 8.2.1. Carbon Nanotubes (CNTs)

The CNTs are further classified into single-walled carbon nanotubes (SWCNTs) by Iijima and Ichihashi (1993) which is just a layer of graphene rolled up into hollow cylindrical product and multi-walled carbon nanotubes (MWCNTs) can be thought of several layers of graphene rolled up forming a similar structure.

### 8.2.2. Carbon Nanofibers (CNFs)

CNFs have lengths measured up to few micrometers, while its diameter ranging from a couple tens of nanometres to several hundreds of nanometres.

### 8.2.3. Carbon Nanohorns (CNHs)

Nevertheless, during synthesis CNH tends to aggregate into spherical clusters having diameter of ~100 nm, thus restricting further functionalization. In contrast to SWCNTs, CNH has one-of-a-kind symmetries that employ chemical reactions to take place in the cavity of CNHs influenced via free molecular movement and clubbing of captured charge carriers, leading to various energy applicative prospects (Karousis, Martinez, et al. (2016).

### 8.2.4. Graphene Nanoribbons (GNRs)

Graphene nanoribbons, a quasi-1-D derivative of graphene, are shown to have two possible edge geometries, namely, armchair and zig-zag having nomenclatures AGNRs and ZGNRs, respectively.

## 8.3. Two-Dimensional Carbon Allotrope

### 8.3.1. Graphene (GR)

Moreover monolayer GR is found to express massless Dirac fermions and Landau level of quantization (Yin, Bai, Wang et al. (2017)).

## 8.4. Three-Dimensional Carbon Allotropes

### 8.4.1. Graphite

Graphite is a 3-D carbon allotrope of $sp^2$ hybridization having each carbon atom bonded with three other atoms.

### 8.4.2. Diamond

Nanodiamonds are an artificial arrangement of carbon atoms having structure and properties analogous to that of a natural diamond.

While most SWNTs are chiral ($m \neq n$), some of them present armchair ($m = n$) or zigzag ($m = 0$) configurations (Figure 10A).

In the most general case, a CNT is composed of a concentric arrangement of several cylinders (Figure 10C). Double-walled carbon nanotubes (DWNTs) are a special case of MWNTs, composed of just two concentric cylinders (Figure 10B).

All possible ways how the graphene sheet can be rolled up to the cylinder can be represented by a pair of integer numbers or chiral indices *n* and *m* that predefine the chiral vector ($c_h \rightarrow$). Chiral angle ($\theta$) and chiral vector ($c_h \rightarrow$) can be classified into three categories as zigzag, armchair, and chiral in terms of the chiral vector $c_h \rightarrow$ as shown in Figure 11(Boumia, Zidour, et al.2014).

Figure 11 shows the three different types of SWCNTs: armchair, zigzag, and chiral. Further details of the structure can be found in reviews by.

# 9. BASIC STRUCTURE AND MORPHOLOGY FEATURES OF CNTS COMPOSITES

Most people call zigzag nanotubes when $m = 0$, armchair nanotubes when $n = m$, and chiral nanotubes when $m > 0$ (Eatemadi, Daraee, Karimkhanloo, et al., 2014).

Also, the tube width d is determined by the chiral vector $C = na1 + ma2$, (wherein, a1 and a2 are the base cell vectors of graphite), as well as the direction of rolling a graphene sheet using this vector (Iijima, 1991, Iijima and Ichihashi 1993 and Eatemadi, Daraee, Karimkhanloo, et al. (2014)) (Figure 12).

*Figure 10. Armchair or zigzag configurations*

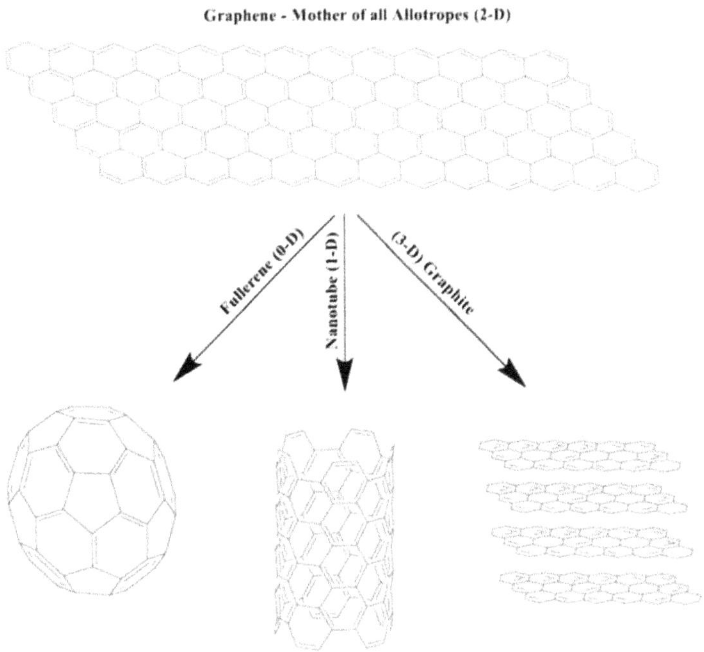


SWCNTs are mono-cylindrical carbon layers with a diameter in the range of 0.4–2.0 nm organized in harmony with chiral, armchair, and zigzag arrangements (Figure 13). Double-walled CNTs (DWCNTs) consist of only two tubes having similar sizes to SWCNTs. However, MWCNTs consist of a number of cylindrical carbon sheets with an average diameter of 1–3 nm for the central cylindrical tubes and 2–100 nm for the external cylindrical tubes (Figure 13).

Few numbers of concentric cylinders were equipped with regular spacing of interlayers that are located around central hollow section and demonstrated as multi walled CNTs Figure 14 (Aravind Kumar Jagadeesan,et al. 2020).

## 9.1 Structure of SWCNT

Based on wrapping mechanism, three different forms of SWCNTs include chiral, armchair, and zigzag pattern.

## 9.2 Formation of Multi Walled Nanotubes

Besides diamond, graphite, and fullerene ($C_{60}$), quasi-one-dimensional nanotube is another form of carbon first reported by Ijima in 1991 when he discovered multiwalled carbon nanotubes (MWCNTs) in carbon soot made by an arc-discharge method (Iijima, 1991).

*Figure 11. Chiral vector $c_h\rightarrow$*

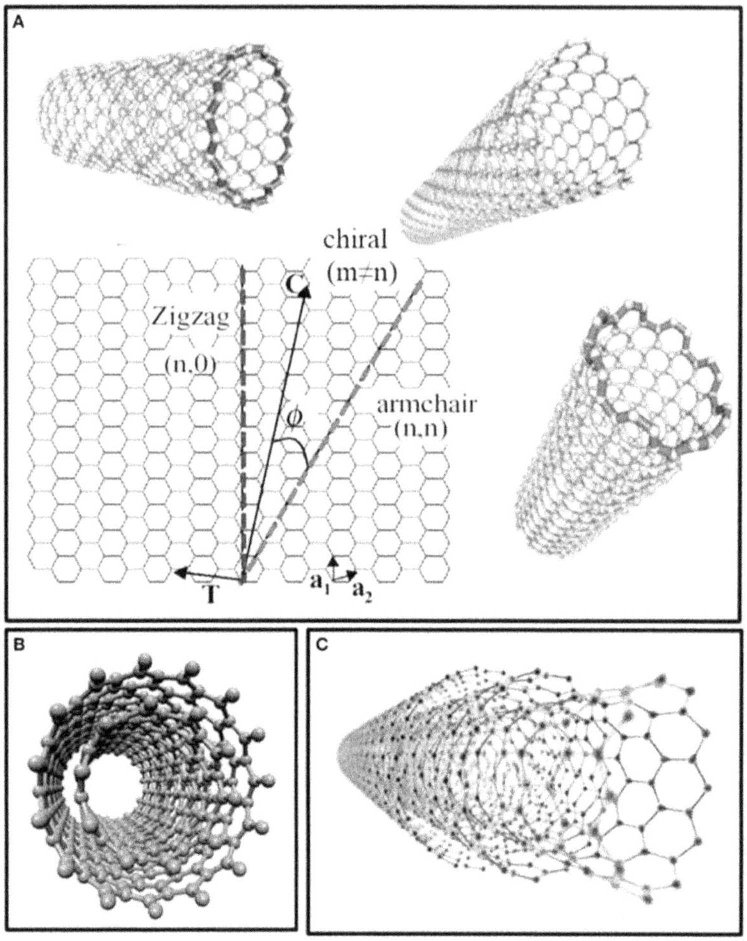


## 9.3. Graphene and Reduced Graphene Oxide

Graphene (G) is defined as a pristine two-dimensional $sp^2$ hybridized carbon sheet. Graphene oxide (GO) is a single graphitic monolayer with $sp^2$ carbon atom and oxygenated aliphatic $sp^3$

## 9.4. Carbon Dots (CDs)

Carbon dots (CDs) are zero-dimensional carbon nanomaterials that have been polymerized. GQDs consist of graphene layers of size less than 10 nm (Kour, Arya, et al.2020).

## 9.5. Structure Morphologies of (SWCNTs)

Figure 15 shows the SWCNT application space (with existing and emerging applications in square and oval boxes, respectively).

*Figure 12. Direction of rolling a graphene sheet*

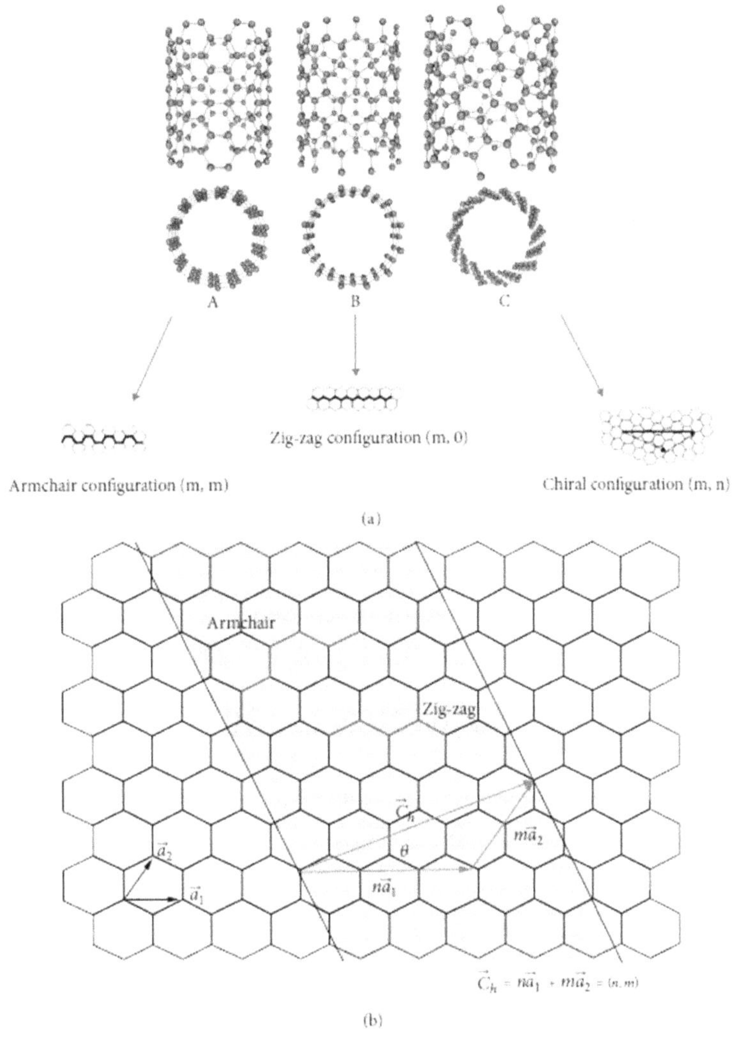


## 9.5.1. About OSCIAL

OCSiAl is the world's largest producer of graphene nanotubes and is the only company with scalable technology for their industrial synthesis (OCSiAl: World's Largest Graphene Nanotube Manufacturer, 2020). "OCSiAl's annual production capacity has now reached 75 tonnes of graphene nanotubes. The growing market demand is closely related to the ongoing revolution in the electric vehicle field. If we take into account the official plans of the largest global car manufacturers, just the Li-ion batteries for them would require 250 tonnes of graphene nanotubes in 2025," said Yury Koropachinskiy. As can be seen in Figure 15, most of the existing technologies such as reinforced composites, Li-ion batteries, and electromagnetic (EM) shielding use randomly aligned, mixed helicity/diameter SWCNTs.

*Figure 13. Cylindrical carbon sheets*

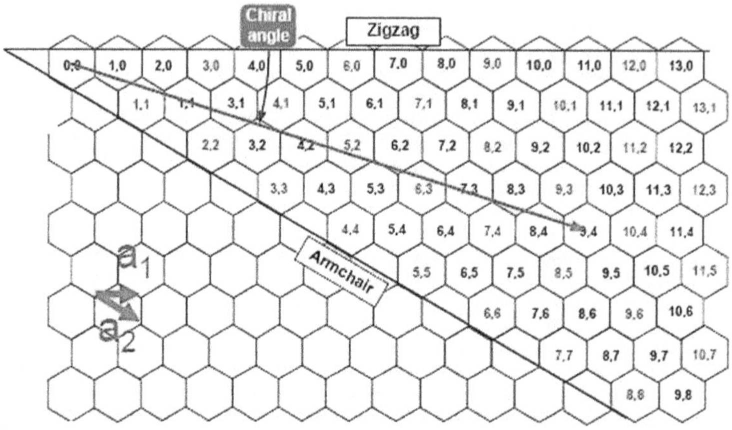


## 9.6. Helicity Control of SWCNTs

Helicity control, which remains the most significant challenge for the synthesis of SWCNTs. Often referred to as the Holy Grail of nanotube synthesis, control over the structure of SWCNTs (Figure 16a)

## 9.7. Catalyst Design for Selective SWCNTs Growth

A recent example is the successful demonstration of selective growth of (12,6), (Yang, Wang, et al. 2014) (16,0), and (14,4) (Yang, Wang, et al. 2017) SWCNTs from W–Co intermetallic compound nanoparticles (Figure 16b) by Li et al., (2018).

*Figure 14. Interlayers*

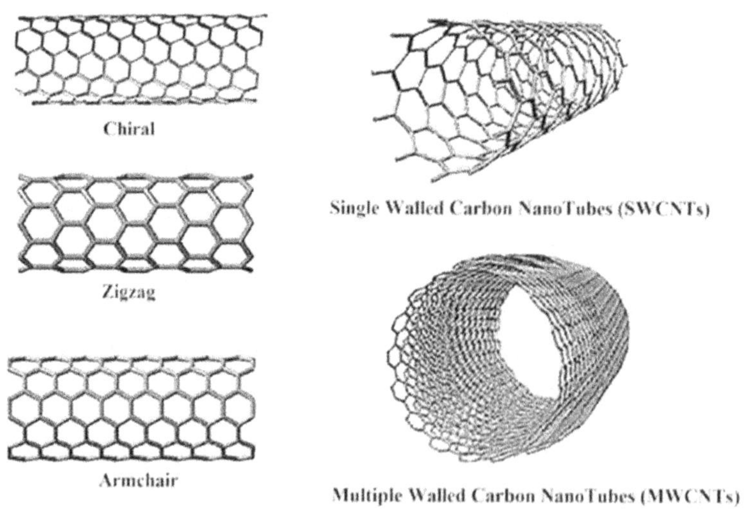

*Figure 15. The SWCNT application space*

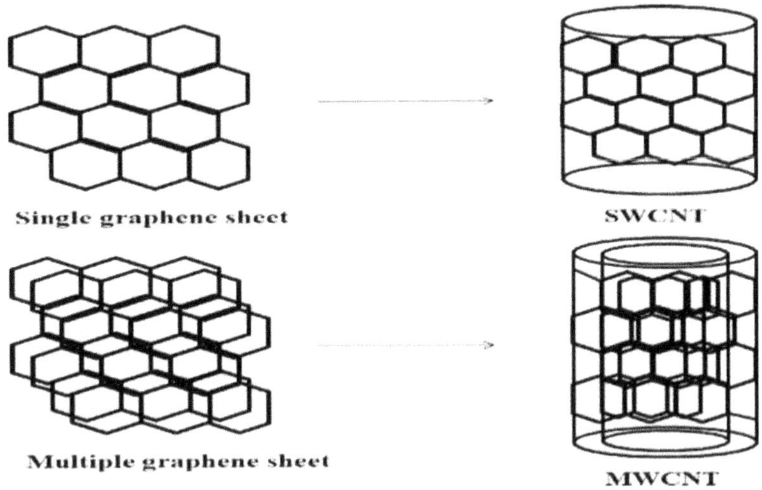


## 9.8. Structure and features of fullerenes ($C_{60}$)

Fullerene offers a high symmetrical lattice in different sizes (C60, C76 and so on) with its unusual structure containing sp$^2$ carbons.

*Figure 16. Helicity control*

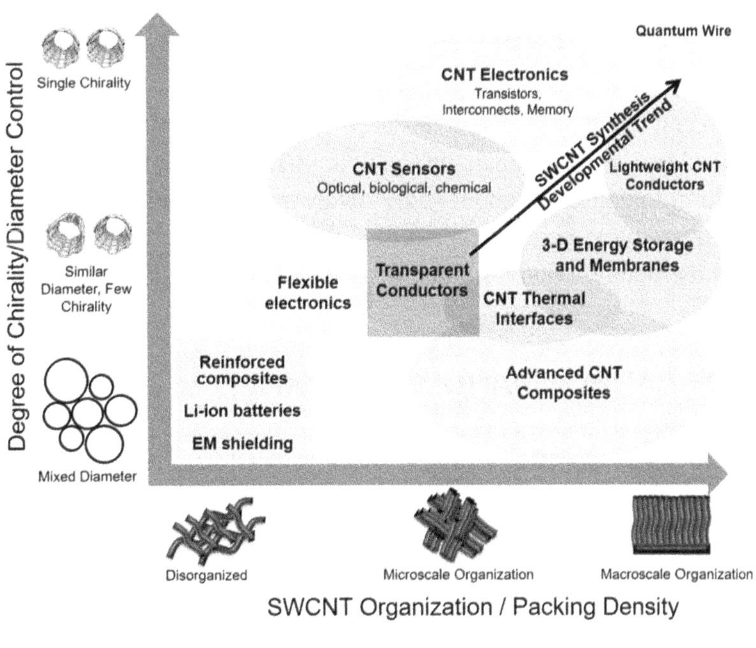

*Figure 17. Chemical approaches*

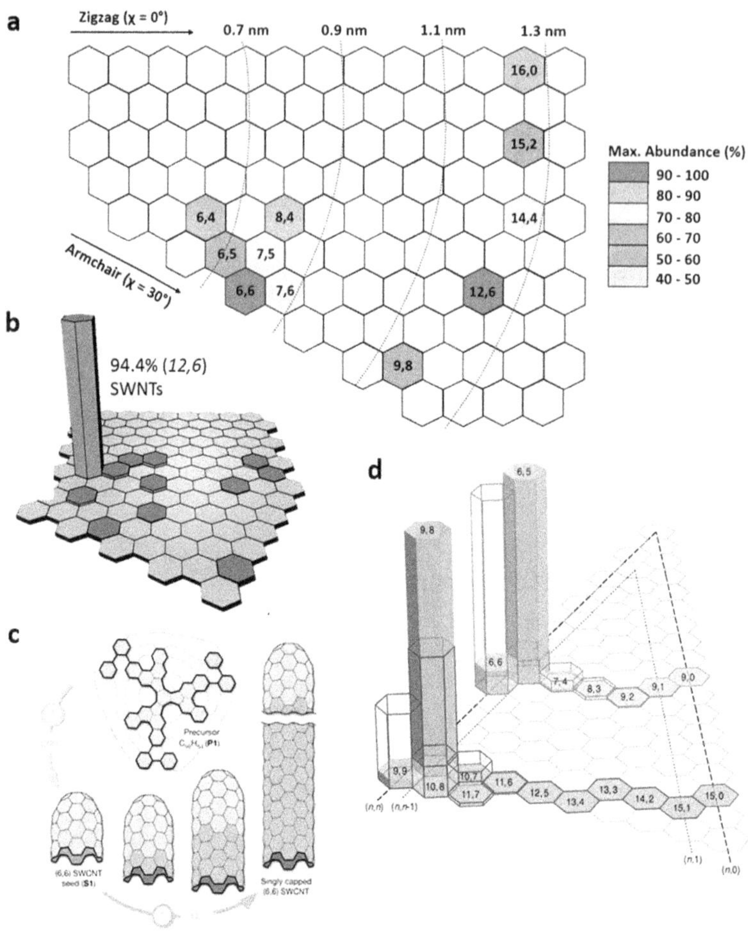


## 10. PRODUCTION AND WORKPLACE EXPOSURE TO CNTs

We present in details recent chemical approaches according to Figure 17 with the functionalization of CNTs including practical uses of functionalized CNTs in many fields of applications. In the light of this fact, we have an intention to do a throughout study reviewing exclusively the recent advances in CNT functionalization which will provide researchers a global insight of the main functionalization strategies of carbon nanotube nanostructures (Lounasvuori, Kelly, Foord, 2018 and Lakhdar Sidi Salah, et al 2021).

## 11. SYNTHESIS OF CARBON NANOTUBES (CNTs)

A brief description of methods contributing to the mass production with their effectiveness is given in Scheme.3 (Anamika Sahu, et al. 2017).

*Figure 18. Vertical thin electrodes installed*

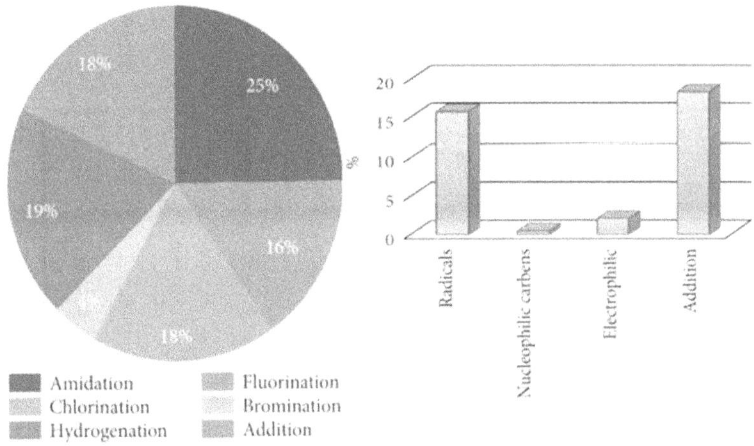


## 12. SCHEME THREE

### 12.1 Physical Processes

These include popular process of carbon nanotubes production such as arc discharge and laser ablation.

#### 12.1.1 Arc Discharge

As shown in Figure 18, while the electrodes are separated by one millimetre. To synthesize multi-walled carbon nanotubes, catalyst nanoparticles of iron, cobalt, and nickel are inserted into the middle region of the positive electrode. (Anamika Sahu, et al. 2017 and Aravind Kumar Jagadeesan, et al. 2020).

Two vertical thin electrodes were installed in the center of the chamber (Figure 18).

#### 12.1.2. Laser Ablation

In the Laser Ablation process, a graphite target is vaporized by a laser source, similar to the Physical Vapor Deposition method shown in Figure 19. This approach is used to manufacture multiwalled carbon nanotubes, and it is also used to synthesise single-walled carbon nanotubes, with the catalyst nanoparticles of Fe, Co, and Ni being employed in both cases (Aravind Kumar Jagadeesan, et al. 2020).

The nanotubes develop on the cooler surfaces of the reactor, as the vaporized carbon condenses. A water-cooled surface is also included in the most prac-tical systems to collect the nanotubes (Figure 19).

### 12.2 Chemical Processes

#### 12.2.1 Chemical Vapor Deposition (CVD)

The term "chemical vapour deposition" (CVD) encompasses many different types of processes, including catalytic chemical vapour deposition (CCVD), which can be either thermal or plasma enhanced

*Figure 19. Physical vapor deposition*

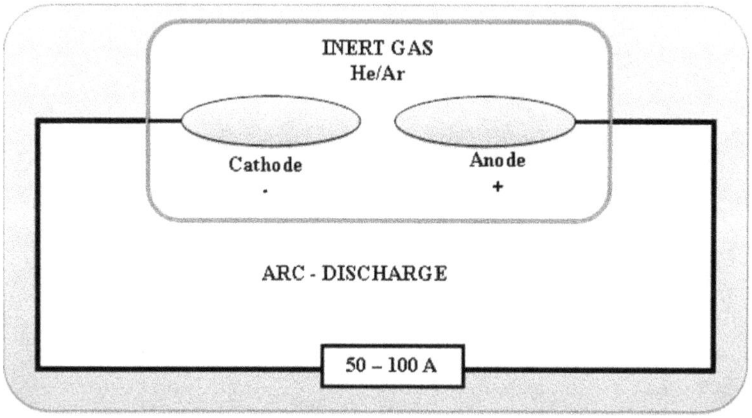


(PE) oxygen assisted CVD (Iijima, 1991) water assisted CVD (Iijima, Ajayan, et al. 1992), microwave plasma (MPECVD), radiofrequency CVD (RF-CVD), and hot-filament (HFCVD). However, catalytic chemical vapour deposition (CCVD) is currently the most widely used approach for the manufacture of carbon nanotubes in industry.

CVD is a technology in which the vaporized reactants react chemically with one another to generate a nanomaterial product that is placed on a substrate during the deposition process (Figure 20). Catalyst nanoparticles of iron, cobalt, and nickel are utilized in the production of single walled Mcarbon nanotubes. (Aravind Kumar Jagadeesan, et al. 2020 and Pham, Jang, et al. 2017).

## 12.2.2 High Pressure Carbon Monoxide Reaction (HiPco®)

This process is called HiPco process. SWNT have also been synthesized in a variant of HiPco process in which a mixture of benzene and ferrocene, $Fe(C_5H_5)_2$ reacts in a hydrogen gas flow to form SWNT.

*Figure 20. Deposition process*

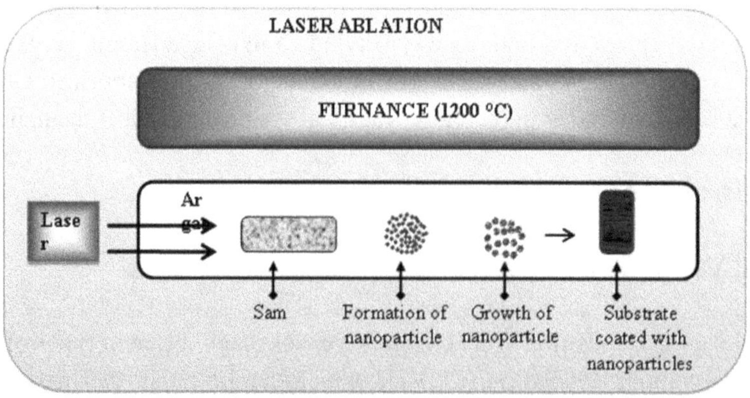

### 12.2.3 CoMoCAT® Process

The secret of the process is in synergistic effect of Co and Mo. Catalyst is most effective when both metals Co and Mo are present at a time on silica substrate with low Co: Mo ratio.

## 12.3 Miscellaneous Processes

Some miscellaneous and relatively less used processes of carbon nanotube production are given below.

### 12.3.1 Helium Arc Discharge Method

It was reported in 2006 by scientists of NASA's Goddard Space Flight Center that they have developed a simple, safe, and very economical process of Single walled car-bon nanotubes production.

### 12.3.2 Electrolysis

In this method carbon nanotubes were produced by de-positing alkali metals on a graphite cathode from a high-temperature molten salt system.

### 12.3.3 Flame Synthesis

This method is based on the synthesis of SWNT in a controlled flame environment that produces the temperature, forms the carbon atoms from the inexpensive hydrocarbon fuels and forms small aerosol metal catalyst islands.

# 13. Growth Mechanism of Carbon Nanotubes (CNTs)

The mechanism comprises the steps shown in (Scheme.4) and the simple diagrams of growth mechanism shown in (Scheme.5). On a growth mechanism of nanotubes, there are several theories one of which describes in (Scheme.4),

## 13.1 Scheme Four

## 13.2 Scheme Five

Generally speaking, when the particle size is a few nm, SWCNT forms, whereas if the particles are a few tens of nm wide, it favors MWCNT growth (Coville, Mhlanga, et al. 2011) (Figure 21).

*Figure 21. MWCNT growth*

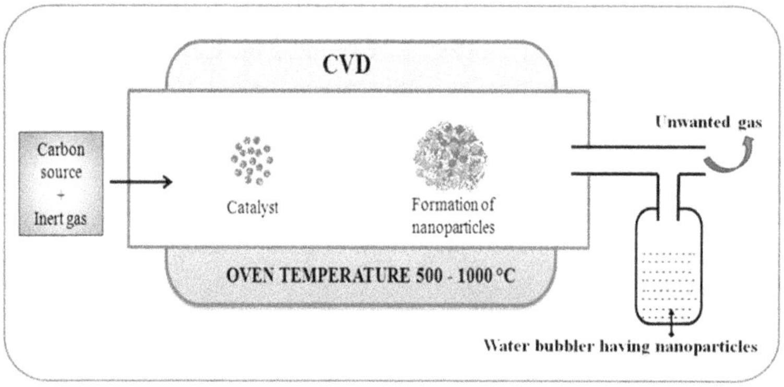


## 13.3. SWCNT Growth from Molecular Seeds

A reduction in growth temperature to 400–500 °C, therefore, helped the growth of (6,6) SWCNTs from (6,6) caps formed by surfacecatalyzed cyclo-dehydrogenation by placing a flat CNT endcap precursor ($C_{96}H_{54}$) on single-crystal platinum catalysts (Figure 16c) (Sanchez-Valencia, Dienel, et al.2014).

## 13.4. Computational Modeling of SWCNT Growth

Models and hypotheses for the origin of SWCNT helicity are just as old as SWCNTs themselves and can be traced back to the seminal work by Iijima (1991). Computational modeling studies targeting various aspects of SWCNT growth mechanisms pertinent to catalytic CVD have since grown considerably in number and complexity and recent reviews (Penev, Ding, et al. 2017). present an ample account of theoretical efforts.

This kinetic route of chiral selectivity, controlled by the R factor, may produce preferentially the fastest growing (2m,m) tubes, corresponding to $\chi \simeq 19°$ where the number of kinks reaches maximum (Figure 16d).

In the context of a possible catalyst-templated chiral selectivity, a catalytic particle "matching" a (n,m) tube edge is more likely to favor "one-index-off" tubes with similar diameter. The recently reported high preference toward (12,6) on W–Co and (8,4) tubes on W–C (Zhang, Kang, et al., (2017) solid catalysts is intriguing. However, very recently, based on large-scale first-principles calculations combined with kinetic Monte Carlo simulations, it was shown that such selectivity in the case of $Co_7W_6$ instead can be a result of complex growth kinetics (Penev, Bets, et al. 2018).

Diameter control in this case would provide additional flexibility for achieving SWCNTs of desired conductivity type. Pushing further the thermodynamic analysis of the growth, an alternative model has recently been proposed (Magnin, Amara, et al. 2018).

Weak interactions between C and Cu have shown to mediate and encourage the formation of C dimers and C–C networks on the surface of Cu nanocatalysts (Rahmani Didar, Balbuena, 2017).

*Figure 22. SWCNT catalysis*

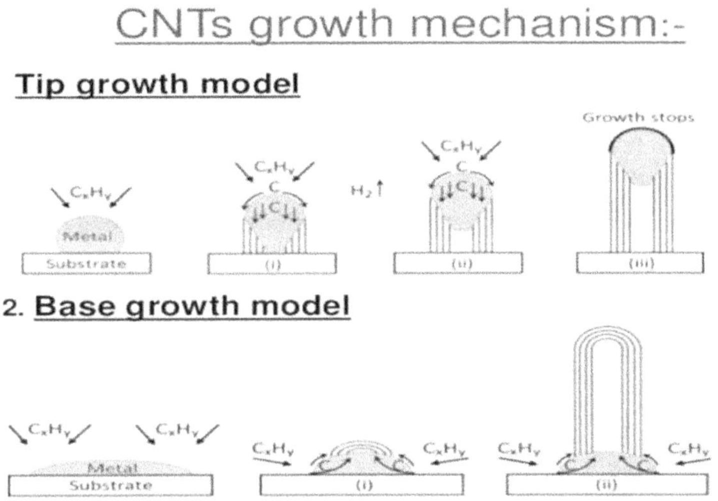


## 13.5. Related Advances in Graphene and Other 2D Material Growth

Indeed, well-defined single crystalline planar substrates serve as simplified model systems for fundamental studies of SWCNT catalysis (Figure 22) that can be readily probed with powerful surface science techniques while avoiding some of the complexities of SWCNT growth such as diverse nanoparticle populations, nanosizing effects, and support interactions.

## 13.6. Graphene Growth Mechanisms

Growth models incorporating kinetic effects have been developed; highlighting in particular that grapheme nucleation simply requires a local carbon supersaturation to be developed at the catalyst surface, while the concentration in the bulk of the catalyst can remain much lower. Centimeter- and even meter-sized domains can be achieved on appropriately treated Cu–Ni alloy foils (Figure 23f) or single crystal Cu foils (Figure 23g, (Xu, Zhang, et al. 2017)) while wafer-scale growth of epitaxially aligned domains has been demonstrated on Ge (110).

## 13.7. In Situ Transmission Electron Microscopy

Early pioneering work focused on the direct observation of SWCNT growth within an ETEM. The in situ ETEM experiments have enabled the observation of catalyst surface and shape reconstruction and step-flow dynamics during SWCNT nucleation, as well as catalyst particle dewetting and glimpses into the early stages of organized growth. Meanwhile, low energy electron microscopy (LEEM) has been used to study graphene growth on a variety of single crystal substrates (Lin, Gomez-Ballesteros, et al. 2017). The ultrahigh vacuum conditions and consequent slower kinetics in these studies can reveal the behavior close to thermodynamic equilibrium but are usually far from the realistic conditions used for large-scale graphene or CNT growth (Figure 22).

*Figure 23. Cu–Ni alloy foils*

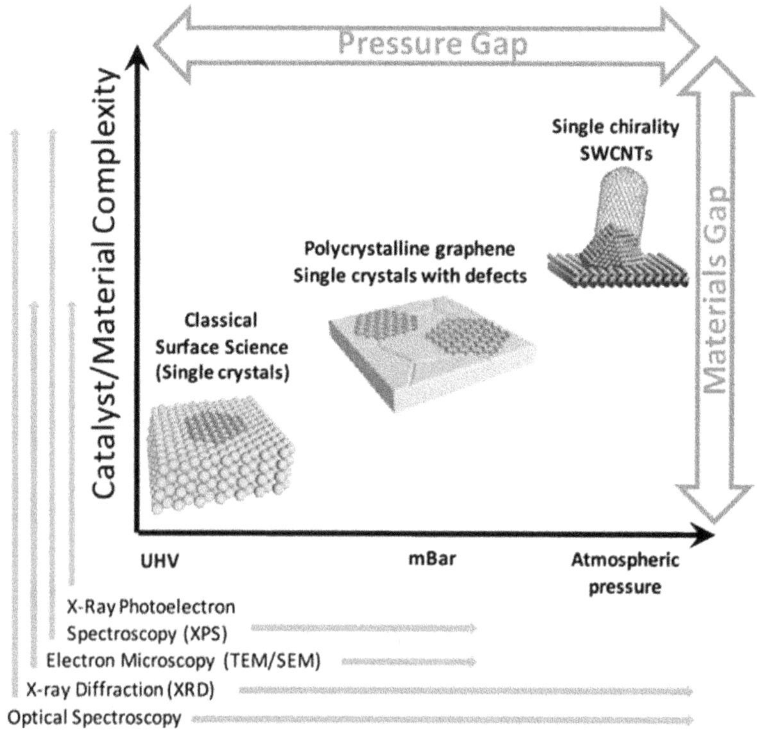


## 13.8. In Situ Spectroscopy

X-ray-based techniques such as X-ray photoelectron spectroscopy (XPS) and X-ray scattering/diffraction have been particularly effective.

Studies performed during the growth of individual SWCNTs have revealed chiral-angle dependent growth rates and defect densities, while in situ spectra from the collective growth of SWCNTs have revealed insights into activation and deactivation mechanisms and into the evolution of tube diameters and helicities (Navas, Picher, et al. 2017).

## 13.9. Growth Challenges

In particular, given the inherent polycrystallinity of the low-cost catalyst foils used for largescale economic growth, improved understanding of the role of different catalyst surface orientations and grain boundaries is required, including how strain is accommodated as graphene grows across these features.

## 14. METHODS USED FOR PURIFICATION AND DISPERSION OF CARBON NANOTUBES

The physicochemical properties (surface topography, solubility, hybridization state, mechanical properties, thermal conductivity) and structural and metallic or carbonaceous impurities, of CNTs, have to be evaluated for biomedical applications (Anamika Sahu, et al. 2017).

Studies have demonstrated that boiling SWNTs in nitric acid or hydrofluoric acid aqueous solutions for purification of SWNTs and removal of amorphous carbon and metal particles is an efficient and simple technique for purification of SWNTs and removal of amorphous carbon and metal particles (Jayendrakumar Pate, Pharm., et al. 2021).

The various steps used in the purification of nanotubes are:

### 14.1. Air Oxidation

Air oxidation treatment of the CNTs is a useful way to eliminate carbon impurities (defect carbon) and metal catalyst.

Carbon based impurities and surface metal can be taken away by the oxidative treatment (Anamika Sahu, et al. 2017).

### 14.2. Acid Refluxing Treatment

Acid treatment: It removes the metal catalyst by exposing the metal surface to oxidation or sonication. The metal will be melted and can be removed by using vacuum having high-temperature.

### 14.3. Surfactant Aided Ultra Sonication, Filtration, and Annealing Purification Methods

Annealing: High temperature between 873K-1873K will result in the rearrangements of nanotubes with the consumption of defects resulting in the pyrolysis of graphitic carbon and short fullerenes. The metal will be melted and can be removed by using vacuum having high temperature.

### 14.4. Ultrasonication

Sonication is well-known as one of the effectual processes to get rid of the amorphous impurities adhering or binding to the walls of CNTs using suitable solvents.

### 14.5. Magnetic Purification

Magnetic Purification-Removal of ferromagnetic (catalytic) particles occurs mechanically from their graphitic shells. Large equipment is not required, and it has been observed that the production of laboratory-sized quantities of SWNTs containing no magnetic impurities (Anamika Sahu, et al. 2017).

## 14.6. Microfiltration

This method is employed to remove carbon nanoparticles by microfiltration, which is completely based on size or particle separation.

## 14.7. Cutting

The nanotubes can be shortened or cut by three steps, chemically, mechanically and the combination of both. (Anamika Sahu, et al. 2017).

## 14.8. Functionalization of Carbon Nanotubes

Functionalization depends on making carbon nanotubes hydrophilic in nature than the contaminations by appending different groups to the tubes.

## 14.9. Chromatography

According to the size of molecules, they get separated with the large size of molecules eluting out first. The only requirement is that the nanotubes should be either solvated or dispersed by means like functionalization and ultrasonication (Anamika Sahu, et al. 2017).

# 15. LIMITATIONS AND MODIFICATION (FUNCTIONALIZATION)

The limitation in the applications of CNT's can be defeated, to some extent, by a process named functionalization (Anamika Sahu, et al. 2017). Functionalization of CNTs can be achieved in two ways shown in (Scheme .6).

## 15.1 Scheme Six

The pioneering works in the functionalization of CNTs have developed three main approaches:

**1.** Covalently linking chemical groups to the skeleton of CNTs,

**2.** Adsorption of various functional molecules to CNTs,

**3.** Endohedral filling of the internal cavity of CNTs.

Thus, a wide range of surface functions can be created on CNTs (Same, Samee, 2018 and Cardenas-Benitez, Djordjevic, et al. 2018).

Figure 24 illustrates the basic techniques accessible for the functionalization of carbon nanotubes (Tobias, Mendoza, Ballesteros, 2016).

## 15.2 Oxidation

The primary method of functionalizing carbon nanotubes includes oxidation by strong acids, resulting in a reduction in their length while generating exposed carboxylic groups, improving dispersibility in aqueous solutions.

*Figure 24. Basic techniques accessible*

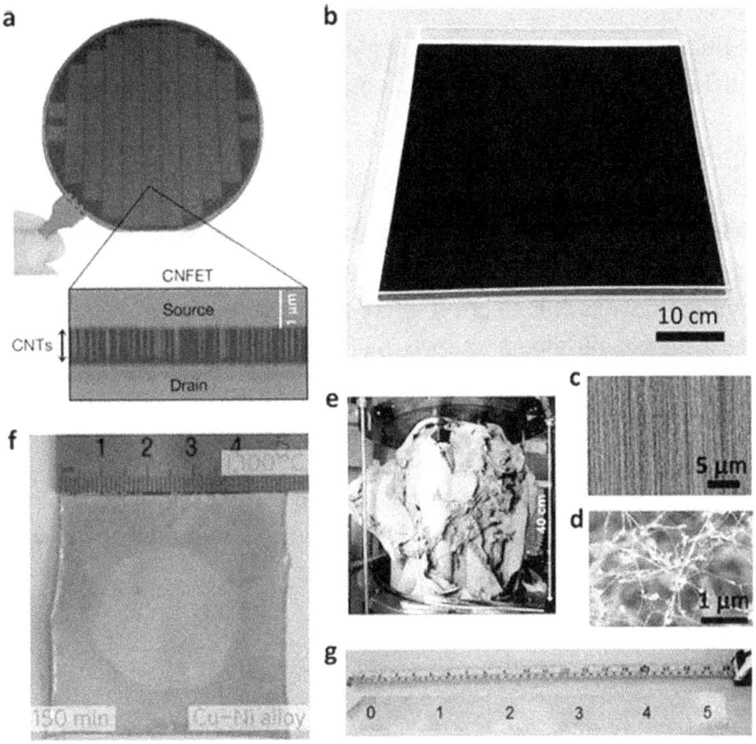


## 15.3 Addition Reactions

A second method is based on addition reactions to the external wall of the carbon nanotube, based on the 1,3-dipolar cycloaddition of azomethine ylides and heating in DMF in the presence of α-amino acids and an aldehyde.

## 15.4 Other Methods

Other functionalization methods include attaching one or more drugs intended to function and constituting a single, powerful payload that can be triggered to release in the target location.

## 15.5 Covalent Functionalization

The attacking species and the carbon atom of the CNT scaffold are covalently linked in direct sidewall functionalization. Both strategies have been used to integrate CNTs into diverse applications, such as tuning electronic properties with active moieties, introducing them into polymeric composites, and fabricating biosensors, drug delivery systems, and tissue engineering scaffolds (Tobias, Mendoza, Ballesteros, 2016).

## 15.6 Noncovalent Functionalization

Depending on their final application, these noncovalently functionalized carbon nanotubes have been modified further with biomolecules, inorganic nanoparticles, and organic compounds.

## 15.7 Endohedral Functionalization

The interest of using inorganic chemicals to decorate carbon nanotubes is twofold. Atomic force microscopy (AFM), scanning tunnelling microscopy (STM), high-resolution transmission electron microscopy (HRTEM), and scanning transmission electron microscopy are the most frequently utilised techniques for characterizing these tagged carbon nanotubes (STEM).

## 15.8 Ends and Defects

Ends of the CNT's are more reactive than the sidewalls. Treating the CNTs with oxidative agents results in introducing oxygenated groups, for instance, carboxylic acid, ketone, alcohol, and ester, at the nanotube ends and defect sites (Anamika Sahu, et al. 2017) (Scheme 6).

## 15.9 Scheme Six

## 15.10 Sidewall Functionalization

To introduces higher concentrations of covalently attached functional groups onto the surface of CNTs with the substitution of significant disruption in the electronic structure or the method used to break down the bonds present on the wall of CNTs which helps to attach the different functional groups.

## 15.11 External decoration with inorganic materials

The biomedical field is one area where the containment of materials in carbon nanotubes is gaining attention, since their interior chamber can be loaded with medicinal or imaging cargos and their outer surface can be changed to improve their dispersibility, biocompatibility, and site-selectivity (Tobias, Mendoza, Ballesteros, 2016).

Thus, functionalization heightens the scope of CNTs as raising nano vectors for the delivery of therapeutics (Anamika Sahu, et al. 2017).

# 16. DRUG LOADING MECHANISM AND CELLULAR UPTAKE OF CNTS

Carbon nanotubes have been incorporated in electrochemical sensors to detect proteins, neurotransmitters, and small bio-molecules for medical applications. Several electrochemical methods are used in these sensors including direct electrochemical detection of DNA biosensors, indirect detection of an oxidation product using enzyme sensors, and detection of conductivity changes using CNT field effect

*Figure 25. CNTs*

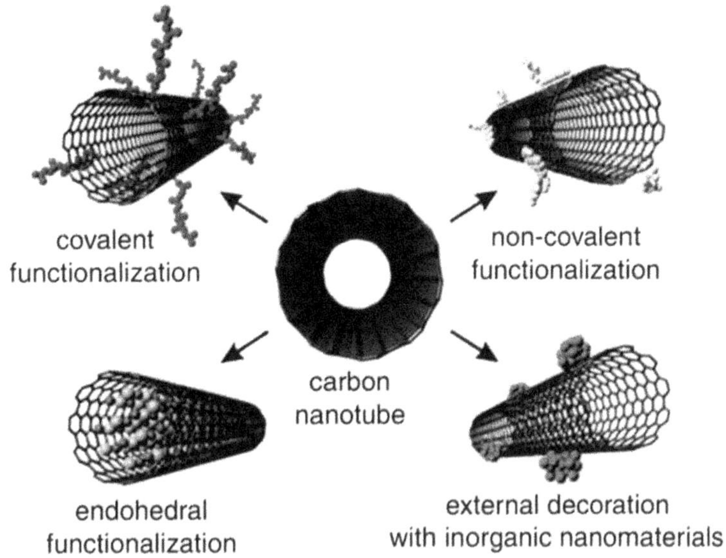


transistors (FETs). A discussion can be found in a recent comprehensive review contributed by Venton and coworkers (Juan et al.2020) .

CNTs possess unique and excellent structural, optical, and electrical properties for the development of advanced drug delivery systems (Anamika Sahu, et al. 2017) (Figure 25).

## 17. UPTAKE OF CNTS BY CELL

CNTs have been aimed and actively explored as multipurpose, innovative nano-carriers for drug-delivery systems. Thanks to their very high aspect ratio, CNTs can penetrate the cell membrane and be uptaken by cells. After entering the cell, CNTs are mainly located inside cell endosomes and lysosomes. These two processes are energy-dependent and are hindered at low temperatures (Figure 26).

## 18. BREAKDOWN MECHANISM OF CNTS IN THE BODY

A group of Swedish and American scientists has shown for the first time that carbon nanotubes can be broken down by alysosomal enzyme-Myeloperoxidase (MPO)--an abundant enzyme of neutrophils granulocytes found in white blood cells and to a lesser extent also in certain macrophages. Shvedova et al., (2012) reported that, in simple biochemical model systems,

*Figure 26. Two processes*

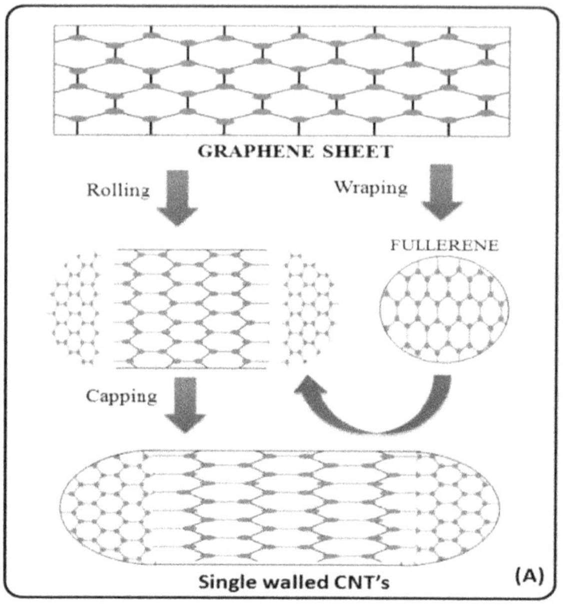


## 19. TOXICITY—DETERMINING PARAMETERS OF CNTS

The toxicity of CNTs may be considered in terms of a number of parameters, in particular their high surface reactivity and high aspect ratio, that is, their tubular structure, which has been shown to enhance cell membrane penetration, thus highlighting the need for extensive biodistribution profiles to be ascertained prior to the systemic use of CNTs. Numerous studies have been conducted both independently and collaboratively around the world to elucidate the toxicology of CNTs. In particular, Kostarelos's group (http://www.kostarelos.gr) have related the CNT characteristics and the parameters that determine pharmacology and toxicology, shown in Scheme 7.the important issue of CNT toxicity is discussed (Chowdhury, 2011) and a detailed review is provided for toxicology work undertaken that is relevant to the pharmaceutical applications of CNTs.

### 19.1 Scheme Seven

It was also observed that CNT/DNA interaction was the preferred route of toxicity in a 3-hour incubation study with 96 µg SWCNT/cm$^2$, which induced DNA damage (through micronucleus generation) in lung fibroblasts (Anamika Sahu, et al. 2017).

## 20. DIFFERENT IMPRESSIVE PROPERTIES OF CARBON NANOTUBES

All the extraordinary features of the CNTs, especially the superior electrical features such as electron mobility (100,000 cm$^2$/Vs), field effect mobility (79,000 cm$^2$/Vs) and electrical conductivity (10$^4$ S/cm) (Rana, Ibrahim, et al. 2017), have made CNTs the perfect candidate for improving bio-sensing tools.

*Table 2. Comparison of mechanical properties of CNTs with other string materials (Yu, Files, et al. 2000)*

| Material | Young's modulus (GPa) | Tensile Strength (GPa) | Density ($g/cm^3$) |
|---|---|---|---|
| Single wall nanotube | 1054 | 150 | N/A |
| Multiwall nanotube | 1200 | 150 | 2.6 |
| Steel | 208 | 0.4 | 7.8 |
| Epoxy | 3.5 | 0.005 | 1.25 |
| Wood | 16 | 0.008 | 0.6 |

Both SWCNTs and MWCNTs are capped at both ends of the tubes in a hemispherical arrangement of carbon networks called fullerenes warped up by the graphene sheet (Figure 25).

## 20.1. Electrical Conductivity Properties of CNTs

Semiconducting SWCNTs are generally p-type semiconductors but MWCNTs comprise of many SWCNT pipes and therefore may not be strictly one-directional conductors.

## 20.2 Strength and Elasticity Property

The elasticity property for both single walled and multi walled CNTs is examined by the term known as modulus of elasticity or elastic modulus.

Instead, day by day investigations on CNTs reveals that the electrical properties of such carbon nanotubes are much sensitive to chemical doping impacts and charge transfer mechanism in spite of exhibiting extreme robustness (Derycke, et al.,2001).

Quasi metallic type's exhibit enhanced electrical conductivity at low temperature levels when subjected to temperature dependent experimental studies (Liu, Lee, et al., 2001).

## 20.3. Mechanical and Electromechanical Characteristics of CNTs

Because of one of the unbelievably solid holdings in nature is the carbon-carbon bonding found inside graphite, CNTs have amazingly brilliant power as they are the hardest game plan still delivered by any analyst. Table 2 represents the comparison of mechanical properties of CNTs with other string materials (Yu, Files, et al. 2000).

## 20.4. Optical Properties of CNTs

As demonstrated as follows, optical assimilation, photo-luminescence, and Raman spectroscopic permit brisk and solid portrayal of this "nanotube quality" as far as non-rounded carbon content, structure (chirality) of the delivered nanotubes, and auxiliary imperfections.

## 20.5. Thermal Conductivity and Expansion Properties of CNTs

Numerous applications of CNTs, such as in nanoscale molecular electronics, sensing and actuating gadgets, or as strengthening-added substance filaments in utilitarian composite materials, have been proposed.

### 20.5.1. Field Emission

Field emission comes about from the tunneling of electrons from a metal tip into vacuum, underneath utilization of a strong electric field.

### 20.5.2. High Aspect Ratio

Carbon nanotubes address an extremely minimal high aspect ratio conductive-added substance for plastics of numerous types.

## 20.6. Dimensional and Small Size

Due to their nanoscale dimensions, electron transport in carbon nanotubes will take place through quantum effects and will only propagate along the axis of the tube.

## 20.7. Chemical Properties and Species Interaction

CNTs have been observed to have enhanced solubility and permeability through the biomembranes when functionalized with lipids [For detailed information on functionalization (Bofan Shen, Shulai Lu, et al. 2022).

### 20.7.1. Solubility

Considerable efforts (Pennetta, Floresta, et al. 2020 and Farghali, Abdel Tawab, et al. 2017) have thus been devoted to the modification of SWCNTs to improve their solubility in solvents and to enhance their compatibility in composite materials. Among various modification methods, sidewall functionalization of SWCNTs has been studied extensively as a route to their separation (Ali, Rehman, et al. 2019 and Haider, Marzoog, et al. 2018), solubilization, and subsequent reactivity.

### 20.7.2. Highly Absorbent

The huge surface area and high absorbency of CNTs make them perfect contender to utilize in air, various gases, and water purification.

## 20.8. Outline Information

While mechanical, electrical, and electrochemical (supercapacitor) properties of the carbon nanotubes are entrenched and have quick applications, the down to earth utilization of optical properties is yet muddled. Specifically, light-discharging diodes (LEDs) and photograph detectors dependent on a solitary nanotube have been created in the lab.

## 21. ADVANTAGE AND DISADVANTAGE OF CARBON NANOTUBES (CNTS)

### 21.1. Advantage of CNTs

1. High electrical along with warm conductivity
2. Very high elasticity
3. Highly adaptable and flexible (~18% lengthening before disappointment)
4. High perspective proportions
5. Good field emanation

### 21.2. Disadvantage of CNTs

The lack of solubility in aqueous and organic media has been a major technological barrier for CNTs. This is due to their stable structure having pure carbon element. In addition, they have very high long range van der Waals forces of attraction due to which they have a tendency to aggregate together and it becomes very difficult to disperse them (Iijima and Ichihashi 1993, Rajwant, et al. 2018 and Beatriz, R.C.M. Karla, F. R. et.al. (2019).

## 22. ADMINISTRATION OF CARBON NANOTUBES

Shorter carbon nanotubes are absorbed into the body more easily through oral routes, passing through columnar cells of the intestines. The biosensors that are based on CNT are acknowledged to be the upcoming generation foundation for ultra-fast and ultra-sensitive bio-sensing systems. CNT has unique expeditious ability to analyze biological samples due to its high surface to volume ratio. CNTs possess numerous advantages which includes high sensitivity, fast response time, high stability and longer lifetime. CNTs provide a very fertile ground for research due to the specific mechanical, electrical, thermal, and chemical properties it holds. There is a particular engrossment in the field of biosensors of these carbon-based nanomaterials based on the biomedical utility of it. In biosensor fabrication CNTs play crucial role which make the biosensor effective to trace and detect target molecules even in trace amount also. The present strong facets of CNTs are instigated from physical chemical interactions transduction and high surface area to volume ratio (Yang et al., 2015). Although the recent advancements in CNTs that have made exciting advances in the field of pathogen detection, but the reality is that there are numerous scientific as well as economic challenges that has to be addressed. To mention a few of them, the first one is cost-effective. CNTs are highly expensive and reducing the cost is very complex challenge. It requires great understanding of the integration of both material science and nanotechnology (Chen et al., 2013). The second one is to have appropriate conditions such as temperature and pH for the functionalization of the CNTs. Moreover, to increase the thermal stability and the lifetime of CNTs is one of the key challenges that the researchers are facing across the globe.

*Figure 27. Applications of CNTs*

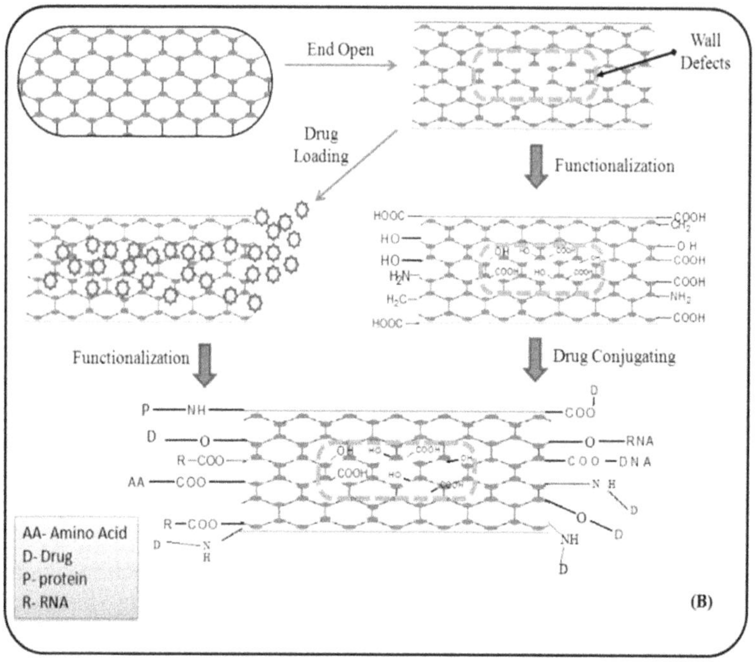


## 23. EXAMPLES OF MEDICAL AND THERAPEUTICS APPLICATIONS FOR CARBON NANOTUBES

A number of biomedical applications of CNTs are proposed including drug vectors, biomolecule, gene delivery to cells or organs, tissue regeneration, and biosensor diagnostics and analysis.

There are two methods, both equally effective-

a) The drug can be attached to the side or behind, or

b) The drug can actually be placed inside the nanotube.

Applications of CNTs encompass major fields and various disciplines, which include nanotechnology, medicine, construction, manufacturing, electronics, peripheral hardware, software and so on Figure 27.

### 23.1. Artificial Implant Scopes

Perhaps, due to its unique material properties such as maximum tensile strength, these can effectively act as implant materials for bone substitutes and if suitably filled with calcium, such implants can be shaped or arranged within the bone structure (Bian, et al. 2004 and Marquis et al. 2003).

### 23.2. Tissue Engineering

Cell tracking/labeling is the specific capacity to identify implanted cell structures and to record the noteworthy improvement of tissue growth *in vivo* as well noninvasive (Abarrategi, et al. 2008).

*Table 3. Release of drugs from CNTs*

| Nucleic acid/drug-CNT type | Cell/tissue | Properties |
|---|---|---|
| Doxorubicin—SWCNT | Colon cancer | Efficient take up of cancer causing cells and then transfers to nucleus |
| Radionuclide-SWCNT | Burkitt lymphocytes | Targeted tumor both in vitro/in vivo |
| siRNA-SWCNT | Human neuro cells | Improves the working of siRNA assisted gastrin peptide receptor |
| Toxic siRNA—MWCNT | Human lung system | Inhibits discrete tumor growth |
| Cisplatin—SWCNT | Neuro Carcinoma | Tumor development—drastic recession |
| Paclitaxel—SWCNT | Breast cancer | Improved efficacy in treatment with minimal effects |

Numerous and popular features involved in the structure of CNTs envisage them to become key elements for nanosensing devices owing to its maximum surface area and DNA or protein immobilizing capacity along with electrical properties (Panini, et al., 2008).

## 23.3. Cancer Cells Tracing

Imaging analysis using single walled CNTs have been explored for the past few decades. Coupling of radioisotopes with single walled CNTs along with imaging techniques based on radio nucleotides can progress advancement in tissue sensitivity, penetration as well as nano medium spatial resolution.

## 23.4. Gene and Drug Delivery Applications (CNT'S in Delivery of Bioactive)

Hydroxyapatite classified as bio-compatible and bio-active materials having biological applications, such as skeletal implants or dental and bone repair scaffolds (Anamika Sahu, et al. 2017).

As discussed above, CNTs possess specific features such as ultrapure/maximum surface distribution, which essentially provoke them to act as a promising tool for drug delivery, nucleic acids and peptides. The releases of drugs using CNTs were highlighted in Table 3.

## 23.5. Sensor-Based Biomedical Applications

Drug delivery, biomolecule distribution, gene transfer to cells or organs, tissue regeneration, and biosensor diagnostics and analysis are the major applications of carbon nanotubes in medicinal field. The drug can either be loaded into the CNT structure or be attached to the CNT surface.

## 23.6. Drug Delivery and Gene Delivery

The drug CNT capsule is ingested by the cell, and eventually the nanotube spills its contents into the cell, resulting in the delivery of the drug (Hua He, Lien Ai Pham-Huy, et al. 2013)and (Madkour, 2019,2020,2022 and 2023).

Figure 28 depicts a diagrammatic illustration of the drug delivery method in action.

*Figure 28. Drug delivery method in action*

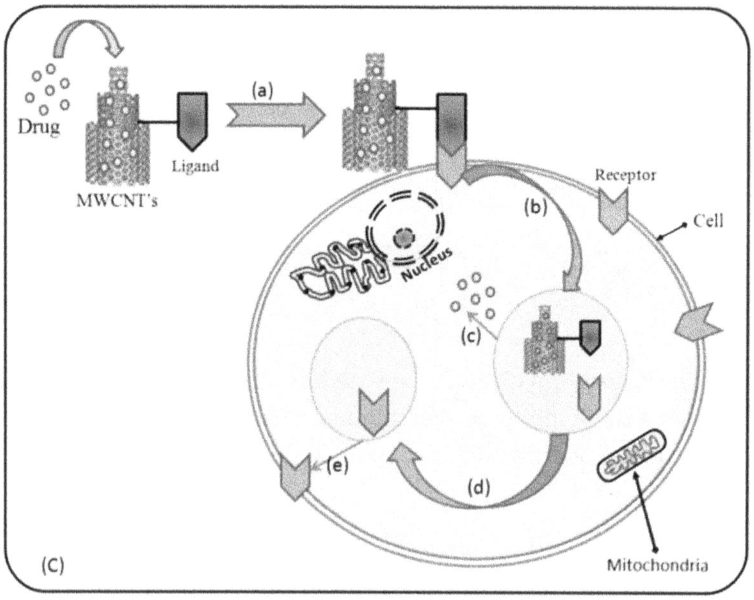


## 23.7. Identification of Cancerous Cells (CNT's in Cancer Therapies)

Several cancer biomarkers that can be detected by CNT-based detection devices are listed in Table 4.

## 23.8. Uses in Anti-Cancer Treatment (CNT's in Cancer Therapies)

The potential application of CNT's in drug delivery has been compiled with special reference to *in vitro* and *in vivo* evaluation.

The variation of a carbon nanotube on a molecular level via biological molecules is effectively an example of the 'bottom-up' fabrication principle of bio-nanotechnology.

## 23.9. Tissue Regeneration

CNTs can be employed as additions to tissue scaffolding to increase the mechanical strength and conductivity of the scaffolding by integrating with the host's body.

## 23.10. Artificial Implant

Table 5 lists some of the applications of nanotubes as artificial implants, with a brief description of each.

## 23.11. CNTs as Biosensor

Although the use of carbon nanotubes in biosensor nanotechnology is relatively new, it represents an extremely promising application area for therapeutic monitoring as well as *in vitro* and *in vivo* diagnostics.

*Table 4. An example of how carbon nanotubes used to identify a cancer biomarker (Jayendrakumar Pate, Pharm., et al. 2021)*

| Carbon nanotube | Biomarker | Cancer Type |
|---|---|---|
| P-type carbon nanotubes | Prostate-specific antigen (PSA) | Prostate cancer |
| Multilabel secondary antibody-nanotube bioconjugates | Prostate-specific antigen (PSA) | Prostate cancer |
| Microelectrode arrays modified with singlewalled carbon nanotubes (SWNTs) | Total prostatespecific antigen (T-PSA) | Prostate cancer |
| Multiwalled carbon nanotubes-thioninechitosan (MWCNTs-THICHIT) nanocomposite film | Chlorpyrifos residues | Many forms |
| Carbon nanomaterial | Carcinoma antigen-125 (CA125) | Carcinoma |
| MWCNT-platinum nanoparticle-doped chitosan (CHIT) | AFP | Many forms |
| Poly-Llysine/ hydroxyapatite/carbo n nanotube (PLL/HA/CNT) hybrid nanoparticles | Carbohydrate antigen 19–9 (CA19-9) | Many forms |
| MWCN-polysulfone (PSf) polymer | Human chorionic gonadotropin (hCG) | Many forms |
| Multiwalled carbon nanotube-chitosan matrix | Human chorionic gonadotropin (hCG) | Many forms |
| MWCNT-glassy carbon electrode (GCE) | Prostate-specific antigen (PSA) | Prostate cancer |
| Nanoparticle (NP) label/immunochromatograp hic electrochemical biosensor | Prostate-specific antigen (PSA) | Prostate cancer |
| SWNT-horseradish peroxidase (HRP) | Prostate-specific antigen (PSA) | Prostate cancer |
| Carbon nanotube field effect transistor (CNT-FET) | Prostate-specific antigen (PSA) | Prostate cancer |
| Carbon nanoparticle (CNP)/poly(ethylene imine) (PEI)-modified screen-printed graphite electrode (CNP-PEI/SPGE) | Carcinoembryonic antigen (CEA), | Urothelial carcinoma |
| Tris(2,2′-bipyridyl)cobalt(III) (Co(bpy)33+)- MWNTs-Nafion composite film | Carcinoma antigen-125 (CA125) | Carcinoma |
| Gold nanoparticles and carbon nanotubes doped chitosan (GNP/CNT/Ch) film | Alpha-fetoprotein (AFP) | Many forms |
| Multiple enzyme layers assembled multiwall carbon nanotubes (MWCNTs) | Alpha-fetoprotein (AFP) | Many forms |

Other CNT-enzyme biosensors, such as CNT-based dehydrogenase biosensors, peroxidase and catalase biosensors, have also been created for a variety of therapeutic monitoring and diagnostic applications.

*Table 5. Application of carbon nanotubes as artificial implants (Jayendrakumar Pate, Pharm., et al. 2021)*

| CNT type | Natural or synthetic Materials type | Cell or tissue type | Properties |
|---|---|---|---|
| Porous SWCNT | Polycarbonate membrane | Osteoblast-like cells | Increase lamellipodia (cytoskeletal) extensions, and Lamellipodia extensions |
| SWCNT-incorporated | Chitosan scaffolds | C2Cl2 cells /C2 Myogenic cell line | Cell growth improvement |
| MWCNT | Collagen sponge Honeycomb scaffold | MC3T3-E1 cells, a mouse osteoblastlike cell line | Increase cellular adhesion and proliferation |
| MWCNT | Polyurethane | Fibroblast cells | Enhance interactions between the cells and the polyurethane surface |
| SWCNT | Alginate | Rat heart endothelial cell | Enhance cellular adhesion and proliferation |
| MWCNT | Poly(acrylic acid) | Human embryonic stem cells | Increase cellular differentiation toward neurons |
| SWCNT | Propylene fumarate | Rabbit tibia | Support cell attachment and proliferation |

## 23.12. Stimuli-Responsive CNT-Based Drug Delivery Systems

Environmentally sensitive drug delivery systems, also known as smart structures, were developed to deliver therapeutic cargo to diseased tissues by leveraging certain elements that are present in the unhealthy tissue.

## 23.13. CNTs as Antioxidants

Recent research has revealed that CNTs, namely carboxylated SWCNTs, are natural antioxidants that may have important medicinal applications in the prevention of chronic illnesses, the slowing of ageing, and the preservation of food. Francisco-Marquez et al. discovered that the addition of –COOH groups increases the free radical scavenging activity of SWCNTs, and that carboxylated SWCNTs are at least as excellent as, if not better than, their nonfunctionalized counterparts in terms of free radical scavenging activity.

## 23.14. CNTs for Infectious Therapy

The use of carbon nanotubes (CNTs) has been investigated as a means of resolving infectious agent resistance to a wide range of antiviral and antibacterial medications, as well as vaccination inefficacy in the body.

### 23.15. CNT for Vaccine Delivery

It has also been demonstrated that functionalized CNTs can be used as vaccine delivery methods.

### 23.16. Antimicrobial Application

Microarray profiling of a monocytic cell line, THP-1, showed that CNTs, both functionalized and non-functionalized, activate several genes involved in monocyte response to infection or vaccination, such as nuclear factor kappa-light-chain-enhancer of activated B cells (NF-κB), interleukin-1β (IL-1β), IL-6, tumor necrosis factor-α (TNF-α), among others.

### 23.17. Treatment of Central Nervous System Disorders

Both pristine and chemically functionalized CNTs have positive effects on neuronal growth, charge on the nanotubes could be manipulated to control neurite outgrowth. It indicated that MWCNTs may possibly act as a safe nano-vector delivery system for immune therapies of brain cancers (e. g. gliomas) (Anamika Sahu, et al. 2017).

## 24. APPLICATIONS OF GRAPHENE AND GRAPHENE DERIVATIVES

Due to the exemplary properties of graphene such as lightweight, electrical conductivity, strong mechanical and thermal strength, graphene is widely involved in different applications such as electrochemiluminescence (ECL) sensor (Han, et al., 2020), transistors, water filtration, energy storage, biosensors and solar cells as shown in the following subtitles:

### 24.1. Energy Storage

Graphene is applied in energy storage devices such as batteries and supercapacitors because of its high surface area (El-Kady, et al. 2016). In Li-ion batteries, graphene is widely used as anode and has a capacity of about 1000 mAh $g^{-1}$ which is three times higher than that of graphite electrode. Graphene also offers longer-lasting batteries and faster recharge time in seconds. Also, due to its flexibility, graphene is used as solid-state supercapacitor printed device in textiles for wearable electronics (Figure 29a) (Abdelkader, et al., 2017). The theoretical specific energy density increases with increase in graphene content in the super-capacitor as illustrated in Figure 29b.

### 24.2. Water Filtration Technology

Graphene nanoporous membranes can ideally be used for water desalination and filtration with an efficiency of 33% to 100% depending on the pore size and the applied pressure.

*Figure 29. Wearable electronics*

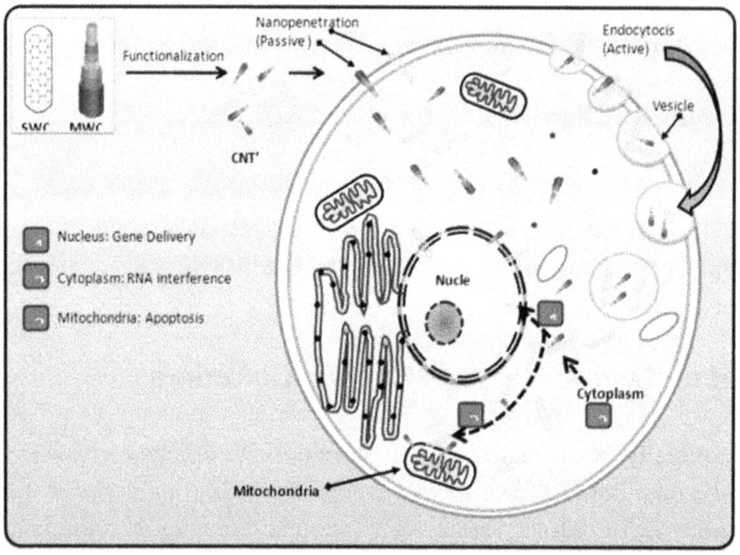


## 24.3. Transistors

At room temperature, the carrier mobility of graphene is very high; this property makes graphene to be applied in transistors. The top gate modulated the drain current and one layer graphene had better field-effect characteristics compared to silicon transistors.

## 24.4. Solar Cells

Solar cells efficiency is an important factor in converting light energy into electricity. Graphene, because of their exemplary optical, mechanical, and electrical properties, have been incorporated in dye-sensitized solar cells (DSSC) and used as electrodes to boost the efficiency of photovoltaic cells.

## 24.5. Other Applications

SWCNTs have strong optical absorption from ultraviolet (UV) to near-infrared (NIR) regions, which can be used for photothermal therapy and photoacoustic imaging from the heat they generate from NIR light absorption. SWCNTs appear to be an excellent platform for biomedical molecular imaging. Scheme 8 represents the applications of enzyme-linked functionalized CNTs.

## 24.6 Scheme Eight

In addition to the feasible drug delivery and unique properties, in study by Vitale et al.,2012, CNTs were used as microelectrodes for electrical simulations that were used in the treatment of neurological disorders, such as PD, due to their biological stability, mechanical strength, and electrical conductivity. This study showed that CNTs are promising candidates, in terms of their use as microelectrodes (Çetin et al.2017). Nano-based drug delivery system (NBDDS) can be used solve potential problems of Central nervous system CNS-targeted therapies, such as passing the Blood-Brain Barrier (BBB). Besides, multifunctional properties, such as bioactivity, targeting, imaging capabilities, and gene delivery, can be accomplished concurrently with these systems (Fig. 30).

Besides these main applications of CNTs, they have been shown as a powerful tool for enantiomer separation of chiral drugs and chemicals in the pharmaceutical industry.

## 25. CONCLUSION

In this chapter, a brief history, the structure, properties, growth mechanisms, characterizations, and applications of carbon nanocomposites are discussed. Graphene is considered as one of the most attractive functional nanomaterials in the world due to its excellent unique properties such as high tensile strength, high electrical conductivity, high carrier mobility, high elasticity, high thermal conductivity and optical transparency of about 97%. Graphene has enticed a wide array of applications in different fields like transistors, solar cells, water purification technologies, sensors, batteries and supercapacitors, which have incredibly attracted a lot of attention among scientists and engineers around the globe. Significant advances have been made in the delivery of anticancer and anti-inflammatory drugs, and bioactive molecules i.e. DNA, RNA and proteins. Drugs and biomolecules can be loaded in CNTs, which can then be utilised as targeted molecules. We have summarized various synthesis techniques and the advanced technological application. The broad scope of these applications suggests that a large number of CNT-based technologies will result from their unique properties of high electrical conductivity, mechanical strength, high aspect ratio, and nanoscale diameter. Until now, it was not realized that NPs as CNTs systems themselves could impose risks to both patients and exposed workers. In fact, as clearly described by Kostarelos and colleagues. 'The use of CNTs in medicine is now at the crossroads between a proof-of-principle concept and an established preclinical candidate for a variety of therapeutic and diagnostic applications. Progress towards clinical trials will depend on the outcomes of efficacy and toxicology studies, which will provide the necessary risk-to-benefit assessments for carbon nanotube, based materials' (Madkour, 2019,2020,2022;2023). CNTs, both SWNT and MWNT, have been proved capable for being applied in water disinfection systems (Indarto et al.2020) owing to their two important properties: antimicrobial activity and adsorption ability. MWNT was reported to hold weaker toxicity than SWNT because of physiochemical differences. However, as MWNT is more affordable than SWNT, there have been a number of efforts trying to improve biocidal effect of MWNTs by decorating them with other antimicrobial agents such as metallic nanoparticles or oxidizing chemicals. CNTs in waterborne pathogen control could be applied either in suspension or immobilized onto hosting media. Immobilized CNTs could benefit from their easy recovery although their antimicrobial effective might be also compromised because of external binding. Meanwhile, direct applications of CNTs in suspension might face difficulties in their retainment system. Although CNT is still expensive when compared to conventional disinfectant, real application of CNTs for water disinfection is very likely to come true in near future because price of CNT has been extremely decreasing for the past few years. Lastly, ecotoxicological concerns of CNT applications are also of great interest as comprehensive study of their life cycle analysis throughout their entire life is still very poorly performed. However, the modified MWNT by $HNO_3$ experienced the structural and property degradation. Besides, there is a slight aggregation at a high content in the matrix, which causes adverse effects to the polybenzoxazine-CNT composite properties. There is still a long way for achieving the optimum modification of CNT, and a better modification of MWNT can do a better performance in the benzoxazine matrix and other matrix. MWNT are reported to be always electrically conductive and to have a high electrical conductivity. Compared with the composites filled with carbon black or carbon fiber as conductive reinforcements, the composite with MWNT can form the conductive path at a very low content, owing to their high aspect ratio and high specific surface areas. Therefore, they can be used as antistatic or electromagnetic shielding materials. Generally speaking, the nanocomposites based on polybenzoxazine and nanocarbon materials are expected to have a good prospect of application and are worth researching further.

## REFERENCES

Abarrategi, A., Gutiérrez, M. C., Moreno-Vicente, C., Hortigüela, M. J., Ramos, V., López-Lacomba, J. L., Ferrer, M. L., & del Monte, F. (2008, January). Multiwall carbon nanotube scaffolds for tissue engineering purposes. *Biomaterials*, *29*(1), 94–102. doi:10.1016/j.biomaterials.2007.09.021 PMID:17928048

Alexey, A. (2003, August 1). Tomchenko, Gregory P. Harmer, Brent T. Marquis, John W. Allen. Semiconducting metal oxide sensor array for the selective detection of combustion gases. *Sensors and Actuators. B, Chemical*, *93*(1–3), 126–134. doi:10.1016/S0925-4005(03)00240-5

Ali, S. (2019). Rehman,S. A. U. Luan, H.-Y. Farid M. U. and Huang, H. *The Science of the Total Environment*, *646*, 1126–1139. doi:10.1016/j.scitotenv.2018.07.348 PMID:30235599

Alim, S., Vejayan, J., Yusoff, M. M., & Kafi, A. K. M. (2018). Recent uses of carbon nanotubes & gold nanoparticles in electrochemistry with application in biosensing: A review. *Biosensors & Bioelectronics*, *121*, 121–136. doi:10.1016/j.bios.2018.08.051 PMID:30205246

Anna, A. (2012, June 1). Shvedova, Antonio Pietroiusti, Bengt Fadeel, Valerian E. Kagan. Mechanisms of carbon nanotube-induced toxicity: Focus on oxidative stress. *Toxicology and Applied Pharmacology*, *261*(2), 121–133. doi:10.1016/j.taap.2012.03.023 PMID:22513272

Ansari, R. (2013). Mirnezhad, M. and Sahmani, S. "An accurate molecular mechanics model for computation of sizedependent elastic properties of armchair and zigzag singlewalled carbon nanotubes,". *Meccanica*, *48*(6), 1355–1367. doi:10.1007/s11012-012-9671-x

Artyukhov, V. I., Penev, E. S., & Yakobson, B. I. (2014). Why Nanotubes Grow Chiral. *Nature Communications*, *5*(1), 4892–4892. doi:10.1038/ncomms5892 PMID:25224858

Beatriz, R. C. M., & Karla, F. R. (2019). Recent advances on the use of carbon nanotubes as smart biomaterials. *Journal of Materials Chemistry. B*, *1*(2), 1–20.

Bian, Z., Ru, J. W., Wei, H. W., & Zhang, T. (2004, January). andAkihisa Inoue.Carbon-Nanotube-Reinforced Zr-Based Bulk Metallic GlassComposites and Their Properties. *Advanced Functional Materials*, *14*(1), 55–63. doi:10.1002/adfm.200304422

Boumia, L. (2014). Zidour, M. Benzair, A. and Tounsi, A. "A Timoshenko beam model for vibration analysis of chiral singlewalled carbon nanotubes,". *Physica E, Low-Dimensional Systems and Nanostructures*, *59*, 186–191. doi:10.1016/j.physe.2014.01.020

Cardenas-Benitez, B., Djordjevic, I., Hosseini, S., Madou, M. J., & Martinez-Chapa, S. O. (2018). Covalent functionalization of carbon nanomaterials for biosensor applications: An update. *Journal of the Electrochemical Society*, *165*(3), B103–B117. doi:10.1149/2.0381803jes

Çetin, M., Aytekin, E., Yavuz, B., & Bozdağ-Pehlivan, S. (2017). Chapter 7 - Nanoscience in Targeted Brain Drug Delivery. Nanotechnology Methods for Neurological Diseases and Brain Tumors Drug Delivery across the Blood-Brain Barrier. Academic Press. doi:10.1016/B978-0-12-803796-6.00007-1

Chatterjee, A., & Deopura, B. L. (2002). Carbon nanotubes and nanofibre: An overview. *Fibers and Polymers*, *3*(4), 134–139. doi:10.1007/BF02912657

Chik, M. W., Hussain, Z., Zulkefeli, M., Tripathy, M., Kumar, S., Majeed, A. B. A., & Byrappa, K. (2019). Polymer-wrapped single-walled carbon nanotubes: A transformation toward better applications in healthcare. *Drug Delivery and Translational Research*, *9*(2), 578–594. doi:10.1007/s13346-018-0505-9 PMID:29594914

Chowdhury, D. F. (2011). *5.49 - Carbon Nanotube for Drug Delivery and Controlled Release* (2nd ed., Vol. 5). Reference Module in Biomedical Sciences Comprehensive Biotechnology., doi:10.1016/B978-0-08-088504-9.00518-3

Demirer, G. S., Zhang, H., Matos, J. L., Goh, N. S., Cunningham, F. J., Sung, Y., Chang, R., Aditham, A. J., Chio, L., Cho, M.-J., Staskawicz, B., & Landry, M. P. (2019). High aspect ratio nanomaterials enable delivery of functional genetic material without DNA integration in mature plants. *Nature Nanotechnology*, *14*(5), 456–464. doi:10.1038/s41565-019-0382-5 PMID:30804481

Deng, S., & Berry, V. (2016). Wrinkled, rippled and crumpled graphene: An overview of formation mechanism, electronic properties, and applications. *Materials Today*, *19*(4), 197–212. doi:10.1016/j.mattod.2015.10.002

Dolez Patricia, I. (2015). *Nanomaterials Definitions, Classifications, and Applications*. Nanoengineering Global Approaches to Health and Safety Issues., doi:10.1016/B978-0-444-62747-6.00001-4

Farghali, A. A. (2017). Abdel Tawab, H. A. Abdel Moaty S. A. and Khaled, R. *Journal of Nanostructure in Chemistry*, *7*, 101–111. doi:10.1007/s40097-017-0227-4

Galluzzi, L., Vitale, I., Abrams, J., Alnemri, E. S., Baehrecke, E. H., Blagosklonny, M. V., Dawson, T. M., Dawson, V. L., El-Deiry, W. S., Fulda, S., Gottlieb, E., Green, D. R., Hengartner, M. O., Kepp, O., Knight, R. A., Kumar, S., Lipton, S. A., Lu, X., Madeo, F., & Kroemer, G. (2012). Molecular definitions of cell death subroutines: Recommendations of the Nomenclature Committee on Cell Death 2012. *Cell Death and Differentiation*, *19*(1), 107–120. doi:10.1038/cdd.2011.96 PMID:21760595

Grushevskaya, H. V., & Krylova, N. G. (2018). Carbon nanotubes as a high-performance platform for target delivery of anticancer quinones. *Current Pharmaceutical Design*, *24*(43), 5207–5218. doi:10.2174/1381612825666190117095132 PMID:30652640

Guan, L., Suenaga, K., & Iijima, S. (2008). Smallest Carbon Nanotube Assigned with Atomic Resolution Accuracy. *Nano Letters*, *8*(2), 459–462. doi:10.1021/nl072396j PMID:18186659

Gul, R. Zainab, N. Asad, M. Salma, B. (2019). An Overview of the Recent Progress in the Synthesis and Applications of Carbon Nanotubes. *J. carbo Res. 5*(3), pp. 2-31.

Han, M., Dong, T. & Hou, D. (2020). Carbon nanotube based Janus composite membrane of oil fouling resistance for direct contact membrane distillation. *Journal of Membrane Science*.

Han, Y. (2020). *A simple and effective flexible electrochemiluminescence sensor for lidocaine detection*.

Hasnain, M. S., & Nayak, A. K. (2019). *Carbon Nanotubes for Targeted Drug Delivery.* Springer Briefs in Applied Sciences and Technology.

He, H. (2013). Carbon Nanotubes: Applications in Pharmacy and Medicine. BioMed Research International.

Hills, G., Lau, C., Wright, A., Fuller, S., Bishop, M. D., Srimani, T., Kanhaiya, P., Ho, R., Amer, A., Stein, Y., Murphy, D., Arvind, Chandrakasan, A., & Shulaker, M. M. (2019). Modern microprocessor built from complementary carbon nanotube transistors. *Nature*, *572*(7771), 595–602. doi:10.1038/s41586-019-1493-8 PMID:31462796

Iijima, S. (1991). Helical microtubules of graphitic carbon. *Nature*, *354*(6348), 56–58. doi:10.1038/354056a0

Iijima, S. (2002). Carbon nanotubes: Past, present and future. Phys. B. *Physica B, Condensed Matter*, *323*(1-4), 1–5. doi:10.1016/S0921-4526(02)00869-4

Iijima, S., Ajayan, P. M., & Ichihashi, T. (1992). Growth model for carbon nanotubes. *Physical Review Letters*, *69*(21), 3100–3103. doi:10.1103/PhysRevLett.69.3100 PMID:10046725

Iijima, S., & Ichihashi, T. (1993). Single-shell carbon nanotubes of 1-nm diameter. *Nature*, *363*(6430), 603–605. doi:10.1038/363603a0

Indarto, A., Ikhsan, N. A., & Wibowo, I. (2020). *Applications of carbon nanotubes for controlling waterborne pathogens*. Waterborne Pathogens Detection and Treatment., doi:10.1016/B978-0-12-818783-8.00020-7

Karousis, N., Martinez, I. S., Ewels, C. P., & Tagmatarchis, N. (2016). Structure, properties, functionalization, and applications of carbon nanohorns. *Chemical Reviews*, *116*(8), 4850–4883. doi:10.1021/acs.chemrev.5b00611 PMID:27074223

Khin Mya Mya, A. (2012). Sreekumaran Nair, V. Jagadeesh Babu, Rajendiran Murugana and Seeram Ramakrishna.A review on nanomaterials for environmental remediation. *Energy & Environmental Science*, *5*, 8075.

Khurshed, A. S., & Bilal, A. T. (2016). Synthesis of carbon nanotubes by catalytic chemical vapour deposition: A review on carbon sources, catalysts and substrates. *Materials Science in Semiconductor Processing*, *41*(5), 67–82.

Kour, R., Arya, S., Young, S.-J., Gupta, V., Bandhoria, P., & Khosla, A. (2020). Recent advances in carbon nanomaterials as electrochemical sensors. *Journal of the Electrochemical Society*, *167*(3), 37555. doi:10.1149/1945-7111/ab6bc4

Kroto, H. W., Heath, J. R., O'Brien, S. C., Curl, R. F., & Smalley, R. E. (1985). C60: Buckminsterfullerene. *Nature*, *318*(6042), 162–163. doi:10.1038/318162a0

Kuila, T., Bose, S., Mishra, A., Khanra, P., Kim, N., & Lee, J. (2012). Chemical functionalization of graphene and its applications. *Progress in Materials Science, 57*(7), 1061–1105. doi:10.1016/j.pmatsci.2012.03.002

Lin, P. A., Gomez-Ballesteros, J. L., Burgos, J. C., Balbuena, P. B., Natarajan, B., & Sharma, R. (2017). Direct Evidence of Atomic-Scale Structural Fluctuations in Catalyst Nanoparticles. *Journal of Catalysis, 349*, 149–155. doi:10.1016/j.jcat.2017.03.009 PMID:28740274

Liu, Z., Pagani, M., Zinniker, D., DeConto, R., Huber, M., Brinkhuis, H., Shah, S. R., Leckie, R. M., & Pearson, A. (2009, February 27). Global Cooling During the Eocene-Oligocene Climate Transition. March (2009). *Science, 323*(5918), 1187–1190. doi:10.1126/science.1166368 PMID:19251622

Lounasvuori, M. M., Kelly, D., & Foord, J. S. (2018). Carbon black as lowcost alternative for electrochemical sensing of phenolic compounds. *Carbon, 129*, 252–257. doi:10.1016/j.carbon.2017.12.020

Madkour, L.H. (2020). *Reactive Oxygen Species (ROS), Nanoparticles, and Endoplasmic Reticulum (ER) Stress-Induced Cell Death Mechanisms.* Imprint: Academic Press.

Madkour, L. H. (2020). *Nanoparticles Induce Oxidativeand Endoplasmic Reticulum Antioxidant TherapeuticDefenses.* Springer International Publishing. doi:10.1007/978-3-030-37297-2

Madkour, L. H. (2020). Nucleic Acids as Gene Anticancer Drug Delivery Therapy. Elsevier.

Madkour, L. H. (2022). *Book: Nanoparticle-Based Drug Delivery in Cancer Treatment.First Published.* Imprint CRC Press., doi:10.1201/9781003229674

Madkour, L. H. (2023). Carbon Nanotubes—Based Intelligent Platform for Cancer Vaccine Co-delivery Nanocarriers Immunotherapy Achievements: Challenges In vitro and In vivo. *International Journal of Nursing Care and Research, 1*(2), 1–55.

Madkour, L.H. (2023). Graphitic Carbon Nitride Quantum Dots (g-C3N4): Fundamentals and Applications.International Research Journal of Basic and Clinical Studies, 8(4), 1-14.

Marquez, F. (2023).*Non-Destructive Testing and Condition Monitoring Techniques in Wind Energy.A volume in Wind Energy Engineering. Copyright.* Elsevier Inc., doi:10.1016/C2021-0-02588-8

Mathur, R., Alver, B., San Roman, A., Wilson, B. G., Wang, X., Agoston, A. T., Park, P. J., Shivdasani, R. A., & Roberts, C. W. M. (2017). ARID1.A loss impairs enhancer-mediated gene regulation and drives colon cancer in mice. *Nature Genetics, 49*(2), 296–302. doi:10.1038/ng.3744 PMID:27941798

Nancy, V. (2008, February 28). Integrated microfluidic systems with an immunosensor modified with carbon nanotubes for detection of prostate specific antigen (PSA) in human serum samples. *Biosensors & Bioelectronics, 23*(7), 1145–1151. doi:10.1016/j.bios.2007.11.003 PMID:18162392

Navas, H., Picher, M., Andrieux-Ledier, A., Fossard, F., Michel, T., Kozawa, A., Maruyama, T., Anglaret, E., Loiseau, A., & Jourdain, V. (2017). Unveiling the Evolutions of Nanotube Diameter Distribution During the Growth of Single-Walled Carbon Nanotubes. *ACS Nano, 11*(3), 3081–3088. doi:10.1021/acsnano.7b00077 PMID:28285520

Negri, V., Pacheco-Torres, J., Calle, D., & López-Larrubia, P. (2020). Carbon Nanotubes in Biomedicine. *Topics in Current Chemistry (Cham)*, *378*(1), 15. doi:10.1007/s41061-019-0278-8 PMID:31938922

Onyancha, R. B., Aigbe, U. O., Ukhurebor, K. E., & Muchiri, P. W. (2021). A facile review on the synthesis and applications of carbon nanotubes in heavy-metal remediation and biomedical fields. *Journal of Molecular Structure*, *1238*, 130462. doi:10.1016/j.molstruc.2021.130462

Penev, E. S., Bets, K. V., Gupta, N., & Yakobson, B. I. (2018). Transient Kinetic Selectivity in Nanotubes Growth on Solid Co–W Catalyst. *Nano Letters*, *18*(8), 5288–5293. doi:10.1021/acs.nanolett.8b02283 PMID:29979600

Penkova, A. V., Acquah, S. F. A., Piotrovskiy, L. B., Markelov, D. A., Semisalova, A. S., & Kroto, H. W. (2017). Fullerene derivatives as nano-additives in polymer composites. *Russian Chemical Reviews*, *86*(6), 530–566. doi:10.1070/RCR4712

Pennetta, C. (2020). Floresta, G. Graziano, A. C. E. Cardile, V. Rubino, L. Galimberti, M. Rescifina A. and Barbera, V. *Nanomaterials (Basel, Switzerland)*, *10*, 1073. doi:10.3390/nano10061073 PMID:32486371

Pham, V. P., Jang, S.-H., Whang, D., & Choi, J.-Y. (2017). Direct growth of graphene on rigid and flexible substrates: Progress, applications, and challenges. *Chemical Society Reviews*, *46*(20), 6276–6300. doi:10.1039/C7CS00224F PMID:28857098

Rahmani Didar, B., & Balbuena, P. B. (2017). Growth of Carbon Nanostructures on Cu Nanocatalysts. *The Journal of Physical Chemistry. C, Nanomaterials and Interfaces*, *121*(13), 7232–7239. doi:10.1021/acs.jpcc.6b12816

Rana, M. M., Ibrahim, D. S., Mohd Asyraf, M. R., Jarin, S., & Tomal, A. (2017). A review on recent advances of CNTs as gas sensors. *Sensor Review*, *37*(2), 127–136. doi:10.1108/SR-10-2016-0230

Ren, J., Li, L., Chen, C., Chen, X., Cai, Z., Qiu, L., Wang, Y., Zhu, X., & Peng, H. (2013). Twisting Carbon Nanotube Fibers for Both Wire-Shaped Micro-Supercapacitor and Micro-Battery. *Advanced Materials*, *25*(8), 1155–1159. doi:10.1002/adma.201203445 PMID:23172722

Rima, K., & Singh, D. P. (2021). *Nano-Biotechnological Approach of Plant Genetic Engineering*. Policy Issues in Genetically Modified Crops.A Global Perspective. doi:10.1016/B978-0-12-820780-2.00021-2

Sahu, A., Gupta, R., Sagar, S., Kumar, M., & Sagar, R. (2017, July-December). A study of psychiatric comorbidity after traumatic limb amputation: A neglected entity. *Industrial Psychiatry Journal*, *26*(2), 228–232. doi:10.4103/ipj.ipj_80_16 PMID:30089974

Sahub, C., Tuntulani, T., Nhujak, T., & Tomapatanaget, B. (2018). Effective biosensor based on graphene quantum dots via enzymatic reaction for directly photoluminescence detection of organophosphate pesticide. *Sensors and Actuators. B, Chemical*, *258*, 88–97. doi:10.1016/j.snb.2017.11.072

Saifuddin, N. Raziah, A. Z. & Junizah, A. R. (2013). Carbon Nanotubes: A Review on Structure and Their Interaction with Proteins. Hindawi Publishing Corporation.

Salah, L. S., Ouslimani, N., Bousba, D., Huynen, I., Danlée, Y., & Aksas, H. (2021). Carbon Nanotubes (CNTs) from Synthesis to Functionalized (CNTs) Using Conventional and New Chemical Approaches. Journal of Nanomaterials.

Salas-Trevino, D., Saucedo-Cardenas, O., & de Jesus Loera-Arias, M. (2018). Carbon nanotubes: An alternative for platinum-based drugs delivery systems. *JBUON*, *23*(3), 541–549. PMID:30003717

Same, S., & Samee, G. (2018). Carbon nanotube biosensor for diabetes disease. *Crescent Journal of Medical and Biological Sciences*, *5*(1), 1–6.

Sanchez-Valencia, J. R., Dienel, T., Groning, O., Shorubalko, I., Mueller, A., Jansen, M., Amsharov, K., Ruffieux, P., & Fasel, R. (2014). Controlled Synthesis of Single-Chirality Carbon Nanotubes. *Nature*, *512*(7512), 61–64. doi:10.1038/nature13607 PMID:25100481

Shen, B., Lu, S., Sun, C., Song, Z., Zhang, F., Kang, J., Cao, Y., & Xiang, M. (2022, June). Effects of Amino Hyperbranched Polymer-Modified Carbon Nanotubes on the Crystallization Behavior of Poly (L-Lactic Acid) (PLLA). *Polymers*, *14*(11), 2188. doi:10.3390/polym14112188 PMID:35683866

Shulaker, M. M., Hills, G., Patil, N., Wei, H., Chen, H.-Y., Wong, H.-S. P., & Mitra, S. (2013). Carbon Nanotube Computer. *Nature*, *501*(7468), 526–530. doi:10.1038/nature12502 PMID:24067711

Sidi Salah, L., Chouai, M., Danlée, Y., Huynen, I., & Ouslimani, N. (2020). Simulation and optimization of electromagnetic absorption of polycarbonate/CNT composites using machine learning. *Micromachines*, *11*(8), 778. doi:10.3390/mi11080778 PMID:32824164

Singh, S., Patel, H., Goyani, M., & Patel, P. (2021, January). *Multidisciplinary International Research Journal of Gujarat Technological University. VOLUME*, *3*(1), 97–120.

Sinha, A., Jain, R., Zhao, H., Karolia, P., & Jadon, N. (2018, January 8). Voltammetric sensing based on the use of advanced carbonaceous nanomaterials: A review. *Mikrochimica Acta*, *185*(2), 89. doi:10.1007/s00604-017-2626-0 PMID:29594390

Sudha, P. N., Sangeetha, K., Vijayalakshmi, K., & Barhoum, A. (2018). *Nanomaterials history, classification, unique properties, production and market.Emerging Applications of Nanoparticles and Architectural Nanostructures.Copyright.* Elsevier Inc., doi:10.1016/B978-0-323-51254-1.00012-9

Tian, P., Tang, L., Teng, K. S., & Lau, S. P. (2018). Graphene quantum dots from chemistry to applications. *Materials Today. Chemistry*, *10*, 221–258. doi:10.1016/j.mtchem.2018.09.007

Tobias, G., Mendoza, E., & Ballesteros, B. (2016). Functionalization of Carbon Nanotubes. In B. Bhushan (Ed.), *Encyclopedia of Nanotechnology* (pp. 1281–1291). Springer. doi:10.1007/978-94-017-9780-1_48

Vashist, A., Kaushik, A., Vashist, A., Sagar, V., Ghosal, A., Gupta, Y. K., Ahmad, S., & Nair, M. (2018). Advances in carbon nanotubes-hydrogel hybrids in nanomedicine for therapeutics. *Advanced Healthcare Materials*, *7*(9), 170–187. doi:10.1002/adhm.201701213 PMID:29388356

Wu, T., Zhang, X., Yuan, Q., Xue, J., Lu, G., Liu, Z., Wang, H., Wang, H., Ding, F., Yu, Q., Xie, X., & Jiang, M. (2016). Fast Growth of Inch-Sized Single-Crystalline Graphene from a Controlled Single Nucleus on Cu–Ni Alloys. *Nature Materials*, *15*(1), 43–47. doi:10.1038/nmat4477 PMID:26595118

Xu, X., Zhang, Z., Dong, J., Yi, D., Niu, J., Wu, M., Lin, L., Yin, R., Li, M., Zhou, J., Wang, S., Sun, J., Duan, X., Gao, P., Jiang, Y., Wu, X., Peng, H., Ruoff, R. S., Liu, Z., & Liu, K. (2017). Ultrafast Epitaxial Growth of Metre-Sized Single-Crystal Graphene on Industrial Cu Foil. *Science Bulletin*, *62*(15), 1074–1080. doi:10.1016/j.scib.2017.07.005 PMID:36659334

Yang, F., Wang, X., Si, J., Zhao, X., Qi, K., Jin, C., Zhang, Z., Li, M., Zhang, D., Yang, J., Zhang, Z., Xu, Z., Peng, L.-M., Bai, X., & Li, Y. (2017). Water-Assisted Preparation of High-Purity Semiconducting (14,4) Carbon Nanotubes. *ACS Nano*, *11*(1), 186–193. doi:10.1021/acsnano.6b06890 PMID:28114760

Yang, F., Wang, X., Zhang, D., Yang, J., Luo, D., Xu, Z., Wei, J., Wang, J.-Q., Xu, Z., Peng, F., Li, X., Li, R., Li, Y., Li, M., Bai, X., Ding, F., & Li, Y. (2014). Chirality-Specific Growth of Single-Walled Carbon Nanotubes on Solid Alloy Catalysts. *Nature*, *510*(7506), 522–524. doi:10.1038/nature13434 PMID:24965654

Yang, N., Chen, X., Ren, T., Zhang, P., & Yang, D. (2015). Carbon nanotube based biosensors.Sensors and Actuators B: Chemical Volume 207. *Sensors and Actuators. B, Chemical*, *207*(February), 690–715. doi:10.1016/j.snb.2014.10.040

Yin, L. J., Bai, K. K., Wang, W. X., Li, S. Y., Zhang, Y., & He, L. (2017). Landau quantization of dirac fermions in graphene and its multilayers. *Frontiers of Physics*, *12*(4), 127208. doi:10.1007/s11467-017-0655-0

Yu, M. F., Files, B. S., Arepalli, S., & Ruoff, R. S. (2000). Tensile loading of ropes of single wall carbon nanotubes and their mechanical properties. *Physical Review Letters*, *84*(24), 5552–5555. doi:10.1103/PhysRevLett.84.5552 PMID:10990992

Yumna, Q. (2023). *Mostapha Tarfaoui, Khalid Lafdi. 13 - Background, advancement, and applications of in situ structural health monitoring based on different modes of failure detection in composites: a review*. Non-Destructive Testing and Condition Monitoring Techniques in Wind Energy Wind Energy Engineering.

Zakaria, M. R., Akil, H. M., Muhammad, H. A. K., Ullah, F., Javed, F., & Nosbi, N. (2019, November). Hybrid carbon fiber-carbon nanotubes reinforced polymer composites: A review. *Composites. Part B, Engineering*, *176*(1), 107313. doi:10.1016/j.compositesb.2019.107313

Zaytseva, O., & Neumann, G. (2016). Carbon nanomaterials: Production, impact on plant development, agricultural and environmental applications. *Chemical and Biological Technologies in Agriculture*, *3*(1), 17. doi:10.1186/s40538-016-0070-8

Zhang, D., Wen, L., Huang, R., Wang, H., Hu, X., & Xing, D. (2018). Mitochondrial specific photodynamic therapy by rare-earth nanoparticles mediated near-infrared graphene quantum dots. *Biomaterials*, *153*, 14–26. doi:10.1016/j.biomaterials.2017.10.034 PMID:29096398

Zhang, F., Peng, W., Yang, Y., Dai, W., & Song, J. (2019). A Novel Method for Identifying Essential Genes by Fusing Dynamic Protein–Protein Interactive Networks. *Genes*, *10*(1), 31. doi:10.3390/genes10010031 PMID:30626157

Zhang, H., Hong, M., Chen, P., Xie, A., & Shen, Y. (2016, April). 3D and ternary rGO/MCNTs/$Fe_3O_4$ composite hydrogels: Synthesis, characterization and their electromagnetic wave absorption properties. *Journal of Alloys and Compounds*, *665*(25), 381–387. doi:10.1016/j.jallcom.2016.01.091

Zhao, Y.-L., & Fraser Stoddart, J. (2009). Noncovalent Functionalization of Single-Walled Carbon Nanotubes. *Accounts of Chemical Research*, *42*(8), 1161–1171. doi:10.1021/ar900056z PMID:19462997

Section 3

# Medical and Biomedical Engineering Technologies

# Chapter 6
# Empowering Informed Healthcare Choices in Rural Areas:
## SIKDA, a Novel Model for Enhancing Patient Access to Reliable Consumer Health Information

**Rita Komalasari**
https://orcid.org/0000-0001-9963-2363
*Yarsi University, Indonesia*

## ABSTRACT

*To bridge the gap between patients' desire for active involvement in healthcare decisions and the availability of trustworthy resources, the chapter introduces SIKDA. This groundbreaking model aims to empower patients and caregivers by facilitating informed patient decision-making. SIKDA represents a transformative approach to consumer health information, providing patients with a single point of access to credible and up-to-date health information online and the knowledge and guidance of a librarian. By leveraging existing NHS online resources and collaborating with public and medical librarians, SIKDA ensures accessibility and inclusivity for diverse demographics, minimizing costs to public services. The study establishes the foundation for a novel paradigm, enhancing patient engagement, healthcare outcomes, information literacy, and collaborative efforts in the digital healthcare era.*

## 1. INTRODUCTION

The application of SIKDA stems from a critical need to address the disparity between patients' eagerness for active participation in healthcare decisions and the limited availability of reliable resources. Recognizing this gap, the chapter unfolds with the primary objective of introducing SIKDA as a transformative model with the explicit purpose of empowering both patients and caregivers. The cornerstone

DOI: 10.4018/979-8-3693-1335-0.ch006

of SIKDA lies in its commitment to facilitate informed decision-making among patients, fostering a paradigm shift in the realm of consumer health information. SIKDA operates as an innovative approach, providing a unified platform that grants patients a singular point of access to credible and real-time health information online, complemented by the invaluable knowledge and guidance of a librarian. This dual-layered approach enhances the quality of information available to patients, positioning them as active participants in their healthcare journey. By strategically leveraging existing NHS online resources and establishing collaborations with public and medical librarians, SIKDA goes beyond mere accessibility, ensuring inclusivity across diverse demographic groups. Notably, this inclusive design is achieved while minimizing costs to public services, demonstrating the model's practical and sustainable impact on healthcare infrastructure. This study serves as the foundational exploration of the SIKDA paradigm, laying the groundwork for a novel approach that transcends traditional boundaries. The overarching motivation behind this work is to contribute significantly to enhanced patient engagement, improved healthcare outcomes, heightened information literacy, and strengthened collaborative efforts within the dynamic landscape of digital healthcare. Through a rigorous examination of the major objectives, this chapter aims to underscore the profound impact of SIKDA on reshaping the healthcare landscape for the benefit of patients, caregivers, and the broader healthcare community. One major challenge in the ever-changing healthcare system is people's growing dependence on online resources for medical information (Gille et al., 2021). Patients are urged to actively engage in making decisions about their health (Swire et al., 2020). Nevertheless, they run the danger of making ill-informed choices due to the enormity of unfiltered and incorrect information available online (Bujnowska et al., 2019). This gap in reliable health information access and the imperative for patient engagement highlights a critical problem. Additionally, the existing literature reveals a need for more research addressing the simultaneous issues of patient participation, information access, and skill utilization (van Schalkwyk et al., 2020). The purpose of this chapter is to fill these gaps and provide a tailored Seven Pillars of Health Information Literacy (SIKDA) model so that people in rural Indonesia, including patients and caregivers, may make better healthcare choices (Verawati et al., 2022). The results of this chapter can substantially impact the world of practice and contribute to the expansion of applied research in several ways: Healthcare practitioners can apply the insights to foster more patient-centred care. Understanding patient information needs and challenges can guide practitioners in tailoring communication strategies, ensuring patients actively participate in decision-making. Patient happiness, treatment adherence, and general health outcomes may be further investigated via applied research into the use of patient-centred care models in various healthcare contexts (Davis et al., 2020). The study's results may help health data suppliers improve the usability of their websites. Improved user interfaces, content curation, and integration of librarian support can create more user-friendly and reliable resources. Future applied research can delve into the effectiveness of different features in online health information platforms, investigating user preferences and the impact on health SIKDA. Healthcare professionals can apply the research findings to improve their communication with patients, recognizing the importance of providing information and support in navigating healthcare choices. Applied research can explore training programs for healthcare professionals, evaluating the effectiveness of interventions to enhance communication skills and patient engagement. Librarians and information specialists can collaborate more effectively within the healthcare ecosystem, supporting users in accessing and understanding health information. Applied research can delve into the dynamics of librarian collaboration in healthcare settings, examining the impact on patient outcomes and the effectiveness of different support models. Policymakers can use the research findings to inform health policies promoting patient engagement, improving health literacy, and fostering collaboration within the healthcare sector. Applied research can explore the implementation and impact of health policies inspired by the research, evaluating their effectiveness in real-world healthcare settings. Health information technology developers can apply research insights to designing platforms catering to diverse user needs and literacy levels. Future applied research can focus on the continuous improvement of health technology, exploring innovations that enhance the user experience, increase accessibility, and promote health literacy. The research contributes to the culture of continuous quality improvement in healthcare, encouraging organizations to reassess their patient engagement and information provision approaches. Applied research can delve into the long-term impact of quality improvement initiatives inspired by the research, examining their sustainability and scalability in different healthcare contexts. The results of this research offer practical insights that can be directly applied to improve various aspects of healthcare practice. Additionally, they provide fertile ground for further applied research to explore and refine interventions, strategies, and models to advance patient-centred care, health information accessibility, and collaborative healthcare efforts.

## 2. BACKGROUND

In the ever-changing healthcare system, it is crucial to closing the gap between patients' desire for active involvement in their healthcare decisions and the trustworthy resources necessary for making informed decisions in an era of patient empowerment, healthcare digitization, and the surge in online health information (Withorn et al., 2021). Patient access to health record summaries, education, and self-management have all been big steps forward for the National Health Service (NHS) (Song et al., 2021). Concurrently, the government's emphasis on shared decision-making underscores the need for accessible, trustworthy health information. However, the internet's expansive landscape poses a challenge: ensuring its reliability and relevance is a pressing concern amidst the wealth of information. Only 20% of the health-related information found online is reliable, despite the fact that 80% of internet users seek it out (Lee et al., 2021). This shows that the majority of individuals are looking for health-related information online but that not all of the sites they encounter are reliable. Feelings of bewilderment, unease, and potential danger may ensue. Help locating trustworthy health information online is needed by 70% of individuals (Ducrot et al., 2021). This shows that individuals need assistance in locating reliable sources of information online and in determining the credibility of various sources. Because of this, individuals may make poor choices about their health, which is a major concern. As much as 60% of Indonesians rely on having an internet connection (Siste et al., 2020). This means that most Indonesians can access health information online without help accessing health information online. This is a significant equity issue, as it means that people who are already disadvantaged are even less likely to have access to reliable health information. These statistics and trends highlight the importance of the study's focus on patient empowerment and access to reliable health information. Patients will be better able to make educated choices about their health as a result of the study's results, which will alleviate these problems. This research addresses this critical gap, focusing on Indonesia, where patient participation in decision-making and challenges in accessing accurate health information intersect. Extensive literature exploring patient involvement and the complexities of online health information forms the foundation of this study. Healthcare providers have jumped on the evidence-based practice bandwagon, but they have yet to figure out how to make patients' decisions and provide them with information based on solid research. Present conditions are inadequate to support patients' active engagement, which necessitates access to information and the ability to assess its quality.

## 3. METHOD

The robust foundation of this chapter is rooted in a meticulous exploration of existing knowledge drawn from an extensive literature review. This comprehensive examination was the compass for navigating the intricate landscape of patient participation and the formidable challenges of obtaining reliable health information (Shukla, 2021). An in-depth literature analysis that included insights from 50 publications set the stage for comprehending the intricacies of patient involvement and the difficulties of obtaining trustworthy health information, which was necessary to guarantee SIKDA's effectiveness. Integrating the SIKDA Seven Pillars of Health Information Literacy with the five-stage process of evidence-based practice laid the groundwork for SIKDA's growth as a Health Information Literacy framework (Fauzan & Noviandi, 2020). The inclusion and exclusion criteria for source selection aim to ensure a focused and comprehensive review of relevant literature. In this context, Google Scholar is the primary source, providing diverse scholarly articles and academic resources. The specified time frame from 2019 to 2023 allows for including recent publications, ensuring that the latest research findings inform the literature review. By limiting the languages to English and Indonesian, the criteria ensure accessibility to a broad audience while excluding studies in other languages that might hinder understanding. The target population focuses on patients, aligning with the research's emphasis on consumer health information, patient empowerment, evidence-based decision-making, and collaborative healthcare. The defined context of a transformative healthcare model guides the selection of studies that align with the overarching theme of innovative healthcare approaches. The inclusive approach to study types encourages examining various scholarly articles and free full-text publications, providing a well-rounded understanding of the chosen concepts within the defined scope. The Unified Theory of Acceptance and Use of Technology (UTAUT) could provide some support for a thorough assessment of the specified SIKDA framework (Tamilmani et al., 2021). This is particularly evident when considering the people who use healthcare services and the development of SIKDA. UTAUT integrates various technology acceptance models, emphasizing factors influencing users' acceptance and adoption of new technologies (Dwivedi et al., 2019). Since SIKDA is about putting an SIKDA framework into healthcare, UTAUT allows us to look at important factors such as effort and performance expectations, social influence, and conditions that make learning easier (Boonsiritomachai & Pitchayadejanant, 2019). This theory is particularly relevant as it considers the impact of social factors and emphasizes facilitating conditions, aligning with the collaborative nature of SIKDA involving public and medical librarians. Analyzing SIKDA through the UTAUT framework can offer insights into the factors influencing its adoption, usability, and effectiveness among healthcare consumers.

### 3.1 Empowering Informed Healthcare Choices

SIKDA stands as a transformative force, providing a novel approach to consumer health information. It furnishes patients with a consolidated platform for accessing credible and real-time health information online, supplemented by the invaluable knowledge and guidance of a librarian. The recent experimental applications of SIKDA underscore its dynamic capabilities in reshaping the patient experience within

*Table 1. Comparing SIKDA to existing health information resources*

| Feature | SIKDA | Existing Health Information Resources |
|---|---|---|
| Access to Credible Information | Centralized platform for credible and up-to-date health information online | Multiple sources with varying degrees of reliability |
| Guidance from Librarians | Knowledge and guidance from dedicated librarians. | Limited interaction with healthcare professionals |
| Collaborative Initiatives | Collaboration with public and medical librarians for comprehensive support. | Limited collaboration and less holistic support. |
| Accessibility | Ensures accessibility and inclusivity for diverse demographics. | Accessibility may vary; not tailored to diverse user needs. |
| Cost Minimization | Minimizes costs to public services through strategic collaborations. | May have associated costs for certain services. |
| Patient Empowerment | Facilitates informed decision-making, actively involving patients in their healthcare. | Limited tools for active patient participation. |
| Information Literacy | Enhances information literacy through curated resources and librarian guidance. | Relies on users' existing information literacy levels. |
| Healthcare Outcomes | Aims to improve healthcare outcomes through empowered patient decision-making. | Outcomes influenced by the quality and accessibility of information. |

healthcare decision-making. Table 1 is an illustrative table comparing SIKDA to existing health information resources, emphasizing its unique features and strengths in promoting patient empowerment:

This section discusses the goal of connecting all current health information products and linking to human help, when necessary, the SIKDA Database seeks to enable access to high-quality consumer health information and librarian support via one gateway. Users can enter as much information as they want, filling in the first box for more information about a condition, the second box for information about a particular treatment, the third box for comparing another treatment to the prescribed treatment, and the final box for all other options or just the condition (Harahap et al., 2021). Additional features include linking to electronic health records (EHR), contacting library staff, learning to search more effectively, and accessing further resources (Daraz et al., 2019). Symptom checkers are not diagnostic tools and should not be used as such (Ennaceur et al., 2023). While only a trained medical expert should make a diagnosis, symptom checks may put patients at ease and provide advice on how to proceed. When it comes to health-related news items, you can rely on Behind the Headlines for objective, evidence-based analysis. Examining the primary sources, the analysts at BTH lay forth the facts in simple English (Madjido et al., 2019). For information on less common diseases, critical evaluation methods such as DiscERN may assess websites without quality assurances (Cuan-Baltazar et al., 2020). Since they are still in their infancy as a concept, generic patient decision aids do not yet exist for all diseases. In order to demonstrate how patients and caregivers may use the model to address information shortages, the SIKDA Database employs four scenarios. In the first case, all the healthcare provider or patient has to do is enter the name of the ailment, and the program will find an index phrase and all its synonyms automatically. Clinical databases like the US National Library of Medicine have developed indexes called MeSH (Dai et al., 2020) specifically for this purpose, and they may be integrated into the program to accomplish this goal. Overcoming low SIKDA and making the search more robust might be achieved via the use of thesaurus mapping, in which the computer automatically detects similar words by mapping them to one index phrase (Pertiwi & Nugrahaeni, 2020). After the caretaker or patient presses the search button, the program will scour NHS Choices and The Information Standard for reputable patient information goods and support organizations. A public librarian will be notified of the patient's or caregiver's request if they want to use the "Ask a Librarian" feature. By creating a central hub, the SIKDA Database hopes to connect patients and their families with reliable health resources and librarians who can answer their questions. By providing symptom checkers, unbiased analysis of health stories, and generic patient decision aids, the system aims to help patients and caregivers find the information they need more quickly. In the rapidly evolving healthcare landscape, the emphasis on patient engagement has emerged as a pivotal factor in enhancing overall healthcare outcomes (Ottosen et al., 2019). Although the trend towards actively involving patients in their healthcare decisions is admirable, it comes with a complex web of difficulties, particularly given the vast amount of online health information (Vassilakaki et al., 2023).

Within the contemporary healthcare setting, there is a discernible shift towards recognizing the value of patient engagement (Esmaeilzadeh et al., 2021). Acknowledging that informed and active patients contribute to improved healthcare outcomes has prompted a reevaluation of traditional healthcare dynamics (Tai-Seale et al., 2019). Empowering patients to play a more active role in decision-making fosters a sense of ownership over one's health journey. In this digital age, the internet is a vast repository of health-related information, offering unprecedented access to a wealth of data. However, this abundance also presents a conundrum wherein individuals navigate an overwhelming sea of information (Robert et al., 2021). The sheer volume and diversity of health-related content on the internet create a challenge for patients striving to make informed decisions about their healthcare. Amidst the wealth of online health information, the critical aspect of patient empowerment lies in the ability to discern reliable and evidence-based information. Patients are faced with distinguishing between credible sources and potentially misleading or inaccurate content (Andersson et al., 2022). This need for discernment is integral to ensuring that the information upon which healthcare decisions are based is trustworthy and aligned with best practices. This study aims to shed light on the difficulties of collecting trustworthy health data and the need for patient involvement in decision-making. The study's overarching goal is to aid in the creation of a solution-oriented model by investigating the difficulties of patient empowerment in the context of the rapidly evolving digital health information ecosystem. This section underscores the importance of understanding the dynamic relationship between patient engagement and decision-making and the formidable challenges posed by the information age. It sets the stage for the subsequent exploration of the SIKDA model, emphasizing the necessity of innovative approaches to navigate these challenges and empower individuals to make informed healthcare decisions (Basri et al., 2021). At the heart of this chapter is the introduction of SIKDA, a groundbreaking model meticulously crafted to address a fundamental challenge in contemporary healthcare—the disparity between patients' aspirations for active participation in decision-making and the availability of reliable resources to facilitate this engagement (Menear et al., 2022). SIKDA transcends the conventional tool definition; it embodies a transformative approach to patient empowerment. In an effort to change the way healthcare decisions are made, SIKDA combines evidence-based patient engagement with information distribution rather than just giving a way to access health records. It gets that giving people knowledge and helping them develop the skills to make good use of it are two parts of empowering them (Akbar et al., 2020). Providing patients and their caregivers with more agency is central to SIKDA's mission. Because healthcare decisions are inherently a team effort, SIKDA broadens its scope to include persons who are actively helping others through the maze of healthcare options. This way, it recognizes that patients and their support systems are part of a larger decision-making ecosystem. Integrating evidence-based patient engagement is a crucial feature of SIKDA's design. With data and research taking centre stage in healthcare decision-making, SIKDA checks that all data is up-to-date and in line with the best standards (Yovadiani et al., 2021). If we want people to make decisions based on up-to-date, relevant information, this integration is vital. As a critical change agent, SIKDA is especially important in the rutal Indonesian setting, where healthcare possibilities and difficulties may vary from other areas. By taking into account the specifics of rural Indonesian healthcare, SIKDA aspires to provide a personalized solution that speaks to the hopes and fears of both patients and caregivers in this context. What makes SIKDA so important is its dedication to closing the gap, making sure that people want to be involved in their healthcare choices and have reliable tools to help them do so (Wattimena et al., 2020). This model, thus, becomes a beacon for individuals seeking a reliable pathway through the complex terrain of health information. SIKDA is positioned not just as a solution but as a pioneering initiative charting the course for informed decision-making in healthcare. By integrating patient empowerment and evidence-based practices, SIKDA stands as a transformative force, promising to reshape the healthcare landscape in Indonesia and beyond. This section sets the stage for a deeper exploration of SIKDA's methodology and the evidence supporting its efficacy in the subsequent chapters. The literature review dissected the multifaceted aspects of patient participation, shedding light on the evolving dynamics between healthcare providers and patients (Hidayat et al., 2020). We learned about the many levels of patient participation, how patients' roles in decision-making are changing, and what this means for healthcare outcomes from a variety of sources. This nuanced exploration forms the bedrock upon which SIKDA is built, ensuring a model that resonates with healthcare consumers' diverse aspirations and needs. Simultaneously, the literature review scrutinized the challenges inherent in the digital age's vast expanse of health information (Aryanto et al., 2020). From misinformation pitfalls to difficulties discerning credible sources, the review meticulously documented the hurdles patients face when navigating the online health information landscape (Senyoni, 2021). This thorough analysis became the basis for devising strategies within SIKDA to address these challenges head-on, ensuring that the model serves as a solution to the identified problems (Sitompul et al., 2019). Building on the insights from the literature review, the research established a robust SIKDA framework. This framework offers a systematic and all-encompassing approach by combining the five-stage process of evidence-based practice with the SIKDA Seven Pillars of Health Information Literacy. It equips individuals with the necessary skills to access health information and critically assess its reliability and relevance in evidence-based decision-making. Incorporating the SIKDA Seven Pillars within the Health Information Literacy framework ensures that individuals engaging with SIKDA possess a holistic set of information competencies. From understanding the information needs to reflecting on one's information-seeking behaviour, these pillars lay the groundwork for an empowered and discerning health information consumer (Sulistyawati et al., 2023). Complementing the Seven Pillars, integrating the five-stage process of evidence-based practice provides a methodological guide for individuals navigating health decisions within the Health Information Literacy framework. From formulating specific questions to critically appraising information, this process instils a systematic approach, fostering informed decision-making grounded in the latest evidence. The amalgamation of the literature review's insights and the Health Information Literacy framework forms a structured foundation for the creation of SIKDA. This methodological approach ensures that the model is informed by existing knowledge and equipped to address the identified challenges and nuances in patient participation and health information access. SIKDA stands at the forefront of consumer health information models, transcending traditional approaches to offer groundbreaking solutions. In a digital age where information is abundant but discerning quality remains challenging, SIKDA serves as a beacon of innovation, aiming to redefine how individuals' access and utilize health information. At the core of SIKDA's innovation is its role as a centralized platform, providing users with a singular access point to credible, up-to-date, and evidence-based health information (Fahmi et al., 2020). This strategic consolidation alleviates the overwhelming nature of navigating multiple sources, ensuring that individuals can seamlessly access the information they need for informed decision-making. SIKDA recognizes that access to information is only part of the equation. The concept takes it a step further by including librarian help, which is crucial since it connects people to the vast amounts of health information that is already accessible. When it comes to helping patrons navigate the platform and learn how to make good use of the information they find there, librarians are indispensable. One important part of health information access, interpretation and usage may be addressed by including library assistance inside SIKDA. Users are equipped with the necessary knowledge and abilities to not only receive information, but also to evaluate, comprehend, and apply it to their own healthcare needs (Ramadani, 2019). A culture of well-informed decision-making cannot be nurtured without this kind of agency. SIKDA's all-encompassing strategy for health literacy is what makes it so revolutionary. It ensures that the model's advantages are available to a varied audience by combining centralized access with librarian help, which caters to persons with various degrees of digital literacy. SIKDA becomes more than a tool; it evolves into an educational resource, enhancing health literacy across the spectrum. In a world inundated with information, SIKDA solves the challenge of information overload. By curating content, ensuring its credibility, and guiding users through the information landscape, SIKDA streamlines the process, making health information not only accessible but also manageable for individuals seeking to make informed healthcare decisions (Musyarofah et al., 2020). SIKDA is more than a model—it catalyzes informed decision-making. SIKDA enables users to navigate the complex healthcare landscape confidently by offering a central hub with support from knowledgeable professionals. This transformative consumer health information model paves the way for a future where individuals have access to information and possess the skills to wield that information effectively for their health and well-being. The subsequent sections will delve into SIKDA's collaborative and inclusive nature, exploring its potential impact on diverse demographics and minimizing costs for public services. SIKDA's strength lies in its collaborative ethos, strategically leveraging the wealth of existing resources the NHS provides. By integrating with established online platforms, SIKDA taps into a reservoir of credible information, ensuring users access a comprehensive array of health-related content. This collaboration not only bolsters the reliability of SIKDA but also aligns it with the broader healthcare infrastructure. One of SIKDA's notable strengths is its commitment to inclusivity. Recognizing the diverse digital literacy levels within the population, SIKDA is designed to be accessible to individuals with varying degrees of technological proficiency (Hasanah et al., 2021). This openness guarantees that SIKDA's advantages reach anyone who may have trouble accessing complicated online health information systems, rather than just the tech-savvy. The digital domain is given a human touch via partnerships with medical and public librarians. In the SIKDA model, librarians play a crucial role as trusted information specialists. Their knowledge and experience improve the overall user experience and help people, particularly those with less familiarity with digital interfaces, connect with technology. Beyond its user interface, SIKDA is a collaborative tool that helps public services save money without sacrificing effectiveness. Instead of building whole new platforms, SIKDA makes use of what the NHS already has. In the long run, this helps the healthcare system save money and makes SIKDA a more viable option for patients (Utari et al., 2022). When it comes to health information, SIKDA's collaborative and inclusive approach is right on target. Through the elimination of obstacles and the maintenance of an intuitive interface, SIKDA transforms into a tool that meets the varied healthcare requirements of the whole community. This tool is designed to empower people to take care of their health, and it is not confined to any certain population. In addition to empowering individuals, SIKDA promotes a community-centred approach to healthcare. A mutually beneficial partnership between digital platforms and human knowledge may be formed via collaboration with librarians, who can serve as a support system. Because of these interdependencies, SIKDA is likely to become a resource that helps the healthcare community share knowledge and work together. SIKDA is poised to revolutionize the healthcare environment due to its collaborative and inclusive character. Collaborating with NHS resources and librarians and making sure that varied user groups can access healthcare information makes SIKDA more than just a model—it becomes a comprehensive solution (Afrizal et al., 2019). In what follows, we will take a closer look at how SIKDA's collaborative design may affect healthcare results in Indonesia and how people there make well-informed decisions. Extensive research has culminated in SIKDA, which is ready to set a new standard for healthcare in Indonesia. This revolutionary methodology is more than just a technical advancement; it heralds a sea change in the way people interact with healthcare data, make choices, and, in the end, manage their health. By tackling patient engagement, healthcare outcomes, and overall well-being, SIKDA's multifaceted approach ushers in a new era in healthcare. It surpasses traditional methods by including patient goals, SIKDA, and teamwork in a seamless manner. This makes SIKDA a comprehensive answer that speaks to people's varied wants and requirements. Raising patient participation to a new level is crucial to this paradigm change. With SIKDA, people are no longer relegated to the role of information receiver and may instead take an active role in making decisions about their healthcare (Ocviyanti & Sari, 2021). Patients are empowered to take an active role in their healthcare decisions when they have access to information that is clear, accurate, and easy to grasp, thanks to SIKDA. The evidence-based basis of SIKDA directly impacts healthcare results. Decisions are grounded on the latest research and data and are in line with industry standards, thanks to SIKDA. Improvements in healthcare quality may be achieved by a dedication to making decisions based on evidence. The impact of SIKDA extends beyond the therapeutic domain and enhances the general welfare of the people. In order to improve preventative care, mental health, and general well-being, SIKDA works to increase health literacy so that people may make informed lifestyle choices. An individual's health encompasses more than just their physical condition, and this holistic approach recognizes that. SIKDA stands out as a trailblazing endeavour that brings together patient engagement, SIKDA, and teamwork in the age of digital healthcare (Meiryani et al., 2022). It represents a sea change from conventional healthcare approaches to ones that take use of digital technology's possibilities, and it is proof that new ideas can solve old problems. This study presents evidence that novel models like SIKDA are critically needed in the modern healthcare system. Transformative solutions are required to address the difficulties of information overload, patient engagement complexity, and healthcare consumers' growing expectations. SIKDA is a model that other healthcare systems throughout the world might follow to meet the challenges of the digital era. SIKDA lays the groundwork for a new kind of healthcare delivery in Indonesia and paves the way for more patient-centred, data-driven healthcare systems throughout the world. By embracing technology, cooperation, and a patient-centric mindset, SIKDA plots a road toward a healthcare environment where people are not simply beneficiaries but active contributors to their well-being (Dilla et al., 2022). This chapter opens the door to investigating how SIKDA might be put into practice, looking at how it could affect healthcare accessibility, patient outcomes, and the health ecosystem in Indonesia as a whole.

## 4. SOLUTIONS AND RECOMMENDATIONS

Given these results, one possible course of action is to push for national health policies to include SIKDA or comparable models. Policymakers should consider supporting initiatives that promote collaborative healthcare, enhance consumer health information accessibility, and empower patients to engage in their healthcare decisions actively. This can contribute to a more patient-centric and efficient healthcare system. In light of the transformative potential of SIKDA and its implications for healthcare, concrete policy solutions and recommendations are imperative for its successful implementation.

1. Integration into National Health Strategies: Advocate for integrating SIKDA into national health strategies, emphasizing its role in enhancing patient empowerment and information. Embedding SIKDA within broader health policies ensures sustained support and resources, facilitating seamless incorporation into existing healthcare frameworks.

2. Incentivizing Collaborative Care Models: Implement policies that incentivize collaborative care models, encouraging healthcare providers to actively engage with librarians in supporting patients through SIKDA. By recognizing and rewarding collaborative efforts, healthcare professionals are more likely to participate actively in the SIKDA model, fostering a culture of interdisciplinary engagement.

3. Health Literacy Education Programs: Introduce health literacy education programs in schools and community settings, emphasizing SIKDA as a tool for informed decision-making. Starting health literacy education promotes a proactive approach to healthcare, and incorporating SIKDA can equip individuals with the skills needed to navigate the healthcare system's complexities.

4. Continuous Training for Healthcare Professionals: Mandate ongoing training for healthcare professionals, ensuring they are proficient in utilizing SIKDA and effectively guiding patients through the platform. Regular training addresses potential barriers to adoption and ensures that healthcare providers remain adept at leveraging SIKDA to benefit their patients.

5. Public Awareness Campaigns: Launch public awareness campaigns highlighting the benefits of SIKDA, its user-friendly interface, and the role of librarians in supporting healthcare decisions. Increasing public awareness fosters a sense of trust in the SIKDA model and encourages individuals to actively seek information through the platform, contributing to improved health outcomes.

6. Research Funding for Implementation Studies: Allocate research funding specifically for implementation studies evaluating the effectiveness of SIKDA in diverse healthcare settings. Investing in research ensures that the real-world impact of SIKDA is rigorously assessed, providing valuable insights for continuous improvement and scalability. By incorporating these policy solutions and recommendations, stakeholders can actively contribute to successfully integrating SIKDA into healthcare practices, fostering a more informed, collaborative, and patient-centric healthcare ecosystem. Beyond the research realm, the successful integration of SIKDA into real-world settings necessitates innovative and practical solutions across clinical, educational, and community contexts.

7. Clinical Decision Support Integration: Integrate SIKDA as a clinical decision support tool within electronic health record systems, ensuring seamless access to reliable consumer health information during patient consultations. Embedding SIKDA directly into clinical workflows empowers healthcare professionals to utilize the platform as an integral part of their decision-making processes, enhancing its practical utility.

8. Patient-Centric Workshops in Healthcare Institutions: Conduct workshops within healthcare institutions, educating patients and healthcare providers on the features and benefits of SIKDA. Encourage collaborative exercises to simulate real-world usage scenarios. Hands-on workshops foster familiarity

with SIKDA, encourage participation from healthcare professionals and patients and provide a practical understanding of its utility.

9. Incorporating SIKDA into the Medical Curriculum: Medical and nursing schools should include SIKDA training modules in their programs to ensure that aspiring doctors and nurses know how to use the platform for patient education and decision support. By introducing SIKDA during formal education, healthcare professionals will be equipped with the necessary skills to leverage the platform effectively in their future practice.

10. Community Health Literacy Events: Organize community health literacy events where SIKDA is introduced and demonstrated. Provide access points for individuals to register and explore the platform with guidance from healthcare professionals. Community events directly link SIKDA and individuals seeking health information, promoting widespread adoption and empowering communities to engage actively in their healthcare.

11. Librarian-Facilitated Health Information Sessions: Collaborate with public libraries to host health information sessions facilitated by librarians. These sessions can showcase SIKDA and its resources, offering personalized assistance to community members. Leveraging librarians as facilitators ensures that individuals receive hands-on guidance in using SIKDA, reinforcing the librarian's role as a valuable resource in health information navigation.

12. Telehealth Integration with SIKDA: Integrate SIKDA into telehealth platforms, allowing healthcare professionals to share information with patients during virtual consultations seamlessly. In the era of telehealth, integrating SIKDA ensures that remote consultations maintain a robust informational component, fostering patient engagement and informed decision-making. These innovative solutions aim to bridge the gap between research and practical application, making SIKDA an integral part of healthcare delivery, education, and community engagement. By weaving these interventions into everyday practices, the transformative potential of SIKDA can be fully realized in diverse and dynamic healthcare environments.

## 5. FUTURE RESEARCH DIRECTIONS

Future research should delve into the practical implementation of SIKDA, assessing its effectiveness in diverse healthcare settings. Additionally, studies can explore the model's scalability to other regions and healthcare systems. Investigating the long-term impact on patient outcomes, healthcare provider practices, and the overall health ecosystem will further enrich our understanding of the transformative potential of consumer health information models like SIKDA. Highlighting Potential Research Directions. While this study lays the foundation for the transformative potential of SIKDA in healthcare, several avenues for future research can deepen our understanding and enhance its applicability in diverse and dynamic healthcare environments:

1. Longitudinal Impact Assessment: Conduct longitudinal studies to assess the sustained impact of SIKDA on patient outcomes, healthcare provider practices, and collaborative healthcare efforts over an extended period. Long-term assessments will provide insights into the durability and scalability of SIKDA's transformative effects.

2. User Experience and Accessibility Studies: Explore user experience and accessibility challenges by conducting in-depth studies on how different demographic groups interact with SIKDA. Investigate

factors influencing usage patterns and satisfaction levels. Understanding user dynamics will inform iterative improvements to enhance SIKDA's accessibility and inclusivity.

3. Integration with Chronic Disease Management: Investigate the integration of SIKDA into chronic disease management programs, assessing its effectiveness in supporting patients with long-term health conditions. Chronic disease management represents a significant healthcare challenge, and exploring SIKDA's role in this context can provide valuable insights.

4. Cross-Cultural Validation Studies: Conduct cross-cultural validation studies to assess the generalizability of SIKDA in diverse cultural and linguistic contexts beyond the Indonesian setting. Adapting SIKDA to different cultural nuances ensures its relevance and effectiveness in a global healthcare context.

5. Healthcare Professional Training Efficacy: Evaluate the efficacy of training programs for healthcare professionals in effectively utilizing SIKDA and supporting patients in information-seeking behaviours. Ensuring that healthcare professionals are proficient in leveraging SIKDA is crucial for its successful integration into clinical practice.

6. Cost-Benefit Analysis of SIKDA Implementation: Analyze the long-term and short-term financial effects of introducing SIKDA in healthcare systems via a thorough cost-benefit analysis. Understanding the economic feasibility of SIKDA implementation aids decision-makers in allocating resources strategically.

7. Exploration of Additional Collaborative Models: Explore additional collaborative models involving different healthcare stakeholders, such as community health workers, patient advocacy groups, and technology developers, to enhance the collaborative nature of SIKDA. Diversifying collaborative partnerships can contribute to a more holistic and practical healthcare ecosystem.

8. User-Generated Content and Feedback Analysis: Analyze user-generated content and feedback within SIKDA to understand how individuals contribute to the platform and the impact of user interactions on information quality and reliability. Incorporating user-generated content analysis ensures that SIKDA remains a dynamic and responsive platform that meets evolving user needs.

These research directions can collectively contribute to a more nuanced understanding of SIKDA's transformative potential and adaptability in varied healthcare settings. Exploring these areas will refine the model and advance our knowledge of how innovative consumer health information frameworks can positively impact healthcare outcomes.

## 6. CONCLUSION

The multifaceted exploration of tailored Seven Pillars of Health Information Literacy (SIKDA), a transformative consumer health information model, has unveiled its potential to revolutionize how individuals' access and utilize healthcare information. From addressing patient empowerment and decision-making challenges to outlining SIKDA's collaborative and inclusive nature, each section has illuminated the model's innovative features and potential impact on Indonesia's healthcare landscape.

## REFERENCES


Afrizal, S. H., Handayani, P. W., Hidayanto, A. N., Eryando, T., Budiharsana, M., & Martha, E. (2019). Barriers and challenges to Primary Health Care Information System (PHCIS) adoption from health management perspective: A qualitative study. *Informatics in Medicine Unlocked*, *17*, 100198. doi:10.1016/j.imu.2019.100198

Akbar, S., Coiera, E., & Magrabi, F. (2020). Safety concerns with consumer-facing mobile health applications and their consequences: A scoping review. *Journal of the American Medical Informatics Association : JAMIA*, *27*(2), 330–340. doi:10.1093/jamia/ocz175 PMID:31599936

Andersson, V., Sawatzky, R., & Öhlén, J. (2022). Relating person-centredness to quality-of-life assessments and patient-reported outcomes in healthcare: A critical theoretical discussion. *Nursing Philosophy*, *23*(3), e12391. doi:10.1111/nup.12391 PMID:35502530

Aryanto, K. Y. E., Abyong, I. W., Seputra, K. A., Pradnyana, G. A., & Paramartha, A. Y. (2020, January). Design of the Integrated Health Information System in the District of Bangli. In *3rd International Conference on Innovative Research Across Disciplines* (ICIRAD 2019) (pp. 240-245). Atlantis Press. https://doi.org/10.2991/assehr.k.200115.039

Basri, M. H., Farmaciawaty, D. A., Rachmania, I. N., Widjaja, F. B., & Wiyana, A. F. (2021, May). Healthcare Ecosystem Mapping in Primary Care: A Case Study in West Java, Indonesia. In *International Conference on Business and Engineering Management* (ICONBEM 2021) (pp. 184-190). Atlantis Press. https://doi.org/10.2991/aebmr.k.210522.025

Boonsiritomachai, W., & Pitchayadejanant, K. (2019). Determinants affecting mobile banking adoption by generation Y based on the Unified Theory of Acceptance and Use of Technology Model modified by the Technology Acceptance Model concept. *Kasetsart Journal of Social Sciences*, *40*(2), 349–358.

Bujnowska-Fedak, M. M., Waligóra, J., & Mastalerz-Migas, A. (2019). The internet as a source of health information and services. *Advancements and Innovations in Health Sciences*, 1-16. https://doi.org/ doi:10.1007/5584_2019_396

Cuan-Baltazar, J. Y., Muñoz-Perez, M. J., Robledo-Vega, C., Pérez-Zepeda, M. F., & Soto-Vega, E. (2020). Misinformation of COVID-19 on the internet: Infodemiology study. *JMIR Public Health and Surveillance*, *6*(2), e18444. doi:10.2196/18444 PMID:32250960

Dai, S., You, R., Lu, Z., Huang, X., Mamitsuka, H., & Zhu, S. (2020). FullMeSH: Improving large-scale MeSH indexing with full text. *Bioinformatics (Oxford, England)*, *36*(5), 1533–1541. doi:10.1093/bioinformatics/btz756 PMID:31596475

Daraz, L., Morrow, A. S., Ponce, O. J., Beuschel, B., Farah, M. H., Katabi, A., Alsawas, M., Majzoub, A. M., Benkhadra, R., Seisa, M. O., Ding, J. F., Prokop, L., & Murad, M. H. (2019). Can patients trust online health information? A meta-narrative systematic review addressing the quality of health information on the internet. *Journal of General Internal Medicine*, *34*(9), 1884–1891. doi:10.1007/s11606-019-05109-0 PMID:31228051

Davis, E. L., Kelly, P. J., Deane, F. P., Baker, A. L., Buckingham, M., Degan, T., & Adams, S. (2020). The relationship between patient-centered care and outcomes in specialist drug and alcohol treatment: A systematic literature review. *Substance Abuse*, *41*(2), 216–231. doi:10.1080/08897077.2019.1671940 PMID:31638870

Dilla, S. M., Tasnim, T., Amraeni, Y., & Selviana, A. (2022). Factors Relating to the Success of the Health Information System at the Buton District Health Center. *Waluya The International Science of Health Journal*, *1*(2), 29–46. doi:10.54883/wish.v1i2.8

Ducrot, P., Montagni, I., Nguyen Thanh, V., Serry, A. J., & Richard, J. B. (2021). Evolution of online health-related information seeking in France from 2010 to 2017: Results from nationally representative surveys. *Journal of Medical Internet Research*, *23*(4), e18799. doi:10.2196/18799 PMID:33851927

Dwivedi, Y. K., Rana, N. P., Jeyaraj, A., Clement, M., & Williams, M. D. (2019). Re-examining the unified theory of acceptance and use of technology (UTAUT): Towards a revised theoretical model. *Information Systems Frontiers*, *21*(3), 719–734. doi:10.1007/s10796-017-9774-y

Ennaceur, S., Bugis, B., & Al-Mohaithef, M. (2023). e-Health information literacy of health science university students: Actual situation and Improvement strategies. *Journal of Electronic Resources Librarianship*, *35*(2), 91–105. doi:10.1080/1941126X.2023.2197749

Esmaeilzadeh, P., Dharanikota, S., & Mirzaei, T. (2021). The role of patient engagement in patient-centric health information exchange (HIE) initiatives: An empirical study in the United States. *Information Technology & People*. Advance online publication. doi:10.1108/ITP-05-2020-0316

Fahmi, A. S., & Ayuningtyas, D. (2020, November). Readiness Analysis of Electronic Information System in Public Health Center as Supporting Tools in Implementation of Fraud Prevention System on UHC Program: Literature Review. In *International Conference of Health Development. Covid-19 and the Role of Healthcare Workers in the Industrial Era* (ICHD 2020) (pp. 100-109). Atlantis Press. https://doi.org/10.2991/ahsr.k.201125.019

Fauzan, A., & Noviandi, N. (2020). Evaluation of Optima Regional Health Information System with HOT-Fit on Technology Aspects Approach in Johar Baru Health Center Jakarta. [JICHI]. *Journal of Intelligent Computing and Health Informatics*, *1*(1), 9–14. doi:10.26714/jichi.v1i1.5397

Gille, S., Griese, L., & Schaeffer, D. (2021). Preferences and experiences of people with chronic illness in using different sources of health information: Results of a mixed-methods study. *International Journal of Environmental Research and Public Health*, *18*(24), 13185. doi:10.3390/ijerph182413185 PMID:34948792

Harahap, N. C., Handayani, P. W., & Hidayanto, A. N. (2021). Barriers in health information systems and technologies to support maternal and neonatal referrals at primary health centers. *Healthcare Informatics Research*, *27*(2), 153–161. doi:10.4258/hir.2021.27.2.153 PMID:34015881

Hasanah, A., Rangkuti, D. N., Pinem, G. Z. D., & Nasution, F. (2021). Evaluation of the Health Information System in Teladan Public Health Center in 2020 Using the Health Metrics Network Instrument Approach. *International Archives of Medical Sciences and Public Health*, 2(1).

Hidayat, I. F., & Hermanto, B. R. (2020, October). A preliminary implementation of hl7 fhir to achieve interoperability in indonesia's local ehr. In 2020 *27th International Conference on Telecommunications* (ICT) (pp. 1-5). IEEE. 10.1109/ICT49546.2020.9239534

Jayatilake, S. M. D. A. C., & Ganegoda, G. U. (2021). Involvement of machine learning tools in healthcare decision making. *Journal of Healthcare Engineering, 2021*, 1–20. Advance online publication. doi:10.1155/2021/6679512 PMID:33575021

Kujala, S., Hörhammer, I., Väyrynen, A., Holmroos, M., Nättiaho-Rönnholm, M., Hägglund, M., & Johansen, M. A. (2022). Patients' Experiences of Web-Based Access to Electronic Health Records in Finland: Cross-sectional Survey. *Journal of Medical Internet Research, 24*(6), e37438. doi:10.2196/37438 PMID:35666563

Lee, H. Y., Jin, S. W., Henning-Smith, C., Lee, J., & Lee, J. (2021). Role of health literacy in health-related information-seeking behavior online: Cross-sectional study. *Journal of Medical Internet Research, 23*(1), e14088. doi:10.2196/14088 PMID:33502332

Madjido, M., Espressivo, A., Maula, A. W., Fuad, A., & Hasanbasri, M. (2019). Health information system research situation in Indonesia: A bibliometric analysis. *Procedia Computer Science, 161*, 781–787. doi:10.1016/j.procs.2019.11.183

Meiryani, M., Aliffiyah, L., Lusianah, L., Lidiyawati, L., Lesmana, T., & Kriswanto, K. (2022, June). Benefits of Information Technology in Improving the Quality of Health Services. In *Proceedings of the 2022 6th International Conference on E-Education, E-Business and E-Technology* (pp. 97-101). ACM.10.1145/3549843.3549858

Menear, M., Girard, A., Dugas, M., Gervais, M., Gilbert, M., & Gagnon, M. P. (2022). Personalized care planning and shared decision making in collaborative care programs for depression and anxiety disorders: A systematic review. *PLoS One, 17*(6), e0268649. doi:10.1371/journal.pone.0268649 PMID:35687610

Musyarofah, S., Muliawati, R., & Mushidah, M. (2020, July). Health Service Information System in Kendal District toward Kendal Smart City. In Proceedings of the Third Andalas International Public Health Conference, AIPHC 2019, Padang, West Sumatera, Indonesia. 10.4108/eai.9-10-2019.2297244

Ocviyanti, D., & Sari, J. M. (2021). Analysis of antenatal care quality in cases of referred pregnant women in emergency rooms based on MCH book records. *Indonesian Journal of Obstetrics and Gynecology, 121-125*, 121–125. doi:10.32771/inajog.v9i3.1339

Ottosen, T., Mani, N. S., & Fratta, M. N. (2019). Health information literacy awareness and capacity building: Present and future. *IFLA Journal, 45*(3), 207–215. doi:10.1177/0340035219857441

Pertiwi, J., & Nugrahaeni, A. (2020). Examination of Technology Acceptance Model on Attitude Toward SIHA in Wonosobo Indonesia. *International Proceedings The 2nd ISMoHIM* 2020. IEEE. doi:10.32585/jikemb.v2i1.810

Ramadani, L. (2019). *Health Information Systems Enabled Transformation of Service Ecosystems: The Case of Indonesian Healthcare*.

Robert, G., Donetto, S., & Williams, O. (2021). Co-designing healthcare services with patients. The Palgrave Handbook of Co-Production of Public Services and Outcomes, 313-333. https://doi.org/doi:10.1007/978-3-030-53705-0_16

Senyoni, W. F. (2021). *Institutional Work in Strengthening Health Information Systems in Fragmented Settings: An Action Research Study of Information Dashboards in East African Community and Indonesia*.

Shukla, R. (2021). The SCONUL Seven Pillars of Information Literacy: Core Model to Test the Skills of LIS Students. Library Philosophy and Practice, 5851, University of Jammu, India..

Siste, K., Hanafi, E., Sen, L. T., Christian, H., Adrian, Siswidiani, L. P., Limawan, A. P., Murtani, B. J., & Suwartono, C. (2020). The impact of physical distancing and associated factors towards internet addiction among adults in Indonesia during COVID-19 pandemic: A nationwide web-based study. *Frontiers in Psychiatry*, *11*, 580977. doi:10.3389/fpsyt.2020.580977 PMID:33101092

Song, T., Liu, F., Deng, N., Qian, S., Cui, T., Guan, Y., Arnolda, L., Zhang, Z., & Yu, P. (2021). A comprehensive 6A framework for improving patient self-management of hypertension using mHealth services: Qualitative thematic analysis. *Journal of Medical Internet Research*, *23*(6), e25522. doi:10.2196/25522 PMID:34152272

Sulistyawati, S., Wibowo, T. A., Rokhmayanti, R., Nugroho, A. S. D., Sukesi, T. W., Hastuti, S. K. W., Mulasari, S. A., & Feletto, M. (2023). Introduction and implementation of an immunization information system in the Indonesian province of Daerah Istimewa Yogyakarta: Lessons for scaling-up. *BMC Health Services Research*, *23*(1), 12. doi:10.1186/s12913-022-08910-6 PMID:36604659

Swire-Thompson, B., & Lazer, D. (2020). Public health and online misinformation: Challenges and recommendations. *Annual Review of Public Health*, *41*(1), 433–451. doi:10.1146/annurev-publhealth-040119-094127 PMID:31874069

Tai-Seale, M., Downing, N. L., Jones, V. G., Milani, R. V., Zhao, B., Clay, B., Sharp, C. D., Chan, A. S., & Longhurst, C. A. (2019). Technology-enabled consumer engagement: Promising practices at four health care delivery organizations. *Health Affairs*, *38*(3), 383–390. doi:10.1377/hlthaff.2018.05027 PMID:30830826

Tamilmani, K., Rana, N. P., Wamba, S. F., & Dwivedi, R. (2021). The extended Unified Theory of Acceptance and Use of Technology (UTAUT2): A systematic literature review and theory evaluation. *International Journal of Information Management*, *57*, 102269. doi:10.1016/j.ijinfomgt.2020.102269

Utari, S. S., Madhumala, R. B., Haryanti, S. C., Rachmawati, U. A., & Atmoko, S. P. U. (2022, December). Enhancing Security on Indonesian E-health Cloud. In *Second International Conference on Advanced Technologies in Intelligent Control, Environment, Computing & Communication Engineering* (ICATIECE) (pp. 1-3). IEEE. 10.1109/ICATIECE56365.2022.10046767

van Schalkwyk, M. C., Bourek, A., Kringos, D. S., Siciliani, L., Barry, M. M., De Maeseneer, J., & McKee, M. (2020). The best person (or machine) for the job: Rethinking task shifting in healthcare. *Health Policy (Amsterdam)*, *124*(12), 1379–1386. doi:10.1016/j.healthpol.2020.08.008 PMID:32900551

Vassilakaki, E., & Moniarou-Papaconstaninou, V. (2023). Librarians' support in improving health literacy: A systematic literature review. *Journal of Librarianship and Information Science*, *55*(2), 500–514. doi:10.1177/09610006221093794

Verawati, F., Abidin, Z., Renaldi, R., Yanthi, D., & Widodo, M. D. (2022). Analysis of the Implementation of the Generic Regional Health Information System (SIKDA) at Simpang Tiga Health Center Pekanbaru City: Analisis Pelaksanaan Sistem Informasi Kesehatan Daerah (SIKDA) Generik Di Puskesmas Simpang Tiga Kota Pekanbaru. *Jurnal Olahraga dan Kesehatan* (ORKES), *1*(2), 229-240. doi:10.56466/orkes/Vol1.Iss2.7

Wattimena, E. S. N., Razak, A., Pallutturi, S., Syafar, D. M., Saleh, L. M., & Rahmadani, S. (2020). Analysis Health Information Systems Quality on the Effectiveness of Public Health Center Officers in Ambon City. *Medico-Legal Update*, *20*(3).

Withorn, T., Eslami, J., Lee, H., Clarke, M., Caffrey, C., Springfield, C., Ospina, D., Andora, A., Castañeda, A., Mitchell, A., Kimmitt, J. M., Vermeer, W., & Haas, A. (2021). Library instruction and information literacy 2020. *RSR. Reference Services Review*, *49*(3/4), 329–418. doi:10.1108/RSR-07-2021-0046

Yovadiani, A., Ramadani, L., & Dewi, F. (2021). Toward National One Health Data: Reconfiguring Local Health Information Systems through Enterprise Architecture. Ultima InfoSys. *Jurnal Ilmu Sistem Informasi*, *12*(2), 101–107. doi:10.31937/si.v12i2.2274

## ADDITIONAL READINGS

Komalasari, R. (2022). A Social Ecological Model (SEM) to Manage Methadone Programmes in Prisons. In Handbook of Research on Mathematical Modeling for Smart Healthcare Systems (pp. 374-382). IGI Global. doi:10.4018/978-1-6684-4580-8.ch020

Komalasari, R. (2022). A Social Ecological Model (SEM) to Manage Methadone Programmes in Prisons. In Handbook of Research on Mathematical Modeling for Smart Healthcare Systems (pp. 374-382). IGI Global. doi:10.4018/978-1-6684-4580-8.ch020

Komalasari, R. (2023). The Relationship Between Cybersecurity and Public Health. In *Handbook of Research on Current Trends in Cybersecurity and Educational Technology* (pp. 78–91). IGI Global., doi:10.4018/978-1-6684-6092-4.ch005

Komalasari, R. (2023). Postnatal Mental Distress: Exploring the Experiences of Mothers Navigating the Healthcare System. In Perspectives and Considerations on Navigating the Mental Healthcare System (pp. 159-181). IGI Global. doi:10.4018/978-1-6684-5049-9.ch007

Komalasari, R. (2023). Designing Health Systems for Better, Faster, and Less Expensive Treatment. In *Exploring the Convergence of Computer and Medical Science Through Cloud Healthcare* (pp. 1–13). IGI Global. doi:10.4018/978-1-6684-5260-8.ch001

Komalasari, R. (2023). The Ethical Consideration of Using Artificial Intelligence (AI) in Medicine. In Advanced Bioinspiration Methods for Healthcare Standards, Policies, and Reform (pp. 1-16). IGI Global. doi:10.4018/978-1-6684-5656-9.ch001

Komalasari, R. (2023). Healthcare for the Elderly With Digital Twins. In Digital Twins and Healthcare: Trends, Techniques, and Challenges (pp. 145-156). IGI Global. doi:10.4018/978-1-6684-5925-6.ch010

Komalasari, R. (2023). Cloud Computing's Usage in Healthcare. In Recent Advancements in Smart Remote Patient Monitoring, Wearable Devices, and Diagnostics Systems (pp. 183-194). IGI Global. doi:10.4018/978-1-6684-6434-2.ch009

Komalasari, R. (2023). History and Legislative Changes Governing Medical Cannabis in Indonesia. In Medical Cannabis and the Effects of Cannabinoids on Fighting Cancer, Multiple Sclerosis, Epilepsy, Parkinson's, and Other Neurodegenerative Diseases (pp. 274-284). IGI Global. doi:10.4018/978-1-6684-5652-1.ch012

Komalasari, R. (2023). Ambient Assisted Living (AAL) Systems to Help Older People. In Exploring Future Opportunities of Brain-Inspired Artificial Intelligence (pp. 84-99). IGI Global. doi:10.4018/978-1-6684-6980-4.ch006

Komalasari, R. (2023). Telemedicine in Pandemic Times in Indonesia: Healthcare Professional's Perspective. In Health Informatics and Patient Safety in Times of Crisis (pp. 138-153). IGI Global. doi:10.4018/978-1-6684-5499-2.ch008

Komalasari, R. (2023). Cloud Computing's Usage in Healthcare. In Recent Advancements in Smart Remote Patient Monitoring, Wearable Devices, and Diagnostics Systems (pp. 183-194). IGI Global. doi:10.4018/978-1-6684-6434-2.ch009

Komalasari, R. (2023). The Relationship Between Cybersecurity and Public Health. In *Handbook of Research on Current Trends in Cybersecurity and Educational Technology* (pp. 78–91). IGI Global. doi:10.4018/978-1-6684-6092-4.ch005

Komalasari, R. (2023). Postnatal Mental Distress: Exploring the Experiences of Mothers Navigating the Healthcare System. In Perspectives and Considerations on Navigating the Mental Healthcare System (pp. 159-181). IGI Global. doi:10.4018/978-1-6684-5049-9.ch007

Komalasari, R. (2023). Designing Health Systems for Better, Faster, and Less Expensive Treatment. In *Exploring the Convergence of Computer and Medical Science Through Cloud Healthcare* (pp. 1–13). IGI Global. doi:10.4018/978-1-6684-5260-8.ch001

Komalasari, R. (2023). The Ethical Consideration of Using Artificial Intelligence (AI) in Medicine. In Advanced Bioinspiration Methods for Healthcare Standards, Policies, and Reform (pp. 1-16). IGI Global. doi:10.4018/978-1-6684-5656-9.ch001

Komalasari, R. (2023). Healthcare for the Elderly With Digital Twins. In Digital Twins and Healthcare: Trends, Techniques, and Challenges (pp. 145-156). IGI Global. doi:10.4018/978-1-6684-5925-6.ch010

Komalasari, R. (2023). Cloud Computing's Usage in Healthcare. In Recent Advancements in Smart Remote Patient Monitoring, Wearable Devices, and Diagnostics Systems (pp. 183-194). IGI Global. doi:10.4018/978-1-6684-6434-2.ch009

Komalasari, R. (2023). History and Legislative Changes Governing Medical Cannabis in Indonesia. In Medical Cannabis and the Effects of Cannabinoids on Fighting Cancer, Multiple Sclerosis, Epilepsy, Parkinson's, and Other Neurodegenerative Diseases (pp. 274-284). IGI Global. doi:10.4018/978-1-6684-5652-1.ch012

Komalasari, R. (2023). Ambient Assisted Living (AAL) Systems to Help Older People. In Exploring Future Opportunities of Brain-Inspired Artificial Intelligence (pp. 84-99). IGI Global. doi:10.4018/978-1-6684-6980-4.ch006

Komalasari, R. (2023). Telemedicine in Pandemic Times in Indonesia: Healthcare Professional's Perspective. In Health Informatics and Patient Safety in Times of Crisis (pp. 138-153). IGI Global. doi:10.4018/978-1-6684-5499-2.ch008

Komalasari, R. (2023). Cloud Computing's Usage in Healthcare. In Recent Advancements in Smart Remote Patient Monitoring, Wearable Devices, and Diagnostics Systems (pp. 183-194). IGI Global. doi:10.4018/978-1-6684-6434-2.ch009

Komalasari, R. (2023). The Relationship Between Cybersecurity and Public Health. In *Handbook of Research on Current Trends in Cybersecurity and Educational Technology* (pp. 78–91). IGI Global. doi:10.4018/978-1-6684-6092-4.ch005

Komalasari, R. (2023). Treatment of Menstrual Discomfort in Young Women and a Cognitive Behavior Therapy (CBT) Program. In Perspectives on Coping Strategies for Menstrual and Premenstrual Distress (pp. 194-211). IGI Global. doi:10.4018/978-1-6684-5088-8.ch011

Komalasari, R. (2023). Healthcare for the Elderly With Digital Twins. In Digital Twins and Healthcare: Trends, Techniques, and Challenges (pp. 145-156). IGI Global. doi:10.4018/978-1-6684-5925-6.ch010

Komalasari, R. (2023). Telemedicine in Pandemic Times in Indonesia: Healthcare Professional's Perspective. In Health Informatics and Patient Safety in Times of Crisis (pp. 138-153). IGI Global. doi:10.4018/978-1-6684-5499-2.ch008

Komalasari, R. (2023). Regulatory Shift Healthcare Applications in Industry 5.0. In Advanced Research and Real-World Applications of Industry 5.0 (pp. 149-165). IGI Global. doi:10.4018/978-1-7998-8805-5.ch008

Komalasari, R. (2023). Exploring Pedagogies Of Affect In Secondary School Physical Education For Enhanced Emotional Well-Being. *Pedagogik: Jurnal Pendidikan Guru Sekolah Dasar*, *11*(2), 117–127. doi:10.33558/pedagogik.v11i2.7374

Komalasari, R. (2023). Integrating Sport education model and the athletics challenges approach for transformative physical education in Indonesian Middle Schools. *Motion: Jurnal Riset Physical Education*, *13*(2), 118–135. doi:10.33558/motion.v13i2.7372

Komalasari, R. (2023). Rancangan Program Bagi Keberhasilan Gerakan Mencuci Tangan Yang Sehat Bagi Para Ibu Dan Anak. *Sulolipu: Media Komunikasi Sivitas Akademika dan Masyarakat*, *23*(1), 55-60. doi:10.32382/sulolipu.v23i1.3103

Komalasari, R. (2023). Culture as A Catalyst: Unveiling the Nexus Between Health Services Performance Management and National Identity. *Journal of Business and Political Economy: Biannual Review of The Indonesian Economy*, *5*(1).

Komalasari, R. (2023). Exploring Pedagogies Of Affect In Secondary School Physical Education For Enhanced Emotional Well-Being. *Pedagogik: Jurnal Pendidikan Guru Sekolah Dasar*, *11*(2), 117–127. doi:10.33558/pedagogik.v11i2.7374

Komalasari, R. (2023). Integrating sport education model and the athletics challenges approach for transformative physical education in Indonesian Middle Schools. *Motion: Jurnal Riset Physical Education*, *13*(2), 118–135. doi:10.33558/motion.v13i2.7372

Komalasari, R. (2023). Efek Ganja Medis pada Pasien Parkinson: A Literature Review of Clinical Evidence. *Journal of Islamic Pharmacy*, *8*(1), 44–48. doi:10.18860/jip.v8i1.17832

Komalasari, R. (2023). Rancangan Program Bagi Keberhasilan Gerakan Mencuci Tangan Yang Sehat Bagi Para Ibu Dan Anak. *Sulolipu: Media Komunikasi Sivitas Akademika dan Masyarakat*, 23(1), 55-60. doi:10.32382/sulolipu.v23i1.3103

Komalasari, R. (2024). Unravelling the Veil: Exploring the Nexus of Insecure Attachment and Functional Somatic Disorders in Adults. In *Discourses, Inquiries, and Case Studies in Healthcare, Social Sciences, and Technology*. IGI Global. doi:10.4018/979-8-3693-3555-0.ch001

Komalasari, R. (2024). Advancing Healthcare: Economic Implications of Immediate MRI in Suspected Scaphoid Fractures - A Comprehensive Exploration. In *Future of AI in Medical Imaging*. IGI Global. doi:10.4018/979-8-3693-2359-5.ch014

Komalasari, R. (2024). Women's perceptions of midlife mothering during perimenopause: the impact on health and well-being through life's transitions. In *Utilizing AI Techniques for the Perimenopause to Menopause Transition*. IGI Global.

Komalasari, R. (2024). Machine Learning in Health Information Security: Unraveling Patterns, Concealing Secrets, and Mitigating Vulnerabilities. In *Enhancing Steganography Through Deep Learning Approaches*. IGI Global.

Komalasari, R. (2024). Navigating Complexity: Unraveling the IVHM Requirements Puzzle in Unmanned Aerial Systems Through Innovative IVHM-RD Methodology and Rigorous Data Analysis. In *Ubiquitous Computing and Technological Innovation for Universal Healthcare*. IGI Global.

Komalasari, R. (2024). Trust Dynamics in Remote Patient-Expert Communication: Unraveling the Role of ICT in Indonesia's Private Healthcare Sector. In *Improving Security, Privacy, and Connectivity Among Telemedicine Platforms*. IGI Global. doi:10.4018/979-8-3693-2141-6.ch010

Komalasari, R. (2024). Cyberbullying in the Health Care Workplace: How to Find Your Way Through the Digital Maze. In *Workplace Cyberbullying and Behavior in Health Professions*. IGI Global. doi:10.4018/979-8-3693-1139-4.ch004

Komalasari, R. (2024). Harmonizing Minds and Machines: A Transformative AI-Based Mental Health Care Framework in Medical Tourism. In *Impact of AI and Robotics on the Medical Tourism Industry*. IGI Global. doi:10.4018/979-8-3693-2248-2.ch002

Komalasari, R. (2024). Navigating the Future of Healthcare: A User-Centric Approach to Designing Lucrative Business Models for the IoMT. In *Lightweight Digital Trust Architectures in the Internet of Medical Things (IoMT)*. IGI Global.

Komalasari, R. (2024). Biospheric Reverie: Unraveling Indoor Air Quality through Bio-Inspired Textiles, Awareness, and Decision-Making. In *Intelligent Decision Making Through Bio-Inspired Optimization*. IGI Global. doi:10.4018/979-8-3693-2073-0.ch013

Komalasari, R. (2024). An Integrated Approach to Next-Generation Telemedicine and Health Advice Systems through AI Applications in Disease Diagnosis. In *AI-Driven Innovations in Digital Healthcare: Emerging Trends, Challenges, and Applications*. IGI Global.

Komalasari, R. (2024). Transformative Insights: Harnessing Artificial Intelligence for Enhanced Ovarian Cancer Prediction and Prognosis. In *Biomedical Research Developments for Improved Healthcare*. IGI Global. doi:10.4018/979-8-3693-1922-2.ch010

Komalasari, R. (2024). Bridging the Gap: Theory of Change Guided Digital Health Implementation in Indonesian Primary Care. In *Analyzing Current Digital Healthcare Trends Using Social Networks*. IGI Global. doi:10.4018/979-8-3693-1934-5.ch002

Komalasari, R. (2024). Navigating the Digital Frontier: A Socio-Technical Review of Indonesia's NHS E-Health Strategy and the Path to Seamless Healthcare Transformation. In *Inclusivity and Accessibility in Digital Health*. IGI Global. doi:10.4018/979-8-3693-1463-0.ch012

Komalasari, R. (2024). *Pursuit of Excellence: A Comprehensive Framework for Ensuring Quality in Digital Health Apps. In Multi-Sector Analysis of the Digital Healthcare Industry*. IGI Global. doi:10.4018/979-8-3693-0928-5.ch005

Komalasari, R. (2024). Fostering Transparency in AI for Disease Detection: Exploring Meaningful Explanations and Interactive Modalities in Healthcare. In *Federated Learning and Privacy-Preserving in Healthcare AI*. IGI Global.

Komalasari, R. (2024). Predictive Pioneers: Bridging the Mental Health Gap in Older Adults through Advanced Digital Technologies. In *Using Machine Learning to Detect Emotions and Predict Human Psychology*. IGI Global. doi:10.4018/979-8-3693-1910-9.ch005

Komalasari, R. (2024). Unravelling the Veil: Exploring the Nexus of Insecure Attachment and Functional Somatic Disorders in Adults. In *Discourses, Inquiries, and Case Studies in Healthcare, Social Sciences, and Technology*. IGI Global. doi:10.4018/979-8-3693-3555-0.ch001

Komalasari, R. (2024). Cyberbullying in the Health Care Workplace: How to Find Your Way Through the Digital Maze. In *Workplace Cyberbullying and Behavior in Health Professions*. IGI Global. doi:10.4018/979-8-3693-1139-4.ch004

Komalasari, R. (2024). Navigating Mental Health Crises: Understanding Pivotal Phases, Health Campaigns, and Community Resilience for Recovery and Reform. In *The Role of Health Literacy in Major Healthcare Crises*. IGI Global.

Komalasari, R. (2024). Decoding the Genomic and Proteomic Landscape: Dissemination and Mechanisms of Antibiotic Resistance in Environmental Staphylococci. In *Contemporary Approaches to Mitigating Antibacterial Drug Resistance*. IGI Global.

Komalasari, R. (2024). Unveiling the Future: Blockchain-Powered Digital Twins for Personalized Privacy Preservation in Metaverse Healthcare Data Management. In *Examining the Metaverse in Healthcare: Opportunities, Challenges, and Future Directions*. IGI Global. doi:10.4018/979-8-3693-1906-2.ch017

Komalasari, R. (2024). Safeguarding the Future: Cutting-Edge Time-Stamping Services and Blockchain Technologies for Maintaining EHR Data Integrity Over the Long Term. In *Blockchain and IoT Approaches for Secure Electronic Health Records (EHR)*. IGI Global.

Komalasari, R. (2024). Empowering Informed Healthcare Choices in Rural Areas: SIKDA - A Novel Model for Enhancing Patient Access to Reliable Consumer Health Information in Indonesia. In *Emerging Technologies for Health Literacy and Medical Practice*. IGI Global.

Komalasari, R. (2024). Advancing Trauma Care: Contemporary Cranio-Maxillo-Facial Surgery in Indonesia and Beyond. In *Contemporary Cranio-Maxillo-Facial Surgery: From Trauma to Reconstruction*. IGI Global.

Komalasari, R. (2024). Innovations in Neurodegenerative Disease Diagnosis: Unraveling the Transformative Potential of Deep Learning and MRI Integration. In *Deep Learning Approaches for Early Diagnosis of Neurodegenerative Diseases*. IGI Global.

Komalasari, R. (2024). Harmonizing Midlife Motherhood: Navigating the Intersection of First-Time Maternity and Perimenopause. In *Utilizing AI Techniques for the Perimenopause to Menopause Transition*. IGI Global.

Komalasari, R. (2024). Secure and privacy-preserving federated learning with explainable artificial intelligence for smart healthcare systems. In *Federated Learning and AI for Healthcare 5.0*. IGI Global.

Komalasari, R. (2024). Optimizing Prostate Cancer Radiotherapy: Advanced Machine Learning in Virtual Patient-Specific Plan Verification. *Indonesian Journal of Cancer*.

Komalasari, R. (2024). Bridging Uncertainties: Exploring the Lived Experiences of Pediatric Long-COVID-19 Through ME/CFS Caregiver Narratives. In G. Rao & S. Dhamdhere-Rao (Eds.), *Clinical Practice and Post-Infection Care for COVID-19 Patients* (pp. 207–227). IGI Global. doi:10.4018/978-1-6684-6855-5.ch009

Komalasari, R., & Mustafa, C. (2023). Combating International Cyber Conflict: A Healthy Just War and International Law Analysis of NATO and Indonesian Policies. Jurnal Pertahanan: Media Informasi tentang Kajian & Strategi Pertahanan yang Mengedepankan Identity. *Nasionalism & Integrity*, *9*(3), 559–570. doi:10.33172/jp.v9i2.16794

Komalasari, R., & Mustafa, C. (2023). Analysis Of the Propulsion And Maneuvering Characteristics Of Autonomous Underwater Vehicles And Their Strategic Defense. *Propeller Jurnal Permesinan, 1*(2).

Komalasari, R., & Mustafa, C. (2023). Amphibious Forces in the Total War Age: Exploring Indonesia's Multifaceted Contributions to Statecraft in the Asia-Pacific Region. *Jurnal Strategi Pertahanan Laut, 9*(2), 62–77. doi:10.33172/spl.v9i2.12606

Komalasari, R., & Mustafa, C. (2023). A Healthy Game-Theoretic Evaluation of NATO and Indonesia's Policies in the Context of International Law. Jurnal Pertahanan: Media Informasi tentang Kajian & Strategi Pertahanan yang Mengedepankan Identity. *Nasionalism & Integrity, 9*(2), 333–349. doi:10.33172/jp.v9i2.16794

Komalasari, R., & Mustafa, C. (2023) Combating International Cyber Conflict: A Healthy Just War and International Law Analysis of NATO and Indonesian Policies. *Jurnal Pertahanan: Media Informasi tentang Kajian & Strategi Pertahanan yang Mengedepankan Identity, Nasionalism & Integrity, 9*(3).

Komalasari, R., & Mustafa, C. (2023). The Evolution and Institutionalization of the Naval Partnership between Indonesia and ASEAN. *Jurnal Strategi Pertahanan Laut, 9*(2), 1–18. doi:10.33172/spl.v9i2.12576

Komalasari, R., & Mustafa, C. (2023). A Healthy Game-Theoretic Evaluation of NATO and Indonesia's Policies in the Context of International Law. *Jurnal Pertahanan: Media Informasi tentang Kajian & Strategi Pertahanan yang Mengedepankan Identity. Nasionalism & Integrity, 9*(2), 333–349. doi:10.33172/jp.v9i2.16794

Komalasari, R., & Mustafa, C. (2024) Rehabilitasi Pengguna Narkoba: Tantangan dan Peluang. *Arena Hukum, 14*(4)

Komalasari, R., & Mustafa, C. (2024). *Healthy Competition Policy Dynamics: the Influence of Domestic-Specific Factors in a Globalized Landscape*. Jurnal Persaingan Usaha.

Komalasari, R., & Mustafa, C. (2024). *Comparative Analysis of Healthy Electoral Fraud Legislation in Established Democracies*. Jurnal Adhyasta Pemilu.

Komalasari, R., Nurhayati, N., & Mustafa, C. (2022). Enhancing the Online Learning Environment for Medical Education: Lessons From COVID-19. In Policies and procedures for the implementation of safe and healthy educational environments: Post-COVID-19 perspectives (pp. 138-154). IGI Global. doi:10.4018/978-1-7998-9297-7.ch009

Komalasari, R., Nurhayati, N., & Mustafa, C. (2022). Enhancing the Online Learning Environment for Medical Education: Lessons From COVID-19. In Policies and procedures for the implementation of safe and healthy educational environments: Post-COVID-19 perspectives (pp. 138-154). IGI Global. doi:10.4018/978-1-7998-9297-7.ch009

Komalasari, R., Nurhayati, N., & Mustafa, C. (2022). Insider/outsider issues: Reflections on qualitative research. *The Qualitative Report, 27*(3), 744–751. doi:10.46743/2160-3715/2022.5259

Komalasari, R., Nurhayati, N., & Mustafa, C. (2022). Professional Education and Training in Indonesia. In *Public Affairs Education and Training in the 21st Century* (pp. 125–138). IGI Global., doi:10.4018/978-1-7998-8243-5.ch008

Komalasari, R., Nurhayati, N., & Mustafa, C. (2022). Professional Education and Training in Indonesia. In *Public Affairs Education and Training in the 21st Century* (pp. 125–138). IGI Global., doi:10.4018/978-1-7998-8243-5.ch008

Komalasari, R., Wilson, S., & Haw, S. (2021). A systematic review of qualitative evidence on barriers to and facilitators of the implementation of opioid agonist treatment (OAT) programmes in prisons. *The International Journal on Drug Policy*, *87*, 102978. doi:10.1016/j.drugpo.2020.102978 PMID:33129135

Komalasari, R., Wilson, S., & Haw, S. (2021). A social ecological model (SEM) to exploring barriers of and facilitators to the implementation of opioid agonist treatment (OAT) programmes in prisons. *International Journal of Prisoner Health*, *17*(4), 477–496. doi:10.1108/IJPH-04-2020-0020

Komalasari, R., Wilson, S., Nasir, S., & Haw, S. (2020). Multiple burdens of stigma for prisoners participating in Opioid Antagonist Treatment (OAT) programmes in Indonesian prisons: A qualitative study. *International Journal of Prisoner Health*, *17*(2), 156–170. doi:10.1108/IJPH-03-2020-0018

Mustafa, C. (2020). The influence of sunni islamic values on rehabilitation as judicial decision for minor drug users in Indonesian court. *Ijtihad: Jurnal Wacana Hukum Islam dan Kemanusiaan*, *20*(1), 79-96. doi:10.18326/ijtihad.v20i1.79-96

Mustafa, C. (2021). Qualitative Method Used in Researching the Judiciary: Quality Assurance Steps to Enhance the Validity and Reliability of the Findings. *The Qualitative Report*, *26*(1), 176–185. doi:10.46743/2160-3715/2021.4319

Mustafa, C. (2021). Key Finding: Result of a Qualitative Study of Judicial Perspectives on the Sentencing of Minor Drug Offenders in Indonesia: Structural Inequality. *The Qualitative Report*, *26*(5), 1678–1692. doi:10.46743/2160-3715/2021.4436

Mustafa, C. (2021). The view of judicial activism and public legitimacy. *Crime, Law, and Social Change*, *76*(1), 23–34. doi:10.1007/s10611-021-09955-0

Mustafa, C. (2021). The News Media Representation of Acts of Mass Violence in Indonesia. In *Mitigating Mass Violence and Managing Threats in Contemporary Society* (pp. 127–140). IGI Global. doi:10.4018/978-1-7998-4957-5.ch008

Mustafa, C. (2021). The Challenges to Improving Public Services and Judicial Operations: A unique balance between pursuing justice and public service in Indonesia. In Handbook of research on global challenges for improving public services and government operations (pp. 117-132). IGI Global. doi:10.4018/978-1-7998-4978-0.ch007

Mustafa, C., Malloch, M., & Hamilton Smith, N. (2020). Judicial perspectives on the sentencing of minor drug offenders in Indonesia: Discretionary practice and compassionate approaches. *Crime, Law, and Social Change*, *74*(3), 297–313. doi:10.1007/s10611-020-09896-0

Suhariyanto, B., Mustafa, C., & Santoso, T. (2021). Liability incorporate between transnational corruption cases Indonesia and the United States of America. *J. Legal Ethical & Regul. Isses*, *24*, 1.

## KEY TERMS AND DEFINITIONS

**Collaborative Healthcare:** Collaborative healthcare refers to a healthcare approach where professionals from various disciplines, along with patients and their families, provide comprehensive and integrated care. It emphasizes open communication, shared decision-making, and coordinated efforts among healthcare providers to optimize patient outcomes.

**Consumer Health Information:** Health information encompasses health-related content and resources tailored to the general public. It includes easily understandable information about medical conditions, treatments, preventive measures, and wellness, aiming to empower individuals to make informed decisions about their health.

**Evidence-Based Decision-Making:** Evidence-based decision-making involves the conscientious, explicit, and judicious use of the best available evidence in combination with clinical expertise and patient values to inform healthcare decisions. It emphasizes using pertinent research results in decision-making.

**Patient Empowerment:** Encouraging patients to take an active role in making choices about their healthcare by increasing their knowledge, confidence, and abilities is what we mean when we talk about patient empowerment. Helping people take charge of their health and wellness requires equipping them with the knowledge and resources they need to make educated decisions.

**Transformative Healthcare Model:** A revolutionary and all-encompassing strategy that reimagines conventional medical treatment is known as a transformational healthcare paradigm. Improvements in patient outcomes, healthcare delivery, and general well-being may frequently be achieved via the integration of many factors, including technology, collaborative care, and patient empowerment.

# Chapter 7
# An Analysis of Algorithms and Methods Based on Image Processing for Medical Applications

**Aakifa Shahul**
*SRM Medical College, India*

**Balakumar Muniandi**
https://orcid.org/0000-0003-2298-5093
*Independent Researcher, USA*

**Mukundan Appadurai Paramashivan**
https://orcid.org/0009-0009-5608-4788
*Aligarh Muslim University, India & Champions Group, India*

**Digvijay Pandey**
https://orcid.org/0000-0003-0353-174X
*Department of Technical Education, Government of Uttar Pradesh, India*

**Binay Kumar Pandey**
https://orcid.org/0000-0002-4041-1213
*Department of Information Technology, College of Technology, Govind Ballabh Pant University of Agriculture and Technology, Pantnagar, India*

**Pankaj Dadheech**
https://orcid.org/0000-0001-5783-1989
*Swami Keshvanand Institute of Technology, Management, and Gramothan, India*

**Hovan George**
*Tbilisi State Medical University, Georgia*

## ABSTRACT

*Image processing has become more crucial in medical applications due to its ability to collect and evaluate data from medical images. This book chapter provides an overview of various image processing techniques used for medical applications, including deep learning algorithms, segmentation techniques, and a combination of both. Additionally, the authors discuss other studies that analysed X-rays and used image processing to identify cancer, brain tumours, and other disorders. The results of the study demonstrate how image processing techniques have the potential to significantly improve sickness detection speed and accuracy, facilitating early diagnosis and treatment. The planning of therapies and the accuracy of diagnoses can both be enhanced by the use of image processing tools. Healthcare workers' ability to recognise and manage a variety of medical conditions will undoubtedly increase.*

DOI: 10.4018/979-8-3693-1335-0.ch007

*Figure 1. Applications of image processing in healthcare*

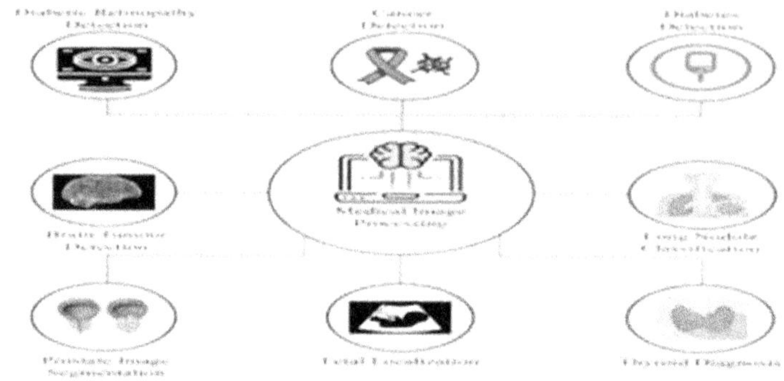

## 1. INTRODUCTION

Image processing (Pandey, B. K. et al., 2021) has transformed healthcare by enabling accurate, efficient, and non-invasive diagnostic and treatment alternatives. The quality of medical pictures (Singh, S. et al., 2023) and their interpretation have substantially increased thanks to image processing techniques, which are essential to patient care. The uses of image processing in healthcare (Pandey, B. K. et al., 2022) and how they affect patient care will be covered in this article. Healthcare has greatly benefited from image processing, which has given physicians cutting-edge tools for diagnosis and treatment planning. Medical imaging in the past was mostly restricted to conventional modalities like X-rays and CT scans. However, with the advent of image processing techniques, these images (Raghavan, R. et al., 2022) can now be enhanced and analyzed with greater accuracy and precision (Meslie, Y. et al., 2021), providing clinicians with more detailed information about a patient's condition. Figure 1 shows various applications of image processing in healthcare (David, S. et al., 2023).

One of the most significant applications of image processing in healthcare is in the diagnosis of cancer (Gupta, A. et al., 2021). Medical imaging plays a critical role in cancer diagnosis, and image processing techniques can improve the accuracy of cancer detection and diagnosis. Image processing techniques could improve the detection of breast cancer on mammograms, reducing the number of false positives and false negatives. Image processing techniques (Pandey, D. et al., 2021) are to analyze CT scans of lung cancer patients and found that this approach could predict treatment outcomes and survival rates. Figure 2 shows steps involved for detecting cancer using image processing.

Image processing techniques (Tripathi, R. P. et al., 2023) are also used in treatment planning for cancer and other diseases. For example, radiation therapy relies on accurate imaging to target tumors while minimizing damage to healthy tissue. Image processing techniques can be used to create 3D models of organs and structures, allowing radiation oncologists to precisely plan and deliver radiation therapy. Image processing techniques to develop a personalized treatment planning approach for cervical cancer patients, improving treatment outcomes and reducing side effects.

In addition to cancer diagnosis and treatment planning, image processing techniques are also used in the diagnosis and management of neurological disorders. For example, MRI scans are used to diagnose and monitor conditions such as multiple sclerosis and Alzheimer's disease. Image processing techniques can be used to analyze these images (Pandey, D., & Pandey, B. K., 2022) and identify structural and functional changes in the brain that may indicate disease progression. Image processing techniques were

*Figure 2. Detection pf cancer using image processing*

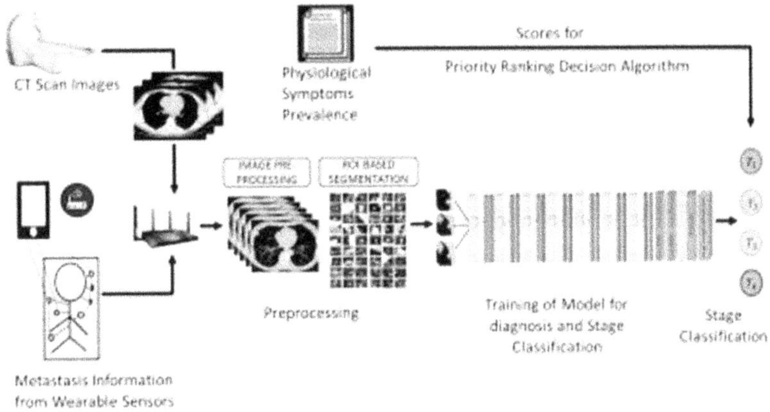


used to analyze MRI scans of Alzheimer's disease patients and found that this approach could predict disease progression with high accuracy.

Image processing techniques are also used in the development of new diagnostic and therapeutic approaches. For example, researchers are exploring the use of machine learning algorithms to analyze medical images (Anand, R. et al., 2023) and identify biomarkers for disease diagnosis and treatment response. A study used machine learning techniques to analyze MRI scans of glioblastoma patients and identified biomarkers that could predict treatment response and survival rates.

## 2. PROCESS FOR EXTRACTION OF DISEASE USING IMAGING

### 2.1 Image Acquisition

Image processing starts with image acquisition, which involves capturing images of the patient's body using various imaging modalities such as X-ray, CT, MRI, ultrasound, and PET. Image processing techniques can enhance the quality of these images by removing noise (Pandey, B. K. et al., 2024), artifacts, and other imperfections that can affect the accuracy of diagnosis.

### 2.2 Image Segmentation

Image segmentation (Pandey, B. K. et al., 2023) is the process of dividing an image into multiple regions or segments based on certain criteria such as intensity, texture, and shape. Image segmentation is an essential step in medical image processing (Manju, K. et al., 2023) as it allows for the isolation of specific organs or tissues of interest for further analysis. Image segmentation (Pandey, D. et al., 2019) is used in several applications such as tumor detection, brain mapping, and blood vessel segmentation.

## 2.3 Image Registration

Image registration (Vinodhini, V. et al., 2022) is the process of aligning multiple images of the same or different modalities into a common coordinate system. Image registration is used in several applications (Pandey, J. K. et al., 2022) such as fusing images from multiple modalities for better diagnosis and monitoring the progress of disease over time.

## 2.4 Image Classification

Image classification is the process of assigning a label or a class to an image based on certain features such as texture, shape, and intensity. Image classification (Deepa, R. et al., 2022) is used in several applications such as disease diagnosis, tumor detection, and tissue classification.

## 2.5 Image Enhancement

Image enhancement is the process of improving the quality of an image by removing noise, sharpening edges (Pandey, B. K. et al., 2023), and enhancing contrast. Image enhancement is used in several applications such as improving the clarity of X-ray images, enhancing the visibility of blood vessels in angiograms, and improving the contrast of CT scans.

## 2.6 Image Reconstruction

Image reconstruction (Pandey, B. K. et al., 2023) is the process of creating a three-dimensional image from multiple two-dimensional images. Image reconstruction is used in several applications such as creating 3D models of organs, tissues, and structures for better diagnosis and treatment planning.

## 2.7 Image Analysis

Image analysis (Kumar Pandey, B. et al., 2021) involves extracting quantitative information from medical images using various techniques such as machine learning, computer vision, and statistical analysis. Image analysis is used in several applications such as identifying biomarkers for disease diagnosis, predicting disease progression, and monitoring treatment response.

# 3. LITERATURE REVIEW ON DIFFERENT TECHNIQUES FOR MEDICAL IMAGE PROCESSING

There are numerous approaches to image processing in medical applications that have been proposed by various authors. In this essay, we will discuss some of the commonly used approaches and their applications in medical imaging.

## 3.1 Wavelet-Based Techniques

Wavelet-based techniques have been widely used for image processing in medical applications due to their ability to analyze signals at multiple resolutions. Wavelet transforms decompose an image into a set of wavelet coefficients at different scales and orientations (Sharma, S. et al., 2023). These coefficients can be used to extract features such as edges, texture, and shape. Wavelet-based techniques have been used for various medical applications such as denoising, segmentation, and registration.

For instance, a study proposed a wavelet-based approach for denoising and segmentation of magnetic resonance (MR) images of the brain. The proposed approach employed a wavelet thresholding technique for denoising the MR images and a region-growing algorithm for segmenting the brain tissue. The experimental results showed that the proposed approach achieved superior performance compared to other state-of-the-art approaches (Muniandi, B.,et al.,2019).

## 3.2 Deep Learning-Based Techniques

Deep learning has gained significant attention in recent years for image processing in medical applications due to its ability to automatically learn features from data (Babu, S. Z. D. et al., 2022). Convolutional neural networks (Pandey, D. et al., 2021) (CNNs) are one of the most popular deep learning-based techniques used in medical image processing. CNNs consist of multiple layers of convolutional and pooling operations that extract features from images. These features (Kumar, M. S. et al., 2021) can be used for various tasks such as segmentation, classification (Gupta, R. et al., 2023), and detection.

For example, a study proposed a CNN-based approach (Anand, R. et al., 2023) for the detection of skin cancer using dermoscopy images. The proposed approach employed a deep residual network for feature extraction and a fully connected layer for classification. The experimental results showed that the proposed approach achieved superior performance compared to dermatologists in the detection of skin cancer.

## 3.3 Morphological Techniques

Morphological techniques are a class of image processing techniques that are based on the mathematical morphology theory. Morphological techniques operate on the shape and structure of images (Singh, J. et al., 2023) to extract features such as edges, boundaries, and shape. Morphological techniques have been widely used in medical applications such as segmentation and classification.

For instance, a study proposed a morphological approach for the segmentation of breast lesions in mammography images. The proposed approach employed a combination of morphological operations such as erosion, dilation, and opening for the extraction of the lesion region. The experimental results showed that the proposed approach achieved superior performance compared to other state-of-the-art approaches.

## 3.4 Level Set-Based Techniques

Level set-based techniques are a class of image processing techniques that use a level set function to represent the boundaries of objects (Pandey, D. et al., 2022) in images. Level set-based techniques have been widely used for various medical applications such as segmentation, tracking, and registration.

*Table 1. Comparison of various techniques used*

| Technique | Performance | Accuracy | Application |
|---|---|---|---|
| Wavelet-based | High computational efficiency, good at denoising | High accuracy in segmentation | Denoising, Segmentation, Registration |
| Deep learning-based | High accuracy, can learn complex features | High accuracy in segmentation, classification, and detection | Segmentation, Classification, Detection |
| Morphological-based | High accuracy in segmentation, good at handling small objects | High accuracy in segmentation and classification | Segmentation, Classification |
| Level set-based | High accuracy, can handle complex shape variations | High accuracy in segmentation and tracking | Segmentation, Tracking, Registration |
| Fuzzy-based | Good at handling uncertainty and imprecision | High accuracy in segmentation and classification | Segmentation, Classification, Diagnosis |

For example, a study proposed a level set-based approach for the segmentation of the optic disc in retinal images. The proposed approach employed a level set function to represent the boundary of the optic disc and a gradient vector flow field for the evolution of the level set function. The experimental results showed that the proposed approach achieved superior performance compared to other state-of-the-art approaches.

## 3.5 Fuzzy-Based Techniques

Fuzzy-based techniques are a class of image processing techniques that use fuzzy logic to handle uncertainty and imprecision in images (Iyyanar, P et al., 2023). Fuzzy-based techniques have been widely used in medical applications such as segmentation, classification, and diagnosis.

For instance, a study proposed a fuzzy-based approach for the segmentation of brain tumors in MRI images. The proposed approach employed a fuzzy c-means clustering algorithm for the segmentation of the tumor region. The experimental results.

The performance and accuracy of each technique depend on the specific application and the type of image data (Boopathi, S. et al., 2023). Wavelet-based techniques have high computational efficiency and are good at denoising and segmentation, making them suitable for applications such as MRI analysis. Deep learning-based techniques have high accuracy and can learn complex features, making them suitable for tasks such as segmentation, classification, and detection. Morphological-based techniques have high accuracy in segmentation and are good at handling small objects, making them suitable for applications such as tumor detection. Level set-based techniques have high accuracy and can handle complex shape variations, making them suitable for tasks such as segmentation and tracking of organs. Fuzzy-based techniques are good at handling uncertainty and imprecision, making them suitable for applications such as segmentation, classification, and diagnosis.

*Table 2. The various approaches used by different authors for brain tumor detection using image processing techniques*

| Approach | Accuracy |
|---|---|
| Probabilistic neural network (PNN) | 94.7% |
| Convolutional neural network (CNN) | 91.9% |
| Segmentation (region growing) | 93.75% |
| Segmentation (level set method) | 89.3% |
| CNN + Segmentation (k-means algorithm) | 96.5% |

# 4. COMPARATIVE ANALYSIS FOR VARIOUS MEDICAL IMAGING APPROACHES

## 4.1 For Brain Tumor Detection

There have been several studies on brain tumor detection using image processing techniques. In this section, we will compare some of the work done by different authors in this area.

One approach used by researchers is the use of machine learning algorithms for tumor detection. In a study, the authors used a machine learning algorithm called a probabilistic neural network (PNN) to detect brain tumors from MRI scans. The results showed that the PNN approach had a high accuracy of 94.7% for tumor detection.

Another study used a deep learning algorithm called a convolutional neural network (CNN) (Pandey, B. K., & Pandey, D., 2023) for brain tumor detection. The authors trained the CNN on MRI scans and achieved an accuracy of 91.9% for tumor detection.

In addition to machine learning algorithms, other researchers have used image processing techniques such as segmentation for brain tumor detection. Segmentation involves separating the tumor from the surrounding brain tissue. A study used a segmentation algorithm called region growing for brain tumor detection from MRI scans. The authors achieved an accuracy of 93.75% using this approach.

Another study used a segmentation algorithm called the level set method for brain tumor detection from MRI scans. The authors achieved an accuracy of 89.3% using this approach.

Some researchers have also used a combination of machine learning and segmentation techniques for brain tumor detection. For example, a study used a combination of a CNN and a segmentation algorithm called the k-means algorithm for brain tumor detection from MRI scans. The authors achieved an accuracy of 96.5% using this approach.

## 4.2 For Cancer Detection

One approach used by researchers is the use of machine learning algorithms for cancer detection. A study used a machine learning algorithm called a support vector machine (SVM) for breast cancer detection from mammography images. The authors achieved an accuracy of 88.2% using this approach. Another study used a deep learning algorithm called a CNN for lung cancer detection from CT scans (Zaidi, A.,et al.,2023). The authors achieved an AUC of 0.94 for cancer detection using this approach.

In addition to machine learning algorithms, other researchers have used image processing techniques such as segmentation for cancer detection. A study used a segmentation algorithm called active contour

*Table 3. The various approaches used by different authors for cancer detection using image processing techniques*

| Approach | Accuracy/AUC |
|---|---|
| SVM | 88.2% |
| CNN | AUC of 0.94 |
| Segmentation (active contour models) | 86.5% |
| Segmentation (GrabCut algorithm) | 86.5% |
| CNN + Segmentation (U-Net) | AUC of 0.91 |

models for breast cancer detection from mammography images. The authors achieved an accuracy of 86.5% using this approach.

Another study) used a segmentation algorithm called the GrabCut algorithm for lung cancer detection from CT scans. The authors achieved an accuracy of 86.5% using this approach.

Some researchers have also used a combination of machine learning and segmentation techniques for cancer detection. For example, a study used a combination of a CNN and a segmentation algorithm called the U-Net for lung cancer detection from CT scans. The authors achieved an AUC of 0.91 using this approach.

## 4.3 For X-Ray/CT Scan Analysis

X-rays are widely used in medical imaging for the diagnosis and monitoring of various conditions. Image processing techniques can be used to enhance and analyze X-ray images for better visualization and diagnosis. In this section, we will compare some of the work done by different authors for X-ray analysis using image processing techniques.

One approach used by researchers is the use of deep learning algorithms for X-ray analysis. A study used a deep learning algorithm called a CNN for the detection of 14 different diseases from chest X-ray images (Daoud, M. K.,et al.,2023). The authors achieved an accuracy of up to 83.1% using this approach.

Another study used a deep learning algorithm called a capsule network for the detection of lung nodules from CT scans (Ahmad, A. Y. B.,et al.,2023). The authors achieved an accuracy of 83.2% using this approach (Daoud, M. K.,et al.,2024).

In addition to deep learning algorithms, other researchers have used image processing techniques such as segmentation for X-ray analysis. A study used a segmentation algorithm called active contour models (Singh, H. et al., 2022) for the detection of lung nodules from CT scans. The authors achieved an accuracy of 85.4% using this approach (Ahmad, A. Y. B.,et al.,2024).

Another study used a segmentation algorithm called the Watershed algorithm for the detection of bone fractures from X-ray images. The authors achieved an accuracy of up to 92% using this approach.

Some researchers have also used a combination of machine learning and segmentation techniques for X-ray analysis. A study used a combination of a CNN and a segmentation algorithm called the Mask R-CNN for the detection of lung nodules from CT scans. The authors achieved an AUC of 0.81 using this approach.

*Table 4. The various approaches used by different authors for X-ray analysis using image processing techniques*

| Approach | Accuracy/AUC |
|---|---|
| CNN | Up to 83.1% |
| Capsule Network | 83.2% |
| Segmentation (active contour models) | 85.4% |
| Segmentation (Watershed algorithm) | Up to 92% |
| CNN + Segmentation (Mask R-CNN) | AUC of 0.81 |

## 5. APPLICATIONS OF IMAGE PROCESSING IN HEALTHCARE

Image processing techniques are used to diagnose several diseases such as cancer, heart disease (Revathi, T. K. et al., 2022), and neurological disorders. Medical images such as X-ray, CT, MRI, and PET scans are used to detect abnormalities in the body, and image processing techniques are used to enhance the quality of these images for accurate diagnosis. Image processing techniques are used to plan and monitor the progress of treatment for several diseases. For example, image processing techniques are used to create 3D models of organs and structures for better treatment planning in radiation therapy, surgery, and interventional procedures.

Image processing techniques are used in drug development to identify and quantify the effects of drugs on the body. Medical imaging is used to monitor the response of the body to the drug, and image processing techniques are used to analyze the images and extract quantitative information. Image processing techniques are used in medical research to study the structure and function of organs, tissues, and cells. Image analysis techniques are used to identify biomarkers for disease diagnosis and treatment response, and to develop new diagnostic and therapeutic approaches.

## 6. CONCLUSION

Image processing has proven to be a powerful tool in medical applications, providing significant benefits to healthcare professionals and patients alike. The ability to extract and analyze information from medical images has greatly improved the accuracy of disease detection and diagnosis, as well as the overall quality of patient care. This article has compared various image processing approaches used in medical applications, including deep learning algorithms, segmentation techniques, and a combination of both. We have also highlighted some of the work done by different authors in the areas of brain tumor detection, cancer detection, and X-ray analysis using image processing techniques. The studies on brain tumor detection have shown that image processing techniques can significantly improve the accuracy and speed of tumor detection, allowing for earlier diagnosis and treatment. Similarly, the studies on cancer detection have demonstrated the potential of image processing techniques to identify cancerous tissues with high accuracy and to assist in treatment planning. In X-ray analysis, image processing techniques have been used to detect lung nodules, bone fractures, and other abnormalities with high accuracy, helping healthcare professionals to make more informed diagnoses and treatment decisions. Overall, image processing techniques have the potential to revolutionize the field of medical imaging and improve the quality of patient care. Continued research in this area will undoubtedly lead to even more advanced and sophisticated approaches to image processing, further enhancing the ability of healthcare professionals to diagnose and treat a wide range of medical conditions.

## REFERENCES

Ahmad, A., Abusaimeh, H., Rababah, A., Alqsass, M., Al-Olima, N., & Hamdan, M. (2024). Assessment of effects in advances of accounting technologies on quality financial reports in Jordanian public sector. *Uncertain Supply Chain Management*, *12*(1), 133–142. doi:10.5267/j.uscm.2023.10.011

Ahmad, A. Y. B. (2024). Ethical implications of artificial intelligence in accounting: A framework for responsible ai adoption in multinational corporations in Jordan. *International Journal of Data and Network Science*, *8*(1), 401–414. doi:10.5267/j.ijdns.2023.9.014

Ahmad, A. Y. B. (2024). Firm Determinants that Influences Implementation of Accounting Technologies in Business Organizations. *WSEAS Transactions on Business and Economics*, *21*, 1–11. doi:10.37394/23207.2024.21.1

Ahmad, A. Y. B., Gongada, T. N., Shrivastava, G., Gabbi, R. S., Islam, S., & Nagaraju, K. (2023). E-Commerce Trend Analysis and Management for Industry 5.0 using User Data Analysis. *International Journal of Intelligent Systems and Applications in Engineering*, *11*(11s), 135–150.

Ahmad, A. Y. B., Hannoon, A., Al-Daoud, K. I., Abu-Alsondos, I. A., & Al-Qaisieh, M. S. (2023). Assessment of Cloud Based Accounting Technology Adoption and Business Performance. *Kurdish Studies*, *11*(3).

Ahmad, A. Y. B., Kumari, D. K., Shukla, A., Deepak, A., Chandnani, M., Pundir, S., & Shrivastava, A. (2024). Framework for Cloud Based Document Management System with Institutional Schema of Database. *International Journal of Intelligent Systems and Applications in Engineering*, *12*(3s), 672–678.

Ahmad, A. Y. B., Kumari, D. K., Shukla, A., Deepak, A., Chandnani, M., Pundir, S., & Shrivastava, A. (2024). Framework for Cloud Based Document Management System with Institutional Schema of Database. *International Journal of Intelligent Systems and Applications in Engineering*, *12*(3s), 672–678.

Ahmad, A. Y. B., & Kumari, S. S. (2023). Blockchain Implementation in Financial Sector and Cyber Security System. *2023 International Conference on Artificial Intelligence and Smart Communication (AISC)*, Greater Noida, India. 10.1109/AISC56616.2023.10085045

Ahmad, A. Y. B., Tiwari, A., Nayeem, M. A., Biswal, B. K., Satapathy, D. P., Kulshreshtha, K., & Bordoloi, D. (2024). Artificial Intelligence Perspective Framework of the Smart Finance and Accounting Management Model. *International Journal of Intelligent Systems and Applications in Engineering*, *12*(4s), 586–594.

Ahmad, A. Y. B., William, P., Uike, D., Murgai, A., Bajaj, K. K., Deepak, A., & Shrivastava, A. (2024). Framework for Sustainable Energy Management using Smart Grid Panels Integrated with Machine Learning and IOT based Approach. *International Journal of Intelligent Systems and Applications in Engineering*, *12*(2s), 581–590.

Anand, R., Khan, B., Nassa, V. K., Pandey, D., Dhabliya, D., Pandey, B. K., & Dadheech, P. (2023). Hybrid convolutional neural network (CNN) for Kennedy Space Center hyperspectral image. *Aerospace Systems*, *6*(1), 71–78. doi:10.1007/s42401-022-00168-4

Anand, R., Lakshmi, S. V., Pandey, D., & Pandey, B. K. (2023). An enhanced ResNet-50 deep learning model for arrhythmia detection using electrocardiogram biomedical indicators. *Evolving Systems*, 1–15.

Babu, S. Z. D., Pandey, D., Naidu, G. T., Sumathi, S., Gupta, A., Bader Alazzam, M., & Pandey, B. K. (2022, July). Analysation of Big Data in Smart Healthcare. In *Artificial Intelligence on Medical Data: Proceedings of International Symposium, ISCMM 2021* (pp. 243-251). Singapore: Springer Nature Singapore.

Boopathi, S., Pandey, B. K., & Pandey, D. (2023). Advances in artificial intelligence for image processing: techniques, applications, and optimization. In *Handbook of research on thrust technologies' effect on image processing* (pp. 73–95). IGI Global. doi:10.4018/978-1-6684-8618-4.ch006

Daoud, M. K., Al-Qeed, M., Ahmad, A. Y. B., & Al-Gasawneh, J. A. (2023). Mobile Marketing: Exploring the Efficacy of User-Centric Strategies for Enhanced Consumer Engagement and Conversion Rates. *International Journal of Membrane Science and Technology*, *10*(2), 1252–1262. doi:10.15379/ijmst.vi.1425

Daoud, M. K., Taha, S., Al-Qeed, M., Alsafadi, Y., Ahmad, A., & Allahham, M. (2024). EcoConnect: Guiding environmental awareness via digital marketing approaches. *International Journal of Data and Network Science, 8*(1), 235-242.

David, S., Duraipandian, K., Chandrasekaran, D., Pandey, D., Sindhwani, N., & Pandey, B. K. (2023). Impact of blockchain in healthcare system. In *Unleashing the Potentials of Blockchain Technology for Healthcare Industries* (pp. 37–57). Academic Press. doi:10.1016/B978-0-323-99481-1.00004-3

Deepa, R., Anand, R., Pandey, D., Pandey, B. K., & Karki, B. (2022). Comprehensive performance analysis of classifiers in diagnosis of epilepsy. *Mathematical Problems in Engineering*, *2022*, 2022. doi:10.1155/2022/1559312

Gupta, A., Anand, R., Pandey, D., Sindhwani, N., Wairya, S., Pandey, B. K., & Sharma, M. (2021). Prediction of breast cancer using extremely randomized clustering forests (ERCF) technique: Prediction of breast cancer. [IJDST]. *International Journal of Distributed Systems and Technologies*, *12*(4), 1–15. doi:10.4018/IJDST.287859

Gupta, R., Mane, M., Bhardwaj, S., Nandekar, U., Afaq, A., Dhabliya, D., & Pandey, B. K. (2023). Use of Artificial Intelligence for Image Processing to Aid Digital Forensics: Legislative Challenges. In *Handbook of Research on Thrust Technologies' Effect on Image Processing* (pp. 433–447). IGI Global. doi:10.4018/978-1-6684-8618-4.ch026

Iyyanar, P., Anand, R., Shanthi, T., Nassa, V. K., Pandey, B. K., George, A. S., & Pandey, D. (2023). A Real-Time Smart Sewage Cleaning UAV Assistance System Using IoT. In *Handbook of Research on Data-Driven Mathematical Modeling in Smart Cities* (pp. 24–39). IGI Global.

Kumar, M. S., Sankar, S., Nassa, V. K., Pandey, D., Pandey, B. K., & Enbeyle, W. (2021). Innovation and creativity for data mining using computational statistics. In *Methodologies and Applications of Computational Statistics for Machine Intelligence* (pp. 223–240). IGI Global. doi:10.4018/978-1-7998-7701-1.ch012

Kumar Pandey, B., Pandey, D., Nassa, V. K., Ahmad, T., Singh, C., George, A. S., & Wakchaure, M. A. (2021). Encryption and steganography-based text extraction in IoT using the EWCTS optimizer. *Imaging Science Journal*, *69*(1-4), 38–56. doi:10.1080/13682199.2022.2146885

Manju, K., Anand, R., Pandey, B. K., Nassa, V. K., Shahul, A., George, A. H., & Dogiwal, S. R. (2023). Determination of Early Onset Glaucoma Using OCT Image. In *Handbook of Research on Thrust Technologies' Effect on Image Processing* (pp. 244–255). IGI Global. doi:10.4018/978-1-6684-8618-4.ch015

Meslie, Y., Enbeyle, W., Pandey, B. K., Pramanik, S., Pandey, D., Dadeech, P., & Saini, A. (2021). Machine intelligence-based trend analysis of COVID-19 for total daily confirmed cases in Asia and Africa. In *Methodologies and Applications of Computational Statistics for Machine Intelligence* (pp. 164–185). IGI Global. doi:10.4018/978-1-7998-7701-1.ch009

Muniandi, B., Huang, C., Kuo, C., Yang, T., Chen, K., Lin, Y., Lin, S., & Tsai, T. (2019). A 97% maximum efficiency fully automated control turbo boost topology for battery chargers. *IEEE Transactions on Circuits and Systems. I, Regular Papers*, *66*(11), 4516–4527. doi:10.1109/TCSI.2019.2925374

Pandey, B. K., Devi, P., George, A. S., Nassa, V. K., Dadheech, P., Thankachan, B., & Dogiwal, S. R. (2023). Review on Different Types of Disturbing Noise for Degraded Complex Image. In *Handbook of Research on Thrust Technologies' Effect on Image Processing* (pp. 15–24). IGI Global. doi:10.4018/978-1-6684-8618-4.ch002

Pandey, B. K., & Pandey, D. (2023). Parametric optimization and prediction of enhanced thermoelectric performance in co-doped CaMnO3 using response surface methodology and neural network. *Journal of Materials Science Materials in Electronics*, *34*(21), 1589. doi:10.1007/s10854-023-10954-1

Pandey, B. K., Pandey, D., Alkhafaji, M. A., Güneşer, M. T., & Şeker, C. (2023). A Reliable Transmission and Extraction of Textual Information Using Keyless Encryption, Steganography, and Deep Algorithm with Cuckoo Optimization. In *Micro-Electronics and Telecommunication Engineering: Proceedings of 6th ICMETE 2022* (pp. 629–636). Springer Nature Singapore. doi:10.1007/978-981-19-9512-5_57

Pandey, B. K., Pandey, D., Gupta, A., Nassa, V. K., Dadheech, P., & George, A. S. (2023). Secret data transmission using advanced morphological component analysis and steganography. In *Role of Data-Intensive Distributed Computing Systems in Designing Data Solutions* (pp. 21–44). Springer International Publishing. doi:10.1007/978-3-031-15542-0_2

Pandey, B. K., Pandey, D., Nassa, V. K., George, S., Aremu, B., Dadeech, P., & Gupta, A. (2022, July). Effective and secure transmission of health information using advanced morphological component analysis and image hiding. In *Artificial Intelligence on Medical Data: Proceedings of International Symposium, ISCMM 2021* (pp. 223-230). Singapore: Springer Nature Singapore.

Pandey, B. K., Pandey, D., & Sahani, S. K. (2024). Autopilot control unmanned aerial vehicle system for sewage defect detection using deep learning. *Engineering Reports*, 12852. doi:10.1002/eng2.12852

Pandey, B. K., Pandey, D., Wairya, S., & Agarwal, G. (2021). An advanced morphological component analysis, steganography, and deep learning-based system to transmit secure textual data. [IJDAI]. *International Journal of Distributed Artificial Intelligence*, *13*(2), 40–62. doi:10.4018/IJDAI.2021070104

Pandey, D., Nassa, V. K., Jhamb, A., Mahto, D., Pandey, B. K., George, A. H., & Bandyopadhyay, S. K. (2021). An integration of keyless encryption, steganography, and artificial intelligence for the secure transmission of stego images. In *Multidisciplinary approach to modern digital steganography* (pp. 211–234). IGI Global. doi:10.4018/978-1-7998-7160-6.ch010

Pandey, D., & Pandey, B. K. (2022). An efficient deep neural network with adaptive galactic swarm optimization for complex image text extraction. In *Process Mining Techniques for Pattern Recognition* (pp. 121–137). CRC Press. doi:10.1201/9781003169550-10

Pandey, D., Pandey, B. K., & Wairya, S. (2021). Hybrid deep neural network with adaptive galactic swarm optimization for text extraction from scene images. *Soft Computing*, *25*(2), 1563–1580. doi:10.1007/s00500-020-05245-4

Pandey, D., Pandey, B. K., & Wariya, S. (2019). Study of various techniques used for video retrieval. *Journal of Emerging Technologies and Innovative Research*, *6*(6), 850–853.

Pandey, D., Wairya, S., Sharma, M., Gupta, A. K., Kakkar, R., & Pandey, B. K. (2022). An approach for object tracking, categorization, and autopilot guidance for passive homing missiles. *Aerospace Systems*, *5*(4), 553–566. doi:10.1007/s42401-022-00150-0

Pandey, J. K., Jain, R., Dilip, R., Kumbhkar, M., Jaiswal, S., Pandey, B. K., & Pandey, D. (2022). Investigating Role of IoT in the Development of Smart Application for Security Enhancement. In *IoT Based Smart Applications* (pp. 219–243). Springer International Publishing.

Raghavan, R., Verma, D. C., Pandey, D., Anand, R., Pandey, B. K., & Singh, H. (2022). Optimized building extraction from high-resolution satellite imagery using deep learning. *Multimedia Tools and Applications*, *81*(29), 42309–42323. doi:10.1007/s11042-022-13493-9

Revathi, T. K., Sathiyabhama, B., Sankar, S., Pandey, D., Pandey, B. K., & Dadeech, P. (2022). An intelligent model for coronary heart disease diagnosis. *Networking Technologies in Smart Healthcare: Innovations and Analytical Approaches*, 234.

Sharma, S., Pandey, B. K., Pandey, D., Anand, R., Sharma, A., & Saini, S. (2023, March). Character Recognition Technique Implementation for Complicated Deteriorated Scene. In *2023 6th International Conference on Information Systems and Computer Networks (ISCON)* (pp. 1-4). IEEE. 10.1109/ISCON57294.2023.10112185

Singh, H., Pandey, B. K., George, S., Pandey, D., Anand, R., Sindhwani, N., & Dadheech, P. (2022, July). Effective Overview of Different ML Models Used for Prediction of COVID-19 Patients. In *Artificial Intelligence on Medical Data: Proceedings of International Symposium, ISCMM 2021* (pp. 185-192). Singapore: Springer Nature Singapore.

Singh, J., Pandey, D., & Singh, A. K. (2023). Event detection from real-time twitter streaming data using community detection algorithm. *Multimedia Tools and Applications*, *83*(8), 1–28. doi:10.1007/s11042-023-16263-3

Singh, S., Madaan, G., Kaur, J., Swapna, H. R., Pandey, D., Singh, A., & Pandey, B. K. (2023). Bibliometric Review on Healthcare Sustainability. Handbook of Research on Safe Disposal Methods of Municipal Solid Wastes for a Sustainable Environment, 142-161.

Tripathi, R. P., Sharma, M., Gupta, A. K., Pandey, D., Pandey, B. K., Shahul, A., & George, A. H. (2023). Timely prediction of diabetes by means of machine learning practices. *Augmented Human Research, 8*(1), 1. doi:10.1007/s41133-023-00062-4

Vinodhini, V., Kumar, M. S., Sankar, S., Pandey, D., Pandey, B. K., & Nassa, V. K. (2022). IoT-based early forest fire detection using MLP and AROC method. *International Journal of Global Warming, 27*(1), 55–70. doi:10.1504/IJGW.2022.122794

Zaidi, A. (2023). Predicting wildfires in Algerian forests using machine learning models. *Heliyon, 9*(7), e18064. Advance online publication. doi:10.1016/j.heliyon.2023.e18064 PMID:37519679

Zaidi, A. (2023). Smart Implementation of Industrial Internet of Things using Embedded Mechatronic System. *IEEE Embedded Systems Letters*. doi:10.1109/LES.2023.3309985

Zaidi, A., Ajibade, S.-S. M., Musa, M., & Bekun, F. V. (2023). New Insights into the Research Landscape on the Application of Artificial Intelligence in Sustainable Smart Cities: A Bibliometric Mapping and Network Analysis Approach. *International Journal of Energy Economics and Policy., 13*(4), 287–299. doi:10.32479/ijeep.14683

Zaidi, A., & Al-Harbi, O. (2023). Statistical Analysis of a Linear Multi-Step Numerical Treatment. *Journal of Statistics Applications & Probability, 12*(1), 26. doi:10.18576/jsap/120126

Zaidi, A., & Al Luhayb, A. (2023). Two Statistical Approaches to Justify the Use of the Logistic Function in Binary Logistic Regression. *Mathematical Problems in Engineering, 2023*, 5525675. doi:10.1155/2023/5525675

# Chapter 8
# Recent Trends in Biomedical Technologies:
## Challenges and Opportunities

**Kannadhasan Suriyan**
https://orcid.org/0000-0001-6443-9993
*Study World College of Engineering, India*

**Nagarajan Ramaingam**
https://orcid.org/0000-0002-4990-5869
*Gnanamani College of Technology, India*

## ABSTRACT

*Telemedicine is the use of immersive multimedia communication to transmit patient knowledge in order to conduct consultations, medical tests and operations, and medical professional partnerships from a distance. The fact that telemedicine is an "accessible and rapidly changing science" is often emphasized. The constant advancement of technology, which paves the way for the expansion of internet connections and the expansion of data processing capability, has increased the possibilities for the global health industry, especially telemedicine, to expand. Information exchange, data mining, the internet of things, wearables, digital computing, and robotics are all emerging as key drivers of creativity in the coming decade.*

## 1. INTRODUCTION

Swarm Intelligence (SI) is a population-based optimization approach that is a sub-domain of soft computing techniques. To solve issues, it typically imitates the social behaviors of many animals seen in nature. In today's world of quadratic equations, Swarm Intelligence (SI) aids in the optimization of quadratic equations with maximum accuracy and in the shortest time possible. Particle Swarm Optimization (PSO) is a kind of swarm intelligence that replicates birds' social behavior while hunting for food. The expanded Kaptur's entropy criteria approach is used as the fitness function in this article to optimize two

DOI: 10.4018/979-8-3693-1335-0.ch008

threshold values utilizing the basic PSO methodology. In computer vision, edge detection is a crucial task. It is still an active study field since it is often employed as a first step in extracting information from pictures. Edge detection is utilized in object identification, picture segmentation, and understanding an image's structure. The edges of an image provide extremely useful information about the items' frontier/boundary in a computer vision application., Rao 2019- M.Han 2010.

Changes in the object's characteristics, such as geometry, shape, size, light intensity, and reflection, create the image's edges. In a gray-level picture, an edge is the line that separates two areas of different grey levels. To put it another way, an edge is a break in the picture intensity or its first derivative. There are two kinds of picture intensity discontinuities. Step discontinuities and line discontinuities are the two types of discontinuities. The rapid variations in picture intensity cause step discontinuities. Line discontinuities, on the other hand, are sudden changes in picture intensity that recover to their previous intensity after a given distance H.Gao 2010- Wu,B 2012. However, because to the smoothing provided by most sensing equipment, low-frequency components in pictures, and external noise, these discontinuities are seldom seen in actual photographs. As a result, the ramp edges were created by step intensity change edges, whereas the roof edges were made by line intensity change edges. It is essential to develop efficient image processing techniques, and clustering is one of them, to offer the proper job of pattern identification. Many current clustering techniques, on the other hand, rely on human tweaking and cannot guarantee correct results when visual input is distorted. One of the most significant difficulties of automatization in many areas of social life is the use of machine vision. The authors investigate the use of machine vision in autonomous cars. The definition of other road users (autonomous vehicles), identification of road infrastructure, and road marking are the main objectives of such study. (Jayakumar D.N 2014- Senthilnath J 2011).

## 2. BIOMEDICAL IMAGE PROCESSING

Most current methods are unique, and it seems that some modifications are required to utilize them in other situations. Some methods for organizing machine vision require a significant amount of computer power. The article focuses on methods to enhance the quality of pattern recognition. The suggested paper's relevance to the computer vision field is explained by the fact that computer vision may be integrated in vehicles, drones, and utilized to automate processes like road traffic and space monitoring. It is critical to offer accurate computer vision work since mistakes may result in fatalities and damages.

The authors claim that their goal is to develop efficient image processing techniques in order to prevent such outcomes. Segmentation and clustering are two terms that may be used to describe them. They do not, however, guarantee reliable findings in the event of visual information distortion or a high degree of noise. The reason for this is because these techniques need manual tweaking. The paper's authors attempted to create a clustering technique based on particle swarm optimization (PSO) that automatically selects parameters. This technique also incorporates the k-means clustering method, which calculates the minimum value of the distance function to arrange pixels into a predefined number of clusters. The particle group motion (pixel-by-pixel transit across the picture) and search for the optimum solution for the whole swarm (search for pixel with the highest average intensity value in a specific area) are utilized in comparison to the PSO.

In contrast to the original k-means technique, there is no requirement for the user to predetermine the cluster quantity, and in addition to the distance function reduction operation, the color function minimiza-

*Figure 1. Biotechnology technology*

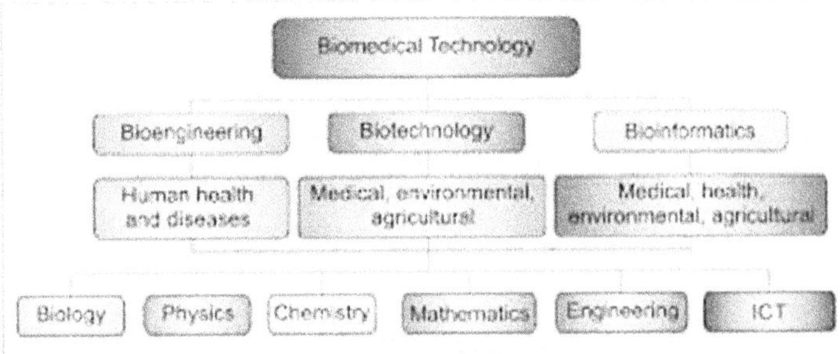


tion operation is introduced. Relative color uniformity describes an object in an image (or a portion of an object), i.e., if the distance function for pixels a and b belonging to the same object goes to zero, their color function likewise tends to zero. Experiments using non-spoiled pictures are given in this article to demonstrate the method's readiness for use in a work process with distorted visual information. In 3D modelling of metropolitan areas, LIDAR has become a common data collecting technique for gathering object surface information. Although these systems offer precise geometric information, they still need data interpretation for object extraction and identification before they can be used for mapping. Several clustering and filtering methods have been developed during the past decade to recognize objects for 3D city modelling or to remove the "artifacts" of barren landscape.

Additional data has been examined in order to enhance the performance of the 3D object extraction procedure. First, LIDAR systems record the first and final pulses of the laser beam, which correspond to the highest and lowest object points struck by the laser beam, respectively. In the presence of vegetation, the first and final pulse values will vary significantly. Furthermore, LIDAR systems record the strength of the returning laser beam, which is mostly in the infrared spectrum. Despite this, the intensity picture is under-sampled and, as a result, noisy. Visible pictures are also a very effective source of information. Due to their intensity and spectrum content, as well as their high spatial resolution, which is still higher than that of laser scanner data, digital pictures may offer essential information. As a result, different types of information have been fused in the context of 3D object extraction in urban areas to overcome the challenges of categorization and identification of complex objects. In the clustering process, collecting this information greatly increases the amount of data sets and correspondingly increases the dimension of feature spaces. The biomedical technology is shown in Figure 1.

As a consequence, most conventional clustering methods that have worked well in other applications with normal data and feature space dimensions are ineffective when used to extract objects from LIDAR data. One of the most common clustering methods is k-means, which may be used to handle large datasets. Because the computational cost of the method increases linearly with the amount of data points, it is effective at clustering big data sets. However, the method may converge on suboptimal solutions. In this article, we propose an alternate clustering technique for solving the k-means algorithm's local optimal issue. The particle swarm optimization technique is used in the approach because it has been shown to perform better than other intelligent optimization methods for a variety of difficult situations.

Contrast enhancement is important in both human and computer vision image processing. It's utilized in medical image processing, texture creation, voice recognition, and a variety of other image/video processing applications as a preprocessing step. For this aim, several methods have previously been devised. Some of these techniques rely on basic linear or nonlinear grey level transformation functions,

*Figure 2. Biomedical engineering*

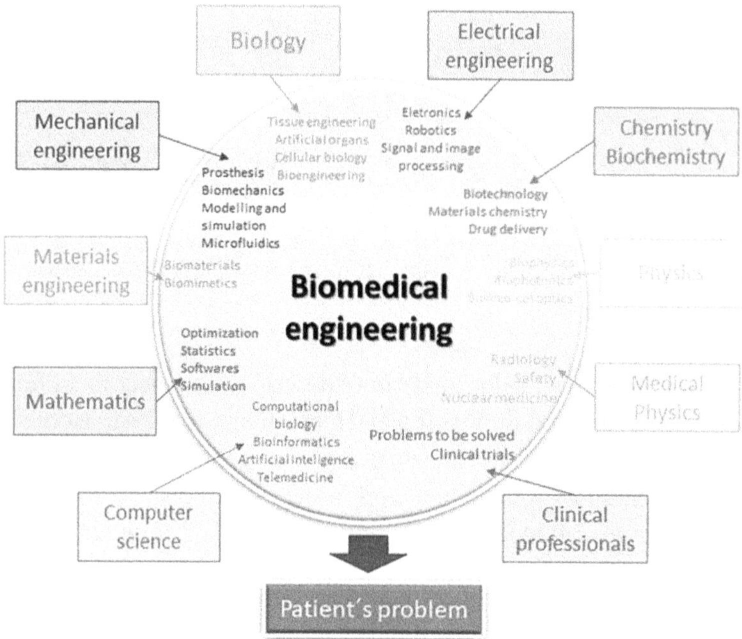


while others rely on a thorough examination of various picture characteristics such as edges, linked component information, and so on. In a digital picture, a histogram is a statistical probability distribution of each grey level. The global and local histogram equalization methods are the two main types of histogram-based equalization approaches.

In its transformation function, Global Histogram Equalization (GHE) utilizes the histogram information of the whole input picture. Despite the fact that this global method is appropriate for overall improvement, it fails to maintain the input image's local brightness characteristics. Normally, high frequency grey levels dominate low frequency grey levels in a picture. In this case, GHE remaps the grey levels in such a manner that the contrast stretching is limited to a few dominant grey levels with bigger picture histogram components, while the remainder suffer from substantial contrast loss.

Local histogram equalization (LHE) aims to solve these issues. It employs a tiny window that progressively glides across each pixel of the picture. The HE is applied to the pixels that are obscured by the window. The grey level mapping for enhancement is therefore done solely for the window's centre pixel. As a result, it makes excellent use of local data. LHE, on the other hand, has a significant computational cost and may result in over-enhancement in certain areas of the picture. Furthermore, this method has the issue of increasing both the sounds and the picture characteristics in the input image. Using non-overlapping block based HE, the significant computational cost of LHE may be reduced. However, on improved pictures, these techniques create an unappealing checkerboard appearance. The biomedical engineering is shown in figure 2.

Histogram Equalization (HE) is a commonly used method for picture contrast enhancement that is relatively successful on nearly all kinds of photos owing to its simplicity. Based on the probability distribution of the input grey levels, HE changes the image's grey levels. Histogram Specification (HS) is a method for controlling the anticipated output of an image histogram by defining the intended output histogram. However, since the output histogram pattern changes with pictures, defining it is a difficult job. The Dynamic Histogram Specification (DHS) technique dynamically produces the required histogram from the supplied picture. Although the original input picture histogram features are preserved, the degree of augmentation is not substantial.

## 3. OPTIMIZATION ALGORITHM

With the introduction of the GPU and numerous strong optimization algorithms that make little or no assumptions about the nature of the issue, the area of Computational Intelligence has experienced remarkable growth in the past two decades. Particle Swarm Optimization (PSO) is one of several such methods that has been extensively utilized to solve issues involving ill-structured continuous/discrete, restricted and unconstrained function optimization. The underlying workings of the PSO make adequate use of probabilistic transition rules to conduct parallel searches of the solution hyperspace without explicit assumption of derivative information, similar to popular Evolutionary Computing paradigms such as Genetic Algorithms and Differential Evolution.

The transition rules are based on an underlying physical model of emergent collective behavior resulting from the social interaction of flocks of birds and schools of fish. PSO has been used in an ever-increasing number of difficult, real-world optimization situations since its debut in 1995, when traditional methods either fail or are of limited use. Its intuitively simple form and small number of changeable parameters make it a popular option for a wide range of situations that need approximate solutions to some extent. However, there are many significant flaws in the basic PSO that create failure modes including stagnation and convergence to local optima, prompting considerable research into mitigation and resolution. Agents in a natural computing paradigm are decentralized creatures that can simulate complicated real-world systems while having no awareness of the high-level objective they are pursuing. This is made feasible by a set of low-level objectives that, when achieved, allow for meaningful collective behavior to emerge from these apparently inept and powerless single actors. Reeves' introduction of particle systems in the context of representing natural things like as fire, clouds, and water in computer-based animations while at Lucasfilm Ltd. provided an early impetus. During the development process, agents or "particles" are created, undergo form changes, and travel about in the modelling environment before being rejected or "died."

According to Reeves, such a model may capture the dynamics and shape of real settings that are impossible to portray using traditional surface-based representations. Reynolds' subsequent work in the developed basic principles that enhanced particle behavior autonomy and set out simple low-level norms that or particles might follow to produce emergent behavior. The Boids Model's complexity is therefore a direct result of the basic interactions among the constituent particles. Separation, alignment, and cohesion are three different flocking principles that a particle must obey, according to Reynolds. While the separation principle enables particles to move away from one another to avoid crowding, the alignment and cohesion principles also need directional updates in order to move towards the average heading and location of adjacent flock members. The boids' intrinsic nonlinearity results in chaotic behavior in emergent group dynamics, while the basic, low-level rules' negative feedback results in orderly activity. The situation where each boid knows where every other boid is is computationally infeasible because of its O(n2) complexity. Reynolds, on the other hand, proposed a neighborhood model in which boids in a general area exchange information, lowering the complexity to O(n) and speeding up the algorithmic implementation.

The flock or swarm prematurely converged in a unanimous manner by integrating local information sharing via closest neighbor velocity matching. As a result, a random disturbance or insanity was introduced in the particle velocities, resulting in sufficient variety and subsequent realistic swarm dynamics. Both of these characteristics were eventually removed since the flock appeared to converge on attractors just as effectively without them. As a result, the paradigm produced a population of agents that resembled the dynamics of a swarm rather than a flock.

One of the most important image processing techniques is image enhancement, which is defined as the transformation of one image into another in order to improve the interpretability or perception of information for human viewers or to provide better input for other automated image processing techniques. Point operation, spatial operation, transformation, and pseudocoloring are the four major types of picture enhancement methods. This paper's research is based on spatial operations. Histogram transformation is one of the basic procedures for gray-level picture contrast enhancement, which helps with subsequent higher-level operations like detection and identification. Linear contrast stretching is a technique that uses a linear transformation to cover the whole range of values in a given picture.

Pseudocoloring is a gray-scale picture enhancement method that falsely "colors" the image based on a color mapping, requiring lengthy interactive attempts to find an appropriate mapping. Separating the picture into its chromaticity and intensity components helps improve color images. The majority of image enhancement work involves using a transformation function to modify the picture histogram in order to get the desired contrast augmentation. As a result, this process also includes the adaptation of the image's color map to meet the needs of the human interpretation. In a real-coded picture, GA is employed in conjunction with a subjective assessment criteria to adjust the gray-level intensity transformation worldwide. GA creates the improved picture by combining various transformation functions with varied parameters.

Noise contamination is more visible during the collection, transmission, storage, or compression of a picture. Various types of noise, such as Gaussian, impulsive, and mixed sound, may influence images. Image denoising is a crucial stage in the image processing process. Pre-processing is a multi-step procedure in image processing. The goal of picture denoising is to eliminate noise from damaged images such that the original image can be estimated while the essential characteristics such as edges, textures, and details are preserved. Different picture denoising techniques have improved the quality of noisy photos damaged by various types of noise in recent years. provide a more comprehensive overview of the denoising methods Despite the fact that there are many types of picture denoising algorithms, their noise removal and run time are not optimum. The transform filtering of domains, spatial filtering, and learning are three kinds of picture denoising methods. Wavelets, Fourier transforms, Block-Matching and 3D filtering (BM3D) methods, and others are examples of transform domain filter picture denoising techniques. The bilateral, Gaussian, guide, and nonlocal mean filters are examples of the second kind of image denoising methods, which are primarily concerned with spatial filtering. The K-Singular Value Decomposition (K-SVD), Large-Scale Sparse Clustering (LSSC) and the Clustering based Sparse Representation (CSR) algorithms are examples of the following third sort focused on learning. The goal of these algorithms is to enhance picture quality.

The compressed sensing technique shows a new way to integrate both internal and exterior data for high-performance magnetic resonance imaging (MRI) data restoration. The Patch Group Prior based Denoising (PGPD) methodology outperforms the Patch Prior based Denoising (PPD) method in terms of

PSNR and picture performance. An energy minimization function is developed to combine the low- and high-frequency denoised results, unlike most of the methods mentioned. A robust method combining gradient histogram with SR is utilized to obtain good evaluations of the sparsely coded coefficients of the residual picture and to denoise images while preserving their textures. An iterative weighted sparse representation (IWSR) may accomplish complete denoising efficiency and excellent picture quality in contrast to other denoising methods. A dual domain filter (DDF) is used to improve the SAGD picture quality. The progressive DDF is used for low-quality optical remote-sensing pictures. DDF is made up of two parts. The employment of a bilateral filter as the guiding filter is the first step toward high-quality pictures. Then, using low-contrast helpful constructs, a Fourier transform is recovered and enhanced using an adaptive adjustment parameter.

Light absorption, back scattering, and forward scattering are common problems with underwater pictures. Back scattering is the amount of light reflected back to the camera by the water before it reaches the picture. Forward scattering occurs when an item emits a random light pattern toward the camera. Floating particles in the water are another issue that degrades the quality of underwater pictures. The amount of noise produced by these impacts is determined by the sea structure, water quality, and other factors. When photographing an underwater picture, the image will darken as the distance between the camera and the subject grows. Another issue that degrades the quality of underwater pictures is the fading of light and the fading of colors. Pre-processing procedures are required to reduce noise from underwater pictures in order to solve these issues. Several researchers presented several picture enhancing techniques. For image enhancement, evolutionary methods such as the genetic algorithm (GA) are employed. Another evolutionary method for improving gray-level and colored pictures are the particle swarm optimization (PSO) algorithm.

The GA-based enhancement methodology was compared to this method. The PSO-based enhancement approach outperformed the other methods in terms of duration, number of pixels on edges, and objective ratings. We apply the same technique here, namely the same enhancement model and objective function, but for underwater pictures, since this enhancement model has shown to be better. Because this method is intended to work for regular pictures, we conduct pre-processing procedures to eliminate noise from the underwater image. Several techniques for improving underwater pictures have been proposed, however evolutionary optimization algorithms have not been applied for underwater image improvement to our knowledge. Improving the brightness, color correction, visibility, or scene contrast may all help to improve the quality of underwater photographs. Optimization is a branch of mathematics concerned with finding the minima and maxima of functions under the influence of constraints. George Dantzig invented optimization in the 1940s, when he utilized mathematical methods to create "programs" (training timetables and schedules) for military usage. Since then, his "linear programming" methods and their descendents have been applied to a broad range of issues, ranging from manufacturing facility scheduling to airline yield management. Today, optimization encompasses a broad range of methods drawn from operations research, artificial intelligence, and computer science, and it is used to enhance business processes in almost every industry.

When the variables in the optimization function can only take a limited number of discrete values, discrete optimization issues emerge. For example, a hospital unit's staff scheduler has a limited number of staff members available, thus staff scheduling entails making discrete choices for each slot in the final schedule. Discrete optimization attempts to make these choices in such a way that a given function is maximized (for example, income) or reduced (for example, cost), while keeping limitations in mind, such as the necessary number of rest days for the staff in a schedule.

A brain tumor is a kind of aberrant cell aggregation that occurs in certain areas of the brain. Primary and metastatic brain tumors may be distinguished based on their incidence. Primary brain tumors begin in the brain, while metastatic brain cancers begin in other regions of the body. Tumors may be cancerous (also known as malignant) or noncancerous (also known as benign) (or benign). A malignant brain tumor expanded quickly to other parts of the brain and spine. The tumor is divided into four classes by a precise thorough classification, with the higher grade tumors being more malignant. Benign tumors may damage the brain and create irreversible consequences due to the presence of brain tumors in the central nervous system. Gliomas are the most common kind of primary brain tumor in adults. Gliomas are graded from I to IV severity according to the World Health Organization's grading system (WHO).

Grade I tumors have benign cells that seem to be normal. Cells in Grade II tumors give them a somewhat odd appearance. Grade III tumors are made up entirely of aberrant and malignant cells. Grade IV cells are thought to be completely aberrant and fast-spreading. Meningioma tumors are caused by glioblastoma multiforme (GBM), a kind of tumor that arises from the tissue layers known as the meninges. Meninges are protective membranes that surround the spinal cord and brain. They are often considered to be benign tumors since they grow slowly and have a minimal chance of spreading. Pituitary tumors are tumors that develop in the pituitary gland and account for approximately 14% of intracerebral cancers; some are caused by inherited genetic abnormalities, while others are caused by continual mutation. It may create serious health problems via tumors, which are believed to be benign due to their presence in sensible brain areas. Early detection plays an important part in the patient's recovery and therapy. Detecting a brain tumor and its grade is a time-consuming and difficult process. In most cases, the patient will be sent to magnetic resonance imaging after the brain tumor has progressed enough and other bothersome symptoms have emerged (MRI). After examining the pictures of the brain, the existence of a tumor is discovered, the patient is submitted to a brain biopsy. In contrast to MR, an accurate result in biopsy may take up to a month. Perfusion is a technique used by MRI experts to evaluate tumors and validate biopsy results. With the exception of biopsy, few novel methods for grading brain tumors have been developed in recent years. The use of perfusion MRI to differentiate low-grade and high-grade glioma has been able to overcome a few limitations in biopsy.

The implication of a computer-aided system is beneficial for detection. An effective and automated method for categorizing brain tumors helps doctors understand medical imaging and assists specialists' decisions in the early phases of tumor development. The grading of brain tumors in this research is done in less time and with greater precision. Furthermore, the whole categorization process is painless. Enough attention has been paid to the goal of diagnosing while evaluating the medical pictures. The existence of current ML methods has proved its effectiveness in addressing many problems, which has sparked interest in the area of health-related procedures and subjects. Several studies have been conducted in different tumor classifications using MRI, primarily evolutionary algorithms, MR brain images, and artificial neural networks (ANN), support vector machine (SVM), and hybrid intelligent techniques are shallow ML algorithms that are used to differentiate the abnormal and normal classes of images in brain MR that are denoted by existing works.

Image analysis is the process of processing pictures on a computer with the goal of determining what things are visible in the image. The process of segmenting a picture into its component pieces is known as image segmentation. It is one of the most crucial jobs in automated image analysis since the outcomes of segmentation will affect all subsequent tasks, such as feature extraction and object classification. Because of its importance, the segmentation process and approach have received a lot of attention in recent decades. This has resulted in a large number of (thousands) of distinct algorithms, and the number is constantly increasing. A few survey articles have also been published, although they only cover a fraction of the techniques that have been created. Because no one segmentation method is suitable for all pictures, and different algorithms are not equally suited for different applications, performance evaluation of segmentation algorithms is essential and therefore a crucial topic in the study of segmentation. Performance evaluation is essential for all computer vision algorithms, from research to implementation, and picture segmentation is a critical and fundamental stage in low-level vision. Fuzzy c-means (FCM) clustering is an unsupervised technique that has been used in astronomy, geology, medical imaging, target identification, and image segmentation to perform feature analysis, clustering, and classifier construction.

An picture may be classified in a variety of feature spaces, and the FCM method categories it by clustering comparable data points in the feature space. This clustering is achieved by repeatedly reducing a cost function in the feature domain that is based on the distance between the pixels and the cluster centers. The pixels in a picture are highly correlated, meaning that pixels in close proximity contain almost identical feature data. As a result, the spatial connection between neighboring pixels is an important feature that may serve as an excellent guide in image segmentation. The traditional FCM method, on the other hand, does not fully use this spatial information. This study makes use of a spatial FCM technique in this manner.

## 4. PARTCLE SWARM OPTIMIZATION ALGORITHM

GAs have been used to develop computer programs for particular tasks as well as to build other computational architectures. The work on Genetic Programming has sparked a recent revival of interest in AP with GA (GP). The GP paradigm enables program induction by exploring a space of potential computer programs for an individual computer program that is well-suited to addressing or approaching the issue at hand. The genetic programming paradigm allows for the evolution of computer programs that can perform alternative computations based on the results of intermediate calculations, that can perform computations on a variety of variables, that can perform iterations and recursions to achieve the desired result, that can define and use computed values and subprograms, and that can perform iterations and recursions to achieve the desired result.

Because such mechanisms create hierarchical structures that would enable the development of new high-level primitives from built-in low-level primitives, GP uses relatively low-level primitives, which are specified individually rather than integrated a priori into high-level primitives. Particle Swarm Optimization (PSO) is a recent heuristic method that may be used to solve problems that aren't linear or continuous. For continuous nonlinear functions, it is a population-based stochastic optimization method. The word "PSO" refers to a group of algorithms that may be used to discover optimum (or near optimal) solutions to numerical and qualitative issues. It's simple to implement in most programming languages, and it's shown to be both effective and fast when applied to a variety of optimization issues. Its connections to artificial life (A-life) in general, and bird flocking, fish schooling, and swarming theory in particular, are perhaps more apparent.

It is, nevertheless, linked to evolutionary computing and has connections to both Genetic Algorithms (GAs) and Evolutionary Programming (EP) (EP). Unlike GAs and EP, PSO is a straightforward idea that is simple to apply. It simply uses basic mathematical operations and is computationally light in terms of both memory and performance. Ant Colony Optimization (ACO) is the outcome of first research into computer intelligence methods to combinatorial optimization. ACO is based on an iterative process in which a population of simple agents construct candidate solutions repeatedly; this process is probabilistically guided by heuristic information on the given problem instance as well as a shared memory containing experience gathered by the ants in previous iterations. ACO has been used to solve a wide variety of difficult combinatorial problems. Components and states, which are sequences of components, are used to define problems.

One of the most important image processing techniques is image enhancement, which is defined as the transformation of one image into another in order to improve the interpretability or perception of information for human viewers or to provide better input for other automated image processing techniques. Point operation, spatial operation, transformation, and pseudocoloring are the four major types of picture enhancement methods, according to. This paper's research is based on spatial operations. Histogram transformation is one of the basic procedures for gray-level picture contrast enhancement, which helps with subsequent higher-level operations like detection and identification. Linear contrast stretching is a technique that uses a linear transformation to cover the whole range of values in a given picture. Pseudocoloring is a grayscale picture enhancement method that artificially "colors" the image based on a color mapping, requiring lengthy interactive trials to find an appropriate mapping. Separating the picture into its chromaticity and intensity components helps improve color images. The majority of image enhancement work involves using a transformation function to modify the picture histogram in order to get the desired contrast augmentation. As a result, this procedure likewise provides the most information from the picture. Image improvement has already been accomplished using evolutionary algorithms. The scientists used a global contrast enhancement method based on genetic programming (GP) to adjust the image's color map to the needs of the human interpretation.

Thresholding is defined as the process of determining an appropriate threshold to separate an item from its surroundings using a set of criteria such as Otsu thresholding, maximum entropy thresholding, minimal error thresholding, 2D maximum entropy thresholding, and Fisher information. The thresholding method must be expanded to multilayer thresholding when segmenting several items from the backdrop. When exploring multilayer thresholds thoroughly, however, a significant amount of computation and a lengthy computation time occur. At this moment, GA has been investigated and used to a variety of optimization problems. GAs is best suited for optimizing large search areas, but not for a complete search procedure. The technique provides an acceptable solution to an applied optimization problem quickly and effectively. The technique provides a trade-off between exploitation and exploration of the search space. GA provides near-optimal solutions suitable for real-time applications. As a result, GA performs a variety of jobs in image processing. A multi-term cost function is given by] which is lowered by using GA-evolved edge detection. In this approach, edge detection is created as a reduction issue as the primary cost function to image segmentation, and edge pictures population is generated by applying particular operators on space of the complete possible edge setup.

An optimum thresholding may be achieved by using GA. Fractional-order Darwinian particle swarm optimization (PSO) and Darwinian PSO and PSO are used to determine the thresholds. The issue of multilevel Otsu is solved using an ant colony optimization (ACO) segmentation technique. Bacterial foraging was utilized to find the thresholds for Kapur's and Otsu's maximizing primary functions. The differential evolution optimization method is used to find multilevel thresholds. Three optimization methods were projected through for selecting multilayer thresholds. The metaheuristic is unable to handle local and global search with increasing thresholds, which may result in inaccurate results and a slower rate of convergence.

Segmentation is one of the optimization problems, and GA effectively assesses the global maxima inside the search space and solves the parameter selection issue in picture segmentation. A new method based on GA was proposed for threshold selection from picture histogram. In this paper, we use entropy to achieve 2-D picture segmentation. The entropy is utilize to calculate the data amounts applied via the dispersion of picture grey levels. Image segmentation is the goal of entropy maximization. The optimization may be accomplished using GA PSO. The projected approach, unlike the traditional thresholding methodology, does not require the pre-organization of picture intensities. Image segmentation is a frequent combinatorial optimization challenge when compared to the traditional thresholding method. The main goal is to increase entropy maximization and the quality of picture segmentation.

Skin cancer is a type of cancer that affects the surface of the skin. Skin cancer can take two forms: malignant and benign. Benign Melanoma is characterized by the appearance of moles on the skin. The appearance of sores that bleed is a symptom of malignant melanoma. Malignant Melanoma is the most dangerous type of skin cancer. It develops as a result of a cancerous growth in a pigmented skin lesion. Malignant melanoma is named after the melanocyte, which is thought to be the cell from which it develops. This disease can be cured if caught early enough. Melanoma is difficult to diagnose and necessitates the use of samples and laboratory tests. Melanoma can spread throughout the body via the lymphatic system or the blood. The main issue to consider when dealing with melanoma is that the first occurrence of the disease can pave the way for subsequent cases. Inflammation or even the spread of a lesion is frequently caused by laboratory sampling. As a result, there has always been a scarcity of less risky and time-consuming methods. Computer-assisted diagnostics, which is based on the signs of the illness, may speed up the detection of skin cancer. The difficulty in diagnosing malignant cells is due to the similarity of skin lesions. Asymmetry, Border irregularity, Color change, and Diameter are some of the distinctive signs of skin cancer. These are often referred to as ABCD parameters. ABCD is a set of parameters. Asymmetry, border irregularity, color, and diameter are all factors to consider. When one half of a tumor does not match the other, it is called asymmetry. The unevenness of pictures is known as border irregularity.

The act of converting various sets of data into a single coordinate system is known as image registration. Multiple pictures, data from various sensors, times, depths, and perspectives are all examples of data. The image registration method is used in a number of fields, including biological imaging, medical imaging, and brain mapping. It may also be used to collect and analyze data and pictures from satellites. The registration procedure is critical for comparing and integrating data from various measures. It may also be described as the process of matching two images of the same scene obtained from various perspectives or by separate sensors geometrically. The goal is to combine several sources of data into a single coordinate system.

Nonalignment of pictures of the same scene may be caused by a variety of factors, including calibration problems, scanning machine setup mistakes, and varied scan geometries, such as slice location, orientation, magnification, and thickness of different sensors. Due to relative motion between the camera and object planes, the non-alignment occurred. Extracted natural or man-made items from Google Earth have been utilised for a variety of purposes, including military, map publication, transportation, and automobile navigation. Automatic extraction of natural objects (roads, buildings, forests, water bodies, and so on) is a difficult issue, and no current program can do it consistently. Automatic methods have the potential to increase the speed and usefulness for military and civic applications, since human extraction of natural objects from images is time intensive. Google Earth was formerly known as Earth Viewer and is a virtual globe, map, and geographic information application. Google Earth can show you satellite pictures of any location. The Google Earth picture is used in this article to demonstrate how to extract natural or man-made objects from satellite images. Simultaneously, a new generation of computer algorithms has evolved from the study of social insect behavior. Self-organization and little or non-existent communication are typical characteristics of social insects. Every social insect is independent in its own right. They may use stigmergy to get information about the surroundings and interact with distant insects or the environment. Swarm Intelligence is defined by all of these characteristics. Ant colonies, bird flocking, mammal herding, and fish schooling are examples of these characteristics in nature. Ant Colony Optimization (ACO) and Particle Swarm Optimization are the two most frequently utilised swarm intelligence methods (PSO).

## 5. CONCLUSION

The technique of dividing a digital picture into several segments set of pixels is known as image segmentation. Color, intensity, and texture are some of the homogeneity criteria used to compare pixels in an area. The objects and borders in a picture are located and identified by these pixels. In the area of image processing, image segmentation is a significant research subject. The majority of computer vision image analysis problems need a segmentation step in order to identify objects and split the picture into homogenous areas based on color, motion, texture, and other factors. Filtering noisy images, medical applications (locate tumors and other pathologies, measure tissue volumes, diagnosis, treatment planning, computer guided surgery, study of anatomical structure), locate objects in satellite images, Face Recognition, Finger Print Recognition, and so on are all examples of practical applications of image segmentation. The choice of one segmentation technique over another, as well as the degree of segmentation, is determined by the kind of picture and the features of the issue at hand. The primary goal of employing a clustering method is to group data into tiny groups with comparable features, despite the fact that the data clusters are different from one another. It is an unsupervised clustering method that has a significant preference for local minima when determining the best solution. As a result, the distribution of initial cluster centers has a significant impact on the clustering process. As a result, determining appropriate starting parameters is a difficult issue, and clustering algorithms need a greater number of experiments to determine the input parameters.

## REFERENCES

Azarbad, M., Ebrahimzadeh, A., & Babajani-Feremi, A. (2010). Brain tissue segmentation using an unsupervised clustering technique based on PSO algorithm. *17th Iranian Conference of Biomedical Engineering (ICBME)*. IEEE.

Broilo, M. & Natale, F.G.B (2010). A Stochastic Approach to Image Retrieval Using Relevance Feedback and Particle Swarm Optimization. IEEE Transactions on Multimedia, 12(4).

Deng-xu, H. E., Hua-zheng, Z. H. U., & Gui-qing, L. I. U. (2011). Glowworm swarm optimization algorithm for solving multi-constrained QoS multicast routing problem. *Seventh International Conference on Computational Intelligence and Security*, (pp. 66-70). IEEE.

Gao, H., Xu, W., Sun, J., & Tang, Y. (2010). Multilevel Thresholding for Image Segmentation Through an Improved Quantum-Behaved Particle Swarm Algorithm. *IEEE Transactions on Instrumentation and Measurement, 59*(4).

Han, M., & Yao, W. (2010). Remote sensing image fusion using particle swam optimization. *International Conference on Intelligent Control and Information Processing, 241*, 13-15

Islam, M. R., Mondal, C., Azam, M. K., & Islam, A. S. M. J. (2016). Text detection and recognition using enhanced MSER detection and a novel OCR technique. In *2016 5th International Conference on Informatics, Electronics and Vision (ICIEV),* (pp. 15-20). IEEE.

Jayakumar, D. N., & Venkatesh, P. (2014). Glowworm swarm optimization algorithm with topsis for solving multiple objective environmental economic dispatch problem. *Applied Soft Computing, 23*, 375–386.

Liao, W. H., Kao, Y., & Li, Y.-S. (2011). A sensor deployment approach using glowworm swarm optimization algorithm in wireless sensor networks. *Expert Systems with Applications, 38*(10), 12180–12188.

Mehdi, H. (2012). Removal of noise patterns in handwritten images using expectation maximization and fuzzy inference systems. *Elsevier journal of pattern recognition*, pp. 4237-4249.

Pramanik, R., & Bag, S. (2018). Shape decomposition-based handwritten compound character recognition for Bangla OCR. *Journal of Visual Communication and Image Representation, 50*, 123–134. doi:10.1016/j.jvcir.2017.11.016

Senthilnath, J., Omkar, S. N., Mani, V., Tejovanth, N., Diwakar, P. G., & Archana Shenoy, B. (2011). Multi-spectral satellite image classification using glowworm swarm optimization. *Geoscience and Remote Sensing Symposium*, (pp. 47-50). IEEE.

Sharma, N., Kumar, B., & Singh, V. (2014). Recognition of Off-line Hand printed English Characters, Numerals, and Special Symbols. *IEEE International Conference- Confluence The Next Generation Information Technology Summit.*

Wu, B., Qian, C., Ni, W., & Fan, S. (2012). The improvement of glowworm swarm optimization for continuous optimization problems. *Expert Systems with Applications, 39*(7), 6335–6342.

Yuli, Z. H. A. N. G., Xiaoping, M. A., & Yanzi, M. I. A. O. (2011). Localization of Multiple Odor Sources Using Modified Glowworm Swarm Optimization with Collective Robots. *Proceedings of the 30th Chinese Control*, (pp. 1899-1904). IEEE.

Zhiwei, Y., Zhengbing, H., Xin, Z., & Bin, X. (2010). Image Segmentation Based on 2-D Fisher and Chaos Particle Swarm Optimization Algorithm. *2nd International Symposium on Information Engineering and Electronic Commerce (IEEC)*. IEEE.

# Chapter 9
# An Industry-Focused Traffic System Utilising Internet of Things

**Binay Kumar Pandey**
https://orcid.org/0000-0002-4041-1213
*Department of Information Technology, College of Technology, Govind Ballabh Pant University of Agriculture and Technology, Pantnagar, India*

**Vinay Kumar Nassa**
https://orcid.org/0000-0002-9606-7570
*Tecnia Institute of Advanced Studies, India*

**A. P. Mukundan**
https://orcid.org/0009-0009-5608-4788
*Champions Group, Singapore & Aligarh Muslim University, India*

**Darshan A. Mahajan**
https://orcid.org/0000-0002-1239-6343
*NICMAR University, India*

**Digvijay Pandey**
https://orcid.org/0000-0003-0353-174X
*Department of Technical Education, Government of Uttar Pradesh, India*

**A. Shaji George**
https://orcid.org/0000-0002-8677-3682
*Business System Department, Almarai Company, TSM, Riyadh, Saudi Arabia*

**Pankaj Dadheech**
https://orcid.org/0000-0001-5783-1989
*Swami Keshvanand Institute of Technology, Management, and Gramothan, India*

## ABSTRACT

*Radio frequency identification technology (RFID) and time series forecasts are used to create a dependable IoT-based traffic system. The system regulates city traffic. The suggested method estimates junction traffic volume over time using LSTM neural networks. RFID technology improves data collection accuracy and reliability. Data preparation includes outlier identification to remove anomalies. Training the LSTM model on preprocessed data reveals traffic volume trends. The trained model predicts traffic volume using historical data. Prediction performance is quantified by MAE, MAPE, and R2. The proposed approach is tested using four intersection traffic data. Results indicate that LSTM-based traffic volume estimation works. The optimal design is determined by evaluating system performance for 12-to-168-time steps. The experimental findings suggest that the proposed method can accurately anticipate traffic volume, helping traffic managers enhance flow. RFID and time series projections bolster traffic system reliability.*

DOI: 10.4018/979-8-3693-1335-0.ch009

## 1. INTRODUCTION

The management of transportation and the control of traffic are essential components of urban infrastructure. These elements have a direct influence on the effectiveness, safety, and sustainability of transportation networks. As a result of the rapid advancement of technology, there has been a rise in the amount of interest shown in the use of innovative solutions to meet the challenges posed by inefficiencies and traffic congestion. One such strategy is the incorporation of Internet of Things (IoT) technology into traffic systems, which enables the collection, evaluation, and use of real-time data with the purpose of producing the best possible results in terms of traffic management. We present an Internet of Things (IoT)-based robust traffic system in this study (Menon, V.,et al.,2022). This system enhances the flow of traffic and increases the overall efficiency of transportation by leveraging the power of IoT and several predictive modelling approaches. Our system utilizes time series forecasts in conjunction with Radio Frequency Identification (RFID) technology in order to enhance its capabilities in the areas of traffic control and monitoring. The development of a comprehensive framework that makes accurate projections of future traffic trends by making use of historical data on traffic and predictive modelling is our key objective. By studying and gaining an understanding of historic traffic patterns, we are able to accurately predict the occurrence of traffic congestion and make well-informed decisions regarding how to best adjust the timing of traffic signals, make effective use of available resources, and successfully alleviate congestion. When it comes to making accurate projections, we rely on a method known as time series analysis. The past traffic data gathered from a variety of crossings is the primary source of input for our forecasting algorithm. The data are preprocessed using stringent procedures to eliminate outliers, which ensures that the results will be accurate and predictable. Following this step, univariate time series forecasting techniques are utilised in order to create forecasts regarding future traffic patterns. Along with improvements to traffic control and monitoring, the RFID technology that our system utilizes also enables us to make more accurate predictions of time series. Vehicles are outfitted with radio frequency identification (RFID) tags, which enables ongoing tracking and identification (Gupta, A. K.,et al.,2023). Traffic managers now have the ability to adjust signal timings, dynamically change traffic flow, and react fast to changing traffic situations thanks to real-time vehicle data and projected traffic patterns. In order to demonstrate that the system we propose actually works, we must first conduct a great deal of research and testing on it (Pandey, B. K.,et al.,2023). We employ performance measures like as mean absolute error (MAE), mean absolute percentage error (MAPE), and coefficient of determination (R-squared) to examine the accuracy and dependability of our predictions. For example, MAE stands for mean absolute error, while MAPE and R-squared stand for mean absolute error and mean absolute percentage error respectively (Pandey, B. K.,et al.,2023).

The results indicate that our Internet of Things–based traffic system is able to accurately forecast traffic patterns and boost the efficiency of traffic management. Utilizing Internet of Things technology and predictive modelling techniques, this research makes a contribution to the field of transportation management. By utilizing the real-time data and insights provided by internet of things and radio frequency identification technologies, traffic managers are able to make more educated decisions that optimize traffic flow, minimize congestion, and increase overall transportation efficiency. The architecture that was proposed has the potential to revolutionize conventional traffic management systems, which would lead to transportation networks that are both more intelligent and more environmentally friendly(Swapna, H. R.,et al.,2023) (Malhotra, P.,et al.,2021).

## 2. APPLICATION OF SENSORS IN TRAFFIC MONITORING

Within the scope of this investigation, an essential factor to take into account is the function that sensors play in the monitoring of traffic. When it comes to the collection of real-time data about traffic conditions including vehicle count, speed, occupancy rate, and flow, sensors play a crucial role. By carefully deploying different kinds of sensors across road networks, it is possible to obtain information on the traffic situation that is both accurate and current. These sensors may take the form of loop detectors that are implanted in the pavement, infrared sensors, acoustic sensors, or even cameras that collect data via visual input. The obtained data is put to use in the construction of prediction models and the development of intelligent solutions for traffic management. By utilizing the power of sensors, traffic engineers and researchers are able to get insights into traffic patterns, locate areas that are prone to high levels of congestion, and make educated decisions on the optimization of traffic signals and route design. To successfully implement the internet-of-things-based traffic system that was proposed in this study, it is therefore essential to have a solid understanding of the role that sensors play in the monitoring of traffic.

## 3. COMMUNICATION PROTOCOLS DESIGNED FOR MONITORING AND CONTROLLING IN REAL TIME

Within the context of this study, the successful implementation of an Internet of Things-based traffic monitoring system calls for the use of communication protocols that deliver real-time monitoring and control of the system. This is a prerequisite for the efficient deployment of the system. Because of these protocols, the movement of data between sensors, traffic management systems, and other linked pieces of equipment may be made to feel very fluid and natural. The Constrained Application Protocol (CoAP) and the Message Queuing Telemetry Transport (MQTT) are both dependable and effective communication protocols that ensure the trustworthy transmission of real-time traffic data through the network. MQTT is an abbreviation that stands for "Message Queuing Telemetry Transport." CoAP is an abbreviation that stands for "Constrained Application Protocol." These protocols provide solutions that are not only lightweight but also scalable, which not only enables more effective utilization of network resources but also reduces the amount of time that is wasted waiting. In addition, the data's integrity and confidentiality are safeguarded by using secure communication protocols such as HTTPS (Hypertext Transfer Protocol Secure) and TLS (Transport Layer Security), which restrict unauthorized access to the data as well as any modifications to the data. HTTPS and TLS are two examples of these kinds of protocols. The Internet of Things-based traffic monitoring system that has been suggested in this work is able to achieve seamless integration of multiple components, as well as effective real-time monitoring and control of the conditions of the traffic, by applying the appropriate communication protocols. This allows the system to monitor and control the conditions of the traffic more effectively.

## 4. PREDICTIVE MODELING

In the context of this article, predictive modelling is essential since it seeks to enhance the functionality of the IoT-based traffic monitoring system by providing insights and forecasts on future circumstances of the traffic. Analyzing historical traffic data gathered from sensors and other sources using predictive

*Figure 1. Histogram of the number of devices in a given year*

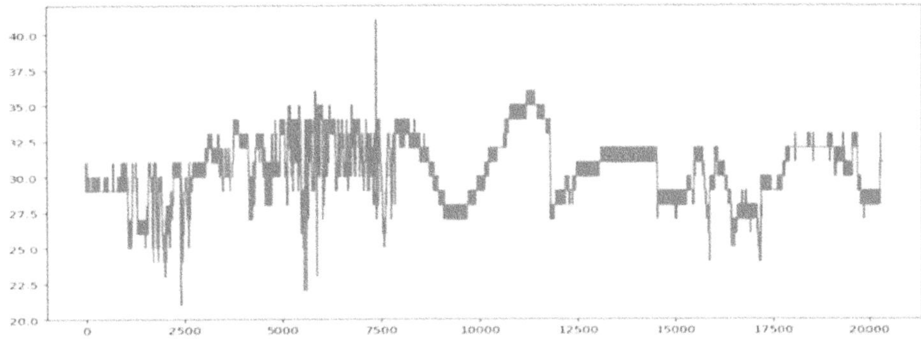


modelling methodologies, such as machine learning algorithms, can help find patterns and trends in the data. The results of these models can then be utilised to provide predictions regarding travel times, traffic flow patterns, and levels of congestion. By applying predictive modelling, traffic management systems are able to make proactive assessments and take preventative steps, thereby reducing the likelihood of anticipated congestion and increasing the efficiency of traffic flow. In addition, these models can assist with resource distribution and traffic planning, which enables authorities to more effectively distribute resources and implement programs to reduce congestion and improve overall traffic management. The incorporation of predictive modelling techniques into the internet-of-things-based traffic monitoring system described in this research provides decision-makers with helpful insights and enables them to make educated decisions to improve the efficiency of traffic and the overall experience of travelling.

## 5. DYNAMIC ADJUSTMENT OF SIGNAL TIMINGS BASED ON TRAFFIC PREDICTIONS

The traffic monitoring system described in this study is based on the Internet of Things (IoT) and places significant emphasis on the use of dynamic signal timing adjustment, which is contingent upon traffic estimates. The system has the ability to predict traffic conditions in real-time and adjust signal timings accordingly, utilizing the predictive modelling skills discussed above (Daoud, M. K.,et al.,2023). The conventional timing of traffic signals is typically predetermined and inflexible, leading to inefficiencies and the possibility of congestion during periods of high demand or unexpected traffic incidents. By including traffic prediction models into the system, the ability to adapt signal timings in response to projected traffic patterns becomes feasible. This adaptive approach enhances the efficiency of traffic movement, mitigates congestion, and decreases travel durations. The system has the capability to assign priority to traffic lights by considering anticipated traffic volumes, levels of congestion, and other relevant parameters. This ensures that traffic signals are able to promptly respond to the prevailing traffic conditions. The proposed system aims to enhance the efficiency of traffic management, reduce delays, and enhance the overall commuting experience for road users by dynamically adjusting signal timings according to traffic estimations (Daoud, M. K.,et al.,2024).

**Previous Work:** An investigation on the use of statistical learning time series algorithms to forecast the traffic that will be produced by devices connected to the internet of things. By concentrating on the

*Figure 2. Histogram of the number of devices in a given year*

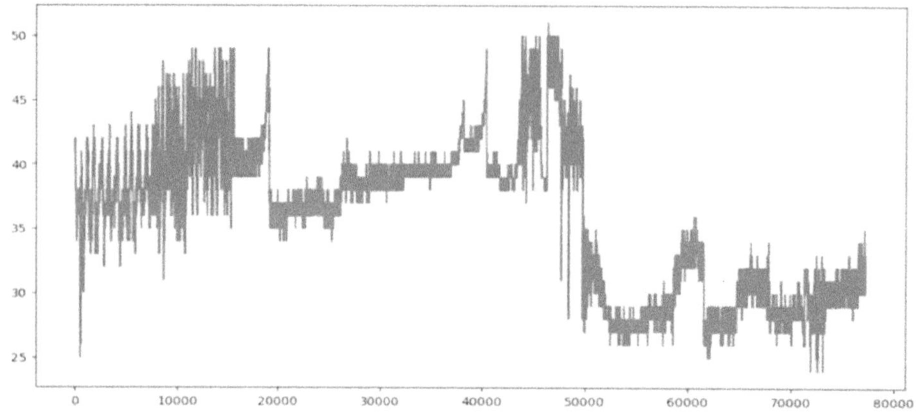


utilization of IoT data to estimate traffic patterns, the authors highlight the significance of statistical learning approaches in the process of processing time series data for the purpose of traffic prediction (Khedkar, S. P.,et al.,2021). A monitoring solution for Internet of Things devices that is built on C5.0 decision trees and time-series analysis. The research places an emphasis on the application of decision tree algorithms and time-series analysis to effectively monitor and manage IoT devices, which will help to contribute to the development of reliable IoT systems (Zhu, B.,et al.,2021). An analysis of the differences and similarities between machine learning and time series forecasting techniques for electric vehicle load. The authors emphasize the necessity of precise load forecasting in the context of electric vehicles and study the application of these methods to anticipate electric car charging patterns (Pandey, D.,et al.,2021). A sophisticated traffic information system that incorporates both Internet of Things and agent technologies. The research is centered on exploiting the capabilities of the internet of things (IoT) as well as agent-based systems in order to construct an intelligent traffic information system that is able to provide efficient traffic management and provide drivers with real-time information (Pandey, B. K.,et

*Figure 3. Histogram of the number of devices in a given year*

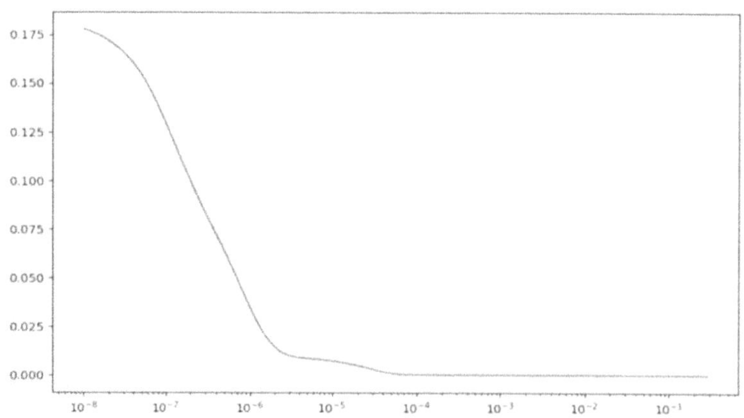

al.,2021). This method anticipates the amount of foot traffic that will enter a shop by utilizing facial recognition, time series, and support vector machines. This study demonstrates how these approaches may be utilised to predict shop foot traffic, offering valuable data that can be utilised by retailers who are aiming to maximize resource allocation and improve customer experience (Sharma, M.,et al., 2022). A smart city monitoring and management system that is capable of effectively handling rubbish. This research demonstrates the usefulness of time-series analysis in the context of smart city applications by optimizing garbage collection and management in urban areas through the application of time-series forecasting (Kumar, M. S.,et al.,2021). The development of a sophisticated transport network using the Internet of Things as its foundation. The authors place a strong emphasis on the incorporation of Internet of Things (IoT) technology into transportation systems, highlighting the potential advantages of these technologies in terms of increased productivity, improved safety, and reduced environmental impact. (Singh, S.,et al.,2023). A Bayesian learning network for estimating the speed of oncoming vehicles while taking into account the degree of uncertainty (Pandey, J. K.,et al.,2022). The goal of the research is to develop a probabilistic method for estimating the speed of traffic, which will make it possible to gain a deeper comprehension of, and a more precise grasp on, the degree of uncertainty associated with traffic forecasts (Sahani, K.,et al.,2023). An in-depth analysis of the several data-driven approaches that can be used to forecast the arrival times of vehicles on highways. The authors emphasize the significance of accurate travel-time prediction for efficient transportation management (Boopathi, S.,et al.,2023), which leads them to study a variety of data-driven approaches and their applications in the process of predicting travel times on highways. A C-ITS data recovery system based on an inductive graph neural network that is aware of its spatial and temporal context. The goal of the project is to recover lost data in connected intelligent transportation systems (C-ITS) by making use of graph neural networks, taking into consideration the data's geographical and temporal linkages. The research contributes to an improvement in the data quality as well as the dependability of C-ITS applications (Abdulkarim, Y. I.,et al.,2023). A back propagation neural network coupled with an adaptive differential evolution approach is utilised for the purpose of time series forecasting. The importance of using neural networks and evolutionary algorithms to make accurate forecasts of time series data is emphasized throughout the study. The research highlights the usefulness of utilizing this hybrid method in a number of different forecasting applications (Raja, D.,et al.,2024). In order to track the development of sugarcane, a monitoring system that uses RFID and ZigBee technology was developed and put into operation. The authors focus on employing RFID and ZigBee to monitor and track the development of sugarcane in order to illustrate the potential of these technologies in agricultural monitoring and management (Kumar Pandey, B.,et al.,2022). An Internet of Things-based system for tracking school buses and other vehicles makes use of RFID technologies and mobile data networks. The study underlines the use of RFID and mobile data networks to track school buses and cars, highlighting the necessity of real-time monitoring and safety in student transportation (Pramanik, S.,et al.,2023). In addition, the study highlights the use of mobile data networks to track school buses and automobiles. A spatial-temporal Chebyshev graph neural network is utilised in IoT-based intelligent transportation systems (ITS) for the purpose of performing traffic flow prediction. This study focuses on employing graph neural networks to anticipate traffic flow while taking into consideration spatial and temporal linkages. It also highlights the potential of the internet of things (IoT) in enhancing the accuracy of traffic prediction [19]. Forecasting traffic congestion with ARIMA requires a lot of modelling. This study demonstrates the usefulness of time series analysis in traffic management and planning (Pandey, D.,et al.,2022) by investigating the application of ARIMA models to forecast traffic congestion. A smart city traffic monitoring system is powered by a combination of a cellular network that supports 5G connections, RFID, and machine learning. The research places a strong emphasis on the incorporation of these technologies to monitor traffic in smart cities, which paves the way for efficient traffic management and decision-making in real time (Pandey, B. K.,et al.,2023). A fusion method for precise vehicle placement that takes use of radio frequency identification and the sensors found within the vehicle itself(B. Muniandi et al.,2019).The authors highlight the use of RFID and in-car sensors as a means to improve the precision of vehicle positioning, contributing in turn to location-based applications and services (Sharma, M.,et al.,2024). A approach for the short-term forecasting of link traffic for urban road networks that is based on GAN. The research is centered on the utilization of generative adversarial networks (GANs) to anticipate short-term link traffic, while also providing a parallel learning framework to improve prediction accuracy and dependability (Dave, P.,et al.,2021). A smart traffic management system for the next generation that combines YOLOv4 with XGBoost in a hybrid form. The research highlights the use of computer vision techniques in conjunction with gradient boosting algorithms in order to develop an advanced traffic management system. The research also highlights the potential of these approaches to improve traffic flow and safety (Khan, B.,et al.,2021). Combining a fruit fly optimization algorithm with a least squares support vector machine to create a method for forecasting traffic flow. This work focuses on enhancing the accuracy of traffic flow predictions by combining support vector machines with a fruit fly optimization method in order to demonstrate the potential of optimization techniques in traffic forecasting (Vinodhini, V.,et al.,2022). A powerful technique based on deep learning that can accurately forecast how well Internet of Things (IoT) communication systems will function in smart cities. This study has an emphasis on the application of deep learning algorithms to forecast the performance of Internet of Things communication networks, which helps to optimize and manage smart city infrastructure (Saxena, A.,et al.,2024). A parking spot-finding technology that works in real time for smart cities (Lelisho, M. E.,et al.,2022). The purpose of the research is to develop a method that can improve parking management and reduce traffic congestion in urban areas by creating a system that can determine the availability of parking spots in real time based on the information collected from sensors and from the past (Devasenapathy, D.,et al.,2024).

## 6. PROBLEM STATEMENT

The problem of inefficiency and congestion in urban traffic management systems is investigated in this paper. Traditional methods of traffic management are based on fixed signal timings and have limited access to real-time information. This leads to inefficient traffic flow, which in turn increases congestion and lengthens travel times. In addition, the absence of comprehensive monitoring and control mechanisms makes it difficult to efficiently respond to shifting traffic circumstances and occurrences. This is a significant barrier. As a solution to these issues, the authors of this study propose an Internet of Things–based traffic monitoring and control system. This system would incorporate a wide variety of sensors, communication protocols, predictive modelling methodologies, and dynamic modifications to signal timing. The objective is to create a traffic management system that is more intelligent and responsive, able to adapt to changing traffic patterns, optimize signal timings, and improve overall traffic efficiency. This will result in less congestion, faster travel times, and a more pleasant commuting experience for people who use roads.

*Figure 4. Histogram of the number of devices in a given year*

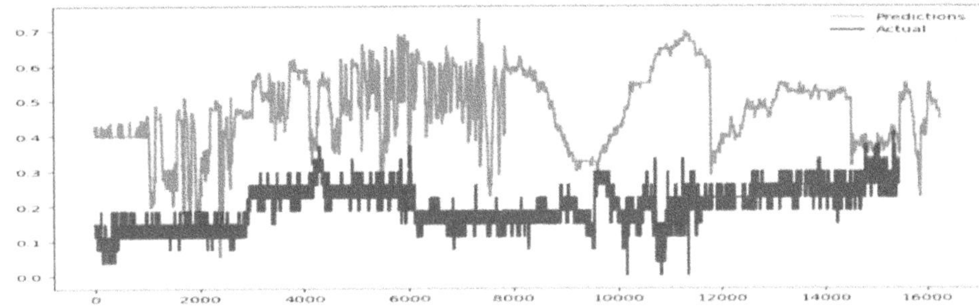


## 7. PROPOSED WORK

The purpose of the suggested study in this paper is to develop an Internet of Things-based traffic monitoring and control system to enhance the management of urban traffic. Using a network of sensors that are spread out across the road network, the system will gather real-time traffic data such as the volume of vehicles, their speeds, and the occupancy levels of the vehicles. This information will be transferred to a centralized control contextualizing effective communication protocols, which will allow for the flow of information to take place in a seamless and prompt manner (Zaidi, A., et al., (2023)). The ability to control traffic will be improved through the application of predictive modelling techniques, which will be used to assess historical and real-time data in order to offer accurate traffic estimates and improve management skills. These estimates will be useful for dynamically changing signal timings at crossings, which will both guarantee optimal traffic flow and reduce congestion. The intended system will be designed to monitor and control traffic conditions in real time, which will allow for quick responses to accident hotspots and areas with heavy congestion. In addition, the system will deliver valuable insights into traffic patterns and trends, which will make it possible to plan for the long term and make decisions regarding urban transport infrastructure. By using sensors, communication protocols, predictive modelling, and dynamic signal modification, the proposed research hopes to usher in a new era of traffic management and boost overall traffic efficiency in urban areas. Within the context of this study, a number of histograms (Figure 1, Figure 2, Figure 3, and Figure 4) are utilised to depict the distribution of the number of devices in a specific year. The frequency of device counts as well as the variability of those counts over time can be gleaned from these histograms. In addition, Table 1 provides an analysis of how accurate the predictions were or how well the model performed by presenting the mean absolute error values for each of the different epochs. Table 2 provides a measure of the overall prediction error by displaying the mean squared error values for the same epochs that were used in Table 1.

## 8. RESULTS AND DISCUSSION

The results of this research indicate that an Internet of Things (IoT)-based traffic monitoring and control system can improve the management of urban traffic. The system acquired accurate data on the current traffic situation in real time through the utilization of a decentralized network of sensors and then successfully transmitted that information to a centralized command and control facility through the ap-

*Table 1. Mean absolute error for epochs*

| Epochs | Mean Squared Error | |
|---|---|---|
| 20 | 0.30431 | |
| 40 | 0.29322 | |
| 60 | 0.13945 | |
| 80 | 0.12344 | |
| 100 | 0.11342 | |

plication of reliable communication protocols. The utilization of predictive modelling methodologies led to accurate traffic estimates, which enabled the real-time modification of signal timings at junctions in accordance with the conditions that were projected to be present in the traffic (Ahmad, A. Y. B.,et al., 2023). As a direct result of these adjustments, traffic flow has been enhanced, congestion has been minimized, and overall traffic efficiency has increased (Muniandi, B.,et al.,2019). Because the system could monitor traffic conditions in real time, it was feasible to react rapidly to events and congestion hotspots, which enabled preventative actions to be taken to reduce any interruptions that may have occurred. In addition to this, the system provided significant insights on traffic patterns and trends, which enabled informed decisions to be made on the long-term planning and development of infrastructure. According to the findings, this Internet of Things-based technology has the potential to transform traffic management practices, which would pave the way for urban transportation systems that are smarter and more efficient. However, additional analysis and testing are need to evaluate the scalability, dependability, and feasibility of the system in more populous urban regions (Ahmad, A. Y. B.,et al., 2024).

## 9. CONCLUSION

This article demonstrated how a traffic monitoring and control strategy that is based on the internet of things has the potential to completely revolutionize the way in which urban traffic is managed. It was discovered that putting into action a strategy that included sensors, communication protocols, predictive modelling, and dynamic signal adjustment was an efficient method for enhancing the flow of traffic, reducing the negative effects of congestion, and making transportation more efficient as a whole. The findings revealed that the system was able to collect real-time traffic data, generate accurate forecasts, and dynamically alter signal timings in accordance with the anticipated conditions of the road traffic.

*Table 2. Mean squared error for epochs*

| Epochs | Mean Squared Error | |
|---|---|---|
| 20 | 0.29524 | |
| 40 | 0.29143 | |
| 60 | 0.26954 | |
| 80 | 0.22925 | |
| 100 | 0.22342 | |

The deployment of this system has the potential to revolutionize the traditional methods of managing traffic by offering preventative responses to problems with traffic flow and facilitating the making of well-informed decisions regarding long-term planning. This system has the ability to revolutionize the traditional methods of managing traffic by delivering these answers. In spite of the fact that the findings are positive, greater study and testing are required in order to solve the problems connected with deploying the system in larger urban areas, including those related to scalability, dependability, and feasibility. This research, despite having some drawbacks, helps pave the way for future improvements in intelligent transport systems and contributes to the development of urban landscapes that are both more intelligent and more environmentally friendly.

The results and implementation of this IoT-based traffic monitoring and control system provide a solid platform for future study and development in this area. There are many other avenues that could be pursued in the course of future research and development. To begin, the system can be expanded such that it covers a larger geographical area, which may include a number of different road networks and intersections. The infrastructure would need to be scaled up, communication protocols would need to be optimized, and predictive modelling techniques would need to be developed so that the additional data volume and complexity can be managed. The application of emerging technologies such as artificial intelligence and machine learning has the potential to enhance the accuracy of traffic forecast algorithms and enable the development of more complex decision-making processes. In addition, the system can be extended to integrate other data sources, such as weather forecasts, public transport timetables, and event information, in order to deliver an approach to traffic management that is more comprehensive and holistic.

## 10. FUTURE SCOPE

In addition, the scalability and interoperability of the system can be researched in order to enable seamless interaction with other smart city initiatives and transportation systems. In general, the scope of this paper will be expanded in the future to include the refining and expansion of the proposed system, the utilization of emerging technologies, and the collaboration with various stakeholders to develop intelligent transportation systems that effectively address the challenges of urban traffic congestion and contribute to the development of smarter, more sustainable cities.

## REFERENCES

Ahmad, A., Abusaimeh, H., Rababah, A., Alqsass, M., Al-Olima, N., & Hamdan, M. (2024). Assessment of effects in advances of accounting technologies on quality financial reports in Jordanian public sector. *Uncertain Supply Chain Management*, *12*(1), 133–142. doi:10.5267/j.uscm.2023.10.011

Ahmad, A. Y. B. (2024). Ethical implications of artificial intelligence in accounting: A framework for responsible ai adoption in multinational corporations in Jordan. *International Journal of Data and Network Science*, *8*(1), 401–414. doi:10.5267/j.ijdns.2023.9.014

Ahmad, A. Y. B. (2024). Firm Determinants that Influences Implementation of Accounting Technologies in Business Organizations. *WSEAS Transactions on Business and Economics*, *21*, 1–11. doi:10.37394/23207.2024.21.1

Ahmad, A. Y. B., Gongada, T. N., Shrivastava, G., Gabbi, R. S., Islam, S., & Nagaraju, K. (2023). E-Commerce Trend Analysis and Management for Industry 5.0 using User Data Analysis. *International Journal of Intelligent Systems and Applications in Engineering*, *11*(11s), 135–150.

Ahmad, A. Y. B., Hannoon, A., Al-Daoud, K. I., Abu-Alsondos, I. A., & Al-Qaisieh, M. S. (2023). Assessment of Cloud Based Accounting Technology Adoption and Business Performance. Kurdish Studies, 11(3).

Ahmad, A. Y. B., Kumari, D. K., Shukla, A., Deepak, A., Chandnani, M., Pundir, S., & Shrivastava, A. (2024). Framework for Cloud Based Document Management System with Institutional Schema of Database. *International Journal of Intelligent Systems and Applications in Engineering*, *12*(3s), 672–678.

Ahmad, A. Y. B., Kumari, D. K., Shukla, A., Deepak, A., Chandnani, M., Pundir, S., & Shrivastava, A. (2024). Framework for Cloud Based Document Management System with Institutional Schema of Database. *International Journal of Intelligent Systems and Applications in Engineering*, *12*(3s), 672–678.

Ahmad, A. Y. B., & Kumari, S. S. (2023). *Blockchain Implementation in Financial Sector and Cyber Security System. 2023 International Conference on Artificial Intelligence and Smart Communication (AISC)*, Greater Noida, India. 10.1109/AISC56616.2023.10085045

Ahmad, A. Y. B., Tiwari, A., Nayeem, M. A., Biswal, B. K., Satapathy, D. P., Kulshreshtha, K., & Bordoloi, D. (2024). Artificial Intelligence Perspective Framework of the Smart Finance and Accounting Management Model. *International Journal of Intelligent Systems and Applications in Engineering*, *12*(4s), 586–594.

Ahmad, A. Y. B., William, P., Uike, D., Murgai, A., Bajaj, K. K., Deepak, A., & Shrivastava, A. (2024). Framework for Sustainable Energy Management using Smart Grid Panels Integrated with Machine Learning and IOT based Approach. *International Journal of Intelligent Systems and Applications in Engineering*, *12*(2s), 581–590.

Banerjee, S., Kaur, D., Pandey, D., Pal, S. R., Ahuchaogu, A. A., Iqbal, M. O., & Pandey, B. K. (2020). Industry 4.0 in the Nutraceutical Industry and Public Health Nutrition. In Robotics and Automation in Industry 4.0 (pp. 396-422). CRC Press.

Daoud, M. K. (2023). The Relationship Between Mobile Marketing and Customer Perceptions in Jordanian Commercial Banks: The Electronic Quality as A Mediator Variable. IJMST, 10(2), 1360-1371.

Daoud, M. K., Al-Qeed, M., Ahmad, A. Y. B., & Al-Gasawneh, J. A. (2023). Mobile Marketing: Exploring the Efficacy of User-Centric Strategies for Enhanced Consumer Engagement and Conversion Rates. *International Journal of Membrane Science and Technology*, *10*(2), 1252–1262. doi:10.15379/ijmst.vi.1425

Daoud, M. K., Taha, S., Al-Qeed, M., Alsafadi, Y., Ahmad, A., & Allahham, M. (2024). EcoConnect: Guiding environmental awareness via digital marketing approaches. *International Journal of Data and Network Science, 8*(1), 235-242.

Dave, P., Chandarana, A., Goel, P., & Ganatra, A. (2021). An amalgamation of YOLOv4 and XGBoost for next-gen smart traffic management system. *PeerJ. Computer Science*, *7*, e586. doi:10.7717/peerj-cs.586 PMID:34239973

Devasenapathy, D., Madhumathy, P., Umamaheshwari, R., Pandey, B. K., & Pandey, D. (2024). Transmission-efficient grid-based synchronized model for routing in wireless sensor networks using Bayesian compressive sensing. *SN Computer Science*, *5*(1), 1–11.

Gupta, A. K., Sharma, R., Pandey, D., Nassa, V. K., Pandey, B. K., George, A. S., & Dadheech, P. (2023). Performance Analysis of Eight-Channel WDM Optical Network with Different Optical Amplifiers for Industry 4.0. In *Innovation and Competitiveness in Industry 4.0 Based on Intelligent Systems* (pp. 197–212). Springer International Publishing. doi:10.1007/978-3-031-29775-5_9

Khan, B., Hasan, A., Pandey, D., Ventayen, R. J. M., Pandey, B. K., & Gowwrii, G. (2021). 12 Fusion of Datamining and Artificial Intelligence. *Machine learning and iot for intelligent systems and smart applications*, 201.

Khedkar, S. P., Canessane, R. A., & Najafi, M. L. (2021). Prediction of traffic generated by IoT devices using statistical learning time series algorithms. *Wireless Communications and Mobile Computing*, *2021*, 1–12. doi:10.1155/2021/5366222

Lelisho, M. E., Pandey, D., Alemu, B. D., Pandey, B. K., & Tareke, S. A. (2022). The Negative Impact of Social Media during COVID-19 Pandemic. *Trends in Psychology*, *31*(1), 1–20. doi:10.1007/s43076-022-00192-5

Malhotra, P., Pandey, D., Pandey, B. K., & Patra, P. M. (2021). Managing agricultural supply chains in COVID-19 lockdown. *International Journal of Quality and Innovation*, *5*(2), 109–118. doi:10.1504/IJQI.2021.117181

Menon, V., Pandey, D., Khosla, D., Kaur, M., Vashishtha, H. K., George, A. S., & Pandey, B. K. (2022). A Study on COVID–19, Its Origin, Phenomenon, Variants, and IoT-Based Framework to Detect the Presence of Coronavirus. In *IoT Based Smart Applications* (pp. 1–13). Springer International Publishing.

Muniandi, B., Huang, C., Kuo, C., Yang, T., Chen, K., Lin, Y., Lin, S., & Tsai, T. (2019). A 97% maximum efficiency fully automated control turbo boost topology for battery chargers. *IEEE Transactions on Circuits and Systems. I, Regular Papers*, *66*(11), 4516–4527. doi:10.1109/TCSI.2019.2925374

Muniandi, B., Huang, C.-J., Kuo, C.-C., Yang, T.-F., Chen, K.-H., Lin, Y.-H., Lin, S.-R., & Tsai, T.-Y. (2019, November). A 97% Maximum Efficiency Fully Automated Control Turbo Boost Topology for Battery Chargers. *IEEE Transactions on Circuits and Systems. I, Regular Papers*, *66*(11), 4516–4527. doi:10.1109/TCSI.2019.2925374

Pandey, B. K., Nassa, V. K., Pandey, D., George, A. S., Gupta, M., Dadheech, P., & George, A. H. (2023). An Analytical Analysis of the Present-Day Procedures for Disposing of Waste. In *Handbook of Research on Safe Disposal Methods of Municipal Solid Wastes for a Sustainable Environment* (pp. 24–34). IGI Global. doi:10.4018/978-1-6684-8117-2.ch002

Pandey, B. K., Pandey, D., Alkhafaji, M. A., Güneşer, M. T., & Şeker, C. (2023). A Reliable Transmission and Extraction of Textual Information Using Keyless Encryption, Steganography, and Deep Algorithm with Cuckoo Optimization. In *Micro-Electronics and Telecommunication Engineering: Proceedings of 6th ICMETE 2022* (pp. 629–636). Springer Nature Singapore. doi:10.1007/978-981-19-9512-5_57

Pandey, B. K., Pandey, D., Dadheech, P., Mahajan, D. A., George, A. S., & Hameed, A. S. (2023). Review on Smart Sewage Cleaning UAV Assistance for Sustainable Development. In *Handbook of Research on Safe Disposal Methods of Municipal Solid Wastes for a Sustainable Environment* (pp. 69–79). IGI Global. doi:10.4018/978-1-6684-8117-2.ch005

Pandey, B. K., Pandey, D., Wairya, S., & Agarwal, G. (2021). An advanced morphological component analysis, steganography, and deep learning-based system to transmit secure textual data. [IJDAI]. *International Journal of Distributed Artificial Intelligence*, *13*(2), 40–62. doi:10.4018/IJDAI.2021070104

Pandey, D., & Pandey, B. K. (2022). An efficient deep neural network with adaptive galactic swarm optimization for complex image text extraction. In *Process Mining Techniques for Pattern Recognition* (pp. 121–137). CRC Press. doi:10.1201/9781003169550-10

Pandey, D., Pandey, B. K., & Wairya, S. (2021). Hybrid deep neural network with adaptive galactic swarm optimization for text extraction from scene images. *Soft Computing*, *25*(2), 1563–1580. doi:10.1007/s00500-020-05245-4

Pandey, D., Wairya, S., Sharma, M., Gupta, A. K., Kakkar, R., & Pandey, B. K. (2022). An approach for object tracking, categorization, and autopilot guidance for passive homing missiles. *Aerospace Systems*, *5*(4), 553–566. doi:10.1007/s42401-022-00150-0

Pandey, J. K., Jain, R., Dilip, R., Kumbhkar, M., Jaiswal, S., Pandey, B. K., & Pandey, D. (2022). Investigating Role of IoT in the Development of Smart Application for Security Enhancement. In *IoT Based Smart Applications* (pp. 219–243). Springer International Publishing.

Saxena, A., Agarwal, A., Pandey, B. K., & Pandey, D. (2024). Examination of the Criticality of Customer Segmentation Using Unsupervised Learning Methods. *Circular Economy and Sustainability*, 1–14. doi:10.1007/s43615-023-00336-4

Sharma, M., Pandey, D., Khosla, D., Goyal, S., Pandey, B. K., & Gupta, A. K. (2022). Design of a GaN-based Flip Chip Light Emitting Diode (FC-LED) with au Bumps & Thermal Analysis with different sizes and adhesive materials for performance considerations. *Silicon*, *14*(12), 7109–7120. doi:10.1007/s12633-021-01457-x

Sharma, M., Talwar, R., Pandey, D., Nassa, V. K., Pandey, B. K., & Dadheech, P. (2024). A Review of Dielectric Resonator Antennas (DRA)-Based RFID Technology for Industry 4.0. *Robotics and Automation in Industry*, *4*(0), 303–324.

Singh, S., Madaan, G., Kaur, J., Swapna, H. R., Pandey, D., Singh, A., & Pandey, B. K. (2023). Bibliometric Review on Healthcare Sustainability. Handbook of Research on Safe Disposal Methods of Municipal Solid Wastes for a Sustainable Environment, 142-161.

Swapna, H. R., Bigirimana, E., Madaan, G., Hasan, A., Pandey, B. K., & Pandey, D. (2023). Impact of Neuromarketing on Consumer Psychology in Digitally Connected Networks. In *Applications of Neuromarketing in the Metaverse* (pp. 193–205). IGI Global. doi:10.4018/978-1-6684-8150-9.ch015

Vinodhini, V., Kumar, M. S., Sankar, S., Pandey, D., Pandey, B. K., & Nassa, V. K. (2022). IoT-based early forest fire detection using MLP and AROC method. *International Journal of Global Warming*, *27*(1), 55–70. doi:10.1504/IJGW.2022.122794

Zaidi, A. (2023). Predicting wildfires in Algerian forests using machine learning models. *Heliyon*, *9*(7), e18064. Advance online publication. doi:10.1016/j.heliyon.2023.e18064 PMID:37519679

Zaidi, A. (2023). Smart Implementation of Industrial Internet of Things using Embedded Mechatronic System. *IEEE Embedded Systems Letters*. doi:10.1109/LES.2023.3309985

Zaidi, A., Ajibade, S.-S. M., Musa, M., & Bekun, F. V. (2023). New Insights into the Research Landscape on the Application of Artificial Intelligence in Sustainable Smart Cities: A Bibliometric Mapping and Network Analysis Approach. *International Journal of Energy Economics and Policy.*, *13*(4), 287–299. doi:10.32479/ijeep.14683

Zaidi, A., & Al-Harbi, O. (2023). Statistical Analysis of a Linear Multi-Step Numerical Treatment. *Journal of Statistics Applications & Probability*, *12*(1), 26.

doi:10.18576/jsap/120126

Zaidi, A., & Al Luhayb, A. (2023). Two Statistical Approaches to Justify the Use of the Logistic Function in Binary Logistic Regression. *Mathematical Problems in Engineering*, *2023*, 5525675. Advance online publication. doi:10.1155/2023/5525675

Zhu, B., Hou, X., Liu, S., Ma, W., Dong, M., Wen, H., Wei, Q., Du, S., & Zhang, Y. (2021). Iot equipment monitoring system based on c5. 0 decision tree and time-series analysis. *IEEE Access : Practical Innovations, Open Solutions*, *10*, 36637–36648. doi:10.1109/ACCESS.2021.3054044

# Chapter 10
# Blockchain's Potential Effect on the Healthcare Industry

**Binay Kumar Pandey**
https://orcid.org/0000-0002-4041-1213
*Department of Information Technology, College of Technology, Govind Ballabh Pant University of Agriculture and Technology, Pantnagar, India*

**Mukundan Appadurai Paramashivan**
https://orcid.org/0009-0009-5608-4788
*Aligarh Muslim University, India & Champions Group, India*

**Vinay Kumar Nassa**
https://orcid.org/0000-0002-9606-7570
*Department of Information Communication Technology, Tecnia Institute of Advanced Studies, India*

**Darshan A. Mahajan**
https://orcid.org/0000-0002-1239-6343
*NICMAR University, India*

**Pankaj Dadheech**
https://orcid.org/0000-0001-5783-1989
*Swami Keshvanand Institute of Technology, Management, and Gramothan, India*

**A. S. Hovan George**
*Tbilisi State Medical University, Georgia*

**A. Shaji George**
https://orcid.org/0000-0002-8677-3682
*Business System Department, Almarai Company, TSM, Riyadh, Saudi Arabia*

## ABSTRACT

*Block chain secures data. Blockchain imitates computerised trading markets. The 2008 cryptocurrency and transaction network Bitcoin leads blockchain technology. When buyers or sellers declare payments, blockchain transfers bitcoin. Blockchain can change finance like the Internet. Block chain technology may affect medicine. This fast-growing issue affects many healthcare providers. Open, irreversible, stable, and collaborative technology are preferred by cryptocurrency over regulated, hidden, proprietary, and changeable ones. Blockchain is used in medical records, security, database management, and biotechnology regulation. Blockchain helps biopharmaceuticals track illicit medications. It monitors medicines and counterfeiters. This book chapter presents a comprehensive block chain technology for health sector decision makers and explores difficulties and constraints using a suitable markup. Here are current and potential medical blockchain applications.*

DOI: 10.4018/979-8-3693-1335-0.ch010

## 1. INTRODUCTION

The data is safeguarded by the block chain. Electronic trading markets that mimic blockchain technology. Blockchain technology was pioneered in 2008 by Bitcoin, a decentralized digital money and transaction network. When a seller or a buyer reveals a payment, the blockchain automatically transfers bitcoin. Similar to the internet, blockchain has the potential to revolutionize the financial industry. There's a possibility that blockchain technology will change medicine. This fast expanding problem is having an effect on a great number of healthcare providers. When compared to technologies that are regulated, concealed, proprietary, and subject to change, cryptocurrencies favor those that are open, irreversible, stable, and collaborative. The administration of medical records, cybersecurity, database management, and the regulation of biotechnology all use blockchain technology. Tracking illegal narcotics is made easier with the use of blockchain technology. It investigates both drug traffickers and counterfeiters. This study provides decision makers in the health industry with a comprehensive overview of block chain technology and utilizes an appropriate markup to conduct an analysis of issues and limits. The article will cover both the present and potential future applications of blockchain technology in the medical field. On a consistent basis, enormous quantities of data are produced, accessed, and exchanged within the healthcare industry (Pramanik, S.,et al.,2023). It is vital to collect and disseminate such a massive amount of data, but doing so presents a formidable challenge due to the delicate nature of the data and the attendant concerns regarding its protection. In order to deliver correct diagnoses and collaborative medical care, the sharing of data in clinical and healthcare settings needs to be scalable, safe, and clean.

In order for clinicians to effectively communicate patient information to the relevant authorities in a timely way, statistics are required. These carers and general practitioners should be able to securely transfer the clinical records of their patients so that their patients can receive information that is accurate and up to date regarding their health. On the other side, clinical evidence is communicated to a distant clinician through telemedicine and e-health in order to acquire an expert opinion. These two online clinical systems send patient data via a technique referred to as "store-and-forward technology" or by real-time clinical surveillance (which may include tele-monitoring, telemetry, and other methods of a similar nature).Sharing clinical results helps medical professionals to establish diagnoses from a distance and provide therapy to patients who are ill in online clinical environments. Issues of confidentiality, sensitivity, and privacy are likely to be major concerns in all therapeutic scenarios since patient data are frequently highly sensitive. In order to facilitate clinical contacts with distant patients, therefore, there is a need for data sharing that is protected, scalable, and consistent. Sharing data in a way that is both secure and efficient enables more effective therapeutic teamwork. This is accomplished by asking for recommendations or validations from a network of clinical experts. This, in turn, leads to a diagnosis that is more accurate and an increase in the treatment's overall effectiveness. Interoperability issues are another problem that the region has to deal with at the moment. The dissemination of scientific information among healthcare institutions and research organizations may present certain difficulties. This phase could be limited by clinical evidence, sensitivity, data interchange agreements, protocols, intricate patient matching procedures, moral policies, or suggestions that are already in place. For the exchange of clinical data, it is necessary to find solutions to a number of fundamental problems first. Researchers have spent years seeking to harness cyberspace, simulated communication, an understanding of equipment, and computer vision in the hopes that this will assist medical professionals in the diagnosis and treatment of a wide variety of chronic illnesses (David, S.,et al.,2023).

## 2. BLOCK CHAINTECHNOLOGY

Block chain technology is one of the most significant technological advances to occur in the business world. The technology behind blockchain is causing a change in the world. Many software updates, particularly in the medical field, have been made possible thanks to blockchain technology (Menon, V.,et al.,2022). The use of blockchain technology was observed in a number of healthcare institutions that attempted data modelling and democracy. This is due, in part, to its versatility as well as its unparalleled capacity to classify, protect, and share health data. Many recent improvements in medicine are supported by blockchain technology. Blockchains are a type of middleware that facilitates collaboration between individuals and organizations working in the healthcare industry.

A blockchain is a distributed digital ledger that is constantly updated and stores digital documents that have been cryptographically signed and timestamped. In the field of healthcare, the term "blockchain" can refer to a number of different uses and repercussions for all involved parties. These technologies can join disparate networks in order to more accurately evaluate the significance of therapies. The use of blockchain technology to store medical records is becoming increasingly common [6]. The use of blockchain is gaining more and more traction. Blockchain, one of the most game-changing technologies of the 21st century, has had an effect on almost every sector of the economy, including the healthcare sector. The immutable public ledger provided by blockchain technology has the potential to revolutionize healthcare in the future (Anand, R.,et al.,2023). Drug Clear and open currency. The components that make up blockchain are called blocks, nodes, and miners.

## 3. BLOCKCHAIN TECHNOLOGY INHEALTHCARE

The technology of block chains is one of the most important technological advancements that has recently occurred in the world of business. The underlying technology of blockchain is having a transformative effect on the world. Because of blockchain technology, numerous software updates have been implemented, most notably in the medical industry (Albermany SA,et al.,2014). Experiments with using blockchain technology to model healthcare organizations' data and to establish democratic processes have been conducted by a wide variety of healthcare organizations. This is partially attributable to its versatility, as well as its unrivalled capacity to categorize, protect, and disseminate health data. The use of blockchain technology has contributed to a number of recent advances in medical care. Blockchains are a sort of middleware that enables individuals and healthcare groups to collaborate on problems pertaining to medicine.

A blockchain is a distributed digital ledger that is constantly updated and maintains digital documents that have been cryptographically signed and timestamped. A blockchain may also be thought of as a digital version of a public ledger. The term "blockchain" can refer to a variety of applications and ramifications for all parties involved in the healthcare industry, depending on the context. These technologies have the ability to connect previously separate networks in order to carry out more precise analyses of the importance of treatments. There has been a recent uptick in the utilization of blockchain technology for the storage of patient medical records (Albermany SA,et al.,2021). The implementation of blockchain technology is getting increasing popularity. The healthcare industry is one of the many areas of the economy that has been impacted by blockchain technology, which is widely regarded as one of the most transformative technologies to emerge in the 21st century. In the not too distant future, the immutable public ledger that blockchain technology makes available has the potential to completely transform the healthcare industry. The tracking of drugs, patient databases, and clinical trials will be the primary focuses of this invention's efforts to solve existing problems.

## 3.1 Increasing the Capacity to Track Out the Origin of Medications

It seems as though the therapeutic supply chains are vulnerable. When it comes to the distribution of medications, the process starts with wholesale suppliers supplying merchandising businesses with supplies, which are subsequently resold to end users by those merchandising organizations. Unfortunately, more illegal narcotics are being added to this loop as it continues. Blockchain technology will not only fix the issues that have been hindering the pharmaceutical supply chain, but it will also eliminate such issues. According to PwC's observations, blockchains are able to record transactions that take place along supply chains. This makes data transportation easier, which in turn gives commodities receivers a stronger sense of confidence in the product's provenance by providing more information about where the product originated. Because of this strategy, recalls can also be made by the regulatory bodies and by the shops. Any entity with a vested interest can use blockchain technology to assure that there are no fake or low-quality goods in the supply chain by using this technology.

## 3.2 Data Control for Patients

Around the course of the previous two years, there have been a number of high-profile cyberattacks that have targeted healthcare facilities located all around the world. The WannaCry ransomware outbreak in 2017 had a significant impact on the United Kingdom's National Health Service (NHS). Three hundred and fifty thousand computers located all around the world were infected by the harmful malware. In 2018, a malicious hacker breached the security of the United States government website HealthCare. Gov, which resulted in the disclosure of sensitive information pertaining to more than 75,000 individuals. More than one and a half million people in Singapore were affected when a data breach occurred in the healthcare industry. In the field of healthcare, there are a significant number of instances of malware attacks and data leaks. The immutability and decentralization of blockchain technology make it possible to develop data processing systems that will persist for an extremely long time. On networks that are based on blockchain technology, hospitals and other healthcare organizations are able to keep the medical records of their patients in the tightest secrecy. The medical records of a patient do not always include all of the relevant information about the patient. As Internet of Things (IoT) advances at an unstoppable rate, patients are able to keep track of their behaviors, establish objectives for themselves, and change their therapies thanks to wearable technology (JayaLakshmi, G.,et al.,2024).

## 3.3 Clinical Trials

Clinical tests can run into the billions of dollars, and if they produce unclear results, everyone involved in the project stands to lose a large amount of money, from the scientist to the consumer. Because of the rise in costs, researchers are now faced with a number of hurdles, including the need to recreate reliable data, effectively communicate scientific and personal data, resolve challenges relating to patient privacy, and enroll patients.

# 4. WHAT DO YOU BELIEVE THE SIGNIFICANCE OF BLOCKCHAIN WOULD BE?

The importance of Block Chain Technology in various sectors becomes clear when its rudimentary components are examined. Forty-four percent of all industries, on average, may have already used blockchain. Block chain technology and its significance.

## 4.1 Immutable

There is no way to alter Blockchain's underlying technology. This paves the way for a plethora of options for systems that require static attributes, allowing them to function to their full potential even in congested settings. Although changes are prohibited in this implementation of Block Chain Technology, users will receive a warning before any modifications are made.

One of blockchain's most valuable features is the transparency it delivers. Due to the widespread availability of information online, most potential points of disagreement have been eliminated. Businesses who adopt and implement this technology may find that their interactions with their target customers become more open and honest, both directly and indirectly.

## 4.2 Digital Autonomy

We are all aware that the internet has granted us a great lot of independence, allowing us to do things like shop, file, and buy goods without leaving the house. Nonetheless, we still do not have any freedoms despite all of this technology. With Block Chain Technology, we are completely independent; we don't have to deal with any governing bodies, and we have complete control over our digital possessions and financial dealings.

## 4.3 Cost-Effective

Blockchain's transaction fees are competitive when measured against other systems. Since we are already aware that no centralization or middle man is occurring during any transactions, we have no need to pay for the services of any person that may be present in the chain. Compared to the many conventional models, which tend to be very expensive, the system that employs this Block Chain technology is far more cost-effective.

## 4.4 Increased Safety Precautions

The duration of each transaction is decreased.

Everyone has reached a point of ease when using today's technologies. All the dealings can be completed in a matter of minutes. There is no longer any need to queue up at the bank to do any kind of business.

## 4.5 Advantages Centered on the Patient

When it comes to patients, blockchain technology has several advantages, including authentication and consent.

Research shows that block chain technology will improve the security of medical information by centralizing previously dispersed cohort routes and treating each person as the system's fundamental building piece.

Information is linked in a decentralized network within a block chain, making it impossible to share incorrect information while yet being simple to move around.

The data in a block chain is linked in a decentralized network, which facilitates distribution and prevents the spread of inaccurate information. Every transaction is timestamped, and every second block is secured by a unique code or biometric. The transaction processing order can be verified with the use of this code or biometric (Pandey, B. K.,et al.,2022).

## 4.6 Industry-Based Benefits and Advantages

Blockchain technology allows for the secure transfer of individual patients' medical records to a network of healthcare providers.

Medical imaging data can be shared safely between doctors and patients if block chain technology is used in its entirety.

For the sake of medical progress, the immutability of blockchain-stored data is the most crucial aspect.

Blockchain technology aided businesses by making it less difficult to comply with statutory drug testing by streamlining the medical testing process (Pandey, J. K.,et al.,2022).

# 5. THREATS OF BLOCK CHAIN IN HEALTHCARE

The underlying reason of the scalability problem was the blockchain platform's inability to handle a large number of transactions in a short amount of time. It was determined that an optimization constraint (Lelisho, M. E.,et al.,2023) was responsible for the trade-off between trading volume and the amount of CPU resources needed to manage the transfers. Authorization and security concerns are intertwined in the context of blockchain technology. Domain name attacks and mempool attacks are two examples of the types of attacks that can be launched against the technology, with the latter allowing an attacker to seize control of a sizable section of blockchain networks. One of the biggest challenges to widespread adoption of blockchain technology was the level of public support for it.

Regulatory organizations are finding it difficult to grant access, citing reasonable concerns about confidentiality, due to the dispersed nature of health information and the lack of a reliable third party. Despite widespread sharing of health information, progress has been stymied by factors including a lack of trust between institutions and the limitations of freely available software.

The challenge of interoperability arises when multiple organizations cannot communicate with one another. One obstacle identified was the absence of necessary IT practitioners with expertise in managing technologies and maintaining the integrated pharmaceutical supply chain across different stakeholders (Sharma, Manvinder,et al.,2024). It was discovered that this was an issue. Even while blockchain technology has the potential to lead to long-term cost reductions, its initial design expenses are rather significant.

## 6. WHAT ARE THE MAJOR ISSUES IN THE HEALTHCARE SECTOR COULD BE RESOLVED WITH BLOCKCHAINTECHNOLOGY?

The technology behind medical block chains seems to have the potential to solve some of the industry's biggest problems, such as conformance, interoperability, and user identification. If this technology is used across the medical care industry, it could lead to an improvement in back-office tasks, a stronger requirement for businesses to handle goods along the supply chain, and more storage space and permissions (Pandey, B. K.,et al.,2022). The medical field faces a number of problems when it comes to collecting and managing data. Now, let's look at how block chain technology can help solve these common problems in the medical field:

### 6.1 The Tracking of Illicit Drugs

In the healthcare business as a whole, forgery has been on the rise for a number of years. The wrong way that drug information is being interpreted is a sad problem that is hurting many parts of the medical field. When medicines are tampered with, their shape and properties are completely changed, making it impossible to tell them apart from the real thing. The patient's illness can't be cured with these medicines. In fact, they are more dangerous and could make the patient's situation even worse.

Block Chain technology can't be changed, which is pretty well known at this point. Because of this, past records of the ingredients in many pharmaceuticals and other private information are protected to a high degree. Before the block chain concept was introduced, unknown users and stakeholders played different parts in putting unapproved or expired pharmaceuticals on the market. This was because the medical field had less protection before the concept was introduced. But in the modern world, we have more ways to check if something is real, and we can easily find medicines that have been lost or stolen, and we can even stop the sale of medicines that haven't been licensed.

Before selling the drug to customers, the company that makes it will make sure it has the right ingredients and get permission from the government. Each of these drugs would have a hash, and the Block Chain Technology would be used to store all of the important information. After the medicines are made, they are shipped to distributors. From there, they are sent to the rest of the world. The wholesaler checks the goods to make sure they are legal and gives the green light for them to be brought into or taken out of any country in the world. The supplier is where the chemist gets the medicines. The state of the patient is looked at, and then the right medicine is confirmed and given by the attending doctor. Figure 4 shows that the block chain will have a record of every single transaction.

### 6.2 Testing in the Clinic

The clinical study is broken up into several stages, and each of these steps needs a lot of information and health data about the patients. Blood test results, performance tracking, estimates, survey data, and diagnostic procedures from a large number of people are all included. It would be harder to track down each person based on their medical information, which would make the whole process harder. During the clinical studies, there would be a lot of room for mistakes to be made. The fact that we can't do good clinical tests is slowing down the progress of medical study, which is bad.

Block chain technology will show us all of the search results when we use a term to look for something on the internet. There may be a study or other information here that fits your search criteria. Every single piece of test data comes with a unique code that matches the data in a report exactly. If a doctor looks at the information that has been saved and wonders if it is real, we can easily show that it is by pointing to the real information that has been stored in the blockchain network. This would make it easier for healthcare professionals, as well as people from outside the healthcare system, to work together (Tripathi, R. P.,et al.,2023).

### 6.3 Management of Medical Records for Patients

The people in healthcare who are in charge of processing patient data are facing not one but two big problems right now. The first important thing to think about is that a person's medical records are.

Everyone has different needs, so each person needs a care plan that is made just for them. For each patient, a different drug would be given (Malhotra, P.,et al.,2021). The second question that needs to be answered is how to store data, make data available, make data easy to find, and share data within management. Patients can only remove or change a few things in their medical records if they have the right permissions, and they can't see their entire medical histories. Another problem that needs to be solved is the question of who owns what.

The answer: Block chain is a type of technology that uses a number of cryptographic methods to make sure that data is as safe as possible (Banerjee, S.et al.,2023). All of the patient's information, such as name, date of birth, age, prescriptions, and so on, is kept in the cloud. Using Block Chain Technology, each patient will be given a user ID to make it easier for them to view their data. With the right permissions and the API software, the other parties can look at the data of a few patients.

## 7. SWOT ANALYSIS ON BLOCK CHAIN TECHNOLOGY IN MDEICAL INDUSTRY

SWOT analysis is a useful strategic tool for analyzing strengths, weaknesses, opportunities, and threats in a competitive market and for strategic growth. Let's talk about a few SWOT analyses of Block Chain in healthcare.

### 7.1 Ways Blockchain Technology is used in healthcare: Electronic Medical Records (EMRs)

From the point of data creation to the point of data retrieval, the accuracy of a digital EMR on a distributed ledger of an allowed blockchain is guaranteed without any human input.

#### 7.1.1 Medical Studies

In terms of clinical study, blockchain gives any knowledge-sharing partnerships a shared, secure platform. Using this method, experts can share information in a safe way.

### 7.1.2 Detecting Medical Fraud

Since blockchain is permanent, it helps stop fraud by keeping transactions from being copied or changed. This makes for a clear and safe exchange.

### 7.1.3 Research in Neurology

As a new technology, blockchain will open the door to many new uses, such as brain augmentation, brain re-enactment, and brain thought. A whole human brain must be digital, which means it must be stored somewhere. This is where blockchain technology comes in.

### 7.1.4 The Drug Business and Research

At any point, blockchain keeps an eye on the medicine supply chain by using its ability to track things precisely: At each point, the medicine's location, its parts, and who owns it are all checked to stop it from being faked or stolen.

### 7.1.5 Challenges and Opportunities

Blockchain is a new platform that is gaining traction in a variety of industries (Singh, S.,et al.,2023) and offers a lot of advantages and prospects . This technology, however, comes with its own collection of issues that must be resolved. A few of these major challenges and opportunities are shown in fig. 5.

### 7.1.6 Protection and Secrecy of Data

Data privacy is the top issue (Pramanik, S.,et al.,2023). Blockchain-based apps eliminate the need for an unknown entity to perform a contract (Parthiban, K.,2021). The blockchain architecture allows the entire population to validate data, making it subject to future secrecy and protection concerns (Babu et al., 2023). All nodes accessing data transmitted by one node compromises data protection (Pandey, D.,et al.,2023). Since all nodes view a single node's data, data protection is weakened. The patient must designate one or two personnel to view their information and medical history in an emergency without third-party approval. This official will give a group access to the same patient's documents. Data transmission is hindered by high-security measures, leaving recipients with little or no information. Blockchain networks also have a 51% attack vulnerability. This attack involves miners with over 50% of a blockchain network's blocks. Miners obtain network power and can block new transactions by rejecting permission. This attack targeted five coins, according to coindesk (Pandey, J. K.,et al.,2022). Not sharing confidential health report information is also possible (B. Muniandi et al.,2019).

### 7.1.7 Storage Capacity management

Storage space management is another concern. Blockchain requires little storage since it stores and processes transaction data with a limited reach. When it entered healthcare, storage issues arose. The healthcare business handles massive amounts of data daily. All data, from patient information, personal histories, and exam results to MRI scans, X-rays, and other diagnostic images, would be accessible to all blockchain nodes, requiring a lot of storage space. Blockchain databases grow quickly because they are transaction-based. When databases become sophisticated,

Records become slower to search and access, making them unsuitable for fast transactions. Therefore, a blockchain system must be flexible and resilient.

### 7.1.8 Issues of Interoperability

Interoperability is another problem with blockchain, which involves allowing blockchains from different connecting providers and utilities speak to one another in a seamless and acceptable manner. This problem makes it difficult to share data effectively.

### 7.1.9 Standardization Challenges

Since blockchain technology is still in its infancy, it will undoubtedly face standardization problems as it moves closer to realistic application in medicine and healthcare. International standardization bodies will be expected to include a range of well-authenticated and accredited specifications. These predefined specifications may be useful in determining the scale, value, and format of data shared in blockchain applications. These specifications will not only scrutinize the shared data, but they will also act as safety precautions.

### 7.1.10 Social Challenges

Since blockchain technology is still in its early stages, it faces social challenges such as cultural shifts in addition to the technical challenges mentioned above. Accepting and implementing a technology that is fundamentally different from conventional work practices is never straightforward. Although the medical industry is gradually moving toward digitization, there is still a long way to go until it fully embraces this technology, especially blockchain, which has yet to be clinically validated. It would take time and effort to get doctors to abandon paper in favor of technology. The technologies and policies offered are largely untrustworthy due to their poor penetration rate in the health sector. We can't yet call it a sustainable and universal option for all healthcare problems because of all of these barriers and risks (Saxena, A.,et al.,2024).

SWOT analysis for blockchain in healthcare.

## 8. THREATS

The included research came up with a total of eight potential challenges to the blockchain technology, which may be broken down into three categories: organizational, sociological, and methodological/scientific. Thirteen studies outlined three different types of methodological and scientific risks, fourteen studies discussed two different types of social dangers, and fourteen investigations uncovered three different types of organizational risks (Boopathi, S.,et al.,2023).

### 8.1 Dangers Posed by Various Forms of Technology

Five separate investigations all reached the conclusion that blockchain technology are scalable. The low number of transactions that could be processed by the network in a single second was the root cause of the scalability issue with blockchain technologies. In addition, according to two separate publications, scalability limitations are the result of a trade-off between the volume of transactions and the amount of computer resources that are required to process the transactions.

## 9. STRENGTHS

This study sheds light on the benefits and limitations of employing blockchain technology in healthcare, specifically with regard to the distribution and storage of health-related data. This study would assist customers, finance sources, and organizations in gaining a clearer grasp of the present good and negative contributions that blockchain technology is making in the field of healthcare. The study collection and data extraction processes were carried out independently in both of the reviews, which contributed to an increase in the validity and dependability of the work. In order to exclude the chance of there being a publication bias, we employed a method called backward-reference list checking to extract grey literature. The benefits and dangers of using blockchain technology that were outlined in this study were broken down into categories that were simpler to comprehend.

### 9.1 Capabilities in Terms of Performance, Throughput, and Scalability

Performance is a popular indicator for software output, and it refers to the amount of data that is sent from one endpoint to another. Efficiency is determined not only by the latency and bandwidth of the network but also, in the context of blockchain technology, by the distribution scale of the consensus nodes. The average output of a platform utilized by a shadowy organization can range anywhere from a few hundred to several thousand blocks per second. There are not many uses of blockchain technology in the actual world at this time.

Data related to healthcare, and the measurements almost certainly will change depending on the different types of data related to healthcare. The current best practices recommend using smaller sample sets whenever possible. The majority of electronic health records (EHR) and clinical data sets contain enormous amounts of data, including historical records and data from a variety of institutions, which can have a detrimental effect on the outcomes. One of the potential applications of blockchain technology in the medical field is to provide patients the ability to retain ownership of their data and share it with whoever they desire. Output characteristics that are based on a demand for speed that is of a lesser importance will have less of an effect on this kind of solution. However, in the event of a crisis, the requirement for critical information regarding a patient would make a solution with high performance essential.

**Throughput:** In general, the throughput of a protocol may be measured in terms of the number of blocks that it adds to the blockchain each second. This transaction metric is established by the consensus method that is currently being used. This algorithm dictates how nodes will cooperate in order to verify that the attached transaction is valid and that each of their mutual ledger copies are accurate. The throughput would be most heavily influenced by the data's design as well as its magnitude and scope, respectively. One of the limiting factors is designing with the least amount of space taken up (differentiating between critical and non-critical information). The standard practice for data these days is to disclose everything and just use what is necessary. However, in order to boost throughput, this technology relies on a different mentality that focuses on transferring a limited range of data. This helps the technology achieve its goal of increasing throughput.

**Scalability -** The best types of information for scaling blockchains include trivial metadata, origin information, contract information, and inspection information. On the other hand, they do not scale well when additional data kinds are included, such as photos or whole genomic datasets. This is because they cannot handle the increased data volume. This is because the data for each block that is committed needs to be repeated across all nodes, and each node stores a full copy of the public ledger, which is a representation of the state of the network. As a result, this causes the network to be in an inconsistent state. When developing software, scalability is still an important element to take into consideration. It is the most major problem that confronts ambitious enterprises that were first founded on a limited group of customers but subsequently need to incorporate a wider number of customers. The same thing will be true for blockchains that hold healthcare data; they are also in the early phases of development and will continue to proliferate swiftly over the world. Multiple design factors are an obvious illustration (Zaidi, A., et al., (2023)).

- With this technique, it would be impossible to take advantage of the potential to refactor all of the knowledge obtained (after the fact) as it develops. If you start with the accurate collection of data from the very beginning (which is challenging), you will boost your ability to scale. The upfront design is what determines how well a technology performs in terms of its efficiency, throughput, and scalability. Implementation patterns for blockchain organizations can exploit the current, traditional models of system deployments by utilizing on-premises environments or cloud-based settings and either Platform-as-a-Service (PaaS) or Infrastructure-as-a-Service (IaaS) hosting models. These deployment models can be used with either on-premises environments or cloud-based environments.

## 10. BLOCKCHAIN PREDICTIONS FOR THE FUTURE ENHANCEMENT IN THE MEDICALFIELD

Emergence of more advanced Block Chain technologies is anticipated, some of which have the potential to be more precise, dependable, and fruitful for the management of health care. Because it makes it possible to find answers to the majority of the company's most pressing issues, the future of Blockchain technology in the healthcare industry looks quite promising. The future in which the technology of block chains is employed as a component of a framework in which healthcare practitioners assume responsibility for their own healthcare records rather than dependent on a centralized system. Utilization of a blockchain like this presents its own unique set of technological challenges. The process of acquiring, recognizing, and retrieving information could be difficult and time-consuming; but, with the assistance of new technologies such as artificial intelligence and the internet of things, such challenges could ultimately be overcome (Pandey, D.,et al.,2021).

Over the past few years, both the public and business sectors have been doing extensive research into the potential applications of block chains in a wide variety of industries. The medical industry can benefit from blockchain technology in the areas of information sharing, the creation of smart contracts, the remediation of security flaws, and cost reduction. To enable transactions easier, blockchain leverages data structures (Daoud, M. K. et al.,2024).

*Table 1. Impact of block chain technology on a wide variety of fields*

| | 2015 | 2016 | 2017 | 2018 | 2019 | 2020 |
|---|---|---|---|---|---|---|
| EHR/MHR | 20 | 20 | 21 | 22 | 23 | 24 |
| TELEMEDICINE | 12 | 14 | 17 | 19 | 22 | 26 |
| MOBILE HEALTH | 8 | 14 | 21 | 28 | 37 | 46 |
| VIRTUAL HEALTH | 39 | 48 | 59 | 73 | 89 | 110 |

The application of Blockchain technology in the medical industry is currently on the cusp of huge paradigm shifts and is on the verge of accomplishing an important landmark. Because technology is developing and becoming more trustworthy, healthcare institutions may be more encouraged to use telemedicine services. This is done in order to establish novel benefits, reduce hospitalizations, improve treatment adherence, and make the facility more comfortable for patients. The application of blockchain technology has a significant possibility to motivate fundamental shifts in the medical care and community-based treatment modalities (Daoud, M. K,et al.,2023).

## 10.1 Monitoring and Data Access Around the Clock

With the assistance of Internet of Things [20, 21] and wearable devices, blockchain for healthcare has the potential to be used to store and update helpful patient data (Meslie, Y.,et al.,2021) in real time. Examples of this type of data include blood pressure and sugar level. This gives doctors the ability to keep an eye on patients who are at high risk around the clock, every day of the week, and to warn and educate the patients' relatives and other caregivers in the event of an emergency.

## 10.2 Efficient Use of Resources

The use of blockchain technology completely does away with any and all processes involving a third party or a mediator, which are presently necessary for the gathering and transferring of data. The healthcare system is able to reach a high level of cost-effectiveness due to the absence of this strategy.

## 10.3 Smart Contracts That Ensure Rules Are Followed Consistently

The use of blockchain technology completely does away with any and all processes involving a third party or a mediator, which are presently necessary for the gathering and transferring of data. The healthcare system is able to reach a high level of cost-effectiveness due to the absence of this strategy.

## 10.4 Patient treatment That is Faster, Cheaper, and of Higher Quality

The distributed ledger technology known as blockchain possesses the potential to offer a unified system for the safe and efficient archiving, retrieval, and use of the medical data contributed by registered users. It is possible to prevent a variety of errors, as well as to expedite diagnosis and treatment, as well as to personalize a patient's medication by ensuring that all of the healthcare personnel involved in providing care for the same patient communicate effectively with one another.

## 10.5 Switching of Patients to Other Providers in a Seamless Manner

Patients can utilize the same information on the blockchain to securely access and trade their health data with other doctors or organizations using a shareable private key. This can be done by using the same details that are stored on the blockchain. Interoperability and collaboration in the use of health information technology (HIT) could be improved as a result of this across the various users (Ahmad, A. Y. B. et al.,2024).

## 10.6 Electronic Medical Records That Are Compatible With One Another

The blockchain has the potential to provide a single transaction layer where organizations can send and share data in a single secure framework, with complete ownership of a particular set of structured data on the record, as well as private scrambled links to information that is stored in a separate location, such as radiographic or other photographs. There is potential for significant gain with the implementation of smart contracts and standardized authorization processes.

## 10.7 Data Security

Over 176 million patient medical records were breached between 2009 and 2017. Health data is already highly secure, but blockchain safety will increase it.

Everyone has a public key and a private key that can only be opened at a certain time and place. Hacking will be harder because hackers must target each user to steal confidential data. Blockchains will have an immutable audit trail of patient records as a result. Access management will shift from "all-or-nothing" to open-ended permissions. The user must create comprehensive transfers that specify who has access, for how long, and what data they can access. Users can change account rights at any time (Ahmad, A. Y. B.,et al.,2023).

Access management will shift from "all-or-nothing" to open-ended permissions. The user must create comprehensive transfers that specify who has access, for how long, and what data they can access. Users can change account rights at any time.

## 10.8 Healthcare Apps for Mobile Devices and Remote Monitoring

Mobile health apps are becoming more significant as technology advances. Blockchains contain encrypted electronic medical records (EMRs).

Network, information can be quickly sent to medical staff and used for home self-monitoring and therapy.

## 10.9 Medical Supplies Are Tracked and Protected

Due to its full accountability, blockchain technology will help preserve and track pharmaceutical supply. It also tracks labor expenses and CO2 emissions from these commodities.

## 10.10 Health Care Claim Statements

The blockchain is ideal for claims processing because it can offer medical situations as they happened without proof modification for fraud. Monitoring disease and epidemic progression. Genomic integrity maintenance

Recently, more companies have offered DNA sequencing to individuals, raising concerns about genomic data theft. Blockchain can overcome this difficulty and set up an online auction where academics can buy genomic data for their study. This might boost sales and eliminate costly intermediaries. Blockchain's unique properties can enable real-time disease reporting and pattern analysis, which can help determine the disease's origin and propagation factors.

Additionally, blockchain technology in healthcare are continually developing. Ethereum and Hyperledger Fabric are limited-scale blockchain application technologies. These systems are Ethereum and Hyperledger Fabric. Blockchain services can improve healthcare data management with better security and coordinated transactions.

# 11. DISCUSSION AND CONCLUSION

Since healthcare sometimes lags behind other industries in technology developments, the dependability of potential data may drive its adoption. Electronic health information management on the blockchain may have several crucial elements. We analyzed followable smart contracts. Innovative agreements would form an unbreakable chain for tailored therapy.

Following clinical guidelines. Combining healthcare operations will use a blockchain smart contract to eliminate duplicate copies in the parent centralized structure. Blockchain helps researchers verify and date scientific results. Smart contracts enable patients access their records, while a documentation blockchain repository lets researchers store their results. Large pharmaceutical companies need blockchain.

Through blockchain blocking, pharmaceutical association research distribution can be decoded. This structure would accelerate data preparation, ensure archive distribution accuracy, and reduce archive destruction, damage, and fabrication. The invention regulates these attractions: Blocks generated cannot be changed or deleted. The block chain prohibits data manipulation. Drug bids that are fake or weak are rejected. Pharmaceutical supply chain manufacturing, shipping, and distribution are controlled.

Block chain technology will enable medical practitioners, healthcare facilities, R&D experts, healthcare agencies, and biomedical researchers to securely and privately share vast amounts of data, clinical information, and recommendations. Successful blockchain deployment in clinical settings would boost biomedical research. Clean, reliable, and scalable clinical data collection, storage, and exchange will aid illness diagnosis and treatment in precision medicine.

Blockchains could store and control neurons. However, few firms have validated blockchain's usefulness in neurotechnology, which is developing. The safety of blockchain storage for brain data is questionable. Decentralized and open blockchains will prevent data from being tampered with or hacked, but many of the general questions about large-scale data collection remain, such as whether personal data will be marketed to third parties for dubious marketing reasons and whether consumers will be indirectly identified (as in bitcoin) by pseudonyms or data trends. This blockchain-based healthcare system will improve quality of life by encouraging patient participation.

Upgrade the system with blockchain technology to maintain data integrity. Security of healthcare and financial data requires blockchain technology.

## REFERENCES

Ahmad, A. Y. B. (2024). Ethical implications of artificial intelligence in accounting: A framework for responsible ai adoption in multinational corporations in Jordan. *International Journal of Data and Network Science*, *8*(1), 401–414. doi:10.5267/j.ijdns.2023.9.014

Ahmad, A. Y. B. (2024). Firm Determinants that Influences Implementation of Accounting Technologies in Business Organizations. *WSEAS Transactions on Business and Economics*, *21*, 1–11. doi:10.37394/23207.2024.21.1

Ahmad, A. Y. B., Abusaimeh, H., Rababah, A., Alqsass, M., Al-Olima, N. H., & Hamdan, M. N. (2024). Assessment of effects in advances of accounting technologies on quality financial reports in Jordanian public sector. *Uncertain Supply Chain Management*, *12*(1), 133–142. doi:10.5267/j.uscm.2023.10.011

Ahmad, A. Y. B., Gongada, T. N., Shrivastava, G., Gabbi, R. S., Islam, S., & Nagaraju, K. (2023). E-Commerce Trend Analysis and Management for Industry 5.0 using User Data Analysis. *International Journal of Intelligent Systems and Applications in Engineering*, *11*(11s), 135–150.

Ahmad, A. Y. B., Hannoon, A., Al-Daoud, K. I., Abu-Alsondos, I. A., & Al-Qaisieh, M. S. (2023). Assessment of Cloud Based Accounting Technology Adoption and Business Performance. Kurdish Studies, 11(3).

Ahmad, A. Y. B., Kumari, D. K., Shukla, A., Deepak, A., Chandnani, M., Pundir, S., & Shrivastava, A. (2024). Framework for Cloud Based Document Management System with Institutional Schema of Database. *International Journal of Intelligent Systems and Applications in Engineering*, *12*(3s), 672–678.

Ahmad, A. Y. B., Kumari, D. K., Shukla, A., Deepak, A., Chandnani, M., Pundir, S., & Shrivastava, A. (2024). Framework for Cloud Based Document Management System with Institutional Schema of Database. *International Journal of Intelligent Systems and Applications in Engineering*, *12*(3s), 672–678.

Ahmad, A. Y. B., & Kumari, S. S. M. S, S. K. Guha, A. Gehlot and B. Pant, "Blockchain Implementation in Financial Sector and Cyber Security System," *2023 International Conference on Artificial Intelligence and Smart Communication (AISC)*, Greater Noida, India, 2023, pp. 586-590, 10.1109/AISC56616.2023.10085045

Ahmad, A. Y. B., Tiwari, A., Nayeem, M. A., Biswal, B. K., Satapathy, D. P., Kulshreshtha, K., & Bordoloi, D. (2024). Artificial Intelligence Perspective Framework of the Smart Finance and Accounting Management Model. *International Journal of Intelligent Systems and Applications in Engineering*, *12*(4s), 586–594.

Ahmad, A. Y. B., William, P., Uike, D., Murgai, A., Bajaj, K. K., Deepak, A., & Shrivastava, A. (2024). Framework for Sustainable Energy Management using Smart Grid Panels Integrated with Machine Learning and IOT based Approach. *International Journal of Intelligent Systems and Applications in Engineering*, *12*(2s), 581–590.

Albermany S, Baqer FM. EEG authentication system using fuzzy vault scheme. *Journal of Discrete Mathematical Sciences and Cryptography*. 2021:1-6.

Albermany, S. A., & Safdar, G. A. (2014). Keyless security in wireless networks. *Wireless Personal Communications*, *79*(3), 1713–1731. doi:10.1007/s11277-014-1954-1

Anand, R., Lakshmi, S. V., Pandey, D., & Pandey, B. K. (2023). An enhanced ResNet-50 deep learning model for arrhythmia detection using electrocardiogram biomedical indicators. *Evolving Systems*, •••, 1–15.

Babu., . . .. SZD. Analysation of Big Data in Smart Healthcare. n: Gupta, M, Ghatak, S, Gupta, A, Mukherjee, AL (eds) Artificial Intelligence on Medical Data Lecture Notes in Computational Vision and Biomechanics. 2023;37. https://doi.org/ doi:10.1007/978-981-19-0151-5_21

Banerjee, S., Ghosh, S., Bhowmick, G. D., Pandey, D., Pandey, B. K., & Dadheech, P. (2023). Biofuels and Health Hazards: An Overview. Handbook of Research on Safe Disposal Methods of Municipal Solid Wastes for a Sustainable Environment, 324-344.

Boopathi, S., Pandey, B. K., & Pandey, D. (2023). Advances in artificial intelligence for image processing: techniques, applications, and optimization. In *Handbook of research on thrust technologies' effect on image processing* (pp. 73–95). IGI Global. doi:10.4018/978-1-6684-8618-4.ch006

Daoud, M. K., . Alqudah, M. . Al-Qeed, B. A. . Al Qaied, and A. Y. A. B. . Ahmad, "The Relationship Between Mobile Marketing and Customer Perceptions in Jordanian Commercial Banks: The Electronic Quality as A Mediator Variable", ijmst, vol. 10, no. 2, pp. 1360-1371, Jun. 2023.

Daoud, M. K., Al-Qeed, M., Ahmad, A. Y. B., & Al-Gasawneh, J. A. (2023). Mobile Marketing: Exploring the Efficacy of User-Centric Strategies for Enhanced Consumer Engagement and Conversion Rates. *International Journal of Membrane Science and Technology*, *10*(2), 1252–1262. doi:10.15379/ijmst.vi.1425

Daoud, M. K., Taha, S., Al-Qeed, M., Alsafadi, Y., Ahmad, A., & Allahham, M. (2024). EcoConnect: Guiding environmental awareness via digital marketing approaches. International Journal of Data and Network Science, 8(1), 235-242.

David, S., Duraipandian, K., Chandrasekaran, D., Pandey, D., Sindhwani, N., & Pandey, B. K. (2023). Impact of blockchain in healthcare system. In *Unleashing the Potentials of Blockchain Technology for Healthcare Industries* (pp. 37–57). Academic Press. doi:10.1016/B978-0-323-99481-1.00004-3

JayaLakshmi. G., Pandey, D., Pandey, B. K., Kaur, P., Mahajan, D. A., & Dari, S. S. (2024). Smart Big Data Collection for Intelligent Supply Chain Improvement. In AI and Machine Learning Impacts in Intelligent Supply Chain (pp. 180-195). IGI Global.

Lelisho, M. E., Pandey, D., Alemu, B. D., Pandey, B. K., & Tareke, S. A. (2023). The negative impact of social media during COVID-19 pandemic. *Trends in Psychology*, *31*(1), 123–142. doi:10.1007/s43076-022-00192-5

Malhotra, P., Pandey, D., Pandey, B. K., & Patra, P. M. (2021). Managing agricultural supply chains in COVID-19 lockdown. *International Journal of Quality and Innovation*, *5*(2), 109–118. doi:10.1504/IJQI.2021.117181

Menon, V., Pandey, D., Khosla, D., Kaur, M., Vashishtha, H. K., George, A. S., & Pandey, B. K. (2022). A Study on COVID–19, Its Origin, Phenomenon, Variants, and IoT-Based Framework to Detect the Presence of Coronavirus. In *IoT Based Smart Applications* (pp. 1–13). Springer International Publishing.

Meslie, Y., Enbeyle, W., Pandey, B. K., Pramanik, S., Pandey, D., Dadeech, P., & Saini, A. (2021). Machine intelligence-based trend analysis of COVID-19 for total daily confirmed cases in Asia and Africa. In *Methodologies and Applications of Computational Statistics for Machine Intelligence* (pp. 164–185). IGI Global. doi:10.4018/978-1-7998-7701-1.ch009

Muniandi, B., Huang, C.-J., Kuo, C.-C., Yang, T.-F., Chen, K.-H., Lin, Y.-H., Lin, S.-R., & Tsai, T.-Y. (2019, November). A 97% Maximum Efficiency Fully Automated Control Turbo Boost Topology for Battery Chargers. *IEEE Transactions on Circuits and Systems. I, Regular Papers*, *66*(11), 4516–4527. doi:10.1109/TCSI.2019.2925374

Pandey, B. K., Pandey, D., Nassa, V. K., George, S., Aremu, B., Dadeech, P., & Gupta, A. (2022, July). Effective and secure transmission of health information using advanced morphological component analysis and image hiding. In *Artificial Intelligence on Medical Data: Proceedings of International Symposium, ISCMM 2021* (pp. 223-230). Singapore: Springer Nature Singapore.

Pandey, B. K., Pandey, D., Wairya, S., Agarwal, G., Dadeech, P., Dogiwal, S. R., & Pramanik, S. (2022). Application of integrated steganography and image compressing techniques for confidential information transmission. *Cyber Security and Network Security*, 169-191.

Pandey, D., Hasan, A., Pandey, B. K., Lelisho, M. E., George, A. H., & Shahul, A. (2023). COVID-19 epidemic anxiety, mental stress, and sleep disorders in developing country university students. *CSI Transactions on ICT*, 1-9.

Pandey, D., Pandey, B. K., & Wairya, S. (2021). Hybrid deep neural network with adaptive galactic swarm optimization for text extraction from scene images. *Soft Computing*, *25*(2), 1563–1580. doi:10.1007/s00500-020-05245-4

Pandey, J. K., Jain, R., Dilip, R., Kumbhkar, M., Jaiswal, S., Pandey, B. K., & Pandey, D. (2022). Investigating Role of IoT in the Development of Smart Application for Security Enhancement. In *IoT Based Smart Applications* (pp. 219–243). Springer International Publishing.

Pandey, J. K., Jain, R., Dilip, R., Kumbhkar, M., Jaiswal, S., Pandey, B. K., & Pandey, D. (2022). Investigating Role of IoT in the Development of Smart Application for Security Enhancement. In *IoT Based Smart Applications* (pp. 219–243). Springer International Publishing.

Parthiban, K., Pandey, D., & Pandey, B. K. (2021). Impact of SARS-CoV-2 in online education, predicting and contrasting mental stress of young students: A machine learning approach. *Augmented Human Research*, *6*(1), 10. doi:10.1007/s41133-021-00048-0

Pramanik, S., Pandey, D., Joardar, S., Niranjanamurthy, M., Pandey, B. K., & Kaur, J. (2023, October). An overview of IoT privacy and security in smart cities. In AIP Conference Proceedings (Vol. 2495, No. 1). AIP Publishing.Albermany S.). doi:10.1063/5.0123511

Pramanik, S., Pandey, D., Joardar, S., Niranjanamurthy, M., Pandey, B. K., & Kaur, J. (2023, October). An overview of IoT privacy and security in smart cities. In AIP Conference Proceedings (Vol. 2495, No. 1). AIP Publishing. doi:10.1063/5.0123511

Saxena, A., Agarwal, A., Pandey, B. K., & Pandey, D. (2024). Examination of the Criticality of Customer Segmentation Using Unsupervised Learning Methods. *Circular Economy and Sustainability*, •••, 1–14. doi:10.1007/s43615-023-00336-4

Sharma, M., Talwar, R., Pandey, D., Nassa, V. K., Pandey, B. K., & Dadheech, P. (2024). A Review of Dielectric Resonator Antennas (DRA)-Based RFID Technology for Industry 4.0. *Robotics and Automation in Industry*, *4*(0), 303–324.

Singh, S., Madaan, G., Kaur, J., Swapna, H. R., Pandey, D., Singh, A., & Pandey, B. K. (2023). Bibliometric Review on Healthcare Sustainability. Handbook of Research on Safe Disposal Methods of Municipal Solid Wastes for a Sustainable Environment, (pp. 142-161). IEEE.

Tripathi, R. P., Sharma, M., Gupta, A. K., Pandey, D., Pandey, B. K., Shahul, A., & George, A. H. (2023). Timely prediction of diabetes by means of machine learning practices. *Augmented Human Research*, *8*(1), 1. doi:10.1007/s41133-023-00062-4

Zaidi, A. (2023). Predicting wildfires in Algerian forests using machine learning models. *Heliyon*, *9*(7), e18064. doi:10.1016/j.heliyon.2023.e18064 PMID:37519679

Zaidi, A. (2023). Smart Implementation of Industrial Internet of Things using Embedded Mechatronic System. *IEEE Embedded Systems Letters*. doi:10.1109/LES.2023.3309985

Zaidi, A., Ajibade, S.-S. M., Musa, M., & Bekun, F. V. (2023). New Insights into the Research Landscape on the Application of Artificial Intelligence in Sustainable Smart Cities: A Bibliometric Mapping and Network Analysis Approach. *International Journal of Energy Economics and Policy.*, *13*(4), 287–299. doi:10.32479/ijeep.14683

Zaidi, A., & Al-Harbi, O. (2023). Statistical Analysis of a Linear Multi-Step Numerical Treatment. *Journal of Statistics Applications & Probability*, *12*(1), 26. doi:10.18576/jsap/120126

Zaidi, A., & Al Luhayb, A. (2023). Two Statistical Approaches to Justify the Use of the Logistic Function in Binary Logistic Regression. *Mathematical Problems in Engineering*, *2023*, 5525675. doi:10.1155/2023/5525675

Section 4

# Engineering Tools and Technologies for Future Challenges

# Chapter 11
# Quantitative Analysis of 3D-Printed Bimetallic Component:
## Wire-Arc Additive Manufacturing and Metal Forming

**Japheth Obiko**
https://orcid.org/0000-0002-3069-1581
*Tshwane University of Technology, South Africa*

**Brendon Shongwe**
*Tshwane University of Technology, South Africa*

**Nicholus Malatji**
*Tshwane University of Technology, South Africa*

**Michael Bodunrin**
*University of the Witwatersrand, South Africa*

**Mariam Kassim**
*Jomo Kenyatta University of Agriculture and Technology, Kenya*

## ABSTRACT

*The book chapter reports on hybrid manufacturing (wire-arc additive manufacturing and metal forming) of 3D-printed bimetallic structure parts. After 3D printing, the metal flow behaviour of the printed material was studied using a Gleeble 3500 thermal-mechanical simulator. The test conditions were temperature range 850-1000°C and strain rate of $1s^{-1}$ and $10s^{-1}$ to strain of 0.6. The results show that the initial microstructure of the bimetallic structure had excellent bonding properties. The flow stress-strain curves increased with a decrease in deformation temperature and vice versa at a given strain rate. The study developed a constitutive equation that predicts the metal flow behaviour of the material studied. From the analysis, the material constants were a stress exponent of 4.76 and an activation energy of 213.7 $kJmol^{-1}$. The study indicates that this method (hybrid manufacturing) is possible in the industrial production process.*

DOI: 10.4018/979-8-3693-1335-0.ch011

## INTRODUCTION

Bimetallic materials offer better performance and numerous advantages than a single material (Sridar, Klecka and Xiong, 2022). For example, heat-resistant 9-12 wt% Cr steel with superior creep strength and excellent corrosion resistance at high temperatures is widely applicable in aerospace, automotive, and fossil power plant applications (Ennis and Czyrska-Filemonowicz, 2003; Czyrska-filemonowicz, Zielińska-lipiec and Ennis, 2006). However, this steel lacks suitability for structure components operating at temperatures higher than 700°C for advanced boiler systems. To achieve high thermal efficiency, hence reducing $CO_2$ emissions, an increase in operating conditions (temperature of up to 700°C and 35 MPa pressure) for modern power plants is paramount. Therefore, joining different steels by welding for the boiler piping system will achieve higher power plant performance. Joining dissimilar materials will take advantage of individual material properties. However, DWJ has associated challenges such as porosity, inclusion and residual stress, thus requiring urgent attention. These structures can be joined using additive manufacturing AM (3D printed) parts to avoid metallurgical challenges associated with dissimilar weld joints.

Additive Manufacturing (AM) has gained popularity as a potential method for reducing production time, cost and material wastage (Balashanmugam, 2021; Krishna, Manjaiah and Mohan, 2021; Srivastava and Rathee, 2022). This method builds a component from computer-readable 3D model data through layer-by-layer deposition. AM saves 50% of production costs compared to conventional manufacturing (Ivanova, Williams and Campbell, 2013; Gibson, Rosen and Stucker, 2015). AM has wide applications in aerospace, construction, manufacturing, automotive, biomedical, oil and gas industries (Ford, 2016; Najmon, Raeisi and Tovar, 2019; Yusuf, Cutler and Gao, 2019; Behera, 2020). The potential of AM as an alternative manufacturing process is growing steadily among the industry players for different applications, hence shaping the future of the manufacturing industry (Gornet, 2017). AM improves sustainability in manufacturing by enabling resource-efficient production and closed-loop material flows, thus improving production through product and process redesign (Ford, 2016). The wide application of AM provides a platform for manufacturing complex geometry components (Mhapsekar, McConaha and Anand, 2018).

AM techniques are mainly classified into powder bed fusion (PBF), sheet lamination, and directed energy deposition (DED) (Friel, 2015; Dinovitzer *et al.*, 2019). Figure 1 summarises the main AM processes and corresponding heat sources. For example, Wire-arc addictive manufacturing WAAM is a DED technique employed as an AM process (Vora *et al.*, 2022). This technique uses a wire feedstock as an energy source. In DED, a concentrated energy source melts metallic feedstock and is deposited onto a substrate to print the component (Ahn, 2021). DED is renowned for its rapid deposition speeds, versatility in materials, and ability to change and repair existing structures (Ahn, 2021). This technique is extensively reported in the literature using wire or powder and heat source such as an electron beam (Sames *et al.*, 2016; Negi *et al.*, 2020; Singh, Kapil and Das, 2020; Svetlizky *et al.*, 2021), plasma arc (Alberti, Bueno and D'Oliveira, 2016; Ivántabernero *et al.*, 2018; Dass and Moridi, 2019; Artaza *et al.*, 2020), laser (Ding *et al.*, 2015; Rumman *et al.*, 2019; Lorenz *et al.*, 2020; Yoo, Lee and Kim, 2023), and electric arc (Sames *et al.*, 2016; Li *et al.*, 2017; Singh, Kapil and Das, 2020; Rosli *et al.*, 2021; Svetlizky *et al.*, 2021). WAAM process uses different heat sources such as gas metal arc welding (GMAW), tungsten arc welding (GTAW), plasma arc welding (PAW) and cold metal transfer (CMT) (Marinelli *et al.*, 2019; Ríos, Colegrove and Williams, 2019). The WAAM basic principle is, as in Figure 2.

The main challenge for all wire-feed AM processes is defect formation, such as residual stresses and distortion generated during the AM process. These phenomena occur due to excessive energy input, high deposition rate, and large temperature gradient during the wire-feed AM processes. Then, the question is, if the component enters into service as-built, what is the influence of these defects on the overall mechanical performance of the WAAM 3D printed parts? Therefore, the said challenges, the hybrid manufacturing process comes in handy. This process incorporates AM followed by the hot deformation process during production. Studies have shown that the hybrid (AM + Hot deformation) process has industrial practicability (Bambach *et al.*, 2020; Liu *et al.*, 2022). Reports on the hot working of additively manufactured parts are available in the literature (Bambach *et al.*, 2020; Liu *et al.*, 2022; Maurya *et al.*, 2022). However, the hot working of bimetallic structures produced via wire-arc additive manufacturing is conspicuously missing in the literature, especially the flow behaviour at the interfacial region of the bimetallic part. Therefore, applying the hybrid method in the processing route of structural or functional components requires a proper understanding of the hot deformation behaviour of WAAM printed parts. This study explored the hot workability of bimetallic 3D-printed structures made from creep-resistant and stainless steel at different deformation conditions. Furthermore, the study developed constitutive equations for predicting the flow stress behaviour of bimetallic components. These equations act as input codes for metal forming process simulation.

## Theory of Metal Forming

Metal forming has two roles: shaping new products and improving the mechanical properties. Materials undergo permanent deformation during shaping, such as rolling, forging and extrusion. The product shape and size depend on the application of external force. Forming is done at room or high temperature. After casting, product manufacturing involves warm or hot working. Warm deformation occurs at $0.4T_m < T < 0.6T_m$, while hot deformation occurs at $T > 0.6T_m$ ($T_m$ is the melting point in absolute temperature). Slip and twinning mechanisms control plastic deformation during forming. Quantitative analysis provides insights into metal flow behaviour at different deformation conditions.

## Constitutive Modelling

Characterisation of metal forming by constitutive equations is the most common method of studying the hot deformation behaviour in metal and alloys. These equations model the relationship between the flow stress and the deformation conditions such as temperature, strain and strain rate. The constitutive equations provide information on the flow stress characteristics that can easily be used in the computational simulation of metal forming (Lin, Chen and Zhang, 2009; Samantaray, Mandal and Bhaduri, 2010; Lin and Chen, 2011; Yin *et al.*, 2013; Wang *et al.*, 2017; Zhu, He and Zhang, 2018). The constitutive models are broadly classified as *phenomenological* (considering continuum mechanics and thermodynamic irreversibility) and *physical* (considering dislocation density theory and dynamic recrystallisation). The classifications depend on the computational parameters for quantifying the flow stress behaviour (He *et al.*, 2015).

The phenomenological models, especially the Arrhenius constitutive model and its modified versions, have been widely used to describe the relationship between flow stress, temperature, strain and strain rate (Wang *et al.*, 2010; Mirzadeh *et al.*, 2011; Mirzadeh, Cabrera and Najafizadeh, 2011; He *et al.*, 2013; Qian *et al.*, 2015; Yang *et al.*, 2017). The Arrhenius type model has been used to characterise the flow stress behaviour of alloys such as:

- Modified 9Cr-1Mo steel (Samantaray, Mandal and Bhaduri, 2010)

*Figure 1. Summarised AM processes*

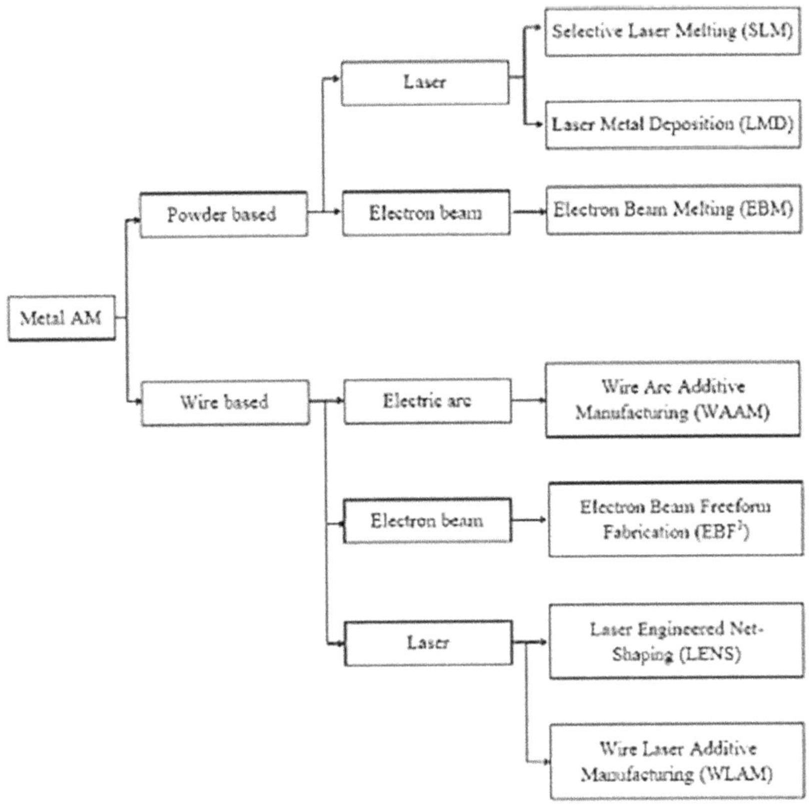


- P92 steel (Liu, Zhang and Yan, 2011; Shi and Liu, 2011; Alsagabi, 2016),
- 35CrMo steel (Huang *et al.*, 2017),
- 20CrMo alloy steel (He *et al.*, 2013),
- Nickel-based superalloys (Zhang *et al.*, 2016),
- N08028 corrosion resistant alloy (Wang *et al.*, 2013),

*Figure 2. WAAM basic principle*
*(Ryan, 2018; Xin et al., 2021)*

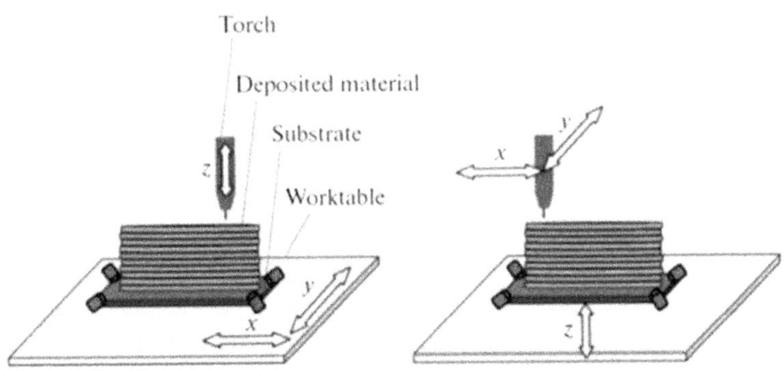

- Aluminium alloys (Gholamzadeh and Karimi Taheri, 2009; Lin *et al.*, 2010; Mostafaei and Kazeminezhad, 2012),
- Magnesium alloys (Luan *et al.*, 2014) and
- Titanium alloys (Jha *et al.*, 2017).

The flow stress behaviour is predicted by fitting experimental data into the constitutive equations to determine the material constants and activation energy.

## Phenomenological Model

The phenomenological constitutive models (Arrhenius-type equations) show the relationship between the flow stress and deformation conditions (temperature and strain rate) using Equation 1 (Yang, Li and Zhang, 2007; Yan *et al.*, 2013; Sun *et al.*, 2016). Experimental data are used in the constitutive equations to determine the material parameters and activation energy during deformation:

$$\dot{\varepsilon} = Af(\sigma)\exp\left(\frac{-Q}{RT}\right) \tag{1}$$

where $\dot{\varepsilon}$ is the strain rate in $s^{-1}$, $T$ is the temperature (K) and $Q$ is the activation energy of deformation in $kJ.mol^{-1}$. The stress function $f(\sigma)$, which is the maximum flow stress value (peak stress $\sigma_p$) can be expressed as:

$f(\sigma) = \sigma^{n'}$ for low flow stress ($\alpha\,\sigma < 0.8$)

$f(\sigma) = \exp(\beta'\sigma)$ for flow high stress ($\alpha\,\sigma > 1.2$)

$f(\sigma)=[\sinh(\alpha\sigma)]^n$ for all flow stress

where $n$, $n'$, $\beta$ and $\alpha$ are material constants and $\alpha \approx \beta'/n'$.

However, the general case (including the low and high flow stress) for analysing hot deformation behaviour developed by Sellars and Tegart (1966) in Equation 2, is commonly used to determine the material constants:

$$\dot{\varepsilon} = A[\sinh(\alpha\sigma)]^n\left[\frac{-Q}{RT}\right] \tag{2}$$

By solving the universal hyperbolic sine function (Equation 2), material constants such as activation energy $Q$ and stress exponent $n$ can be determined from linear graphs (Figure 4). At a constant temperature and taking the natural logarithm of Equation 2 and its partial derivative with $\ln\dot{\varepsilon}$:

$$\frac{1}{n} = \frac{\partial \ln[\sinh(\alpha\sigma)]}{\partial \ln\dot{\varepsilon}} \tag{3}$$

*Figure 3. a) ln $\dot{\varepsilon}$ vs. ln $\sigma$ to determine $n'$, b) ln ( $\dot{\varepsilon}$ ) vs. $\sigma$ to calculate $\beta'$*
*(Li et al., 2018)*

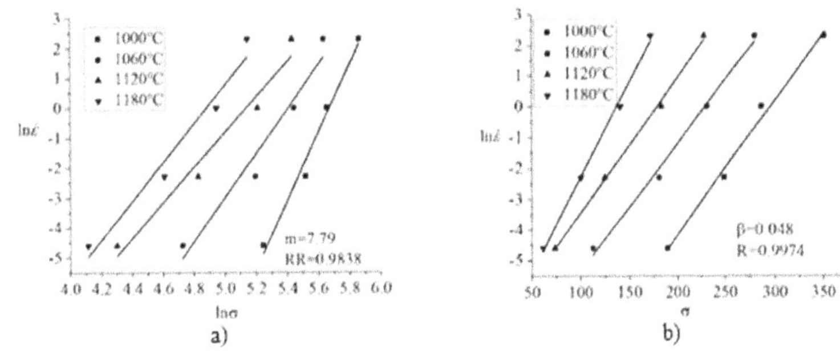


Similarly, at constant strain rate $\dot{\varepsilon}$ :

$$Q = Rn\frac{\partial \ln\left[\sinh\left(\alpha\sigma\right)\right]}{\partial \frac{1}{T}} \tag{4}$$

Similarly, the material constants $n'$ and $\beta'$ can be determined using the power law and exponential law equations for high and low flow stress and $\alpha \approx \beta'/n'$. The linear graphs used to calculate the material constants $n'$ and $\beta'$ are shown in Figure 3. Substituting the material constants and activation energy into Equation 2, an equation for predicting the flow stress during deformation can be developed.

The Zener-Hollomon parameter (*Z*) can describe the combined effects of the deformation temperature and strain rate (Wang *et al.*, 2008; Alsagabi, 2016). The relationship between the Zener-Hollomon parameter and the peak flow stress is described by He *et al.* (2017):

$$Z = \dot{\varepsilon}\exp\left[\frac{Q}{RT}\right] = f\left(\sigma\right) = B\exp\left(\beta\sigma_p\right) = A\left[\sinh\left(\alpha\sigma_p\right)\right]^n. \tag{5}$$

*Figure 4. Relationship between a) ln (strain rate) and ln (sinh (ασ)) to determine stress exponent n, and b) ln (sinh (ασ)) and 100 0/T ($K^{-1}$) to determine activation energy Q*
*(Nayan et al., 2014)*

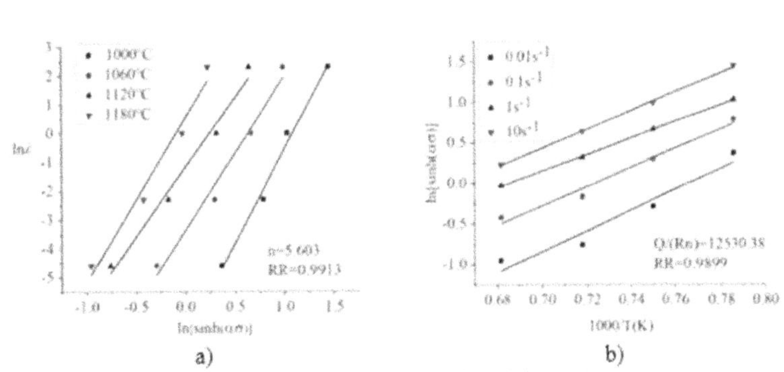

*Table 1. Chemical composition (%wt) of electrodes*

| Electrode | C | Si | P | Mn | S | Cr | Mo | Ni |
|---|---|---|---|---|---|---|---|---|
| ER308L | 0.02 | 0.32 | 0.011 | 1.7 | 0.009 | 20 | | 10 |
| E9018-B3 | 0.08 | 0.55 | 0.02 | 0.68 | 0.01 | 2.39 | 1.05 | - |

The natural logarithm of Equation 5 and a plot of *sinh* ($\alpha\sigma_p$) *vs.* ln *Z* can be used to determine the material constants *A* and *n* in the hyperbolic sine law by solving Equation 6:

$$\ln Z = \ln A + n\sinh(\boldsymbol{\alpha\sigma}p) \tag{6}$$

The peak flow stress $\sigma p$ can then be expressed as:

$$\sigma = \frac{1}{\alpha}\ln\left\{\left(\frac{Z}{A}\right)^{1/n} + \left[\left(\frac{Z}{A}\right)^{2/n} + 1\right]^{1/2}\right\} \tag{7}$$

## WAAM Processing

The study used welding electrodes having a similar chemical composition to creep-resistant steel (E9018-B3) and austenitic steel (ER308L) with the chemical composition given in Table 1. These electrodes were purchased from the local merchants and used as consumable material. Figure 5 shows the tungsten-inert gas welding machine that fabricated the bimetallic WAAM structure. The substrate was a mild steel plate of 5 mm thick. The bimetallic structure had 15 layers (for the first material-SS308L), cooled for 20 minutes, and then another 15 layers (for the second material) deposited on the substrate. The interpass cooling time was ~1 minute for the first and second layers. The WAAM-produced bimetallic structure was heat-treated at 740°C/1h before further processing. Figure 6 shows a section of the bimetallic component studied.

## Uniaxial Compression Test

Specimens of bimetallic WAAM structure measuring 8 mm diameter and 12 mm height were sectioned (Figure 5) for compression test. The uniaxial compression tests were done using Gleeble® 3500 equipment under vacuum. Hot deformation conditions were: temperature range of 800-1000°C and strain rate of $1s^{-1}$ and $10s^{-1}$ up to 60% reduction. The K-type thermocouples were spot welded at the centre monitor temperature between the test and preset temperature during testing. The test sample and anvil were pasted with graphite and nickel paste to minimise friction. All the tested samples were first heated to a temperature of 1100°C at a heating rate of 5°C/s and held for 180 seconds for microstructure homogenisation. Then, the specimen was cooled to the deformation temperature at 10°C/s and soaked for 60 seconds to reduce the thermal gradient before compression. After deformation, the samples were rapidly air-cooled to room temperature.

*Figure 5. TIG welding machine*

## RESULTS AND DISCUSSION

### Microstructure Characterisation

The overall performance of the bimetallic structure depends on the interfacial characteristics, especially the bond strength and microstructure evolution. The microstructure evolution of the bimetallic structure after heat treatment is in Figure 7. Generally, CrMo steel (material Zone 1) exhibits martensitic microstructure having precipitates along the grain and lath boundaries. The precipitates are the $M_{23}C_6$ carbides and MX carbonitrides. $M_{23}C_6$ (M = Fe, Mo, W, Cr) carbides precipitate along the prior austenite grain

*Figure 6. Sample of 3D printed bimetallic structure*

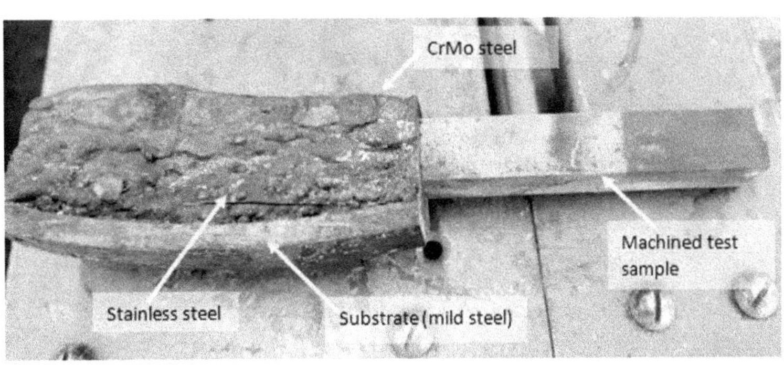

*Figure 7. a) Macrostructure, b) Microstructure of bimetallic structure showing the material zones*

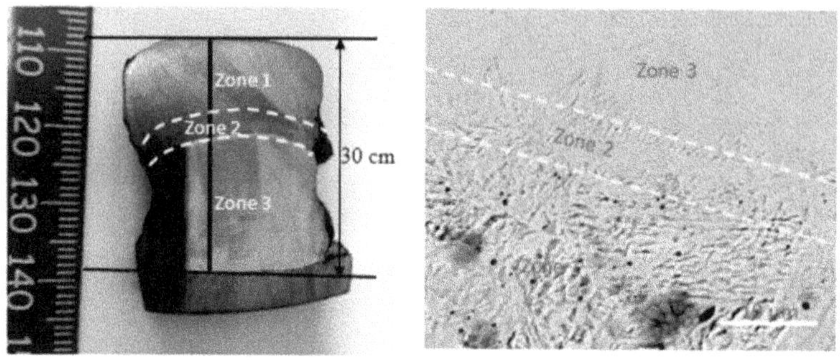


a) b)

boundaries PAGBs (Pandey *et al.*, 2019), while MX (M = V, Nb; X = C, N) precipitates are usually randomly distributed in the matrix (Barbadikar *et al.*, 2015). Precipitation of Coarse carbides occurs along the prior austenite grain boundaries, hindering the movement of the PAG boundaries and the subgrain dislocations, thus improving the creep resistance (Pandey *et al.*, 2018). $M_{23}C_6$ carbides improve the creep strength by pinning the grain boundary movement (Abson and Rothwell, 2013), while the MX precipitates impede dislocation movement during deformation (Nagode *et al.*, 2011).

Zone 1 (topmost zone) exhibited CrMo steel microstructure as in Figure 8. The microstructure had well-defined grain and lath boundaries decorated with carbides along the grain boundaries (bright features along the grain boundaries). Inclusions observed in the microstructure (Figure 8a) were due to slag trapped during deposition. These inclusions are detrimental to the mechanical properties of the bimetallic structure. Therefore, when using a cored electrode for the built-up process, more attention is required to eliminate slags before adding a layer to the existing deposited layer. The region near Zone 2 for the CrMo steel had a well-defined martensitic microstructure (Figure 8b). At this region, the continuous reheating process during deposition causes the material to undergo tempering, hence causing precipitation.

Figure 9 shows the microstructure of SS308L steel deposited on the substrate (mild steel). This region exhibited cellular and columnar structures formed due to repeated heating and cooling during the built-up process. The substrate and the environment are at room temperature during deposition, and the solidifi-

*Figure 8. SEM-BSE image of a) Topmost (Zone 1), b) near the interface (Zone 1)*

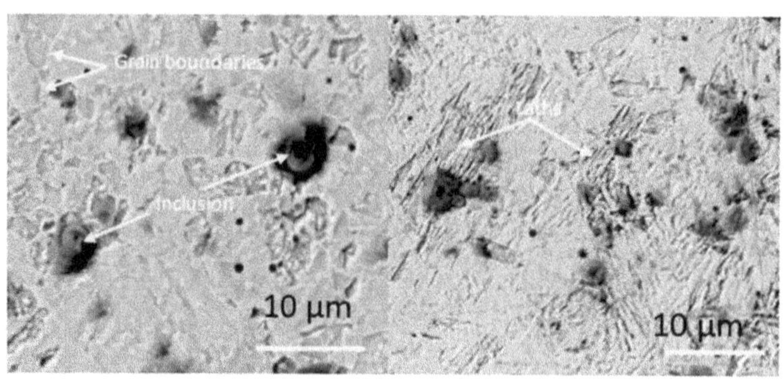


a) b)

*Figure 9. a) SEM-BSE image of Zone 3, b) Quantitative analysis*

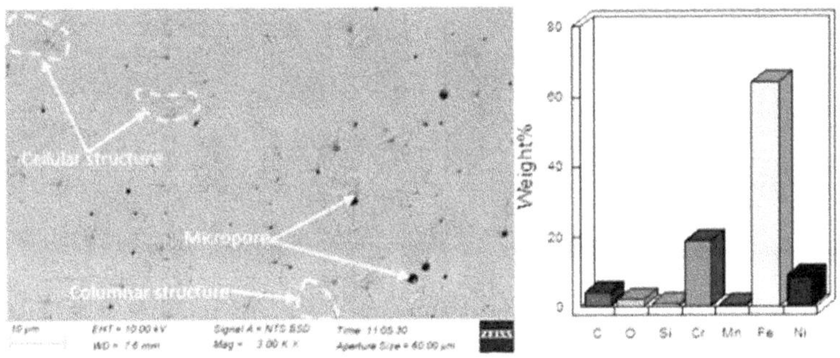


cation rate is high, thus causing the formation of finer grains. Gel'atko *et al.* (2023) reported that finer grains form due to a faster cooling rate accompanied by undercooling. A higher cooling rate limits the cooling time and thus restricts dendrite formation. The cellular and columnar structures occur naturally due to morphology variation at the molten pool during cooling along the grain boundaries (Gel'atko *et al.*, 2023). Figure 9 shows that the SS308L steel had micropores, which might have developed due to entrapped gases caused by a high solidification rate. Therefore, to improve the structural integrity, the printed parts require post-processing techniques such as forging and rolling during production.

## Flow Stress-Strain Behaviour

Figure 10 shows the flow stress-strain curves of the bimetallic structure manufactured using WAAM. The flow stress-strain curves were obtained during hot deformation at 1 $s^{-1}$ and 10 $s^{-1}$ and a temperature range of 800 to 1000°C. The flow stress-strain curve shape shows the material response to the deformation conditions (Wang, Jin and Wang, 2014). At different strain rates, the flow stress-strain curves exhibited similar metal flow behaviour: the flow stress increased with an increase in strain up to a specific strain value before stress saturation occurred, as defined by Laasraoui and Jonas (1991). In the initial stages, the flow stress increased rapidly until a strain $\varepsilon$ of ~0.2, representing a work-hardening stage. At higher $\varepsilon > 0.2$, the flow stress-strain curves slope decreased slightly to attain a steady-state condition. This characteristic behaviour is due to the balance between work hardening and dynamic recovery. The flow stress at this stage is called saturated flow stress (Laasraoui and Jonas, 1991). At this stage, the flow stress-strain curves show the characteristic behaviour of a material experiencing work hardening at strains $< 0.2$, followed by softening due to dynamic recovery at $\varepsilon > 0.2$. This flow behaviour occurs in commercial metals and alloys.

Figure 6 shows that the stress-strain curves are sensitive to the deformation conditions: the temperature, strain and strain rate. For a given strain rate, the flow stress increased with a decrease in the deformation temperature. This phenomenon occurs due to the diffusional mobility of atoms and dislocation mobility as a function of temperature (Zhu, 2018). For a given temperature, the flow stress increases as the strain rate increases.

Generally, flow stress is lower at higher deformation temperatures and lower strain rates due to the generation of atomic vacancies and dislocation diffusion. Carsí *et al.* (2011) reported that at temperatures higher than the dissolution temperatures of carbides, the carbides dissolve in the matrix, thus reducing precipitation hardening. The absence of carbides enhances dislocation motion, hence lowering the

*Figure 10. Flow stress-strain curves under different deformation conditions*

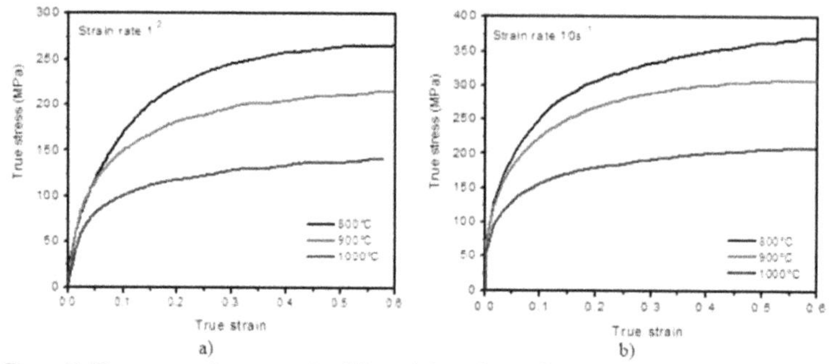


flow stress. Contrarily, deformation at lower temperatures and higher strain rates results in higher flow stresses. Higher flow stress is due to increased dislocation density. Forming at lower temperatures and higher strain rates occurs rapidly. Hence, there is no time for softening to occur. Therefore, the study recommends higher deformation temperatures above the dissolution temperature of carbides and lower strain rates for industrial metal forming processes.

## Kinetic Analysis

The thermal activation controls the metal flow behaviour pattern during deformation, thus controlling the final product microstructure (Liu *et al.*, 2020). During deformation, the softening mechanisms occur. These mechanisms' derivation is by comparing the deformation and self-diffusion activation energy. For dynamic recovery DRV, the activation energy is closely equal to the self-diffusion activation energy. However, the activation energy for a material undergoing dynamic recrystallisation DRX is higher than the self-diffusion activation energy. These softening mechanisms, DRX and DRV, are identified by comparing their activation energy and self-diffusion. The activation energy Q depends on the flow stress at different deformation conditions. The empirical constitutive equations derive this parameter Q. Arrhenius-type equations developed by Sellars and Tegart (1966) are used to determine constitutive material constants and activation energy. The peak stress or the steady-state stress is used to determine material properties (McQueen and Ryan, 2002). However, the flow stress curves in this study (Figure 6) did not exhibit DRX flow behaviour. Therefore, the kinetic analysis was done using the saturation flow stress $\sigma_{ss}$ (Laasraoui and Jonas, 1991). The stress values are read directly from the flow stress-strain curves, as reported by Jonas *et al.*(2009). Therefore, material constants *n'* and $\beta$ were obtained by solving the Arrhenius-type constitutive equation given in Equation 1, written as Equations 8 and 9:

$$\text{Low stress: } \dot{\varepsilon} = A_1 \sigma_{ss}^{\ n'} \exp\left(\frac{-Q}{RT}\right). \tag{8}$$

$$\text{High stress: } \dot{\varepsilon} = A_2 \exp\left(\beta \sigma_{ss}\right) \exp\left(\frac{-Q}{RT}\right). \tag{9}$$

Taking the natural logarithms of the equations:

$$\ln\dot{\varepsilon} + \frac{Q}{RT} = \ln A_1 + n'\ln\sigma_{ss} \tag{10}$$

$$\ln\dot{\varepsilon} + \frac{Q}{RT} = \ln A_2 + \beta\sigma_{ss} \tag{11}$$

Taking the partial derivatives of Equations 10 and 11 at constant deformation temperature:

$$n' = \frac{\partial \ln\dot{\varepsilon}}{\partial \ln\sigma_{ss}} \tag{12}$$

$$\beta' = \frac{\partial \ln\dot{\varepsilon}}{\partial \sigma_{ss}} \tag{13}$$

Averages of the material constants $n'$ and $\beta'$ were obtained by solving Equations 12 and 13 and then using linear regression of the plotted results for each steel. Substituting the saturation flow stress into the hyperbolic sine equations:

$$\dot{\varepsilon} = A\left[\sinh\left(\alpha\sigma_{ss}\right)\right]^n \left[\frac{-Q}{RT}\right] \tag{14}$$

The equations for solving the stress exponent $n$ and activation energy $Q$ are:

$$\frac{1}{n} = \frac{\partial \ln\left[\sinh\left(\alpha\sigma_{ss}\right)\right]}{\partial \ln\dot{\varepsilon}} \tag{15}$$

$$Q = Rn\frac{\partial \ln\left[\sinh\left(\alpha\sigma_{ss}\right)\right]}{\partial \frac{1}{T}} \tag{16}$$

The $n'$ value was obtained by taking the average slope of $\ln\dot{\varepsilon}$ *vs.* $\ln\sigma_{ss}$ (Figure 11a). The value of $\beta'$ was determined from the slope of $\ln\dot{\varepsilon}$ *vs.* $\sigma_{ss}$ plots (Figure 11b).

$n' = 6.43$, $\beta' = 0.0279$, $\alpha \approx \beta'/n' = 0.00434$.

*Figure 11. Plot of ln ( $\dot{\varepsilon}$ )vs. ln σ to obtain n′ and (b) ln ( $\dot{\varepsilon}$ ) vs. σ to obtain β.*

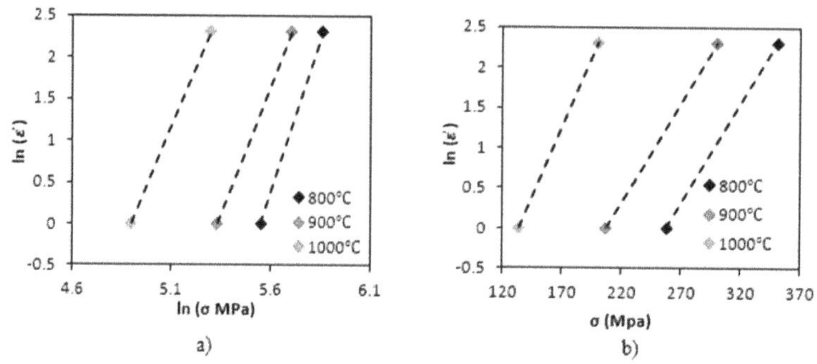


The stress exponent *n value* is an average of the slope of the plots of $\ln\dot{\varepsilon}$ *vs.* ln sinh ($\alpha\,\sigma_{ss}$) at different deformation temperatures (Figure 12a). The activation energy *Q* was determined from the plot of ln sinh ($\alpha\,\sigma_{ss}$) *vs.* 1000/*T* at different strain rates as shown in Figure 12b.

The stress exponent $n = 4.76$, $Q = 213.71$ kJ.mol$^{-1}$, and $A = 8.56 \times 10^9$ (coefficient of the hyperbolic stress parameter).

From the obtained material constants and activation energy, the developed constitutive equation to predict the flow stress for the 3D printed bimetallic material:

$$\dot{\varepsilon} = 8.56\times10^{9}\sinh\left(0.00434\sigma_{ss}\right)^{4.76}\exp\left[\frac{-213710}{RT}\right] \tag{17}$$

The steady-state flow stress for the bimetallic steel:

$$\sigma_{ss} = \frac{1}{0.00434}\ln\left[\left(\frac{Z}{8.56\times10^{9}}\right)^{\frac{1}{4.76}} + \left(\left(\frac{Z}{8.56\times10^{9}}\right)^{\frac{2}{4.76}} + 1\right)^{\frac{1}{2}}\right] \tag{18}$$

$$Z = \dot{\varepsilon}\left[\frac{616000}{RT}\right] = 4.89\times10^{32}\left(\sinh\left(0.00220\sigma_{ss}\right)\right)^{32.0} \tag{19}$$

## CONCLUSION

WAAM 3D printing and the Gleeble® 3500 thermal-mechanical simulator Equipments were employed to study the industrial production process. The study observed the following:

*Figure 12. (a) Plot of ln ( $\dot{\varepsilon}$ ) vs. ln (sinh (ασ)) to obtain stress exponent n and (b) ln (sinh (ασsat)) vs. 1000/T to obtain apparent activation energy Q.*

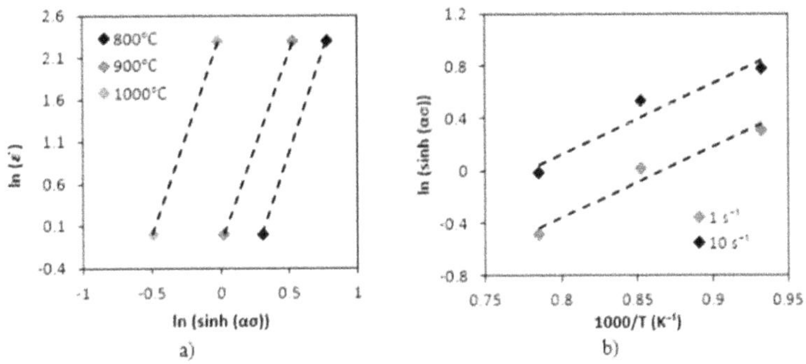


1. The initial microstructure of the WAAM 3D printed bimetallic structure had excellent bonding properties with a narrow gradient zone. The microstructure had defects such as micropores and inclusion. The optimisation of the WAAM process parameter will ensure complete suppression of inherent WAAM defects. The study advises the application of optimisation tools, such as Artificial Neural Network (ANN), to predict the optimal process parameters.

2. The flow stress-strain curves of the bimetallic structure exhibited an increase in flow stress with a decrease in deformation temperature at a given strain rate and vice-versa. The results indicate that the flow stress value is sensitive to deformation conditions. This behaviour is similar to wrought metal and alloys. Thus, the printed parts meet the requirements for industrial application.

3. The material constants values for the bimetallic structure were stress exponent *n* of 4.76 and activation energy *Q* of 213.71 kJmol-1. The equation for predicting flow stress behaviour for the studied material was:

$$\dot{\varepsilon} = 8.56 \times 10^{9} sinh\left(0.00434\sigma_{ss}\right)^{4.76} exp\left[\frac{-213710}{RT}\right]$$

This equation acts as input code in computer simulation software for analysing metal flow behaviour.

4. The study revealed that hybrid manufacturing is possible for the industrial production process. The processing route incorporating WAAM and metal forming enhances the quality of 3D printed parts.

*Table 2. Material constants calculated using the constutive equations*

| $n$ | $Q$ | $ln\,A$ | $n'$ | $\beta'$ | $\alpha$ |
|---|---|---|---|---|---|
| **4.76** | 213.71 | 22.87 | 6.43 | 0.0279 | 0.00434 |

## REFERENCES

Abson, D. J., & Rothwell, J. S. (2013). Review of type IV cracking of weldments in 9–12%Cr creep strength enhanced ferritic steels. *International Materials Reviews*, *58*(8), 437–473. doi:10.1179/1743280412Y.0000000016

Ahn, D. G. (2021). *Directed Energy Deposition (DED) Process: State of the Art, International Journal of Precision Engineering and Manufacturing - Green Technology*. Korean Society for Precision Engineering., doi:10.1007/s40684-020-00302-7

Alberti, E. A., Bueno, B. M. P., & D'Oliveira, A. S. C. M. (2016). Additive manufacturing using plasma transferred arc. *International Journal of Advanced Manufacturing Technology*, *83*(9), 1861–1871. doi:10.1007/s00170-015-7697-7

Alsagabi, S. (2016). High Temperature Deformation Behavior of P92 Steel. *Transactions of the Indian Institute of Metals*, *69*(8), 1513–1518. doi:10.1007/s12666-015-0725-3

Artaza, T., Suárez, A., Veiga, F., Braceras, I., Tabernero, I., Larrañaga, O., & Lamikiz, A. (2020). Wire arc additive manufacturing Ti6Al4V aeronautical parts using plasma arc welding: Analysis of heat-treatment processes in different atmospheres. *Journal of Materials Research and Technology*, *9*(6), 15454–15466. doi:10.1016/j.jmrt.2020.11.012

Bambach, M., Sizova, I., Sydow, B., Hemes, S., & Meiners, F. (2020). Hybrid manufacturing of components from Ti-6Al-4V by metal forming and wire-arc additive manufacturing. *Journal of Materials Processing Technology*, *282*, 116689. Advance online publication. doi:10.1016/j.jmatprotec.2020.116689

Bambach, M., Sizova, I., Sydow, B., Hemes, S., & Meiners, F. (2021). Perspectives on additive manufacturing in Industry 4.0. In M. Manjaiah & ... (Eds.), *Woodhead Publishing Reviews: Mechanical Engineering Series* (pp. 127–150). Woodhead Publishing. https://doi.org/https://doi.org/10.1016/B978-0-12-822056-6.00001-1

Barbadikar, R., Deshmukh, S., Maddi, L., Laha, K., Parameswaran, P., Ballal, R., Peshwe, R., Paretkar, K., Nandagopal, M., & Mathew, D. (2015). Effect of normalizing and tempering temperatures on microstructure and mechanical properties of P92 steel. *International Journal of Pressure Vessels and Piping*, *132–133*, 97–105. doi:10.1016/j.ijpvp.2015.07.001

Behera, A. (2020). Processes and Application in Additive Manufacturing: Practices in Aerospace, Automobile, Medical, and Electronic Industries. In *Additive Manufacturing applications for metals and composites* (pp. 25–47). IGI Global. doi:10.4018/978-1-7998-4054-1.ch002

Czyrska-filemonowicz, A., Zielińska-lipiec, A., & Ennis, P. J. (2006). Modified 9% Cr Steels for Advanced Power Generation : Microstructure and Properties. *Journal of Achievements in Materials and Manufacturing Engineering*, *19*(2), 43–48.

Dass, A., & Moridi, A. (2019). State of the Art in Directed Energy Deposition: From Additive Manufacturing to Materials Design. *Coatings*, *9*(7), 418. doi:10.3390/coatings9070418

Ding, D., Pan, Z., Cuiuri, D., & Li, H. (2015). Wire-feed additive manufacturing of metal components: Technologies, developments and future interests. *International Journal of Advanced Manufacturing Technology*, *81*(1), 465–481. doi:10.1007/s00170-015-7077-3

Dinovitzer, M., Chen, X., Laliberte, J., Huang, X., & Frei, H. (2019). Effect of wire and arc additive manufacturing (WAAM) process parameters on bead geometry and microstructure. *Additive Manufacturing*, *26*, 138–146. doi:10.1016/j.addma.2018.12.013

Ennis, P.J. & Czyrska-Filemonowicz, A. (2003). Recent advances in creep-resistant steels for power plant applications', *Sadhana - Academy Proceedings in Engineering Sciences*, *28*(3–4), 709–730. . doi:10.1007/BF02706455

Ford, S. (2016). Additive manufacturing and sustainability : an exploratory study of the advantages and challenges. *Journal of Cleaner Production*. . doi:10.1016/j.jclepro.2016.04.150

Friel, R. J. (2015) 'Power ultrasonics for additive manufacturing and consolidating of materials', Power Ultrasonics: Applications of High-Intensity Ultrasound, (pp. 313–335). Springer. doi:10.1016/B978-1-78242-028-6.00013-2

Gel'atko, M., Hatala, M., Botko, F., Vandžura, R., Hajnyš, J., Šajgalík, M., & Török, J. (2023). Stress Relieving Heat Treatment of 316L Stainless Steel Made by Additive Manufacturing Process. *Materials (Basel)*, *16*(19), 6461. doi:10.3390/ma16196461 PMID:37834599

Gholamzadeh, A., & Karimi Taheri, A. (2009). The prediction of hot flow behavior of Al-6%Mg alloy. *Mechanics Research Communications*, *36*(2), 252–259. doi:10.1016/j.mechrescom.2008.06.011

Gibson, I., Rosen, D., & Stucker, B. (2015). *3D Printing*. Rapid Prototyping, and Direct Digital Manufacturing, Additive Manufacturing Technologies.

Gornet, T. (2017). History of Additive Manufacturing. Research Gate. doi:10.4018/978-1-5225-2289-8.ch001

He, A., Xie, G., Yang, X., Wang, X., & Zhang, H. (2015). A physically-based constitutive model for a nitrogen alloyed ultralow carbon stainless steel. *Computational Materials Science*, *98*, 64–69. doi:10.1016/j.commatsci.2014.10.044

He, A., Xie, G., Zhang, H., & Wang, X. (2013). A comparative study on Johnson-Cook, modified Johnson-Cook and Arrhenius-type constitutive models to predict the high temperature flow stress in 20CrMo alloy steel. *Materials & Design*, *52*, 677–685. doi:10.1016/j.matdes.2013.06.010

He, J., Zhang, D., Zhang, W., Qiu, C., & Zhang, W. (2017). Constitutive equation and hot compression deformation behavior of homogenized Al-7.5Zn-1.5Mg-0.2Cu-0.2Zr alloy. *Materials (Basel)*, *10*(10), 1–12. doi:10.3390/ma10101193 PMID:29057825

Huang, Y., Wang, S., Xiao, Z., & Liu, H. (2017). Critical Condition of Dynamic Recrystallization in 35CrMo Steel. *Metals*, *7*(5), 161. doi:10.3390/met7050161

Ivanova, O., Williams, C., & Campbell, T. (2013). Additive manufacturing (AM) and nanotechnology: Promises and challenges. *Rapid Prototyping Journal*, *19*(5), 353–364. doi:10.1108/RPJ-12-2011-0127

Ivántabernero, P. A., Álvarez, P. & Suárez, A. (2018). Study on Arc Welding Processes for High Deposition Rate Additive Manufacturing. Procedia CIRP, 68, 358–362. doi:10.1016/j.procir.2017.12.095

Jha, S., Tewari, A., Mishra, S., & Toppo, S. (2017). Constitutive Relations for Ti-6Al-4V Hot Working. *Procedia Engineering*, *173*, 755–762. doi:10.1016/j.proeng.2016.12.089

Jonas, J., Quelennec, X., Jiang, L., & Martin, É. (2009). The Avrami kinetics of dynamic recrystallization. *Acta Materialia*, *57*(9), 2748–2756. doi:10.1016/j.actamat.2009.02.033

Krishna, R., Manjaiah, M., & Mohan, C. B. (2021). Developments in additive manufacturing. In M. Manjaiah, (Eds.), *Woodhead Publishing Reviews: Mechanical Engineering Series* (pp. 37–62). Woodhead Publishing.

Laasraoui, A., & Jonas, J. J. (1991). Prediction of steel flow stresses at high temperatures and strain rates. *Metallurgical Transactions. A, Physical Metallurgy and Materials Science*, *22*(7), 1545–1558. doi:10.1007/BF02667368

Li, F., Chen, S., Shi, J., Tian, H., & Zhao, Y. (2017). Evaluation and Optimization of a Hybrid Manufacturing Process Combining Wire Arc Additive Manufacturing with Milling for the Fabrication of Stiffened Panels. *Applied Sciences (Basel, Switzerland)*, *7*(12), 1233. doi:10.3390/app7121233

Li, Y., Ji, H., Li, W., Li, Y., Pei, W., & Liu, J. (2018). Hot deformation characteristics-Constitutive equation and processing maps-of 21-4N Heat- resistant steel. *Materials (Basel)*, *12*(1), 89. doi:10.3390/ma12010089 PMID:30591698

Lin, C., Xia, C., Chen, M., & Chen, S. (2010). Constitutive descriptions for hot compressed 2124-T851 aluminum alloy over a wide range of temperature and strain rate. *Computational Materials Science*, *50*(1), 227–233. doi:10.1016/j.commatsci.2010.08.003

Lin, Y. C., Chen, M. S., & Zhang, J. (2009). Modeling of flow stress of 42CrMo steel under hot compression. *Materials Science and Engineering A*, *499*(1–2), 88–92. doi:10.1016/j.msea.2007.11.119

Lin, Y. C., & Chen, X. M. (2011). A critical review of experimental results and constitutive descriptions for metals and alloys in hot working. *Materials & Design*, *32*(4), 1733–1759. doi:10.1016/j.matdes.2010.11.048

Liu, C. Y., Zhang, R. J., & Yan, Y. N. (2011). Hot deformation behaviour and constitutive modelling of P92 heat resistant steel. *Materials Science and Technology*, *27*(8), 1281–1286. doi:10.1179/026708310X12683158443323

Liu, F., Ma, J., Peng, L., Huang, G., Zhang, W., Xie, H., & Mi, X. (2020). Hot deformation behavior and microstructure evolution of Cu-Ni-Co-Si alloys. *Materials (Basel)*, *13*(9), 2042. doi:10.3390/ma13092042 PMID:32349437

Liu, Y., Zhang, C., Wang, Y., Xu, X., Zhu, H., & Jiang, J. (2022). Reveal the hot deformation behaviour and microstructure evolution in additively manufactured 316L stainless steel. *Materials Science and Engineering A*, *861*(August), 144290. doi:10.1016/j.msea.2022.144290

Lorenz, A., Jones, B., Wimpenny, I., & Jackson, R. (2020). A review of hybrid manufacturing. *Proceedings - 26th Annual International Solid Freeform Fabrication Symposium - An Additive Manufacturing Conference,* (pp. 96–108).

Luan, J., Sun, C., Li, X., & Zhang, Q. (2014). Constitutive model for AZ31 magnesium alloy based on isothermal compression test. *Materials Science and Technology*, *30*(2), 211–219. doi:10.1179/1743284713Y.0000000341

Marinelli, G., Martina, F., Ganguly, S., Williams, S., Lewtas, H., Hancock, D., Mehraban, S., & Lavery, N. (2019). Microstructure and thermal properties of unalloyed tungsten deposited by Wire + Arc Additive Manufacture. *Journal of Nuclear Materials*, *522*, 45–53. doi:10.1016/j.jnucmat.2019.04.049

Maurya, K., Yeom, T., Kang, W., Park, H., Hong, K., & Reddy, S. (2022). Optimization of hybrid manufacturing process combining forging and wire-arc additive manufactured Ti-6Al-4V through hot deformation characterization. *Journal of Alloys and Compounds*, *894*, 162453. doi:10.1016/j.jallcom.2021.162453

McQueen, H. J., & Ryan, N. D. (2002). Constitutive analysis in hot working. *Materials Science and Engineering A*, *322*(1–2), 43–63. doi:10.1016/S0921-5093(01)01117-0

Mhapsekar, K., McConaha, M., & Anand, S. (2018). Additive Manufacturing Constraints in Topology Optimization for Improved Manufacturability. *Journal of Manufacturing Science and Engineering*, *140*(5), 051017. doi:10.1115/1.4039198

Mirzadeh, H., Cabrera, J. M., & Najafizadeh, A. (2011). Constitutive relationships for hot deformation of austenite. *Acta Materialia*, *59*(16), 6441–6448. doi:10.1016/j.actamat.2011.07.008

Mirzadeh, H., Cabrera, M., Prado, M., & Najafizadeh, A. (2011). Hot deformation behavior of a medium carbon microalloyed steel. *Materials Science and Engineering A*, *528*(10–11), 3876–3882. doi:10.1016/j.msea.2011.01.098

Mostafaei, M. A., & Kazeminezhad, M. (2012). The prediction of hot flow behavior of Al-6%Mg alloy. *Materials Science and Engineering A*, *535*, 216–221. doi:10.1016/j.msea.2011.12.067

Nagode, A., Kosec, L., Ule, B., & Kosec, G. (2011). Review of creep resistant alloys for power plant applications. *Metalurgija*, *50*(1), 45–48.

Najmon, J. C., Raeisi, S., & Tovar, A. (2019). *Review of additive manufacturing technologies and applications in the aerospace industry*. Additive Manufacturing for the Aerospace Industry., doi:10.1016/B978-0-12-814062-8.00002-9

Nayan, N., Singh, G., Murty, N., Jha, K., Pant, G., Koshy, M., & Ramamurty, U. (2014). Hot deformation behaviour and microstructure control in AlCrCuNiFeCo high entropy alloy. *Intermetallics*, *55*, 145–153. doi:10.1016/j.intermet.2014.07.019

Negi, S., Nambolan, A., Kapil, S., Joshi, P. R. M., Karunakaran, P., & Bhargava, P. (2020). Review on electron beam based additive manufacturing. *Rapid Prototyping Journal*, *26*(3), 485–498. doi:10.1108/RPJ-07-2019-0182

Pandey, C., Mahapatra, M., Kumar, P., & Saini, N. (2018). 'Some studies on P91 steel and their weldments', *Journal of Alloys and Compounds. Journal of Alloys and Compounds*, *V*, 332–364. doi:10.1016/j.jallcom.2018.01.120

Pandey, C., Mohan, M., Kumar, P., Thakre, G. and Saini, N. (2019) 'Role of evolving microstructure on the mechanical behaviour of P92 steel welded joint in as-welded and post weld heat treated state', *Journal of Materials Processing Technology*, *263*(June 2018), 241–255. . doi:10.1016/j.jmatprotec.2018.08.032

Qian, L., Fang, G., Zeng, P., & Wang, X. (2015). Correction of flow stress and determination of constitutive constants for hot working of API X100 pipeline steel. *International Journal of Pressure Vessels and Piping*, *132–133*(November), 43–51. doi:10.1016/j.ijpvp.2015.05.008

Ríos, S., Colegrove, P. A., & Williams, S. W. (2019). Metal transfer modes in plasma Wire + Arc additive manufacture. *Journal of Materials Processing Technology*, *264*, 45–54. doi:10.1016/j.jmatprotec.2018.08.043

Rosli, A., Alkahari, A., Mohd, M., Shajahan, R., Faiz, H., & Safarudin, G. (2021). Review on effect of heat input for wire arc additive manufacturing process. *Journal of Materials Research and Technology*, *11*, 2127–2145. doi:10.1016/j.jmrt.2021.02.002

Rumman, R., Lewis, A., Hascoet, Y., & Quinton, S. (2019). Laser metal deposition and wire arc additive manufacturing of materials: An overview. *Archives of Metallurgy and Materials*, *64*(2), 467–473. doi:10.24425/amm.2019.127561

Ryan, E. M. (2018) *On Wire and Arc Additive Manufacture of Aluminium.* [Thesis\.

Samantaray, D., Mandal, S., & Bhaduri, A. K. (2010). Constitutive analysis to predict high-temperature flow stress in modified 9Cr-1Mo (P91) steel. *Materials & Design*, *31*(2), 981–984. doi:10.1016/j.matdes.2009.08.012

Sames, J., List, A., Pannala, S., Dehoff, R., & Babu, S. (2016). The metallurgy and processing science of metal additive manufacturing. *International Materials Reviews*, *61*(5), 315–360. doi:10.1080/09506608.2015.1116649

Sellars, C. M., & McTegart, W. J. (1966). On the mechanism of hot deformation. *Acta Metallurgica*, *14*(9), 1136–1138. doi:10.1016/0001-6160(66)90207-0

Shi, R. X., & Liu, Z. D. (2011). Hot deformation behavior of P92 steel used for ultra-super-critical power plants. *Journal of Iron and Steel Research International*, *18*(7), 53–58. doi:10.1016/S1006-706X(11)60090-3

Singh, A., Kapil, S., & Das, M. (2020). A comprehensive review of the methods and mechanisms for powder feedstock handling in directed energy deposition. *Additive Manufacturing*, *35*, 101388. doi:10.1016/j.addma.2020.101388

Sridar, S., Klecka, M. A., & Xiong, W. (2022). Interfacial characteristics of P91 steel - Inconel 740H bimetallic structure fabricated using wire-arc additive manufacturing. *Journal of Materials Processing Technology*, *300*, 117396. doi:10.1016/j.jmatprotec.2021.117396

Srivastava, M., & Rathee, S. (2022). Additive manufacturing: Recent trends, applications and future outlooks. *Progress in Additive Manufacturing*, *7*(2), 261–287. doi:10.1007/s40964-021-00229-8

Sun, C., Zuo, X., Xiang, Y., & Yang, J. (2016). Investigation on Hot Deformation Behavior and Hot Processing Map of BSTMUF601 Super-Alloy. *Metals*, *6*(3), 70. doi:10.3390/met6030070

Svetlizky, D., Das, M., Zheng, B., Vyatskikh, L., Bose, S., Bandyopadhyay, A., Schoenung, M., Lavernia, J., & Eliaz, N. (2021). Directed energy deposition (DED) additive manufacturing: Physical characteristics, defects, challenges and applications. *Materials Today*, *49*, 271–295. doi:10.1016/j.mattod.2021.03.020

Vora, J., Parmar, H., Chaudhari, R., Khanna, S., Doshi, M., & Patel, V. (2022). Experimental investigations on mechanical properties of multi-layered structure fabricated by GMAW-based WAAM of SS316L. *Journal of Materials Research and Technology*, *20*, 2748–2757. doi:10.1016/j.jmrt.2022.08.074

Wang, B., Fu, W., Lv, Z., Jiang, P., Zhang, W., & Tian, Y. (2008). Study on hot deformation behavior of 12%Cr ultra-super-critical rotor steel. *Materials Science and Engineering A*, *487*(1–2), 108–113. doi:10.1016/j.msea.2007.10.007

Wang, L., Liu, F., Zuo, Q., & Chen, F. (2013). Prediction of flow stress for N08028 alloy under hot working conditions. *Materials & Design*, *47*, 737–745. doi:10.1016/j.matdes.2012.12.074

Wang, M., Jin, P., & Wang, J. (2014). Hot deformation and processing maps of 7005 aluminum alloy. *High-Temperature Materials and Processes*, *33*(4), 369–375. doi:10.1515/htmp-2013-0083

Wang, S., Huang, Y., Xiao, Z., Liu, Y., & Liu, H. (2017). A Modified Johnson-Cook Model for Hot Deformation Behavior of 35CrMo Steel. *Metals*, *7*(9), 337. doi:10.3390/met7090337

Wang, Z., Fu, W., Wang, B., Zhang, W., Lv, Z., & Jiang, P. (2010). Study on hot deformation characteristics of 12%Cr ultra-super-critical rotor steel using processing maps and Zener-Hollomon parameter. *Materials Characterization*, *61*(1), 25–30. doi:10.1016/j.matchar.2009.10.001

Xin, H., Tarus, I., Cheng, L., Veljkovic, M., Persem, N., & Lorich, L. (2021). Experiments and numerical simulation of wire and arc additive manufactured steel materials. *Structures*, *34*, 1393–1402. doi:10.1016/j.istruc.2021.08.055

Yan, P., Liu, Z., Liu, W., Bao, H., & Weng, Y. (2013). Hot deformation behavior of a new 9%Cr heat resistant steel G115. *Journal of Iron and Steel Research International*, *20*(9), 73–79. doi:10.1016/S1006-706X(13)60159-4

Yang, H., Li, Z.H. and Zhang, Z.L. (2007) 'Investigation on Zener-Hollomon parameter in the warm-hot deformation behavior of CF53', *Journal of Shanghai Jiaotong University (Science)*, *12 E*(3), 352–358. . doi:10.1631/jzus.2006.A1453

Yang, N., Dai, Q., Chu, H., Zhang, C., Wang, W., & Xiao, P. (2017). Effect of Aluminum Alloying on the Hot Deformation Behavior of Nano-bainite Bearing Steel. *Journal of Materials Engineering and Performance*, *26*(12), 5954–5962. doi:10.1007/s11665-017-3018-7

Yin, F., Hua, L., Mao, H., & Han, X. (2013). Constitutive modeling for flow behavior of GCr15 steel under hot compression experiments. *Materials & Design*, *43*, 393–401. doi:10.1016/j.matdes.2012.07.009

Yoo, S. W., Lee, C. M., & Kim, D. H. (2023). Effect of Functionally Graded Material (FGM) Interlayer in Metal Additive Manufacturing of Inconel-Stainless Bimetallic Structure by Laser Melting Deposition (LMD) and Wire Arc Additive Manufacturing (WAAM). *Materials (Basel)*, *16*(2), 535. Advance online publication. doi:10.3390/ma16020535 PMID:36676271

Yusuf, S. M., Cutler, S., & Gao, N. (2019). Review: The Impact of Metal Additive Manufacturing on the Aerospace Industry. *Metals*, *9*(12), 1286. doi:10.3390/met9121286

Zhang, P., Yi, C., Chen, G., Qin, H., & Wang, C. (2016). Constitutive Model Based on Dynamic Recrystallization Behavior during Thermal Deformation of a Nickel-Based Superalloy. *Metals*, *6*(7), 161. doi:10.3390/met6070161

Zhu, L., He, J., & Zhang, Y. (2018). A two-stage constitutive model of X12CrMoWVNbN10-1-1 steel during elevated temperature. *Materials Research Express*, *5*(2), 1–11. doi:10.1088/2053-1591/aaa911

# Chapter 12
# Advanced Digital Data Processing Using Cloud Cryptography:
## Industrial Applications

**Digvijay Pandey**
https://orcid.org/0000-0003-0353-174X
*Department of Technical Education, Government of Uttar Pradesh, India*

**Binay Kumar Pandey**
https://orcid.org/0000-0002-4041-1213
*Department of Information Technology, College of Technology, Govind Ballabh Pant University of Agriculture and Technology, Pantnagar, India*

**Mukundan Appadurai Paramashivan**
https://orcid.org/0009-0009-5608-4788
*Aligarh Muslim University, India & Champions Group, India*

**Darshan A. Mahajan**
https://orcid.org/0000-0002-1239-6343
*National Institute of Construction Management and Research, India*

**Pankaj Dadheech Dadheech**
https://orcid.org/0000-0001-5783-1989
*Swami Keshvanand Institute of Technology, Management, and Gramothan, India*

**A. Shaji George**
https://orcid.org/0000-0002-8677-3682
*Business System Department, Almarai Company, TSM, Riyadh, Saudi Arabia*

**A. Shahul Hameed**
*Department of Telecommunication, Consolidated Techniques, Saudi Arabia*

## ABSTRACT

*Cloud computing uses the internet instead of discs or memory. Computing services include servers, databases, networks, and programmes. The primary benefit of cloud computing is easy and cheap data backup and access from anywhere. Cloud storage doesn't store consumer data, raising safety concerns. Cloud backup and storage users may not know how data is transported. The user is unaware a third party is secretly accessing their data. For safety, we offer numerous encryption algorithms. This book chapter covered cryptography and cloud computing.*

DOI: 10.4018/979-8-3693-1335-0.ch012

## 1. INTRODUCTION

By reorganizing a large number of resources and making them accessible to consumers in accordance with the criteria defined by the customers (Saxena, A. et al., 2021), the model of cloud computing brings in an entirely new way for the delivery of services. The cloud computing model (Anand, R. et al., 2023) makes this new technique possible (M. A. Vouk, et al. (2008)). The introduction of cloud computing also marks the start of an entirely new era in the annals of computer history. The concept that drives cloud computing is analogous, in its most generic form, to that of a virtualized software programmer (Kirubasri, G. et al., 2022). In addition to this, it is an essential component of the investigation and development of the cellular networks (Kirubasri, G. et al., 2021) and services of the next generation, which makes it an essential component of the entire picture (P. S. Wooley, et al. (2011)).

This is because it is an essential aspect of the investigation and development of the next generation of cellular networks and services (Anand, R. et al., 2022). One of the most essential aspects of cloud computing is the practice of storing customers' data on the cloud rather than on their local devices. This helps customers significantly cut down on the amount of data (Kumar, M. S. et al., 2021) they need to store and gives them easier access to their data (Pandey, B. K. et al., 2024). Because of cloud computing, it is no longer necessary for a user to install software on their local computer system in order to access a service that is supplied via the internet and made available through cloud computing. This is because cloud computing eliminates the need for such software installations. The use of cloud computing enables the availability of this capability. Users, both individually and collectively in the form of organisations, are eligible to take advantage of this benefit. The most significant benefits of utilising this technology include the relatively low cost of utilising it, increased storage capacity, and numerous application opportunities that are given by cloud computing.

The use of cloud computing also offers However, the most significant risk that cloud computing poses is to the users' safety and privacy (Pramanik, S. et al., 2023), and most industry experts agree that this is becoming a key obstacle that prevents cloud computing from reaching its full potential (i.e., by storing sensitive data on a server that is owned by someone else and located in an unknown location). Cloud computing has the potential to revolutionize many industries, but it has yet to reach its full potential because of this risk. Computing in the cloud has the potential to change many different industries, but because of the risk involved, it has not yet reached its full potential (S. Subashini et al. (2011)). Cloud computing has the potential to transform many different businesses, but because of the risk involved, it has not yet realized its full potential. Cloud computing has the potential to change many different industries (David, S. et al., 2023). The phrase "safety in the cloud" refers to the infrastructure and processes that are necessary to safeguard cloud computing services and commodities against the danger of being compromised by cybercriminals. This is important since cloud computing services and commodities are becoming increasingly popular. The fact that cloud computing services and items are kept in off-site locations introduces this potential safety concern. In this context, the fact that cloud computing services and items are stored in remote locations offers a possible security risk (Pandey, J. K. et al., 2022).

As a direct result of this reality, a substantial amount of cryptography is applied in cloud computing in order to secure (Pandey, B. K. et al., 2021) the validity, integrity, and secrecy of the information that is kept there. This is done in order to keep sensitive data (Khan, B. et al., 2021) from falling into the wrong hands. Encryption ensures that this objective can be accomplished with success. However, the solutions that are now being investigated are not only inadequate, but they are also inefficient. As a consequence of this, it is not possible to put any of these ideas into action because they do not satisfy

*Figure 1. Different cloud models*

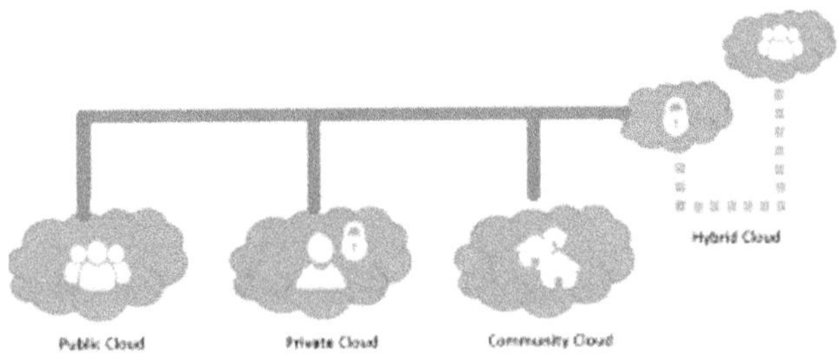


the requirements. When information is encrypted not only before it is stored in the cloud but also while it is stored in the cloud, it makes auditing and exercising control over the information more difficult. This is because the information cannot be decrypted until it is removed from the cloud.

Despite the fact that encryption significantly lessens the likelihood that private information will be compromised by unauthorized (Pandey, B. K. et al., 2023) individuals, we find ourselves in the current predicament (Daoud, M. K., et al.,2024).. This is so despite the fact that the aforementioned fact is true. This special issue's goal is to bring together cryptography and data security (Kumar Pandey, B. et al., 2021) experts from academia and industry for the purpose of having a debate about the various features of cloud computing that are important to cryptography and data security (C. Modi, et al. (2013)). These experts will then discuss the various aspects of cloud computing that are relevant to cryptography and data security (Daoud, M. K., et al.,2023). The multiple properties of cloud computing that are connected to encryption and data security will be the focus of the conversation that will take place later in Figure 1.

**Private Cloud**: One organisation or one group at a time is the only one who is permitted to use a private cloud. A non-affiliated third party or organisation is in charge of managing it. Because of its high level of security and its great degree of flexibility, the private cloud is frequently utilised by larger organisations as well as by government sectors.

**Public Cloud:** Any user who has access to the internet and is willing to pay for the storage they use will be able to use a public cloud, which stores files on servers that are managed by a third party. Amazon, the Windows Azure Service Platform, and the sales force are some examples.

**Community Cloud**: Any user who has access to the internet and is willing to pay according to their consumption can access the files that are hosted in a public cloud by a third party. Examples include Amazon, the Azure Service Platform from Microsoft, and sales force.

Based on a service the cloud model (Singh, H. et al., 2022) is offering, we are speaking of either:

- IaaS (Infrastructure-as-a-Service)
- PaaS (Platform-as-a-Service)
- SaaS (Software-as-a-Service)
- or, Storage, Database, Information, Process, Application, Integration, Security, Management, Testing-as-a-service in Figure 2.

The cloud provider issuer will give the buyer with more or less control over their cloud, depending on the buyer's desires for how to use the space and resources associated with the cloud. These desires can be communicated by the buyer through the cloud provider issuer (V. J. Winkler (2011)). For instance, the cloud need can take on a variety of forms depending on whether it will be utilised for business pur-

*Figure 2. Different service cloud model*

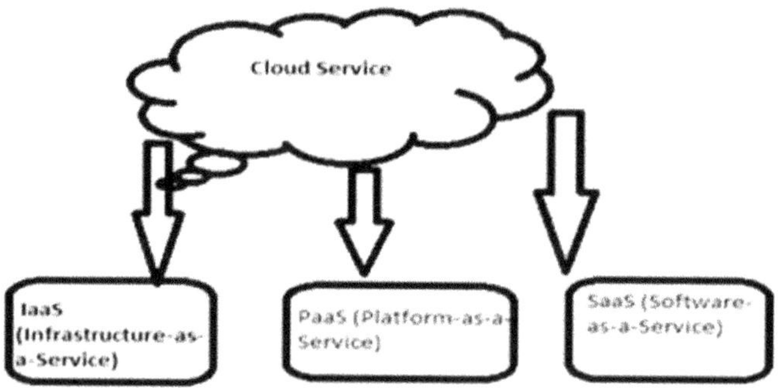


poses or for personal, domestic purposes. There are 3 forms of cloud that provide software program as a Service (SaaS), Infrastructure as a provider (IaaS), platform as a provider (PaaS).

**1. Software as a provider: –** Also known as "cloud software services," SaaS stands for software as a service. The management of SaaS is accomplished by the engagement of a third party. SaaS is used most often used in commercial business due to the fact that it does now now not require to the established of the software at once withinside the purchaser device, the software is immediately launched through an internet browser. SaaS is used most commonly used in commercial enterprise due to the fact that it does now not require to the founded of the software at once withinside the purchaser device. GoToMeeting and Google Apps are two instances of popular software as a service programs.

**2. Infrastructure as a provider**: – IaaS makes a variety of computer (Bessant, Y. A. et al., 2023) resources, including hardware, software, and storage tools, available to customers on demand. Customers (Saxena, A. et al., 2024) of an IaaS can gain access to the provider by utilizing the internet in their operations. Amazon, three Tera, and Go Grid are a few examples of well-known IaaS providers.

**3. Platform as a service:–** The platform as a service (PaaS) machine performs significantly better than the code as a service arrangement. A platform as a service (PaaS) provider grant subscribers access to the elements that are necessary for them to develop and run applications on top of software. J2EE, Ruby, and LAMP are some of the instances for PaaS that are available.

On cloud cryptography refers to the application of cryptographic methods for the purpose of safeguarding data that is stored on the cloud (A. U. Khan, (2012)). It is an essential component of cloud security since it helps to ensure that only authorized users may access and view sensitive data. This is an essential aspect of cloud security (Ahmad, A. Y. B, et al.,2024).

The use of cloud cryptography can safeguard data while it is stored, while it is being sent, and while it is being accessed.

- Data that is saved in the cloud, such as through the use of a cloud storage service, is referred to as "data at rest." Even if the cloud provider's systems are breached, it is still possible to safeguard data from unauthorized access by encrypting it while it is stored in the cloud using cloud cryptography (Ahmad, A. Y. B.,et al.,2023).
- Data that is travelling between on-premises systems and cloud-based services or between multiple cloud services is considered to be data that is in transit. Cloud cryptography can be used to encrypt data while it is being transferred (Raja, D. et al., 2024), which protects it from being tampered with or eavesdropped on.

• Data that is being processed by a cloud service is referred to as "data that is in use." The utilization of cloud cryptography enables data in use to be protected from unauthorized access as well as modification.

The cloud environment lends itself well to the implementation of a wide range of cryptographic protocols. The use of a single key for both encrypting and decrypting data is the basis of symmetric encryption, which is one of the most prevalent approaches. Although symmetric encryption is quick and effective, it is essential that the encryption key be kept a secret at all times.

• Asymmetric encryption is characterized by the utilization of two distinct keys, namely a public key and a private key. The data that is to be encrypted can be done so using the public key, and the data can then be decrypted using the private key. Asymmetric encryption is more secure than symmetric encryption; yet, it is slower than symmetric encryption. This is due to the fact that the private key does not need to be exchanged (Muniandi, B.,et al.,2019).

• The application of hash functions is what leads to the creation of a data item's one-of-a-kind fingerprint. It is possible to utilize hash algorithms in order to check the authenticity of data and identify any unauthorized modifications (F. Yahya, (2014)).

Cloud service companies often make available a comprehensive range of cryptographic features and services to their customers. For instance, the majority of cloud service providers offer default encryption for both the data while it is at rest and while it is in transit. In addition, suppliers of cloud computing services offer key management services, which enable businesses to better manage and safeguard their encryption keys.

## 2. IMPLICATIONS OF EMPLOYING CLOUD CRYPTOGRAPHY

The use of cloud cryptography has a number of advantages, including the following:

• Data protection: The use of encryption in the cloud helps to prevent sensitive data from being accessed, stolen, or modified without authorization.

• Compliance: Cloud cryptography can assist businesses in complying with data protection standards such as GDPR and HIPAA. This is a benefit of using cloud cryptography.

• Confidence that one's data is safe even when it's hosted in the cloud Cryptography in the cloud can provide businesses with the confidence that their data is safe even when it's hosted in the cloud.

## 3. HOW TO IMPLEMENT CLOUD CRYPTOGRAPHY

There are several essential steps that businesses need to take in order to successfully adopt cloud cryptography, including the following:

• Select the appropriate encryption algorithms Companies and other organizations should select encryption methods that are suitable for their particular requirements. For instance, businesses that place a high priority on performance may prefer to employ symmetric encryption, whereas businesses that place a high priority on security may want to use asymmetric encryption (T. Mather, (2009)).

• Ensure the proper management of encryption keys: Organizations are responsible for ensuring the proper management of their encryption keys. This includes maintaining the confidentiality of encryption keys and rotating them on a regular basis.

• Take advantage of the many cloud cryptography features and services that are available Cloud service providers offer a wide range of cloud cryptography features and services. To ensure the safety of their data, businesses should make use of the capabilities and services provided by this platform.

## 4. CHALLENGES IN CLOUD CRYPTOGRAPHY

Cryptography in the cloud is a strong tool for protecting data stored in the cloud; nevertheless, it also raises a number of issues. The following are some of the most frequently encountered difficulties:

• **Key management** is one of the most difficult difficulties in cloud cryptography, and you'll need to be prepared for it. Key management is the process of producing, storing, distributing, rotating, and revoking encryption keys in a manner that is both secure and effective. If the encryption keys are misplaced, stolen, or otherwise compromised in any way, the data will either become inaccessible or revealed. As a result, effective key management needs meticulous planning and implementation, in addition to consistent audits and backups.

• **Data sovereignty and compliance**: Organizations have a responsibility to be aware of the rules of data sovereignty and compliance that are applicable to their data. For some businesses, it may be mandatory to back up their information in a particular part of the world or to work with a cloud service provider that has been independently validated as complying with a predetermined set of safety requirements. Because of these criteria, selecting and putting into practice appropriate cloud cryptography solutions might be difficult.

• **Dangers to one's security and privacy** Despite the fact that cloud cryptography can assist in the protection of data, there are still some dangers to one's security and privacy that are linked with the utilization of cloud computing environments. In the event that a cloud provider is hacked, for instance, cybercriminals may have access to data that has been encrypted. In the event that cloud providers require access to encrypted data in order to carry out maintenance or troubleshooting, this access may be granted to them. It is imperative that organizations give careful consideration to these dangers and immediately begin taking preventative measures.

Impact on performance Cloud cryptography may have an effect on the performance of both applications and operating systems. This is due to the requirement of computing power for both encryption and decryption. Organizations have a responsibility to be aware of the potential impact on performance and to select cloud cryptography solutions that meet the performance standards they have established.

• **Cost**: Cloud cryptography services can be pricey, especially for businesses that have to encrypt big (Babu, S. Z. D. et al., 2022) volumes of data. This is especially true for businesses that use public clouds. Before implementing cloud cryptography, businesses need to give great consideration to both the costs and the benefits of doing so.

In addition to these more basic issues, particular organizations may also be confronted with a wide variety of additional challenges. Organizations, for instance, that make use of more than one cloud provider can be required to handle encryption keys across more than one platform. In addition, businesses who run cloud-native apps can find it necessary to locate cloud cryptography solutions that are compatible with their software.

On general, cloud cryptography is a powerful tool that may be used to protect data that is stored on the cloud. Nevertheless, it is essential to be aware of the difficulties connected with cloud cryptography and to take measures to address these difficulties (Vijaya Pinjarkar, (2016)).

Cloud cryptography offers a number of advantages, including:

• Data protection: The use of encryption in the cloud helps to prevent sensitive data from being accessed, stolen, or modified without authorization. This is of utmost significance for businesses that keep private information on the cloud, such as records of their customers and clients, financial information, and intellectual property.

• Compliance: Cloud cryptography can assist businesses in complying with data protection standards such as GDPR and HIPAA. This is a benefit of using cloud cryptography. Because of these requirements, organizations are required to put in place particular security precautions in order to safeguard sensitive data. The use of cloud cryptography may be able to assist organizations in fulfilling these requirements.

• Confidence that one's data is safe even when it's hosted in the cloud Cryptography (Revathi, T. K. et al., 2022) in the cloud can provide businesses with the confidence that their data is safe even when it's hosted in the cloud. This is of utmost significance for businesses that conduct mission-critical activities through the cloud.

In addition to these benefits, cloud cryptography also has the potential to provide the following advantages:

• **Scalability**: Cloud cryptography solutions are often scalable, which means that they can easily be scaled up or down to meet the requirements of expanding organizations. This allows cloud cryptography solutions to better accommodate these organizations. This is something that may be useful for businesses who have a significant amount of data that needs to be protected.

• **Cost reductions**: Cloud cryptography solutions can help businesses reduce the money they spend on information technology. One use case for cloud cryptography is helping businesses reduce the amount of on-premises hardware and software they require, as an example.

• **Ease of use**: Solutions for cloud cryptography are often simple to use and manage. This may be useful for companies that do not have the in-house competence to run their own encryption solutions and hence must outsource this function. The use of cloud cryptography generally confers a number of benefits upon businesses of varying sizes and scopes. The use of cloud cryptography enables businesses to safeguard their sensitive data, bring themselves into compliance with data security rules, and experience a sense of calm. Cloud cryptography has a number of advantages, but it also has some

**Disadvantages**. Here are a few of the most common disadvantages:

**Cost**: Cloud cryptography services can be expensive, especially for organizations that need to encrypt large amounts of data (Pandey, B. K., & Pandey, D., 2023).

**Complexity**: Cloud cryptography can be complex to implement and manage. Organizations need to have the expertise to choose the right cloud cryptography services (Pandey, D., & Pandey, B. K., 2022) and to configure them correctly.

**Performance impact**: Cloud cryptography can impact the performance of applications and systems. This is because encryption and decryption require processing power.

**Vendor lock-in**: Once an organization has encrypted its data using a cloud cryptography service, it can be difficult to switch to a different provider.

**Security risks:** Even though cloud cryptography can help to protect data, there are still some security risks associated with using cloud computing environments. For example, if a cloud provider is hacked, attackers could gain access to encrypted data (Boopathi, S. et al., 2023)..

Organizations need to carefully consider the advantages and disadvantages of cloud cryptography before deciding whether to use it. Organizations should also choose a cloud provider that has a strong track record of security and compliance.

Here are some additional **disadvantages** that may be relevant to specific organizations:

**Legal and regulatory requirements**: Some organizations may be subject to legal or regulatory requirements that restrict how they can encrypt and manage their data. For example, some organizations may be required to keep their encryption keys on-premises.

**Compliance requirements**: Some organizations may need to comply with industry-specific regulations that have specific requirements for data encryption and key management. For example, organizations in the healthcare industry may need to comply with HIPAA regulations.

**Operational requirements**: Some organizations may have operational requirements that make it difficult to use cloud cryptography. For example, organizations that need to access their data in real time may not be able to afford the performance impact of encryption.

## 5. CONCLUSION

The use of cryptography in the cloud is an important component of cloud security. The use of cloud cryptography enables organizations to safeguard their sensitive data from being accessed, stolen, or modified without their permission. The most important objective is to safely store and retrieve information in the cloud, which is managed by a third party rather than the data's owner. It is common for software structures to have multiple endpoints, often more than one client, and at least one and perhaps more give up servers. These connections between the customer and the server are carried out across networks that cannot be relied upon. Communication takes place across open, public networks such as the internet as well as private networks that have the potential to be hacked by either malevolent outsiders or malicious people working within the organization. The use of cryptography can protect communications that go via networks that cannot be trusted. An adversary can try to carry out a number of different sorts of attacks on a network, but the most common ones are listed here (Zaidi, A., et al., (2023)).

An attacker really listens in on a community phase and makes an effort to study sensitive records as they flow. This type of attack is known as a passive attack. On-line passive attacks (in which an attacker reads traffic in real-time) and offline passive attacks (in which an attacker definitely captures site traffic in realtime and views it later — maybe after spending some time decrypting it) are both considered to fall under the category of passive attacks. An active attack consists of the attacker pretending to be a customer or server, intercepting communications while they are in transit, examining and/or changing the contents of the communication before sending it directly to its intended destination (or dropping it completely), and then passing on the communication.

In this work, a clear picture of cloud computing and the security challenges associated with it utilizing cryptographic methods has been presented. In conclusion, there are many reasons to have a positive outlook regarding the foreseeable future of cloud cryptography. Cloud cryptography is a powerful instrument for safeguarding data that is stored in the cloud, and it is likely that in the future, it will play an increasingly important role in the context of cloud security. Cloud computing refers to the practice of storing data on remote servers that are connected to the internet.

## 6. FUTURE SCOPE

The implementation of cryptography in cloud computing is poised for significant growth in the years to come. As a direct consequence of the growing acceptance of cloud computing, an increasing number of companies are shifting the location where they store their data to the cloud. As a consequence of this, there will be a rise in demand for cloud-based cryptographic solutions.

The following is a summary of some of the most significant recent advancements that are having an impact on the direction that cloud cryptography may go in the future:

• The emergence of computing based on quantum mechanics: Quantum computers have the potential to decipher a significant number of the encryption techniques that are now utilised to keep cloud-based data secure. This poses a serious risk to the confidentiality of data that is stored on the cloud. As research into quantum computing continues to expand, organizations will be compelled to adopt new encryption methods that are immune to the effects of quantum computing.

• The proliferation of applications that make use of artificial intelligence (AI): At this very moment, AI is being put to use in the process of developing fresh and significantly more involved strategies for cloud cryptography. AI is now being utilized, for instance, in the design of novel encryption algorithms, the improvement of key management systems, the detection and response to security threats, and a variety of other applications.

• The development of a larger consumer base for cloud computing In the years to come, it is projected that the market for cloud computing will continue to enjoy significant increase. As a direct consequence of this trend, there is likely going to be an increased need for cloud-based cryptography solutions.

In addition to these general tendencies, there are a number of additional aspects that are contributing to the growing significance of cloud cryptography. The ever-increasing number of data breaches and the ever-increasing awareness of the need to protect data privacy and security are only two examples of the factors that contribute to this problem.

The following are some examples of applications that could be developed in the future using cloud cryptography:

• Fully holomorphic encryption, also referred to as FHE, is a form of encryption that, in contrast to other forms of encryption, does not require the data to be decrypted before computations may be carried out on it. This distinguishes fully homomorphic encryption from other forms of encryption. As a consequence of this advancement, the manner in which data is processed on the cloud may undergo significant transformations. For instance, FHE might be used to enable third-party firms to carry out analysis on encrypted data (Singh, S. et al., 2023) without being required to view the data in its original, unencrypted form in order to do so.

• The term "post-quantum cryptography" refers to a type of cryptography that is secure against breaches carried out by quantum computers. The term "post-quantum" cryptography refers to this particular method of encoding data. The usage of post-quantum cryptography will become increasingly crucial for the purpose of securing data that is stored on the cloud as the field of quantum computing continues to make advancements in the coming years.

• Cryptography that is based on the technology behind blockchains: The use of blockchain technology enables the development of key management systems that are both trustworthy and unassailable by unauthorized parties. It is feasible that this would make it easier for businesses to manage the encryption keys that they use to protect their data while it is kept in the cloud. This would be beneficial for both the firms and their customers.

## REFERENCES


Ahmad, A., Abusaimeh, H., Rababah, A., Alqsass, M., Al-Olima, N., & Hamdan, M. (2024). Assessment of effects in advances of accounting technologies on quality financial reports in Jordanian public sector. *Uncertain Supply Chain Management*, *12*(1), 133–142. doi:10.5267/j.uscm.2023.10.011

Ahmad, A. Y. B. (2024). Ethical implications of artificial intelligence in accounting: A framework for responsible ai adoption in multinational corporations in Jordan. *International Journal of Data and Network Science*, *8*(1), 401–414. doi:10.5267/j.ijdns.2023.9.014

Ahmad, A. Y. B. (2024). Firm Determinants that Influences Implementation of Accounting Technologies in Business Organizations. *WSEAS Transactions on Business and Economics*, *21*, 1–11. doi:10.37394/23207.2024.21.1

Ahmad, A. Y. B., Gongada, T. N., Shrivastava, G., Gabbi, R. S., Islam, S., & Nagaraju, K. (2023). E-Commerce Trend Analysis and Management for Industry 5.0 using User Data Analysis. *International Journal of Intelligent Systems and Applications in Engineering*, *11*(11s), 135–150.

Ahmad, A. Y. B., Hannoon, A., Al-Daoud, K. I., Abu-Alsondos, I. A., & Al-Qaisieh, M. S. (2023). Assessment of Cloud Based Accounting Technology Adoption and Business Performance. *Kurdish Studies, 11*(3).

Ahmad, A. Y. B., Kumari, D. K., Shukla, A., Deepak, A., Chandnani, M., Pundir, S., & Shrivastava, A. (2024). Framework for Cloud Based Document Management System with Institutional Schema of Database. *International Journal of Intelligent Systems and Applications in Engineering*, *12*(3s), 672–678.

Ahmad, A. Y. B., Kumari, D. K., Shukla, A., Deepak, A., Chandnani, M., Pundir, S., & Shrivastava, A. (2024). Framework for Cloud Based Document Management System with Institutional Schema of Database. *International Journal of Intelligent Systems and Applications in Engineering*, *12*(3s), 672–678.

Ahmad, A. Y. B., & Kumari, S. S. (2023). Blockchain Implementation in Financial Sector and Cyber Security System. *2023 International Conference on Artificial Intelligence and Smart Communication (AISC)*, Greater Noida, India. 10.1109/AISC56616.2023.10085045

Ahmad, A. Y. B., Tiwari, A., Nayeem, M. A., Biswal, B. K., Satapathy, D. P., Kulshreshtha, K., & Bordoloi, D. (2024). Artificial Intelligence Perspective Framework of the Smart Finance and Accounting Management Model. *International Journal of Intelligent Systems and Applications in Engineering*, *12*(4s), 586–594.

Ahmad, A. Y. B., William, P., Uike, D., Murgai, A., Bajaj, K. K., Deepak, A., & Shrivastava, A. (2024). Framework for Sustainable Energy Management using Smart Grid Panels Integrated with Machine Learning and IOT based Approach. *International Journal of Intelligent Systems and Applications in Engineering*, *12*(2s), 581–590.

Anand, R., Lakshmi, S. V., Pandey, D., & Pandey, B. K. (2023). An enhanced ResNet-50 deep learning model for arrhythmia detection using electrocardiogram biomedical indicators. *Evolving Systems*, 1–15.

Anand, R., Singh, J., Pandey, D., Pandey, B. K., Nassa, V. K., & Pramanik, S. (2022). Modern technique for interactive communication in LEACH-based ad hoc wireless sensor network. In *Software Defined Networking for Ad Hoc Networks* (pp. 55–73). Springer International Publishing. doi:10.1007/978-3-030-91149-2_3

Babu, S. Z. D., Pandey, D., Naidu, G. T., Sumathi, S., Gupta, A., Bader Alazzam, M., & Pandey, B. K. (2022, July). Analysation of Big Data in Smart Healthcare. In *Artificial Intelligence on Medical Data: Proceedings of International Symposium, ISCMM 2021* (pp. 243-251). Singapore: Springer Nature Singapore.

Bessant, Y. A., Jency, J. G., Sagayam, K. M., Jone, A. A. A., Pandey, D., & Pandey, B. K. (2023). Improved parallel matrix multiplication using Strassen and Urdhvatiryagbhyam method. *CCF Transactions on High Performance Computing*, *5*(2), 1–14. doi:10.1007/s42514-023-00149-9

Boopathi, S., Pandey, B. K., & Pandey, D. (2023). Advances in artificial intelligence for image processing: techniques, applications, and optimization. In *Handbook of research on thrust technologies' effect on image processing* (pp. 73–95). IGI Global. doi:10.4018/978-1-6684-8618-4.ch006

Daoud, M. K., Al-Qeed, M., Ahmad, A. Y. B., & Al-Gasawneh, J. A. (2023). Mobile Marketing: Exploring the Efficacy of User-Centric Strategies for Enhanced Consumer Engagement and Conversion Rates. *International Journal of Membrane Science and Technology*, *10*(2), 1252–1262. doi:10.15379/ijmst.vi.1425

Daoud, M. K., Taha, S., Al-Qeed, M., Alsafadi, Y., Ahmad, A., & Allahham, M. (2024). EcoConnect: Guiding environmental awareness via digital marketing approaches. *International Journal of Data and Network Science, 8*(1), 235-242.

David, S., Duraipandian, K., Chandrasekaran, D., Pandey, D., Sindhwani, N., & Pandey, B. K. (2023). Impact of blockchain in healthcare system. In *Unleashing the Potentials of Blockchain Technology for Healthcare Industries* (pp. 37–57). Academic Press. doi:10.1016/B978-0-323-99481-1.00004-3

Khan, A. U., Oriol, M., Kiran, M., Jiang, M., & Djemame, K. (2012). Security risks and their management in cloud computing. *4th IEEE Int. Conf. Cloud Comput. Technol. Sci. Proc.* (pp. 121–128). IEEE. 10.1109/CloudCom.2012.6427574

Khan, B., Hasan, A., Pandey, D., Ventayen, R. J. M., Pandey, B. K., & Gowwrii, G. (2021). 12 Fusion of Datamining and Artificial Intelligence. *Machine learning and iot for intelligent systems and smart applications*, 201.

Kirubasri, G., Sankar, S., Pandey, D., Pandey, B. K., Nassa, V. K., & Dadheech, P. (2022). Software-defined networking-based Ad hoc networks routing protocols. In *Software defined networking for Ad Hoc networks* (pp. 95–123). Springer International Publishing. doi:10.1007/978-3-030-91149-2_5

Kirubasri, G., Sankar, S., Pandey, D., Pandey, B. K., Singh, H., & Anand, R. (2021, September). A recent survey on 6G vehicular technology, applications and challenges. In *2021 9th International Conference on Reliability, Infocom Technologies and Optimization (Trends and Future Directions)(ICRITO)* (pp. 1-5). IEEE.

Kumar, M. S., Sankar, S., Nassa, V. K., Pandey, D., Pandey, B. K., & Enbeyle, W. (2021). Innovation and creativity for data mining using computational statistics. In *Methodologies and Applications of Computational Statistics for Machine Intelligence* (pp. 223–240). IGI Global. doi:10.4018/978-1-7998-7701-1.ch012

Kumar Pandey, B., Pandey, D., Nassa, V. K., Ahmad, T., Singh, C., George, A. S., & Wakchaure, M. A. (2021). Encryption and steganography-based text extraction in IoT using the EWCTS optimizer. *Imaging Science Journal*, *69*(1-4), 38–56. doi:10.1080/13682199.2022.2146885

Mather, T., Kumaraswamy, S., & Latif, S. (2009). Cloud. *Security and Privacy*, 299.

Modi, C., Patel, D., Borisaniya, B., Patel, A., & Rajarajan, M. (2013). A survey on security issues and solutions at different layers of Cloud computing. *The Journal of Supercomputing*, *63*(2), 561–592. doi:10.1007/s11227-012-0831-5

Muniandi, B., Huang, C., Kuo, C., Yang, T., Chen, K., Lin, Y., Lin, S., & Tsai, T. (2019). A 97% maximum efficiency fully automated control turbo boost topology for battery chargers. *IEEE Transactions on Circuits and Systems. I, Regular Papers*, *66*(11), 4516–4527. doi:10.1109/TCSI.2019.2925374

Pandey, B. K., & Pandey, D. (2023). Parametric optimization and prediction of enhanced thermoelectric performance in co-doped CaMnO3 using response surface methodology and neural network. *Journal of Materials Science Materials in Electronics*, *34*(21), 1589. doi:10.1007/s10854-023-10954-1

Pandey, B. K., Pandey, D., Gupta, A., Nassa, V. K., Dadheech, P., & George, A. S. (2023). Secret data transmission using advanced morphological component analysis and steganography. In *Role of Data-Intensive Distributed Computing Systems in Designing Data Solutions* (pp. 21–44). Springer International Publishing. doi:10.1007/978-3-031-15542-0_2

Pandey, B. K., Pandey, D., & Sahani, S. K. (2024). Autopilot control unmanned aerial vehicle system for sewage defect detection using deep learning. *Engineering Reports*, 12852. doi:10.1002/eng2.12852

Pandey, B. K., Pandey, D., Wairya, S., & Agarwal, G. (2021). An advanced morphological component analysis, steganography, and deep learning-based system to transmit secure textual data. [IJDAI]. *International Journal of Distributed Artificial Intelligence*, *13*(2), 40–62. doi:10.4018/IJDAI.2021070104

Pandey, D., & Pandey, B. K. (2022). An efficient deep neural network with adaptive galactic swarm optimization for complex image text extraction. In *Process Mining Techniques for Pattern Recognition* (pp. 121–137). CRC Press. doi:10.1201/9781003169550-10

Pandey, J. K., Jain, R., Dilip, R., Kumbhkar, M., Jaiswal, S., Pandey, B. K., & Pandey, D. (2022). Investigating Role of IoT in the Development of Smart Application for Security Enhancement. In *IoT Based Smart Applications* (pp. 219–243). Springer International Publishing.

Pramanik, S., Pandey, D., Joardar, S., Niranjanamurthy, M., Pandey, B. K., & Kaur, J. (2023, October). An overview of IoT privacy and security in smart cities. In AIP Conference Proceedings (Vol. 2495, No. 1). AIP Publishing. doi:10.1063/5.0123511

Raja, D., Kumar, D. R., Santhiyakumari, N., Kumarganesh, S., Sagayam, K. M., Thiyaneswaran, B., Pandey, B. K., & Pandey, D. (2024). A compact dual-feed wide-band slotted antenna for future wireless applications. *Analog Integrated Circuits and Signal Processing*, *118*(2), 1–15. doi:10.1007/s10470-023-02233-0

Revathi, T. K., Sathiyabhama, B., Sankar, S., Pandey, D., Pandey, B. K., & Dadeech, P. (2022). An intelligent model for coronary heart disease diagnosis. *Networking Technologies in Smart Healthcare: Innovations and Analytical Approaches*, 234.

Saxena, A., Agarwal, A., Pandey, B. K., & Pandey, D. (2024). Examination of the Criticality of Customer Segmentation Using Unsupervised Learning Methods. *Circular Economy and Sustainability*, 1–14. doi:10.1007/s43615-023-00336-4

Saxena, A., Sharma, N. K., Pandey, D., & Pandey, B. K. (2021). Influence of tourists satisfaction on future behavioral intentions with special reference to desert triangle of Rajasthan. *Augmented Human Research*, *6*(1), 1–9. doi:10.1007/s41133-021-00052-4

Singh, H., Pandey, B. K., George, S., Pandey, D., Anand, R., Sindhwani, N., & Dadheech, P. (2022, July). Effective Overview of Different ML Models Used for Prediction of COVID-19 Patients. In *Artificial Intelligence on Medical Data: Proceedings of International Symposium, ISCMM 2021* (pp. 185-192). Singapore: Springer Nature Singapore.

Singh, S., Madaan, G., Kaur, J., Swapna, H. R., Pandey, D., Singh, A., & Pandey, B. K. (2023). Bibliometric Review on Healthcare Sustainability. Handbook of Research on Safe Disposal Methods of Municipal Solid Wastes for a Sustainable Environment, 142-161.

Subashini, S., & Kavitha, V. (2011, January). A survey on security issues in service delivery models of cloud computing. *Journal of Network and Computer Applications*, *34*(1), 1–11. doi:10.1016/j.jnca.2010.07.006

Vouk, M. A. (2008). Cloud computing - Issues, research and implementations. *Proc. Int. Conf. Inf. Technol. Interfaces*. ITI. 10.2498/cit.1001391

Winkler, V. J. (2011). *"Securing the Cloud," Cloud Comput. Secur. Tech. tactics*. Elsevier.

Wooley, P. S. (2011). Identifying Cloud Computing Security Risks. *Continuity in Education*, *1277*(February).

Yahya, F., Chang, V., Walters, J., & Wills, B. (2014). *Security Challenges in Cloud Storage*. doi:10.1109/CloudCom.2014.171

Zaidi, A. (2023). Predicting wildfires in Algerian forests using machine learning models. *Heliyon*, *9*(7), e18064. doi:10.1016/j.heliyon.2023.e18064 PMID:37519679

Zaidi, A. (2023). Smart Implementation of Industrial Internet of Things using Embedded Mechatronic System. *IEEE Embedded Systems Letters*. doi:10.1109/LES.2023.3309985

Zaidi, A., Ajibade, S.-S. M., Musa, M., & Bekun, F. V. (2023). New Insights into the Research Landscape on the Application of Artificial Intelligence in Sustainable Smart Cities: A Bibliometric Mapping and Network Analysis Approach. *International Journal of Energy Economics and Policy.*, *13*(4), 287–299. doi:10.32479/ijeep.14683

Zaidi, A., & Al-Harbi, O. (2023). Statistical Analysis of a Linear Multi-Step Numerical Treatment. *Journal of Statistics Applications & Probability*, *12*(1), 26. doi:10.18576/jsap/120126

Zaidi, A., & Al Luhayb, A. (2023). Two Statistical Approaches to Justify the Use of the Logistic Function in Binary Logistic Regression. *Mathematical Problems in Engineering*, *2023*, 5525675. doi:10.1155/2023/5525675

# Chapter 13
# Internet of Things-Based Service-Oriented Architecture for Industrial Applications

**Kamalendu Pal**
https://orcid.org/0000-0001-7158-6481
*University of London, UK*

## ABSTRACT

*TThe industrial world has witnessed rapid development and popularity of the Internet of Things (IoT) based information systems that provide various services to business applications (e.g., manufacturing, supply chain management). These information systems use pervasive computing, ubiquitous computing and unique data communication protocols to sense industrial applications in an environment where the physical and digital worlds work together. The new information system can serve various business service categories through continuous symbiotic interactions. These interactions exchange data and information to create services that benefit the industrial world and help smooth business operations. This chapter describes the Apparel Business Decentralized Data Integration (ABDDI) information system architecture that requires a description of a logic-based knowledge representation scheme for information modelling and reasoning to provide web services semantic interoperability with the help of ontology. The chapter also presents an example of the service composition ontology's similarity assessment based on an algorithm.*

## 1. INTRODUCTION

Humanity dwells on the Earth with ambitious goals to manage and mitigate unprecedented social, economic, and environmental challenges. Science, technology, and innovation play an enormous role in realizing these ambitious goals to improve human living conditions. The process of creative progress started by technological, and its evolution can help change economies and improve living conditions by increasing productivity, reducing production costs and goods prices, and helping to raise real benefits

DOI: 10.4018/979-8-3693-1335-0.ch013

for the world. Thus, one of the essential ingredients to create a better world is using technology to move forward and unprecedented change in the scope and pace of daily human life.

Harnessing the frontier of technologies can help create and deliver innovations (including non-technological and new forms of social innovation) that could be transformative in creating sustainable development goals and producing more prosperous, inclusive, and healthy human societies. In this way, society gets the required solutions and chances for future sustainable development that are essential, cheaper, faster, scalable, and easy to use. The extent of technological development impact has already ushered in the transformative implications of information and communication technologies (ICTs) in many countries worldwide. However, these new technologies often threaten to outpace the capability of societies and policymakers to adapt to the changes they can create, giving rise to widespread anxiety and ambivalence or hostility to some technological advances.

One of the central questions of where ideas come from is on the mind of researchers visiting a research laboratory, a painter's workshop, or an inventor's experimental laboratory. It is the secret human society hopes to see – the magic that happens when new things are born. In this way, it is possible to provide creativity, like the discovery of millimetre radio waves by Professor Jagadish Chandra Bose (Sarkar, 2006).

Today, the world is witnessing the tremendous influence of wireless communication technology on daily working activities. Three great scientific minds heavily influence modern wireless telecommunications - James Clerk Maxwell (Mahon, 2004), Jagadish Chandra Bose (Sarkar et al., 2006), and Tim Berners-Lee (Berners-Lee, 2000). James Clerk Maxwell provided the theoretical foundation of electromagnetic wave propagation; Jagadish Chandra Bose showed his colleagues the transmission of millimetre waves by transmitting this new type of waves in the Presidency College (Kolkata, India) laboratory; and Tim Berners-Lee created the World Wide Web at CERN (Geneva, Switzerland). Today's Computer data communication network is at once intangible and constantly mutated, growing larger and more complex with each passing second. Many of the world's business community uses this incredible network of networks for daily business-related work and relevant services.

The advent of development and adoption of new technologies in recent decades is continuing, and this continuation is driven by: (i) the cumulative nature of technological change; (ii) the exponential nature of technologies such as microchips that are doubled in power every two years for more than half a century; (iii) the convergence of technologies into new combinations; (iv) drastic reduction in costs of production; (v) the emergence of digital "platforms of platforms" – most prominently the Internet; and (v) adoption of artificial intelligence (AI) techniques, the Internet of Things (IoT), and cognitive technologies have successfully been applied to various industrial applications (Zhao & Kumar, 2021). In addition, IoT has paved the way for many industrial application domains while posing several challenges as many devices, protocols, communication channels, architectures, and middleware exist. Big data generated by these devices calls for advanced machine learning (ML) and data mining techniques to understand, learn effectively, and reason with this volume of information, such as cognitive technologies. Cognitive technologies play a significant role in developing successful cognitive systems which mimic "cognitive" functions associated with human intelligence, such as "learning" and "problem-solving".

AI-based applications now help to solve real-world problems, it includes image recognition, problem-solving, and logical reasoning that sometimes exceed human performance. AI applications, particularly robotics, can transform production processes and business activities, specially manufacturing. Big data technologies are opening new opportunities and enabling breakthroughs related to, among others, manufacturing data analytics addressing different perspectives: (i) descriptive to answer what happened, (ii) diagnostic to answer the reason why it happened, (iii) predictive to understand what will happen and (iv) prescriptive to detect how human operators can make it happen.

Without any doubt, the potential impact of big data technology can bring huge changes, the economy, and society is relevant, boosting innovations in organizations and improving business models. Besides, today the Internet has become ubiquitous, has influenced almost every corner of the world, and affects human life unimaginably. However, the journey is far from over. Human society is entering an era of even more pervasive connectivity where many service applications will be connected to the Web. Human society is entering an era of the IoT. Different academics and practitioners have defined this term in many ways. The IoT is a things-connected network wirelessly connected via smart sensors; IoT can interact without human intervention. Some preliminary IoT applications have already been developed in the manufacturing, transportation, and automotive industries (He et al., 2014) (Pretz, 2013). The evolution of IoT involves many development issues, such as infrastructure, communications, interfaces, protocols, and standards. The IoT refers to a new world where almost all devices and appliances are connected to a network. In addition, industrial application service providers can use them collaboratively to achieve complex tasks requiring high intelligence.

IoT devices are geared with embedded sensors, actuators, processors, and transceivers for this intelligence and interconnection. IoT is not a single technology but an agglomeration of various technologies that work together. In simple terms, sensors and actuators help interact with the physical environment. The data collected by the sensors must be stored and processed intelligently to derive valuable inferences from it. This chapter broadly defines the term *sensor*; a handheld mobile phone or even a microwave oven can be a sensor if it provides inputs about its current state (internal state + environment). An *actuator* is a device used to effect a change in the environment, such as the temperature controller of an air conditioner.

Data storage and processing can be done on the network's edge, or a server located in a different place. If any data preprocessing is done, it simply happens at some other proximate device or sensor. In the end, data is then typically sent to a remote server after processing. An IoT object's storage and processing capabilities are also restricted by the available resources, which are often very constrained due to size, energy, power, and computational capability limitations. Consequently, the central research challenge is ensuring that users can get the appropriate data type at the desired level of correctness.

In addition to data collection and handling challenges, there are also challenges in processed data communication to the relevant service entities. The collaboration and communication between IoT devices are generally wireless because these devices are simply installed at different work locations. Wireless communication channels are often suffering from high rates of disturbance. This way, reliably communicating data without too many retransmissions is a central research issue. Moreover, actions must be taken based on the derived inferences after processing the received data. The nature of action can be diverse. One can directly modify the physical world through actuators. Alternatively, one may do something virtually. For example, one may send raw data and information to other smart objects. Creating a change in the physical world depends on its state now, called *context awareness*. Each action is taken, considering the context and why an application can behave differently in separate business contexts. For example, an office member may not like messages from office colleagues to interrupt when that individual is on vacation.

Sensors, actuators, data storage, and the communication network form the central infrastructure of an IoT framework. However, many software aspects need to be considered. First, one needs middleware that can be used to interconnect and manage all these heterogeneous components. This way, the IoT industry needs much standardization to connect many different devices. However, the conceptual realization of IoT is far from achieving a full deployment of converged IoT services and technology. Current information technology and communication (ITC) research focuses on providing integrated solutions, primarily on the feature that enables convergence or "*interoperability*". Interoperability can be simplified as the feature for providing a seamless exchange of information to personalize services automatically or simply exchanging information so that other systems can use it to improve performance, enable and create services, and control operations information processing. This type of scenario requires increased interoperability in service management operations.

This chapter reviews the recent trends and challenges on interoperability in the IoT applications domain, discuss physical versus virtual sensors, and, while addressing technology interoperability challenges in parallel, discusses how, with the growing importance of data understanding and processing, semantic web technologies, frameworks, and information models can support interoperability in the design of services in the industrial information system design. Finally, this chapter aims to identify relevant issues and challenges that need to be considered in the coming and future information system projects.

The remainder of this chapter is described as follows. Section 2 presents the background of IoT-based industrial applications. Section 3 reviews representative research on semantic modelling methods for general IoT-based applications. Section 4 describes the background knowledge about the proposed system. Section 5 presents the knowledge representation and reasoning approach. Section 6 explains some mathematical concepts using a business case and includes a concept similarity assessment. Finally, section 7 puts the ideas of future research work forwards. Finally, section 8 concludes the chapter by discussing relevant research issues.

## 2. BACKGROUND KNOWLEDGE OF IoT ARCHITECTURE

Reviewing Industrial IoT applications combines many emerging technologies, from sensing and actuating to networking and application-specific service provisions. In this way, the paradigm of industrial IoT application covers a broad range of technological issues (e.g., protocol, architecture, and hardware), network and connectivity technologies (e.g., cellular and Wi-Fi), infrastructure and platform technologies (e.g., standards, and gateway), and users' application-specific technologies (e.g., predictive maintenance, and extended reality). Moreover, there is no unique consensus on architecture for IoT, which is agreed upon universally. Different research groups have proposed different architectures. From the broad scope, this chapter discusses two critical aspects of an industrial IoT ecosystem: (i) architecture and (ii) interoperability of service provision. In the following subsections, the chapter explains these concepts.

### 2.1 Architectural Issues

### 2.2 Three-Layer and Five-Layer Architectures

The most straightforward architecture is a three-layer architecture (Mashal et al., 2015) (Said & Masud, 2013) (Wu et al., 2010), as shown in Figure 1. The presented layered IoT system is initially applied to the Mobile Business Architecture (MBA) entities, and Mobile Internet Technical Architecture (MITA) fundamentals are mapped to these principles and concept models for MIBA and MITA.

*Figure 1. Layered architectures of IoT system*

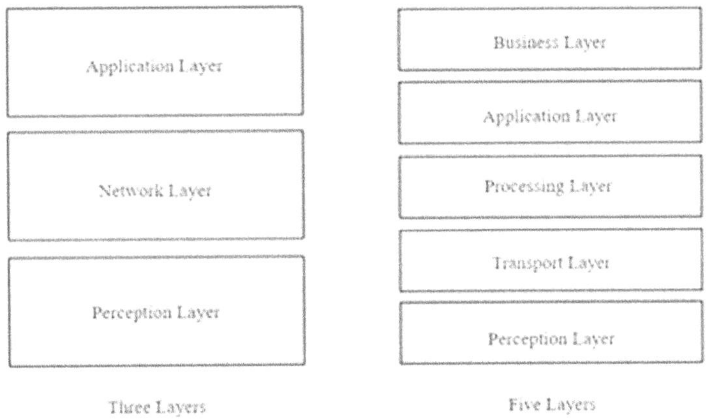


The middleware layer in digital Mobile Internet Technical Architecture has been named the Network Layer, as illustrated in Figure 1 (Three Layers) model. The above three-element layers are used for each of the MIBA segments and their elements. In addition, an interaction between the elements is provided with protocols and content delivery within the payloads of the protocols.

This three-layer architecture was introduced in the early research stage in this area. It has got three layers, namely, the perception, network, and finally, application layers.

1. The *perception layer* is the physical layer, with sensors for sensing, gathering data, and creating information about the business operating environment. It senses physical arguments or identifies other smart objects in the business environment.
2. The *network layer* is the sole authority for making other smart objects (e.g., servers, network devices, and physical servers). Its characteristics are also used for transmitting and processing sensor data.
3. The *application layer* is responsible for providing business-specific services to the user. However, there are different applications in which the IoT can be deployed, such as smart cities, factories, smart health, and smart homes.

A three-layer architecture defines the main idea of the IoT. However, this is not enough for research on IoT due to its focus on more delicate aspects of the IoT. That is why many more layered architectures are presented in the academic article. One is the five-layer architecture that includes the processing and business layers (Mashal et al., 2015) (Said & Masud, 2013) (Wu et al., 2010) (Khan et el., 2012). The five-layers architecture includes perception, transport, processing, application, and business layers (see Figure 1). The purpose of the perception and application layers is the same as the architecture with three layers. This chapter outlines the function of the remaining three layers.

- The *transport layer* transfers the gathered sensor data from the perception layer to the processing layer and vice versa through data communication networks (e.g., wireless, 3G, LAN, Bluetooth, RFID, and NFC).

*Figure 2. Fog architecture of an innovative IoT gateway*

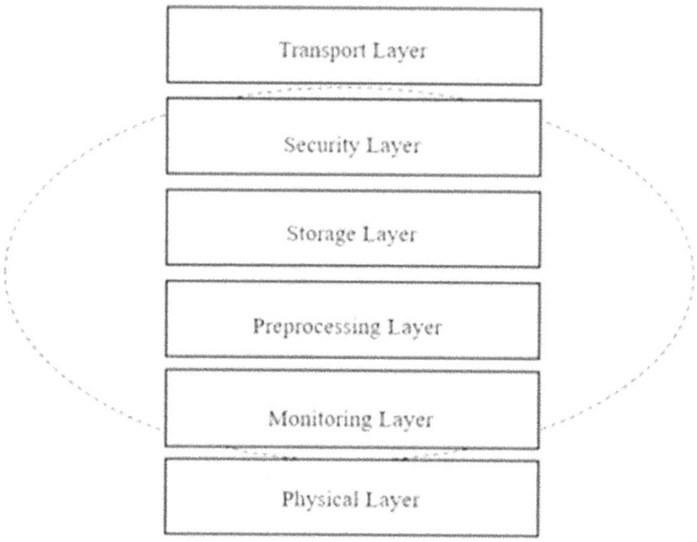


- The *processing layer* is also called the middleware layer. It stores, analyses, and processes vast data from the transport layer. It can control and serve diverse services to the lower layers. In addition, it uses many technologies such as databases, cloud computing, and big data processing modules.
- The *business layer* controls the IoT system environment, including applications, profit models, and business and end-users' privacy. However, the business layer details are out of the scope of this chapter.

Another architecture proposed by Ning and Wang (Ning & Wang, 2011) is influenced by the processing layers in a typical human brain. The human intelligence characteristics also inspire it (e.g., think, feel, remember, make decisions, and react to the physical environment). It is constituted of three parts. First is the human brain, analogous to the processing and data management unit or the data centre. The second is the spinal cord, analogous to the distributed network of data processing nodes and smart gateways. The third is the network of nerves, corresponding to the networking components and sensors.

## 2.3 Cloud and Fog Based Architectures

Cloud and fog computing architectures have played a significant role in recent years. In some system architectures, cloud computers do the data processing in a significant, centralized fashion. Such a cloud-centric architecture keeps the cloud at the centre, applications above it, and the network of smart things below it (Gubbi et al., 2013). Cloud computing is given main importance because it gives enormous scalability and flexibility. It provides services (e.g., platform, core infrastructure, storage, and software). Developers can use their storage tools, software tools, data mining, machine learning, and visualization tools through the cloud.

Lately, there has been a move towards another system architecture, namely, *fog computing* (Bonomi et al., 2014) (Bonomi et al., 2012) (Stojmenovic & Wen, 2014), where the sensors and network gateways do a part of the data processing and analytics. A fog architecture (Aazam & Huh, 2014)is shown in Figure 2, including storage, monitoring, pre-processing, and security layers between the transport and physical layers. The monitoring layer looks after power, resources, responses, and services. The processing layer does sensor data filtering, processing, and analytics. The short-time storage layer provides storage functionalities (e.g., data replication, distribution, and storage). Lastly, the security layer uses encryption/decryption to ensure data integrity and privacy. Controlling and pre-processing are done on the edge of the data communication network before sending data to service-oriented computing.

Moreover, "fog computing" and "edge computing" are used interchangeably in different business contexts. The latter term relates to the former and is constructed to be more generic. Fog computing, originally called Cisco, refers to smart gateways and smart sensors, whereas edge computing is slightly more penetrative. This paradigm envisions smart data pre-processing capabilities for physical devices such as motors, pumps, or lights. The aim is to do data processing on these devices located at the *edge* of the network. The diagram is not appreciably different from Figure 2, and this chapter does not describe edge computing details separately.

## 2.4 Interoperability and Related Issues for Business Applications

IoT-based service oriented architecture plays an essential role in modern industrial systems automation. Two central factors, (i) interoperability of business services, and (ii) quality of service (QoS), lay the foundation for system automation. Different ways industrial systems interoperability are contextualized (e.g., technical interoperability, syntactical interoperability, semantic interoperability, pragmatic interoperability, platform interoperability, and conceptual interoperability) for system development purpose. Moreover, academic literature states that the interoperability between IoT devices and computing servers depends on different operational characteristics (e.g., business models, hardware and software composition, communication protocols, and business policies). A research group (Pal & Karakostas, 2014) tried to introduce a hardware and software interoperability issue with an ontology-based system architecture. In this chapter, interoperability represents the capability of heterogeneous things or devices to connect and exchange meaningful information among themselves (Pal, 2023).

The QoS values of a composite service are determined by the QoS values of its component services and by the composition structure used (e.g., sequential, parallel, conditional and/or loops). Here, we focus on the sequential composition model. Other models may be reduced or transformed to the sequential model, using for example techniques for handling multiple execution paths and unfolding loops (Cardoso et al., 2004). Importance of business process orchestration and design patterns are highlighted by academics and practitioners (Courbis & Finkelstein, 2005) (Yoshioka et al., 2004) for service aggregation purpose. In a typical commercial IoT setting, this chapter treats the devices and services as bots to set up relationships between them and change them over time. This will allow the situation to let the devices cooperate and achieve a complex task seamlessly.

One needs to have many interoperating components to make such a model work. So let us look at some of the significant components of such a system.

- Object Identification (ID): One needs a unique way to identify an object. An ID can be allocated to an object based on traditional parameters such as the MAC ID, IPv6ID, a universal product code, or other custom methods.
- Metainformation: Along with an ID, one needs some extra information about the device that describes its characteristics and operational methods.
- Security Controls: This way, an industrial owner of a device might restrict the kinds of devices that can connect to it. These are referred to as *owner control.*
- Service Discovery: Such a system is like a service cloud, where one needs to store details of devices providing certain kinds of services. It becomes essential to keep these directories updated so that devices can learn about other devices.
- Relationship management: This module manages relationship information. For example, it also stores the types of devices that a given device should connect with based on the services provided.
- Service Composition: This module takes the IoT model to a new level. The goal of having such a system is to provide better-integrated services to users.

The rest of the chapter presents a manufacturing service integration example with the help of an ontology-concept similarity assessment algorithm.

## 3. INDUSTRIAL MANUFACTURING APPLICATIONS

Today, all manufacturing businesses value the consequence of building an effective supply chain as part of enterprise proliferation and profitability (Pal, 2018). Different industry-specific supply chain types (e.g., automotive, pharmaceutical, apparel, agriculture), and a supply chain consists of a system with organization, people, technology, activity, information, and resource involved in delivering a service or product from suppliers to customers.

Supply chain business activities transform resources into ultimate products and deliver them to customers. The supply chain network comprises the enterprises and departments involved in this process. The essential requirements of supply chain operation are minimizing the inventory, creating seamless material and information flow, effective communication among the business partners, market, sale, purchase, manufacturing plan, and control, customer delivery service, and after-sales service. Therefore, a supply chain is a network of facilities and distribution options that performs material procurement functions, transforming these materials into intermediate and finished products and delivering these finished products to customers. This definition, or a modified version of it, has been used by several researchers (e.g. (Lee & Billington, 1993) (Swaminathan, 2001a) (Keskinocak & Tayur, 2001) (Pal, 2017)). Supply Chain Management (SCM) aims to improve logistical resource allocation, management, and control. In this way, manufacturing SCM is a set of synchronized activities for integrating suppliers, manufacturers, transporters, and efficient customer service so that the right product or service is delivered in the right quantities, at the right time, to the right places (Pal, 2020) (Pal & Ul-Haque, 2020).

The ultimate objective of SCM is the efficient management of the end-to-end process, which starts with the design of the product or service and ends with the time when it has been sold, consumed, and, finally, discarded by the consumer. This process includes product design, procurement, planning and forecasting, production, distribution, fulfilment, after-sales support, and end-of-life disposal. Supply chain management issues can be classified into two broad categories: configuration (design-oriented) issues that relate to the basic infrastructure on which the supply chain executes and coordination (execution-oriented) issues that relate to the actual execution of the supply chain. Configuration-level issues include the following topics:

1. Procurement and Supplier Decisions: Procurement involves making buying decisions under conditions of scarcity. At the same time, the requirements criteria for selecting suppliers and the number of suppliers need to be decided. If sound data is available, using economic analysis methods such as cost-benefit or cost-utility analysis is good practice. Procurement ensures the buyer receives goods, services, or works at the best possible price when aspects such as quality, quantity, time, and location are compared.
2. Production Decisions: This is a multi-criteria decision activity. Does it include the decisions regarding production network design (e.g., Where, and how many manufacturing sites should be used for production purposes? How much capacity should be installed at each of these sites? What kind of products and services will be supported through the supply chain network?).
3. Distribution Decisions: It is based on infrastructure design decisions (e.g., What kind of distribution channels should a manufacturing company have? How many and where should the distribution centers and retail outlets be situated? What types of transportation services and routes should be used? What types of environmental issues does the distribution infrastructure need to be considered?).
4. Information Support Decisions: Managing a manufacturing supply chain involves numerous decisions about the flow of information, product, funds, and coordination. SCM has been instrumental in connecting and smoothing business activities and forming various kinds of business relationships (e.g., Customer Relationship Management, Supplier Relationship Management) among supply chain stakeholders.

Coordination-level issues include the following topics:

1. Material Flow Decisions: These decisions include – How much inventory of different product types should be stored to provide the target service levels? Should inventory be carried in finished form or semi-finished form? How often should inventory be replenished? And so many other issues need to be considered.
2. Information Flow Decisions: SCM systems utilize modern Information and Communication Technologies (ICT) to acquire, interpret, retain, and distribute information. The software applications are ready-made packages, usually targeting a set of tasks, e.g., tracking product-related information during the transportation process. These ready-made package-software applications are mass-customized products that ignore the specific requirements of a particular business sector, so they are problematic. However, the problem of the appropriate IT solutions for supporting collaboration between supply chain business partners is not new, and it has been approached with several standards and protocols implemented in numerous enterprise information systems.

Application like ERP (Enterprise Resource Planning), CRM (Customer Relationship Management), and WMS (Warehouse Management System) contains valuable data that can be utilized by the decision support systems (DSS). Moreover, the digital transformation of business and society presents enormous growth opportunities offered by technologies such as the Internet of Things (IoT), Big Data, advanced manufacturing, blockchain technologies, and artificial intelligence. This digital transformation is char-

*Figure 3. Diagrammatic representation of the supply chain business process*

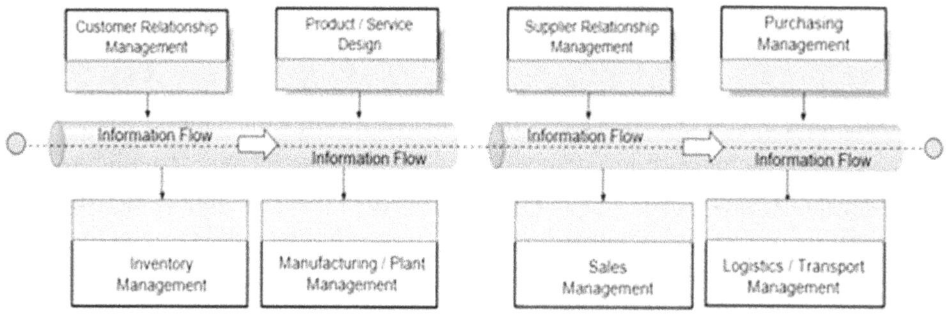


acterized by a fusion of advanced technologies and the integration of physical and digital systems, the predominance of innovative business models and new processes, and intelligent products and services.

The main goal was to reduce inventory levels drastically and regulate the suppliers' interaction with the production line more effectively. It consisted of material and information flow through the supply chain organizations. The scope of the supply chain begins with the source of supply and ends at the point of consumption. It extends beyond simply a concern with the physical movement of materials. Equal emphasis is given to supplier management, purchasing, inventory management, manufacturing management, facilities planning, customer service, information flow, transport, and physical distribution. Some of the critical business processes along the supply chain are shown in Figure 3.

Also, the critical challenges in supporting massive heterogeneous data integration in global supply networks are: (i) increasing number of business alliance partners due to the globalization of business processes, (ii) different business practice and infrastructure facilities within participating business partners, and (iii) differences in data exchange formats and standards among business-partners. Moreover, data capture and transmission mechanisms (e.g., barcoding, radio frequency identification technique, electronic data interchange, wireless networking infrastructure and protocols, global positioning system's capability) produce vast amounts of supply chain transportation data that, if properly controlled and shared, can enhance performance and agility of global supply chain networks. A single representation data format is essential to harness the value-added service.

As a result, much global manufacturing (e.g., textile and clothing) businesses invest in new ICT to harness smooth information-sharing ability in their supply chain operations (Monteneggro et al., 2007). With the recent progress in Radio Frequency Identification (RFID) technology, low-cost wireless sensor hardwires, and world comprehensive web technologies, the Internet of Things (IoT) advance has attracted attention in connecting global apparel business activities and sharing operational business information. In this context, IoT technology supports connecting and integrating both digital and physical business entities, enabling the provision of a new type of information system (IS) applications and services. Also, the Semantic Web of Things (SWoT) is ushering in a new opportunity in the business community by integrating the Semantic Web and IoT technologies. It aims to associate semantically rich and easily accessible information with real-world objects, locations, and events using inexpensive, disposable, and unobtrusive micro-devices, such as RFID tags and wireless sensors. This opportunity provides new IS applications and services in many apparel business areas (e.g., manufacturing, inventory, and transportation management).

In order to facilitate this vision, information technologies and software system frameworks must mitigate typical pervasive computing application-related issues: platform heterogeneity, appropriate resource utilization, intelligent intractability of the user, device volatility, dependence on context, and limitation of device-specific computation power. Hence, the SWoT vision requires pervasive knowledge-based systems with higher degrees of automatic capability in information storage, management, and discovery and transparent access to information sources for processing.

Most IoT technology solutions inherit support from stable data communication infrastructures and assistance from centralized brokers for service management and information discovery. Also, optimization of alternative information provision has been recently getting prominence, for example, 6LoWPAN (IPv6 over Low Power Wireless Personal Area Networks) (Monteneggro et al., 2007) and the Constrained Application Protocol (CoAP) (Colitti et al., 2011). In parallel attempts, device and data annotation ontologies were presented, for example - OntoSensor (Russomanno et al., 2005) and the SSN-XG ontology of the World Wide Web Consortium for semantic sensor networks (Lefort et al., 2005). Current research projects, such as UBIWARE (Katasonov et al., 2008), and Sense2Web (Barnaghi et al., 2010), integrate data communication networking and semantic technologies to make software frameworks for semantically enriched IoT application services.

These services are composed of different data sources from real-world objects related to apparel business processes. With many devices in the textile and clothing world, many physical parameters and real-world entities can collaborate in a business service through data communication networks or the Internet. Thus, there is a need for seamless integration of the physical world with the digital world in IoT. Also, the progress in IoT-enabled device abstraction and integrating different data sources in the apparel business is challenging. The traditional web service discovery mechanisms are incapable of producing the appropriate result. Also, the heterogeneous nature of IoT-generated data requires semantic modelling. Thus, IoT entities must be formally represented and managed to achieve interoperability.

An essential step towards the vision of IoT-based information systems interoperability is reusing data collected from widely distributed sensor-enabled devices. Ontology-based semantic modelling helps capture entities' capability to represent information and its relationships among other entities to enable efficient information exchange. In this way, semantic modelling in conjunction with service-oriented computing and ontology, ushers a scalable means of accessing IoT entities. An ontology describes a vocabulary modelling a domain of interest and specifying the meaning of terms in that vocabulary. Depending on the accuracy of this specification, the notion of ontology encompasses different data or conceptual models (e.g., classifications). In other words, ontology is a shared conceptualization, and it is used to represent knowledge as a set of concepts related to each other. The structural part comprises four key components: *classes*, *relations*, *attributes*, and *individuals*. Classes represent the concepts in ontology design, and individuals are the basic, '*ground level*' components or instances of an ontology. Attributes or characteristics represent the features of the classes, and relations describe how the classes and individuals are related to one another.

This chapter presents semantic knowledge-based on IoT-related entities in an apparel manufacturing business. Since most of the IoT data in a textile and clothing industry setting needs to be made available homogeneously to allow integration from various sources. A unified machine-understandable representation of world knowledge must put things into an everyday semantic context. The integrated Apparel Business Decentralised Data Integration (ABDDI) framework uses semantic modelling. Ontologies are used to form a unified knowledge base to support: (a) semantic definition and representation of IoT entities; (b) dynamic service discovery and matching based on user request; and (c) service composition and orchestration in dynamic environments. This chapter considers ontologies as the vital component for automatic service representation, composition, discovery, and orchestration for IoT in dynamic environments. The proposed knowledge base hides the heterogeneity of entities and consequently enables semantic searching and querying capabilities. Finally, the presented knowledge base integrates several existing ontologies related to sensor resources and web services and extends them for IoT.

## 4. RELATED RESEARCH WORKS

In recent decades automatic identification technologies are gaining more attraction for industrial applications. These applications are combined with RFID, Electronic Product Code (EPC) technologies, and sensor-based data communication networks to share application-specific data. Advances in ICT are bringing into reality the vision of many uniquely identifiable, interconnected objects and things that gather data from diverse physical environments and deliver the information to various innovative applications and services. In this way, the network of objects (e.g., devices, vehicles, machines, containers), embedded with sensors and software can collect and communicate data over the Internet. The inter-and intra-organizational communication and information exchange are perceived to be facilitated by IoT capability. This, IoT-based technology adoption can be viewed as an additional capability that may add value to the supply chain industries.

While IoT's many applications are used in different supply chain industries (Jandl et el., 2019), industrial IoT (IIoT) focuses primarily on the industrial usage of IoT devices (Jakl et al., 2018) (Jeschke et al., 2017). Thomas Moser's research group recently presented an overview of industrial IoT applications (Jandl et al., 2019) for the smart industry (also known as Industry 4.0). With the advent of the IoT, new opportunities and capabilities emerge in real-time monitoring, management, and optimizing goods distribution and supply chain. As more physical objects in supply chain industries are equipped with barcodes, RFID tags, or sensors, transport, and logistics companies can monitor the movement of physical objects from one location to another. The goal is to track a product along the entire supply chain, including material management, production, transportation, and distribution (Karakostas, 2013).

Manufacturing companies are changing their IT infrastructure towards the fourth industrial revolution (Lasi et al., 2014). The utilization of artificial intelligence (AI) techniques in cyber-physical systems (CPSs) (Broy et al., 2012) is a typical characteristic of this change (Lee et al., 2014). In this context, flexibility is one of the essential criteria for manufacturing companies, mainly because of ever-shorter market launch times and increasing customer demands for individualization (Cheng et al., 2017; Lasi et al., 2014). To conduct industry 4.0 research, researchers use the factory simulation model because companies are often unwilling to provide data from and access to their production lines for research purposes.

More researchers have started concentrating on techniques enabling machines to understand IoT data better intelligently from various sources. In order to use AI applications, contextual operational knowledge must be available in formal and machine-readable representation (Humm et al., 2020). Semantic web services address the issues of automatic discovering, composing, and executing by providing a declarative, ontological framework for describing them. Using AI methods (e.g., automated planning such as (Marrella, 2018) (Marrella & Mecella, 2018), multi-agent systems for decentralized manufacturing control such as (Ciortea et al., 2018), Case-Based Reasoning (CBR) such as Minor et al., 2014) (Muller, 2018)) to enhance flexibility in cyber-physical production workflows (Bordel Sanchez et al., 2018) (Seiger et al., 2018)) inevitably require such semantic annotations.

Much related research work (Puttonen et al., 2010) (Puttonen et al., 2013) exists that already highlights these issues by using semantic web services (SWS). Moreover, the currently available approaches that use semantic web services in the context of Industry 4.0 focus only on specific aspects and do not consider the entire context of a manufacturing environment. Also, the complex reasoning within the knowledge base makes real-time execution and monitoring of manufacturing processes difficult. In this way, managing interoperable industrial applications is the key to success. One of the key components of interoperability is the specific information system's architecture. A reference architecture (RA) represents a robust and ground-level understanding of the scenarios that help recognize problems and difficulties. The key idea of designing a RA (e.g., service-oriented architecture (SOA)) is to highlight modularity, scalability, adaptiveness, and interoperability among the connected heterogeneous devices in a real-time environment (Sarkar et al., 2021) (Schneider, 2018) (Hazra et al., 2021) (Nagarajan et al., 2021).

Several research works propose using SWSs for smart manufacturing in Industry 4.0 but focus only on partial aspects and do not consider the entire shop floor context. For instance, Puttonen et al. (2013) present an approach to using SWSs for executing manufacturing processes using three software agents represented as web services. One of these agents, Service Monitor, is a specialized web service that performs semantic web service composition by using planning techniques concerning a given production goal and the current state of the world provided by a domain ontology. Therefore, they use OWL to describe the state of the production system and OWL-S and SPARQL expressions for semantically describing the available web services that offer production capabilities.

Since modern Cyber-Physical Production Systems (CPPSs) (Monostori, 2014) consist of many different components and therefore, many stakeholders are involved in their development process up to the later use in the manufacturing of products, Lobov et al. (2008) investigated the application of SWSs for orchestration of flexible control. They propose OWL for modelling a Process Taxonomy, Product Ontology, Equipment Ontology, and Service Ontology and discuss the responsibilities of involved persons for knowledge acquisition and maintenance rather than present their detailed semantic specification.

Many academics and practitioners, work-like Henson et al. (Cory et al., 2009), experiment a semantically enhanced sensor service application, known as SemSOS, having the capability to query both high-level knowledge and low-level environmental reading by sensors. Concerning the classical sematic-matchmaking approach, different research groups distinguish among full (subsume), potential (intersection-satisfiable), and partial (disjoint) match types to represent the relevant knowledge of different types of entities (Colucci et al., 2007) and (Li & Horrocks, 2004) respectively. Similarly, queries in ubiquitous infrastructure allow only exact matches with facts derived from a support knowledge base. Non-standard inferences like abduction and contraction are needed to support approximate matches, semantic ranking, and explanations of outcomes (Colucci et al., 2007).

This chapter highlights the requirement for a unified semantic knowledge base for automatic service representation, discovery, modelling, and composition in dynamic environments. Different research projects try to address these issues using ontologies in prototype system design. For example, semantic commonalities in the RFID semantic streams project use DLs for system modelling (Ruta et al., 2011). This work is of considerable interest, but so far, it is sui generis.

All these schemes are worthy of considerable study, and together they represent the wealthiest characterization of a knowledge-based approach in pervasive service computing so far produced. Moreover, the systems are implemented and can be shown to generate relevant services using appropriate data streams.

## 5. OVERVIEW OF THE FRAMEWORK

The proposed framework uses model-theoretic semantics modelled in ontologies for IoT-generated data modelling purposes. It helps to gather detailed information regarding the characteristics of IoT devices based on their technical requirements. The advantages of this encoding are – (i) interconnecting different classification systems to represent capabilities and properties of constituent parts, (ii) translating characteristics or properties among compound constituent parts, and (iii) aggregating basic properties into complex properties based on the constituents of a superordinate system. Those concepts can be used and adapted for the IoT to enhance the uses of IoT devices in connecting them as a group, create coordination between IoT devices, and improve their interoperability.

One of the main objectives of this framework is to define an IoT architecture, which can also be used for other applications. The design principle in Apparel Business Semantic Data Management (ABDDI) is that any physical/real-world object in the global textile and clothing business can have a virtual representation through a Virtual Object (VO). AVO uses a semantic representation of the functionality and conceals the varied identity of the real-world object. Multiple VOs can be combined to form a Composite Virtual Object (CVO) that provides more compact and reliable services. In simple, CVOs are combined to form a service request. Thus, the ABDDI architecture has three layers: VO, CVO, and Service, as shown in Figure 4.

The functionalities of the three layers are presented below:

1. VO layer: Real-world objects are represented in the digital format as VO. End-users can search semantically and retrieve information from any existing VO. Also, actuation can be done through the VO.
2. CVO layer: In this layer, VOs are combined to form a service request. This layer caters functionality to search and query categories of CVOs for service provision semantically.
3. Service layer: This layer gets the request from a user and analyses the service requests to determine the categories of CVOs needed for service accomplishment. This layer also performs service composition and orchestration in a dynamic cloth and textile business environment.

The ABDDI system also has got other components: *a registry* and a *control unit*. Every layer in this framework has a registry referencing the available VOs, CVOs, and services. These registries provide methods to semantically search and query existing VOs, CVOs, and services. In addition, the control unit regulates access to the VOs, CVOs, and services based on the level of the end-user requirements.

## 6. KNOWLEDGE REPRESENTATION AND REASONING APPROACH

One of the motivations for this chapter stems from the Description Logics (DL) [1] based knowledge representation approach in pervasive computing applications along the global apparel supply chain. This DL-based knowledge representation systems play a role very much like Database Management Systems (DBMS). In DL, elementary descriptions are *atomic concepts* and *atomic roles*. Complex descriptions can be built from them inductively with *concept constructors*. In abstract notation, one can use the letters A and B for atomic concepts, R for atomic roles, and C and D for concept descriptions. Possible

*Figure 4. RFID tagging level at different stages in the apparel manufacturing network*

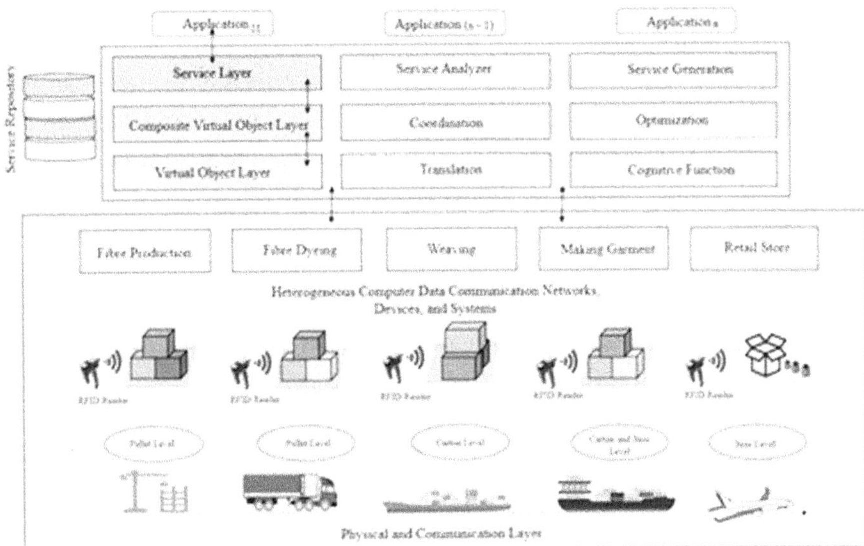


DL constructors and the related examples are shown in Table 1, and these constructors are used in the DLs investigated in this chapter.

## 6.1 Resource Representation and Reasoning in Description Logics

The most important and well-known service characterizing reasoning in DL checks for specificity hierarchies by determining whether a concept description is more specific than another one or, formally, if there is a *subsumption* relation between them.

*Table 1. DLs set of constructors*

| Constructor Name | Syntax | Explanation |
|---|---|---|
| Top concept | ⊤ | Universal concept. All the objects in the domain. |
| Bottom concept | ⊥ | The empty set. |
| Atomic concept | A | All objects belong to set A. |
| Atomic negation | ¬A | All the objects are not belonging to set A. |
| Conjunction | C⊓D | The objects belong to both C and D sets. |
| Disjunction | C⊔D | The objects that are in the extension of either C or D or both |
| Value restriction | ∀RC | All the objects are participating in the R relation, whose range is all the objects belonging to the C set. |
| Existential restriction | ∃RC | At least one object participating in the relation R. |
| Concept definition | A ≡ C | Concepts represent sets of elements and can be viewed as unary predicates. |

**Definition 1 (Subsumption):** *Give two concept descriptions, C and D, and a TBox* $\tau$ *in a DL L; one can say that D subsumes C concerning if for every model of* $\tau$, $C^I$, $D^I$. *In a particular case, two concepts are equivalent if they subsume each other.*

For example, consider the following concept descriptions, referring to different garment types in an apparel supply chain network: $G_1$ = *SweaterBodyGarment* ⊓ ∀ *hasMain.Colour.Red*, and $G_2$ = *UpperBodyGarment* ⊓ ∀ *hasMain.Colour.Red*. Then using TBox reasoning – the concept inclusion can be achieved, and the output will be *Sweater* $\sqsubseteq_T$ *UpperBodyGarment*. Hence, given the model, knowledge expressed by $G_1$ is more specific than the one required by $G_2$ concerning the reasoning mechanism and the definition of $G_2$ subsumes $G_1$.

Based on subsumption, new reasoning mechanisms can be defined in DLs. The ABDDI system development uses several non-standard reasoning mechanisms (e.g., Least Common Subsumer – LCS).

**Definition 2 (Least Common Subsumer):** *Let* $C_1, \dots, C_p$ *be p concept descriptions in a DL L. A Least Common Subsumer (LCS) of* $C_1, \dots, C_p$, *denoted by LCS* ($C_1, \dots, C_p$) *is a concept description E in L. state that the following conditions hold: (i)* $C_h \sqsubseteq E$ *for h = 1, …, p; (ii) E is the least L–concept description satisfying (iii), i.e., if* $E'$ *is an L–concept satisfying* $C_i \sqsubseteq E'$ *for all i = 1, …, n, then* $E \sqsubseteq E'$.

It is worth showing how to model concept collections formalized in ALN (D) according to a compact lossless representation. Such modelling allows for finding commonalities in resource annotations formalized in DL.

**Definition 3 (Concept Components):** *Let C be a concept described in a DL L, with C formalized as* $C^1 \sqcap \dots C^m$. *The Concept Components of C are defined as follows: if* $C^j$, *with j = 1,…, m is either a concept name, a negated concept name, or a concrete feature or a number restriction, then* $C^j$ *is a Concept Component of C; if* $C^j = \forall R. E$, *with j = 1 …, m, then* $\forall R.E^k$ *is a Concept Component of C, for each* $E^k$ *Concept Component of E.*

**Definition 4 (Subsumption):** *Give two concept descriptions C and D, and a TBox* $\tau$ *in a DL L; one can say that D subsumes C concerning if for every model of* $\tau$, $C^I$, $D^I$. *In a particular case, two concepts are equivalent if they subsume each other.*

**Definition 5 (Least Common Subsumer):** *Let* $C_1, \dots, C_p$ *be p concept descriptions in a DL L. A Least Common Subsumer (LCS) of* $C_1, \dots, C_p$, *denoted by LCS* ($C_1, \dots, C_p$) *is a concept description E in L. state that the following conditions hold: (i)* $C_h \sqsubseteq E$ *for h = 1, …, p; (ii) E is the least L–concept description satisfying (iii), i.e., if* $E'$ *is an L–concept satisfying* $C_i \sqsubseteq E'$ *for all i = 1, …, n, then* $E \sqsubseteq E'$.

**Definition 6 (r-Common Subsumer, Informative r-Common Subsumers):** *Let* $C_1, \dots, C_p$ *be p concept descriptions in a DL L, and let be* $k \leq p$. *An r-Common Subsumer (r-CS) of* $C_1, \dots, C_p$ *is a concept* $_{D \neq} \top$ *such that D is an LCS of at least r = k/p concepts among* $C_1, \dots, C_{p.}$ *One can define a particular case as Informative r-Common Subsumers (IrCS) that specific r-CSS for which r < 1.*

It is worth showing how to model concept collections formalized in ALN (D) according to a compact lossless representation. Such a modelling framework allows for finding commonalities in resource annotations formalized in DL.

**Definition 7 (Concept Components):** *Let C be a concept described in a DL L, with C formalized as* $C^1 \sqcap \dots C^m$. *The Concept Components of C are defined as follows: if* $C^j$, *with j = 1,…, m is either a concept name, a negated concept name, or a concrete feature or a number restriction, then* $C^j$ *is a Concept Component of C; if* $C^j = \forall R. E$, *with j = 1 …, m, then* $\forall R.E^k$ *is a Concept Component of C, for each* $E^k$ *Concept Component of E.*

**Definition 8 (Aggregate Collection Matrix):** *Let* $S_1, \dots, S_n$ *be an aggregate collection, with* $S_j = C_{1i},$ *..,* $C_{pi}$ *for i = 1 … n. Let* $D \in \{D_1, \dots, D_m\}$ *be the Concept Components deriving from all the concepts in* *the aggregate collection. The Aggregate Subsumers Matrix is defined as A = ($a_{ij}$), with i = 1 … n and j = 1 … m, such that for each i, $a_{ij}$ = v, with $0 \leq v \leq p_i$, where v is the number of concept descriptions in $S_i$ subsumed by the component $D_j$.*

**Definition 9 (Aggregate Model):** *Let $S_1$, …, $S_n$ be an aggregate of concept collections; for i = 1 … n, $S_i$ is a concept collection descriptions $_{Ski}$ with k = 1 … $p_i$. An Aggregate Model for $S_1$, …, $S_n$ and each of this element consists of the pair of items - <E, G> with the following characteristics: (i) E represents the subsumers matrix deriving from the collection $C_1$,..,$C_p$ = ($C_{ki}$), with i = 1 … n and k = 1 … $p_i$, whose elements $e_{kj}$ are calculated by using prognostications to subsumption; and (ii) G is the collection subsumers matrix deriving from the input collection $S_1$,..,$S_n$, whose elements $a_{ij}$ are calculated by using information stored in E. In this computation, each row i in G is related to an aggregate collection $S_i$, defined as a collection of description $C_{ki}$ whose subsumption relationship with components deriving from $S_1$,.., Sn is stored in E. To this modelling, values $a_{ij}$, for each component $D_j$, are determined as Concept Component Relative Cardinality $RC_{D_j}^{S_i}$ .*

## 6.2 Semantic Similarity Assessment

Before describing the proposed approach's theoretical framework, the employed reasoning services will be shortly recalled in the following subsection to make the chapter self-contained. Furthermore, the proposed algorithmic concept of similarity measurement is presented in this section.

In ABSDM, the similarity between concepts $C_i$, $C_j$ can be expressed by a number, and its values can fall between 0 and 1. It may be viewed as a one-directional relation, and its larger values imply a higher similarity between the concepts. The concept similarity is described as follows:

**Definition 10 (Concept Similarity):** An ontological concept (C) similarity ($\partial$) is considered a *relation*, and it can be defined as $\partial C \times C \rightarrow [0, 1]$. In simple, it is a function from a pair of concepts to a real number between *zero* and *one*, expressing the degree of similarity between two concepts such that:

1. $\forall C_1 \in G$, $\partial(C_1,C_1)= 1$
2. $\forall C_1,C_2 \in G$, $0 \leq \partial(C_1,C_2) \leq 1$
3. $\forall C_1,C_2,C_3 \in G$, IF $Sim_d(C_1,C_2) > Sim_d(C_1,C_3)$ THEN $\partial(C_1,C_2) < \partial(C_1,C_3)$

The above properties provide the range of semantic similarity functions $\partial(C_i,C_j)$. For exactly similar concepts, the similarity is $\partial(C_1,C_1)=1$; when two concepts have nothing in common, their similarity is $\partial(C_1,C_2)=0$. In this way, the output of the similarity function should be in the closed interval [0, 1]. Here $Sim_d$ represents the semantic distance, and ($C_1$, $C_2$, $C_3$) represent three concepts of graph G. In CSIA, the following semantic similarity ($\partial$) function has been used for computation purposes:

$$\partial\left(C_1,C_2\right)=\frac{1}{\deg * Sim_d\left(C_1,C_2\right)+1}$$

Where $C_1$ and $C_2$ represent two concepts, and 'deg' represents the impact of semantic distance on semantic similarity, it should be between $0 < \deg \leq 1$. A weight allocation function is used, as shown below, to compute the semantic similarity between concepts:

$$w(C_m, C_n) = \left[ \max(depth(C_m)) + \frac{OrderNumber(C_n)}{TNodes(G) \ \ +1} + 1 \right]^{-1}$$

Where $C_m$ and $C_n$ represents two nodes directly connected, max(*depth*($C_m$)) represents the maximum depth of the node $C_m$ (the depth of the root node is equal to 0 and 1 for the nodes directly connected to the root node and so on), TNodes(G) and OrderNumber($C_n$) represent the total number of nodes in concept graph G and the order number of the node ($C_n$) between their siblings. The detailed description of these mathematical formalizations is beyond the scope of this chapter.

## 6.3 Semantic Similarity Assessment

In ABDDI, the similarity between two concepts $C_i$, a number can express Cj, and its values can fall between 0 and 1. It may be viewed as a one-directional relation, and its larger values imply a higher similarity between the concepts. The concept similarity is described as follows:

# 7. EXAMPLE OF A BUSINESS SCENARIO

A simple apparel manufacturing scenario is used to present a part of ABDDI algorithmic computation. Semantic IoT-based product flow in a retail outlet is considered. Each product is described using semantic-enhanced IoT as an ALN (D) concept expression in OWL language. As the retail apparel product arrive or depart the shop, they are scanned by the gate RFID readers; reading events, including semantic annotation extracted from tags, are fed to a semantic Data Service Management Service (DSMS), which computes Concept Components and subsumption test through a reasoning mechanism.

Let us consider a situation that allows a user to purchase a sweater from an online business. This example considers how a request is matched with the service advertised for wool garments selling service. An algorithm (i.e., ALGORITHM 1) tries to perform semantic matching for a relevant sweater.

**Algorithm 1.** Algorithm for semantic similarity computation
input: two concepts ($C_1$, $C_2$), the root node (root), concepts graph (G)
output: semantic similarity value between two concepts
begin
if $C_1$ and $C_2$ are same concept then $Sim_d = 0$
else
if $C_1$ and $C_2$ are directly connected then $Sim_d = w(C_1, C_2)$
else
if idirect path connection exist then
$S_{path01}$ = ShortestPath (G, $C_1$, $Root_N$)
$S_{path02}$ = ShortestPath (G, $C_2$, $Root_N$)
$Sim_d = w(S_{path01}) + w(S_{path02}) - 2*w(CSPath]$
end if
$\partial(C_1, C_2) = \dfrac{1}{\deg * Sim_d + 1}$

*Figure 5. The hierarchical concept relationships*

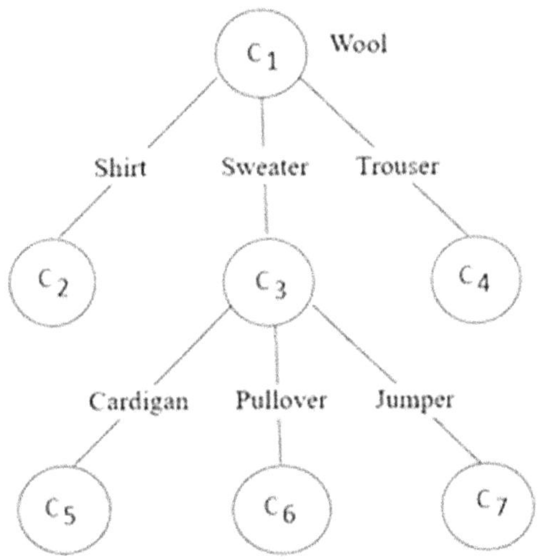


```
end if
end if
return ∂
end
```

The algorithm takes two ontological concepts, the root node (root), and the concepts graph (G), as input and computes a semantic similarity between the concepts as output.

The part of the concept hierarchy used in this example is shown in Figure 5. Each node of this hierarchy represents a concept. The experimental comparison considers semantic similarity among Wool, Shirt, Sweater, Trousers, Cardigan, Pullover, and Jumper. The proposed algorithm (i.e., Algorithm-I) provides semantic similarity between concepts with a higher score than the path similarity algorithm.

In Table 2, (a) is the result of path similarity [18], and (b) tabulates the results of the proposed Algorithm-I used in ABDDI. In ABDDI, the similarity measure is flexible and customizable, allowing the consideration of user preferences. This refers to two aspects. Firstly, the user may determine some of the similarity assessment parameters' relative importance using the advanced search interface. Sec-

*Table 2. The results of various similarity measures*

| | $C_1$ | $C_2$ | $C_3$ | $C_4$ | $C_5$ |
|---|---|---|---|---|---|
| $C_1$ | 1.00 | 0.25 | 0.50 | 0.20 | 0.20 |
| $C_2$ | 0.25 | 1.00 | 0.50 | 0.33 | 0.16 |
| $C_3$ | 0.50 | 0.50 | 1.00 | 0.25 | 0.16 |
| $C_4$ | 0.20 | 0.33 | 0.25 | 1.00 | 0.20 |
| $C_5$ | 0.20 | 0.16 | 0.16 | 0.20 | 1.00 |

(a) Path similarity

| | $C_1$ | $C_2$ | $C_3$ | $C_4$ | $C_5$ |
|---|---|---|---|---|---|
| $C_1$ | 1.00 | 0.48 | 0.65 | 0.51 | 0.38 |
| $C_2$ | 0.48 | 1.00 | 0.65 | 0.51 | 0.38 |
| $C_3$ | 0.65 | 0.65 | 1.00 | 0.71 | 0.48 |
| $C_4$ | 0.51 | 0.51 | 0.71 | 1.00 | 0.59 |
| $C_5$ | 0.38 | 0.38 | 0.48 | 0.59 | 1.00 |

(b) The proposed method

ond, apart from presenting a single rank for each candidate service, more detailed results may also be provided (e.g., different values for recall, and the degree of match) to facilitate the user in identifying the timelier service.

## 8. FUTURE RESEARCH DIRECTIONS

Based on the findings, future studies, and experiments must be conducted on semantic reasoning with more diverse information content, complex scenarios, and more detailed contextual knowledge (e.g., rules and other forms). Reasoning engines certainly influence reasoning performance. Thus, different reasoning engines should be evaluated. For example, integrating real-time reasoned knowledge with background knowledge by utilizing federated RDF databases would be valuable, as they can provide background reasoning and knowledge integration services on specific platforms. In the future, the current research plans to use many context scenarios to evaluate the proposed approach's benefit more appropriately.

Although several academic and industry proposals address IoT-based information systems interoperability issues, no reasonable ground can cover some related research issues. The lack of standards and the absence of cutting-edge technologies slow the development of IoT. Proving semantically interoperable platforms across the different IoT domains requires research improvements. Hence, there is still significant room for future work on this topic.

## 9. CONCLUSION

This chapter presents some of the uses of IoT technologies (e.g., RFID tag-based technology, sensor) in the apparel industry. RFID technology can interact with items (i.e., transport carts, trolleys, keys, and valuable products) without physical contact in the textile and clothing retail industry. Thus, item-level IoT infrastructures provide item handling efficiency and offer a promising way to capture customers' in-store behavioural data and then gain insight into these data using data mining technology. In this way, these sensing objects and things form the Internet of Things (IoT), which are used to automate manufacturing business processes (e.g., material management, transportation management, and logistics management). In addition, ioT applications rely on real-time context data and allow sending information to drive users' behaviours in intelligent supply chain environments. These IoT-based solutions are mostly tailored for vertical applications and systems, utilizing knowledge only from business areas. However, to realize IoT's full potential, these specialized silo applications must be replaced with horizontal collaborative applications, including knowledge acquisition and sharing capabilities.

One of the main challenges in realizing an IoT system lies in dealing with device and data heterogeneities and fostering the interoperability between devices, information, and services built on top of it. Ontologies and semantic description frameworks play an essential role in the IoT. They operate on top of a standardized data interoperability infrastructure and enforce the concept of data uniformity that allows data and data semantics to be described in application-independent ways. Semantic technologies and ontologies decouple data semantics from application logic, facilitating information exchange, adaptability, and interoperability among tools and systems.

This chapter describes a unified semantic knowledge base (ABDDI) for IoT-related apparel supply chain management technologies. Semantic modelling is a critical component to address issues related to interoperability among different entities to realize the grand vision of IoT. ABDDI's knowledge base comprises ontologies to model different aspects of the apparel business (e.g., IoT resources, location information, contextual information, domain knowledge, policies for the dynamic environment, and IoT services). Most of the current work focuses on IoT resources and services; however, modelling contextual information in a dynamic environment assists in more accurate knowledge representation for IoT entities.

Real-time data gathering and processing are the main ingredients of current supply chain operations. To further enhance the potential of this promising application, in the future, this research will try to propose a unified framework for IoT-based path analytics, which uses both in-store shopping paths and IoT-based purchasing data for actionable navigation patterns. In the pre-processing data module, the critical problem of capturing the mainstream shopping path sequences while wiping out redundant and repeated details must be addressed in detail.

## REFERENCES

Aazam, M., & Huh, E. N. (2014). Fog computing and smart gateway based communication for cloud of things. *Proceedings of the 2nd IEEE International Conference on Future Internet of Things and Cloud (FiCloud' 14)*. IEEE. 10.1109/FiCloud.2014.83

Atzori, L., Iera, A., & Morabito, G. (2011). IoT: Giving a social structure to the internet of things. *IEEE Communications Letters*, *15*(11), 1193–1195. doi:10.1109/LCOMM.2011.090911.111340

Baader, F., Calvanese, D., McGuinness, D., Nardi, D., & Patel-Schneider, P. (Eds.). (2002). *The Description Logic Handbook*. Cambridge University Press.

Barnaghi, P., Presser, M., & Moessner, K. (2010). Publishing Linked Sensor Data. *Proceedings of the 3rd International Workshop on Semantic Sensor Networks*. IEEE.

Berners-Lee, T. (2000). *Weaving the Web: The Original Design and Ultimate Design of the World Wide Web by its inventor*. Harper Business.

Bonomi, F., Milito, R., Natarajan, P., & Zhu, J. (2014). *"Fog computing: a platform for internet of things and analytics," in Big Data and Internet ofThings: A Road Map for Smart Environments*. Springer.

Bonomi, F., Milito, R., Zhu, J., & Addepalli, S. (2012). Fog computing and its role in the internet of things. *Proceedings of the 1st ACM MCC Workshop on Mobile Cloud Computing*, (pp. 13–16). 10.1145/2342509.2342513

Bordel Sánchez, B., Alcarria, R., Sánchez de Rivera, D., & Robles, T. (2018). Process execution in CyberPhysical Systems using cloud and Cyber-Physical Internet services. *The Journal of Supercomputing*, *74*(8), 4127–4169. doi:10.1007/s11227-018-2416-4

Broy, M., Cengarle, M. V., & Geisberger, E. (2012). Cyber-Physical Systems: Imminent Challenges. In *Large-Scale Complex IT Syst. Dev., Operat. and Manag. - 17th Monterey Workshop*. IEEE.

Cardoso, J., Sheth, A., Miller, J., Arnold, J., & Kochut, K. (2004). Quality of service for workflows and web service processes. *Journal of Web Semantics*, *1*(3), 281–308. doi:10.1016/j.websem.2004.03.001

Cheng, H., Xue, L., Wang, P., Zeng, P., & Yu, H. (2017). Ontology-based web service integration for flexible manufacturing systems. In *15th Int. Conf. on Ind. Inf.*, (pp. 351–356). IEEE. 10.1109/INDIN.2017.8104797

Ciortea, A., Mayer, S., & Michahelles, F. (2018). Repurposing Manufacturing Lines on the Fly with Multi-agent Systems for the Web of Things. In *Proc. of the 17th Int. Conf. on Autonomous Agents and Multi-Agent Systems*, (pp. 813–822). Int. Found. for Autonomous Agents and Multiagent Systems / ACM.

Colitti, W., Steenhaut, K., Caro, N. De., (2011). Integrating Wireless Sensor Networks with the Web", Extending the Internet to Low power and Lossy Networks (IP+SN).

Colucci, S., Di Noia, T., Pinto, A., Ruta, M., Ragone, A., & Tinelli, E. (2007). A Nonmonotonic Approach to Semantic Matchmaking and Request Refinement in E-Marketplaces. *International Journal of Electronic Commerce*, *12*(2), 127–154. doi:10.2753/JEC1086-4415120205

Cory, A., Henson, J., Pschorr, K., Sheth, A. P., & Thirunarayan, K. (2009). SemSOS: Semantic sensor Observation Service. *Proceedings of the International Symposium on Collaborative Technologies and Systems*. IEEE.

Courbis, C., & Finkelstein, A. (2005). Weaving Aspects into Web Service Orchestrations. *Proceeding of International Conference of Web Services (ICWS '05),* IEEE. 10.1109/ICWS.2005.129

De Virgilio, R., Di Sciascio, E., Ruta, M., Scioscia, F., & Torlone, R. (2011). Semantic-based rfid data management. In *Unique Radio Innovation for the 21st Century* (pp. 111–141). Springer. doi:10.1007/978-3-642-03462-6_6

Gubbi, J., Buyya, R., Marusic, S., & Palaniswami, M. (2013). Internet of Things (IoT): A vision, architectural elements, and future directions. *Future Generation Computer Systems*, *29*(7), 1645–1660. doi:10.1016/j.future.2013.01.010

Hazra, A., Adhikari, M., Amgoth, T., & Srirama, S. N. (2021, November). A Comprehensive Survey on Interoperability for IIoT: Taxonomy, Standards, and Future Directions. *ACM Computing Surveys*, *55*(1), 1–35. doi:10.1145/3485130

Jakl, A., Schoffer, L., Husinsky, M., & Wagner, M. (2018), Augmented Reality for Industry 4.0: Architecture and User Experience. In *Proceeding of the 11$^{th}$ Forum Media Technology, CER-WS,* 38-42.

Jandl, C., Nurgazina, J., Schoffer, L., Reichl, C., Wagner, M., & Moser, T. (2019). SensiTrack – A Privacy by Design Concept for Industrial IoT Applications, In *Proceeding of the 24th IEEE International Conference on Emerging Technologies and Factory Automation*, (pp. 1782-1789). IEEE. 10.1109/ETFA.2019.8869186

Jeschke, S., Brecher, C., Meisen, T., Ozdemir, D., & Eschert, T. (2017), Industrial Internet of Things and Cyber Manufacturing Systems. In Industrial Internet of Things. Springer.

Karakostas, B. (2013). A DNS architecture for the Internet of things: A case study in transport logistics. *Procedia Computer Science*, *19*, 594–601. doi:10.1016/j.procs.2013.06.079

Katasonov, A., Kaykova, O., Khriyenko, O., Nikitin, S., & Terziyan, V. (2008). Smart Semantic Middleware for the Internet of Things. *Proceedings of the 5th International Conference of Informatics in Control, Automation and Robotics*, (pp. 11-15). IEEE.

Keskinock, P., & Tayur, S. (2001). Quantitive analysis of Internet-enabled supply chain. *Interfaces, 31*(2), 70–89. doi:10.1287/inte.31.2.70.10626

Khan, R., Khan, S. U., Zaheer, R., & Khan, S. (2012). Future internet: the internet of things architecture, possible applications and key challenges. *Proceedings of the 10th International Conference on Frontiers of Information Technology (FIT '12).* IEEE. 10.1109/FIT.2012.53

Lasi, H., Fettke, P., Kemper, H.-G., Feld, T., & Hoffmann, M. (2014). Industry 4.0. *BISE, 6*(4), 239–242.

Lee, H. L., & Billington, C. (1992). Managing supply chain inventories: Pitfalls and opportunities. *Sloan Management Review, 33*(3), 65–77.

Lee, J., Kao, H.-A., & Yang, S. (2014). Service Innovation and Smart Analytics for Industry 4.0 and Big Data Environment. *Procedia CIRP, 16*, 3–8. doi:10.1016/j.procir.2014.02.001

Lefort, L., Henson, C., Taylor, K., Barnaghi, P., Compton, M., Corcho, O., Garcia-Castro, R., Graybeal, J., Herzog, A. & Janowicz, K. (2005). *Semantic Sensor Network XG Final Report.* W3. https://www.w3.org/2005/Incubator/ssn/XGR-ssn/

Li, L., & Horrocks, I. (2004). A software framework for matchmaking based on semantic web technology. *International Journal of Electronic Commerce, 8*(4), 39–60. doi:10.1080/10864415.2004.11044307

Lobov, A., Lopez, F. U., Herrera, V. V., Puttonen, J., & Lastra, J. L. M. (2008). Semantic Web Services framework for manufacturing industries. In *Int. Conf. on Rob. and Biomim.*, (pp. 2104–2108). IEEE.

Maass, W., Filler, A. (2006). Towards an infrastructure for semantically annotated physical products. INFORMATIK 2006–Informatik für Menschen–Band 2, Beiträge der 36. *Jahrestagung der Gesellschaft für Informatik eV* (GI).

Mahon, B. (2004). *The Man Who Changed Everything: The Life of James Clerk Maxwell.* John Wiley & Sons Ltd.

Marrella, A. (2018). *Automated Planning for Business Process Management.*

Mashal, I., Alsaryrah, O., Chung, T. Y., Yang, C. Z., Kuo, W. H., & Agrawal, D. P. (2015). Choices for interaction with things on Internet and underlying issues. *Ad Hoc Networks, 28*, 68–90. doi:10.1016/j.adhoc.2014.12.006

Minor, M., Montani, S., & Recio-García, J. A. (2014). Process-oriented Case-based Reasoning. *Information Systems, 40*, 103–105. doi:10.1016/j.is.2013.06.004

Monostori, L. (2014). Cyber-physical Production Systems: Roots, Expectations and R&D Challenges. *Procedia CIRP, 17*, 9–13. doi:10.1016/j.procir.2014.03.115

Montenegro, G., Kushalnagar, N., Hui, J., & Culler, D. (2007). *Transmission of IPv6 packets over IEEE 802.15.4 networks.*

Müller, G. (2018). Workflow Modeling Assistance by Casebased Reasoning. Springer.

Nagarajan, S. M., Muthukumaran, V., Vinoth Kumar, V., Beschi, I. S., & Magesh, S. (2021). *Fine Tuning Smart Manufacturing Enterprise Systems: A Perspective of Internet of Things-Based Service-Oriented Architecture, Handbook of Research on Innovations and Applications of AI, IoT, and Cognitive Technologies*. IGI Global Publication. doi:10.4018/978-1-7998-6870-5.ch006

Ning, H., & Wang, Z. (2011). Future internet of things architecture: Like mankind neural system or social organization framework? *IEEE Communications Letters*, *15*(4), 461–463. doi:10.1109/LCOMM.2011.022411.110120

Pal, K. (2017). Supply Chain Coordination Based on Web Services. In H. K. Chan, N. Subramanian, & M. D. Abdulrahman (Eds.), *Supply Chain Management in the Big Data Era* (pp. 137–171). IGI Global Publication. doi:10.4018/978-1-5225-0956-1.ch009

Pal, K. (2018). A Big Data Framework for Decision Making in Supply Chain. IGI Publication.

Pal, K. (2020). Information Sharing for Manufacturing Supply Chain Management Based on Blockchain Technology. Cross-Industry Use of Blockchain Technology and Opportunities for the Future. IGI Global Publication. doi:10.4018/978-1-7998-3632-2.ch001

Pal, K., & Karakostas, B. (2014). A Multi Agent-Based Service Framework for Supply Chain Management, In the proceeding of International Conference on Ambient Systems, Networks and Technology. *Procedia Computer Science*, *32*, 53–60. doi:10.1016/j.procs.2014.05.397

Pal, K., & Ul-Haque, A. (2000). *Internet of Things and Blockchain Technology in Apparel Manufacturing Supply Chain Data Management*. In the proceeding of 11th International Conference on Ambient Systems, Networks and Technologies (ANT-2020), Procedia Computer Science, Warsaw, Poland.

Pal, K. (2023). Advancing Towards Sustainable Supply Chain Management Using IoT and Blockchain Technology. In Y. Ramakrishna & S. N. Wahab (Eds.), *Handbook of Research on Designing Sustainable Supply Chains to Achieve a Circular Economy* (pp. 224–246). IGI Global.

Pal, K. (2024). Information Technology-Driven Industrial Revolution and Supply Chain Operations Sustainability. In M. R. Khan, N. R. Khan, & N. Z. Jhanihi (Eds.), *Digital Transformation for Improved Industry and Supply Chain Performance* (pp. 108–132). IGI Global.

Pena-opez, I. (2005). Internet Report 2005. *Internet of Things : Engineering Cyber Physical Human Systems*, 2005.

Puttonen, J., Lobov, A., & Lastra, J. L. M. (2013). Semantics-Based Composition of Factory Automation Processes Encapsulated by Web Services. *IEEE Transactions on Industrial Informatics*, *9*(4), 2349–2359. doi:10.1109/TII.2012.2220554

Puttonen, J., Lobov, A., Soto, M. A. C., & Lastra, J. L. M. (2010). A Semantic Web Services-based approach for production systems control. *Advanced Engineering Informatics*, *24*(3), 285–299. doi:10.1016/j.aei.2010.05.012

Russomanno, D. J., Kothari, C. R., & Thomas, O. A. (2005). Building a Sensor Ontology: A Practical Approach Leveraging ISO and OGC Models. *2005 International Conference on Artificial Intelligence,* (pp. 637-643). IEEE.

Ruta, M., Colucci, S., Scioscia, F., Di Sciascio, E., & Donini, F. M. (2011). Finding commonalities in RFID semantic streams. *Procedia Computer Science, 5,* 857–864. doi:10.1016/j.procs.2011.07.118

Said, O., & Masud, M. (2013). Towards internet of things: Survey and future vision. *International Journal of Computer Networks, 5*(1), 1–17.

Sarkar, I., Adhikari, M., Kumar, N., & Kumar, S. (2021). Dynamic task placement for deadline-aware IoT applications in federated for networks. *IEEE Internet of Things Journal, 1,* 2021.

Sarkar, T. K., Mailloux, R. J., Oliner, A. A., Salazar-Palma, M., & Sengupta, D. L. (2006). *History of Wireless.* John Wiley & Sons Inc. doi:10.1002/0471783021

Schneider, M., Hippchen, B., Abeck, S., Jacoby, M., & Herzog, R. (2018). Enabling IoT platform interoperability using a systematic development approach by example. In *Proceedings of the Global Internet of Things Summit (GIoTS'18).* IEEE. 10.1109/GIOTS.2018.8534549

Seiger, R., Huber, S., & Schlegel, T. (2018). Toward an execution system for self-healing workflows in cyber-physical systems. *Software & Systems Modeling, 17*(2), 551–572. doi:10.1007/s10270-016-0551-z

Stojmenovic, I., & Wen, S. (2014). "The fog computing paradigm: scenarios and security issues," in Proceedings of the Federated Conference on Computer Science and Information Systems (FedCSIS' 14), 1–8, IEEE, Warsaw, Poland, September 2014. 10.15439/2014F503

Swaminathan, J. M. (2000). *Supply chain management.* International Encyclopedia of the Social and Behavioural Sciences, Elsevier Sciences.

Varelas, G., Voutsakist, E., Raftopoulout, P., Petrakis, E. G. M., & Milios, E. (2005). Semantic Similarity methods in WordNet and their application to information retrieval on the Web. *Proceedings of the 7th annual ACM international workshop on web information and data management.* ACM. 10.1145/1097047.1097051

Vermesan, O., Friess, P., & Guillemin, P. (2011). "Internet of things strategic research roadmap," in Internet of Things. *Global Technological and Societal Trends, 1,* 9–52.

Weyrich, M., & Ebert, C. (2016). Reference architectures for the internet of things. *IEEE Software, 33*(1), 112–116. doi:10.1109/MS.2016.20

Wu, M., Lu, T. J., Ling, F. Y., Sun, J., & Du, H. Y. (2010). Research on the architecture of internet of things. *Proceedings of the 3rd International Conference on Advanced Computer Theory and Engineering, 10*(5). IEEE.

Yoshioka, N., Honiden, S., & Fnkelstein, A. (2004). Security Patterns: A Method for Constructing Secure and Efficient Inter-Company Coordination Systems. *Proceedings of Enterprise Distributed Object Computing Conference 2004 (EDOC'04).* IEEE.. 10.1109/EDOC.2004.1342507

Zhao, J., & Kumar, V. V. (2021). *Handbook of Research on Innovations and Applications of AI, IoT, and Cognitive Technologies*. IGI Global Publication. doi:10.4018/978-1-7998-6870-5

## KEY TERMS AND DEFINITIONS

**Description Logic:** Description logics (DL) are a family of formal knowledge representation languages. Many DLs are more expressive than propositional logic, but less expressive than first-order logic.

**EPC:** Electronic Product Code (EPC), is a low-cost RFID tag designed for consumer products as a replacement for the universal product code (UPC).

**Internet of Things:** Internet of Things (IoT) means networks of things, software, sensors, network connectivity, and embedded 'things or physical objects. It collects or exchanges data. IoT makes objects sensed or controlled through a network infrastructure, supports integration between physical real world and automated information systems, and brings various effects such as improved productivity or economy in manufacturing industries.

**Ontology:** Information sharing among supply chain business partners using information systems is an important enabler for supply chain management. There are diverse types of data to be shared across the supply chain, namely – *order, inventory, shipment*, and *customer service*. Consequently, information about these issues needs to be shared to achieve efficiency and effectiveness in supply chain management. In this way, information-sharing activities require that human and / or machine agents agree on common and explicit business-related concepts (the shared conceptualization among hardware / software-agents, customers, and service providers) are known as explicit ontologies; and this help to exchange data and derived knowledge out of the data to achieve collaborative goals of business operations.

**RFID Reader:** An RFID transceiver, providing real and access to RFID tags information.

**RFID Tag:** An RFID tag (or transponder), typically consisting of an RF coupling element and a microchip that carries identifying data. Tag functionality may range from simple identification to being able to form an ad hoc network.

**Semantic Web Service:** A Semantic Web Service, like conventional web services, is the server end of a client-server system for machine-to-machine interaction via the Web. Semantic services are a component of the semantic Web because they use mark-up which makes data machine-readable in a detailed and sophisticated way (as compared with human-readable HTML which is usually not easily "understood" by computer programs).

**Supply Chain Management:** Supply chain management encompasses the planning and management of all activities involved in sourcing, procurement, manufacturing, and distribution. Importantly, it also includes coordination and collaboration with channel partners, which can be suppliers, intermediaries, third-party service providers, and customers. Supply chain management integrates supply and demand management within and across companies.

**Web Ontology Language (OWL):** The Web Ontology Language (OWL) is a semantic mark-up language for publishing and sharing ontologies on the Web. OWL is developed as a vocabulary extension of RDF (the Resource Description Framework) and is derived from the DAML + OIL Web Ontology Language.

# Chapter 14

# An Analysis of Internet-of-Things-Based Fire Detection and Alert Systems

**Digvijay Pandey**
https://orcid.org/0000-0003-0353-174X
*Department of Technical Education, Government of Uttar Pradesh, India*

**Vinay Kumar Nassa**
https://orcid.org/0000-0002-9606-7570
*Department of Information Communication Technology, Tecnia Institute of Advanced Studies, India*

**Binay Kumar Pandey**
https://orcid.org/0000-0002-4041-1213
*Department of Information Technology, College of Technology, Govind Ballabh Pant University of Agriculture and Technology, Pantnagar, India*

**Darshan A. Mahajan**
https://orcid.org/0000-0002-1239-6343
*NICMAR University, India*

**Pawan Kumar Patidar**
https://orcid.org/0009-0008-8115-9094
*Swami Keshvanand Institute of Technology, Management, and Gramothan, India*

**Pankaj Dadheech**
https://orcid.org/0000-0001-5783-1989
*Swami Keshvanand Institute of Technology, Management, and Gramothan, India*

**A. Shaji George**
https://orcid.org/0000-0002-8677-3682
*Business System Department, Almarai Company, TSM, Riyadh, Saudi Arabia*

## ABSTRACT

*One of the most valuable resources is the forest, home to many animals and plants. Forest fire agencies worldwide have studied forest fire prevention and detection. Worldwide, natural and man-made calamities occur. Forest fires are environmental tragedies. The dense forest fire devours everything in its path. This research examines the forest fire detection and alert system to detect fires early. This research identifies forest fires before they spread to safeguard wildlife and natural resources. An Arduino microcontroller, flame sensor, ultrasonic sensor, thermistor, smoke sensor, buzzer, and GPRS are in every IoT (internet of things) device. Each IoT sensor records sensor values in the thing speak cloud. The cloud storage may pick and map forest fire threats by eliminating features from the input. MLP mapping maps forest fire danger, while AROC maps forest fire hazard. GPRS delivers cloud-based SMS warnings. Finally, forest department officials may interact.*

DOI: 10.4018/979-8-3693-1335-0.ch014

## INTRODUCTION

The forest, which is home to a wide variety of plant and animal species, is one of the most precious resources. Research on forest fire detection and prevention has been conducted by forest fire departments all over the world. Disasters, both natural and those caused by humans, can strike everywhere in the world. Forest fires are a terrible tragedy for the ecology. The raging blaze in the dense forest consumes everything in its path. This research helps to protect wildlife and natural resources by locating forest fires before they spread throughout the surrounding area (Pandey, B. K.,et al.,2023). The research being conducted looks into the forest fire detection and alert system to determine how early fires can be detected. Every device that is part of the Internet of Things (IoT) has the following components: an Arduino microcontroller, flame sensor, ultrasonic sensor, thermistor, smoke sensor, buzzer, and GPRS. Every Internet of Things sensor uploads its readings to the cloud storage service Thing Speak(Raja, D.,et al.,2024). The cloud storage might identify and map potential forest fire hazards by filtering out certain elements of the incoming data. MLP Mapping maps the risk of forest fires, whereas AROC maps the potential for forest fires. SMS alerts can be delivered across the cloud using GPRS. Finally, officials from the forest department may engage with a fire alarm that pinpoints the location of the forest.

The Internet of Things (IoT) refers to a network of sensors and other internet-connected devices that exchange data with one another. In the future, the Internet of Things (IoT) will have the most significant impact on the logic of businesses. The Internet of Things is used in a variety of contexts, including smart homes, agriculture, industries, wearables, cars, forests, and healthcare (Gomez-Chabla, R.,et al.,2019). Both the administration of forests and the Internet of Things are quite interesting to us (Pandey, B. K.,et al.,2011).

The term "forest" refers to a large area that is home to a diverse collection of plant and animal life. Over thirty percent of the globe is covered in forest. Our economy is dependent on a great number of different species, and a large number of people find employment in the forest (Meslie, Y.,et al.,2021). The forest's most important resource is the air that circulates throughout it. It rids our bodies of harmful substances and shields us from harm (Jaya Lakshmi, G.,et al.,2024). People put themselves in danger when they carelessly set forest fires ablaze. Forest fires that are started by accident might have a negative impact on the food chain (Kumar, M. S.,et al.,2021). Destruction of forest habitat is harmful to both people and wildlife.

There are two primary causes of forest fires: Reasons related to nature: Lightning strikes in locations with low humidity have the potential to start forest fires. Rain is the only thing that can put an end to it and repair the damage done to the ground if it spread across a vast area. 2. Causes that are not natural Accidents caused by carelessness or ignorance, such as fires, cigarette and electrical sparks, can result in extensive property damage (Saxena, A.,et al.,2021).

The density of the forest, the average wind speed, and the temperature all had a high correlation with the daily forest fire rates. Even at low temperature densities, the forest has a substantial impact on the amount of area that has burned (Daoud, M. K.,et al.,2024). According to the National Interagency Centre's report from 2016, forest fires had an impact on 65,575 different species of animals. In 2017, the fire was responsible for the destruction of about 10 million acres and 71,499 additional structures (Jadon, A.,et al.,2019).

Deforestation in the Amazon rainforest served as the impetus for our idea. The majority of fires in the Amazon were caused by mining, forestry, and agriculture. After cutting down trees, farmers often practise slash-and-burn agriculture in order to help the soil regain its nutrients (Daoud, M. K.,et al.,2023).

There were hundreds of people fighting flames in the southern Amazon rainforest. The smoke and soot produced by forest fires contribute to air pollution, endangering species and causing damage to one of the most important carbon sinks on the world (Boopathi, S.,et al.,2023).

A Comparison of Different Techniques for the Detection and Emergency Warning of Forest Fires Based on the Internet of Things

An unforeseeable catastrophe that needs to be contained as soon as possible is a forest fire since it has the potential to spread. Several authors, each with their own approach and set of limitations (Pandey, D.,et al.,2022), recommend investigations and trials to prevent problems, identify them, and alert people to their existence. A review of ten articles focused on forest firefighting strategies was conducted (Kumar Pandey, B.,et al.,2021).

It was proposed by (Jayaram et al., 2019) that the Arduino was the most crucial part of the system because it was the one that controlled everything. After that, sensors for temperature and smoke were connected. Temperature sensors detect the temperature, while smoke sensors measure the amount of smoke in the forest. The apparatus that measures temperature indicates that the flames will continue to get higher. This Arduino board comes equipped with Wi-Fi and GPS modules. In this scenario, having GPS to track position and Wi-Fi to upload data to the cloud could prove to be useful. In order to connect to the cloud and receive blaze notifications, the user needs to have access to the internet. Concerning a crisis, this system is ineffective.

Monitoring and warning were two methods that Saikumar and his colleagues (T. Saikumar,et al.,2019) utilised to protect trees from being destroyed by forest fires. A system for detecting fires that is comprised of a buzzer, temperature and smoke sensors, and an Arduino Board. The initial text message was delivered to the GSM phone number that was provided. The temperature sensor showed both extremely high and extremely low temperatures. These pieces of information were represented by icons. If the blaze range is high, the display will recognise a bushfire; however, it will not identify a forest fire if the range is medium. Flame sensors are used to measure the range of the flame. One of the many advantages of using this strategy is how quickly SMS warnings can be delivered. On the other hand, the position of the node is irrelevant. It may be difficult for teams to identify opportunities (Pandey, D., & Pandey, B. K. (2022)).

Divya et al., 2019 recommended employing wireless devices in conjunction with sensors that were anchored to structures in order to detect fires. The base station was responsible for data reception and replication. The approach identifies forest fires using data collected via wireless mesh reticulation. Temperature and smoke sensors are placed at varying distances from one another to conceal the forest while simultaneously evaluating $CO_2$ levels and monitoring temperatures. Data is transmitted to the microcontroller by the sensors (Abdulkarim, Y. I.,et al.,2023). Due to the fact that the flame sensor's reactions showed it was more sensitive to regular light, it was put to use in the detection of fire alarms (Vinodhini, V.,et al., 2022). It is pointless to transfer the data to the cloud once again if there is a fire occurring. Because it is individualised, this technology provides information in a timely manner. Because not all microcontrollers offer analogue I/O, the transfer of sensor data can be difficult (Pramanik, S.,et al.,2023).

In Dubey et al.,2019 study, the authors interfaced the Node MCU module with temperature and smoke sensors. The smoke sensor picked up on any traces of combustion or smoke from a fire, while the temperature sensor picked up on any signs of heat. The buzzer on the Arduino sounded the alarm. When it discovered that there was a fire, it immediately notified the customer via the Ethernet module. Even though the Wi-Fi module is fantastic for connecting to the cloud, it does not offer SMS notifications in the event of a fire. A significant advantage was the noise-resistant nature of the Ethernet cable that was utilised in the connection of the systems. Due to its resistance to noise, data quality is maintained at a high level. The most significant negative is that it is difficult to determine whether an Ethernet problem is being caused by a cable or a network node. Particularly in regions that are heavily covered in forest.

In their study, Kameswari et al.,2018 suggested using an Arduino Uno equipped with a temperature sensor, a humidity sensor, and an LCD display. This model was able to detect fire as well as humidity. When the predetermined levels of temperature and humidity were attained, an alarm for fire was triggered. Data is transmitted from Arduino Uno sensors to the controller. The system cannot detect fires because it does not have a smoke or fire sensor. This is a limitation. Sensors that measure temperature and humidity have the potential to trigger false fire alarms (Pandey, D.,et al.,2021).

An Arduino Uno, a temperature sensor, a smoke sensor, and a buzzer are the components that should be used in order to construct a smoke alarm, as recommended by (Sharma et al., 2017). This model was responsible for creating the fire alarm. Temperature sensors detect the heat, while smoke sensors detect the presence of fire smoke. When there is a problem, the buzzer on the Arduino board will sound. If the temperature is at an unsafe level, the alert will sound. The warning will be turned down as the temperature returns to normal and the amount of smoke diminishes. There are no cloud and remote warning solutions currently available (Pandey, J. K.,et al.,2022).

IoT research and applications in the flame sector were discussed in (Vijayalakshmi et al, 2017) article. This study makes environmental monitoring and business practises more applicable in the real world. In the sensing layer, wireless sensor network (WSN) sensors, radio frequency identification (RFID) tags on products, and video nodes tracked fires and products (Muniandi, B.,et al.,2019). The network layer was responsible for connecting the mobile and sensor networks. The research presented various potential solutions to forest fires. However, it seems as though the implementation will be expensive.

The improved fire detection and alarm system developed by (Singh et al., 2017) was predicated on a real-time surveillance system that detected fire, captured webcam photographs of the scene, and displayed them on a screen (Ahmad, A. Y. B.,et al.,2023). When the feedback from this sensor reaches its maximum value, whether it is heat, moisture, carbon dioxide, or fire, the control system will transmit signals to GSM, and the second controller will activate a relay to turn on the screen. The use of the wireless camera for live surveillance is now being tested (Ahmad, A. Y. B.,et al.,2024). This smoke detector was made with an Arduino UNO. The capability of this system to issue remote fire warnings is essential. Even in the event of a fire, the cameras are still capable of losing data and video.

Patel and his colleagues (Patel, D. et al., 2016) presented a system with a primary focus on security and automation. The need of designing for privacy and automating both the home and the office was emphasised. In this piece, the authors make an attempt to enhance safety by combining novel approaches with conventionally acceptable levels of automated home and commercial security. Users will have access to smoke, gas, temperature, and fire sensors in their homes and places of employment if a basic circuit board is configured in the appropriate manner. Thermocouples are utilised extensively inside fire suppression systems. This strategy is successfully utilised by the local law enforcement. To store the massive volumes of data generated by a forest, a cloud-based storage solution is required.

Wang et al., 2015 presented their idea for a GIS-WSN system. It was utilised in the planning and carrying out of the project. In order to monitor CO and smog concentrations, identify flames, maintain temperature and relative humidity, and create numerous sensing nodes for regional wireless sensor networks, sensors were utilised. In addition to these tasks, the sensors also maintained temperature and humidity. These particulars and the location are transmitted to the monitoring station (Zaidi, A., et al., (2023)). Because of this, using a GIS will be more difficult and expensive. The existing approaches for

*Table 1. Various forest fire detection system*

| Author's name | Problem statement | Proposed method | Advantages | Limitations |
|---|---|---|---|---|
| Jayaram.K et al., 2019 | Fire spreads low humidity days which destroyed trees. | A solar panel with a battery served as the power source. | Measuring the rainfall, humidity. | Any time the solar panel is used, it must be charged. |
| T. Saikumar et.al., 2019 | The problem of movement of animals from forest area to residential area | Magnetic transducer, involving an iron focus with damage twist. | The SMS alert was quick. | This might be difficult for the fire fighters to track the location. |
| Divya.A et al., 2019 | Forest fires pose a significant threat to the environment on a regular basis | The strategy was devised using information from wireless sensor reticulation. | Information is sent in a flexible manner. | The drawback was interfacing issues will occur in micro-controllers. |
| Dubey .V et al., 2019 | According to a poll, the forest has lost about 80% of its value | Node MCU module was interfaced with the sensors. | The cable used to interconnect the systems in Ethernet was robust to noise. | There is problem in Ethernet during the passing of information; it is difficult to troubleshoot in forest area. |
| Kameswari.G et al., 2018 | One of the most dangerous applications was forest fire detection | Fire and humidity was sensed using Arduino Uno. | Ability of accessing the database from any area. | By using only temperature and humidity sensor, there can be false alarms of a fire. |
| Sharma.A.K et al., 2017 | It has been found in a survey that 80% losses caused due to fire would have been avoided if the fire was detected immediately | Detector of fire was built using Arduino Uno which was interfaced with sensors. | Fire surveillance was quick and reliable. | It may give a false alarm. |
| Vijayalakshmi.S et al., 2017 | Prevention of fire and fire risk level control difficulty are increased day by day. | Features of IoT technology and Wireless Sensor Network technology. | It promotes army construction development. | The overall implementation seems costly. |
| Singh.D et al., 2017 | The goal is to detect fire as soon as possible. | The fire was detected and images were captured using a camera. | It was low cost, power efficient. | During a fire, the cameras might lose connection. |
| Patel.D et al., 2016 | A fire outbreak was a major tragedy that must be avoided due to the potential loss of lives and property. | Thermistors were the most commonly used sensing device in fire detection. | It was cost effective and medium coverage method. | A cloud platform was needed to store enormous volume of information sensed. |
| Wang.M et al., 2015 | Burning weeds in the field start forest fires. | CO and smog concentrations are measured using wireless sensor networks. | The process of sending information was more flexible. | The usage of GIS becomes complex and costly. |

detecting forest fires are outlined in Table 1, together with an analysis of their advantages and disadvantages (Kirubasri, G.,et al.,2021).

## CONCLUSION

Each Internet of Things (IoT) device in the system contains a variety of components, some of which are an Arduino microcontroller, a flame sensor, an ultrasonic sensor, a thermistor, a smoke sensor, and a buzzer. Other components include a smoke sensor and an ultrasonic sensor. GPRS is also included as an additional component. Every sensor that is a component of the Internet of Things (IoT) collects sensor data and then transmits that data to the cloud storage that is provided by Thing Speak. The feature selection and mapping of potential forest fire threats can be done in the cloud storage after the features have been removed from the input. MLP Mapping and Advanced Relative Operating Characteristics (AROC), which stands for "Advanced Relative Operating Characteristics," are two techniques that are used in the mapping of forest fire danger and hazard, respectively. AROC stands for "Advanced Relative Operating Characteristics." The application of MLP Mapping may be seen here for the function of mapping potential forest fire dangers. GPRS ensures that SMS notifications received from the cloud are delivered to the device that the user is using. A fire alarm that specifically identifies the area of the forest will make it possible for authorised members of the forest department to be contacted, which is surely not the least important aspect of the situation.

## REFERENCES

Abdulkarim, Y. I., Awl, H. N., Muhammadsharif, F. F., Saeed, S. R., Sidiq, K. R., Khasraw, S. S., ... Pandey, D. (2023). Metamaterial-based sensors loaded corona-shaped resonator for COVID-19 detection by using microwave techniques. *Plasmonics*, 1–16.

Ahmad, A., Abusaimeh, H., Rababah, A., Alqsass, M., Al-Olima, N., & Hamdan, M. (2024). Assessment of effects in advances of accounting technologies on quality financial reports in Jordanian public sector. *Uncertain Supply Chain Management*, *12*(1), 133–142. doi:10.5267/j.uscm.2023.10.011

Ahmad, A. Y. B. (2024). Ethical implications of artificial intelligence in accounting: A framework for responsible ai adoption in multinational corporations in Jordan. *International Journal of Data and Network Science*, *8*(1), 401–414. doi:10.5267/j.ijdns.2023.9.014

Ahmad, A. Y. B. (2024). Firm Determinants that Influences Implementation of Accounting Technologies in Business Organizations. *WSEAS Transactions on Business and Economics*, *21*, 1–11. doi:10.37394/23207.2024.21.1

Ahmad, A. Y. B., Gongada, T. N., Shrivastava, G., Gabbi, R. S., Islam, S., & Nagaraju, K. (2023). E-Commerce Trend Analysis and Management for Industry 5.0 using User Data Analysis. *International Journal of Intelligent Systems and Applications in Engineering*, *11*(11s), 135–150.

Ahmad, A. Y. B., Hannoon, A., Al-Daoud, K. I., Abu-Alsondos, I. A., & Al-Qaisieh, M. S. (2023). Assessment of Cloud Based Accounting Technology Adoption and Business Performance. *Kurdish Studies, 11*(3).

Ahmad, A. Y. B., Kumari, D. K., Shukla, A., Deepak, A., Chandnani, M., Pundir, S., & Shrivastava, A. (2024). Framework for Cloud Based Document Management System with Institutional Schema of Database. *International Journal of Intelligent Systems and Applications in Engineering*, *12*(3s), 672–678.

Ahmad, A. Y. B., Kumari, D. K., Shukla, A., Deepak, A., Chandnani, M., Pundir, S., & Shrivastava, A. (2024). Framework for Cloud Based Document Management System with Institutional Schema of Database. *International Journal of Intelligent Systems and Applications in Engineering*, *12*(3s), 672–678.

Ahmad, A. Y. B., & Kumari, S. S. (2023). Blockchain Implementation in Financial Sector and Cyber Security System. *2023 International Conference on Artificial Intelligence and Smart Communication (AISC)*, Greater Noida, India. 10.1109/AISC56616.2023.10085045

Ahmad, A. Y. B., Tiwari, A., Nayeem, M. A., Biswal, B. K., Satapathy, D. P., Kulshreshtha, K., & Bordoloi, D. (2024). Artificial Intelligence Perspective Framework of the Smart Finance and Accounting Management Model. *International Journal of Intelligent Systems and Applications in Engineering*, *12*(4s), 586–594.

Ahmad, A. Y. B., William, P., Uike, D., Murgai, A., Bajaj, K. K., Deepak, A., & Shrivastava, A. (2024). Framework for Sustainable Energy Management using Smart Grid Panels Integrated with Machine Learning and IOT based Approach. *International Journal of Intelligent Systems and Applications in Engineering*, *12*(2s), 581–590.

Boopathi, S., Pandey, B. K., & Pandey, D. (2023). Advances in artificial intelligence for image processing: techniques, applications, and optimization. In *Handbook of research on thrust technologies' effect on image processing* (pp. 73–95). IGI Global. doi:10.4018/978-1-6684-8618-4.ch006

Daoud, M. K. (2023). The Relationship Between Mobile Marketing and Customer Perceptions in Jordanian Commercial Banks: The Electronic Quality as A Mediator Variable. *IJMST, 10* (2).

Daoud, M. K., Al-Qeed, M., Ahmad, A. Y. B., & Al-Gasawneh, J. A. (2023). Mobile Marketing: Exploring the Efficacy of User-Centric Strategies for Enhanced Consumer Engagement and Conversion Rates. *International Journal of Membrane Science and Technology*, *10*(2), 1252–1262. doi:10.15379/ijmst.vi.1425

Daoud, M. K., Taha, S., Al-Qeed, M., Alsafadi, Y., Ahmad, A., & Allahham, M. (2024). EcoConnect: Guiding environmental awareness via digital marketing approaches. *International Journal of Data and Network Science, 8*(1), 235-242.

Divya, A., Kavithanjali, T., & Dharshini, P. (2019, March). IoT Enabled Forest Fire Detection and Early Warning System. In *2019 IEEE International Conference on System, Computation, Automation and Networking (ICSCAN)* (pp. 1-5). IEEE. 10.1109/ICSCAN.2019.8878808

Dubey, V., Kumar, P., & Chauhan, N. (2019). Forest fire detection system using IoT and artificial neural network. In *International Conference on Innovative Computing and Communications* (pp. 323-337). Springer, Singapore. 10.1007/978-981-13-2324-9_33

Gomez-Chabla, R., Real-Aviles, K., Moran, C., Grijalva, P., & Recalde, T. (2019, January). IoT applications in agriculture: A systematic literature review. In *2nd International Conference on ICTs in Agronomy and Environment* (pp. 68-76).Springer, Cham. 10.1007/978-3-030-10728-4_8

Jadon, A., Omama, M., Varshney, A., Ansari, M. S., & Sharma, R. (2019). Firenet: A specialized lightweight fire & smoke detection model for real-time iot applications. arXiv preprint arXiv:1905.11922.

Jaya Lakshmi. G., Pandey, D., Pandey, B. K., Kaur, P., Mahajan, D. A., & Dari, S. S. (2024). Smart Big Data Collection for Intelligent Supply Chain Improvement. In AI and Machine Learning Impacts in Intelligent Supply Chain (pp. 180-195). IGI Global.

Jayaram, K., Janani, K., Jeyaguru, R., Kumaresh, R., & Muralidharan, N. (2019, March). Forest Fire Alerting System With GPS Co-ordinates Using IoT. In 2019, *5th International Conference on Advanced Computing & Communication Systems (ICACCS)* (pp. 488-491). IEEE. 10.1109/ICACCS.2019.8728383

Kameswari, G., Himaja, B., & Tirumala, N. (2018). Wild Fire Detection using. *Internet of Things : Engineering Cyber Physical Human Systems.*

Kirubasri, G., Sankar, S., Pandey, D., Pandey, B. K., Singh, H., & Anand, R. (2021, September). A Recent Survey on 6G Vehicular Technology, Applications and Challenges. In *2021 9th International Conference on Reliability, Infocom Technologies and Optimization (Trends and Future Directions) (ICRITO)* (pp. 1-5). IEEE.

Kumar, M. S., Sankar, S., Nassa, V. K., Pandey, D., Pandey, B. K., & Enbeyle, W. (2021). Innovation and creativity for data mining using computational statistics. In *Methodologies and Applications of Computational Statistics for Machine Intelligence* (pp. 223–240). IGI Global. doi:10.4018/978-1-7998-7701-1.ch012

Kumar Pandey, B., Pandey, D., Nassa, V. K., Ahmad, T., Singh, C., George, A. S., & Wakchaure, M. A. (2021). Encryption and steganography-based text extraction in IoT using the EWCTS optimizer. *Imaging Science Journal*, *69*(1-4), 38–56. doi:10.1080/13682199.2022.2146885

Meslie, Y., Enbeyle, W., Pandey, B. K., Pramanik, S., Pandey, D., Dadeech, P., & Saini, A. (2021). Machine intelligence-based trend analysis of COVID-19 for total daily confirmed cases in Asia and Africa. In *Methodologies and Applications of Computational Statistics for Machine Intelligence* (pp. 164–185). IGI Global. doi:10.4018/978-1-7998-7701-1.ch009

Muniandi, B., Huang, C., Kuo, C., Yang, T., Chen, K., Lin, Y., Lin, S., & Tsai, T. (2019). A 97% maximum efficiency fully automated control turbo boost topology for battery chargers. *IEEE Transactions on Circuits and Systems. I, Regular Papers*, *66*(11), 4516–4527. doi:10.1109/TCSI.2019.2925374

Pandey, B. K., Pandey, D., Alkhafaji, M. A., Güneşer, M. T., & Şeker, C. (2023). A Reliable Transmission and Extraction of Textual Information Using Keyless Encryption, Steganography, and Deep Algorithm with Cuckoo Optimization. In *Micro-Electronics and Telecommunication Engineering: Proceedings of 6th ICMETE 2022* (pp. 629–636). Springer Nature Singapore. doi:10.1007/978-981-19-9512-5_57

Pandey, B. K., Pandey, S. K., & Pandey, D. (2011). A survey of bioinformatics applications on parallel architectures. *International Journal of Computer Applications*, *23*(4), 21–25. doi:10.5120/2877-3744

Pandey, D., Ogunmola, G. A., Enbeyle, W., Abdullahi, M., Pandey, B. K., & Pramanik, S. (2021). COVID-19: A framework for effective delivering of online classes during lockdown. *Human Arenas,* 1-15.

Pandey, D., & Pandey, B. K. (2022). An efficient deep neural network with adaptive galactic swarm optimization for complex image text extraction. In *Process Mining Techniques for Pattern Recognition* (pp. 121–137). CRC Press. doi:10.1201/9781003169550-10

Pandey, D., Wairya, S., Sharma, M., Gupta, A. K., Kakkar, R., & Pandey, B. K. (2022). An approach for object tracking, categorization, and autopilot guidance for passive homing missiles. *Aerospace Systems*, *5*(4), 553–566. doi:10.1007/s42401-022-00150-0

Pandey, J. K., Jain, R., Dilip, R., Kumbhkar, M., Jaiswal, S., Pandey, B. K., & Pandey, D. (2022). Investigating Role of IoT in the Development of Smart Application for Security Enhancement. In *IoT Based Smart Applications* (pp. 219–243). Springer International Publishing.

Patel, D., Gosh, A., Banerjee, S. J., & Jana, D. (2016). GSM Based Fire Sensor Alarm Using Arduino. *International Journal of Scientific and Engineering Research*, *7*(4).

Pramanik, S., Pandey, D., Joardar, S., Niranjanamurthy, M., Pandey, B. K., & Kaur, J. (2023, October). An overview of IoT privacy and security in smart cities. In AIP Conference Proceedings (Vol. 2495, No. 1). AIP Publishing. doi:10.1063/5.0123511

Raja, D., Kumar, D. R., Santhiyakumari, N., Kumarganesh, S., Sagayam, K. M., Thiyaneswaran, B., Pandey, B. K., & Pandey, D. (2024). A compact dual-feed wide-band slotted antenna for future wireless applications. *Analog Integrated Circuits and Signal Processing*, *118*(2), 1–15. doi:10.1007/s10470-023-02233-0

Saikumar, T., & Sriramya, P. (2019). *IoT Enabled Forest Fire Detection and Altering the Authorities. International Journal of Recent Technology and Engineering*. IJRTE.

Savitha, N., & Malathi, S. (2018, October). A survey on fire safety measures for industry safety using IOT. In *2018 3rd International Conference on Communication and Electronics Systems (ICCES)* (pp. 1199-1205).IEEE. 10.1109/CESYS.2018.8723930

Saxena, A., Sharma, N. K., Pandey, D., & Pandey, B. K. (2021). Influence of tourists satisfaction on future behavioral intentions with special reference to desert triangle of Rajasthan. *Augmented Human Research*, *6*(1), 1–9. doi:10.1007/s41133-021-00052-4

Sharma, A. K., Ansari, M. F. R., Siddiqui, M. F., & Baig, M. A. (2017). IOT enabled forest fire detection and online monitoring system. [IJCTER]. *Int. J. Curr. Trends Eng. Res.*, *3*, 50–54.

Shinde, S. R., Karode, A. H., & Suralkar, D. S. (2017). Review on IOT Based Environment Monitoring System. *International Journal of Electronics and Communication Engineering and Technology*, *8*(2).

Singh, D., Sharma, N., Gupta, M., & Sharma, S. (2017). Development of system for early fire detection using ArduinoUNO. *International Journal of Engineering Science*, 10857.

Vijayalakshmi, S. R., & Muruganand, S. (2017, February). A survey of Internet of Things in fire detection and fire industries. In *2017 International Conference on I-SMAC (IoT in Social, Mobile, Analytics and Cloud)(I-SMAC)* (pp. 703-707). IEEE. 10.1109/I-SMAC.2017.8058270

Vinodhini, V., Kumar, M. S., Sankar, S., Pandey, D., Pandey, B. K., & Nassa, V. K. (2022). IoT-based early forest fire detection using MLP and AROC method. *International Journal of Global Warming*, *27*(1), 55–70. doi:10.1504/IJGW.2022.122794

Wang, M., Liu, H., Chen, F., & Liu, J. (2015, August). Forest fire warning system based on GIS and WSNs. In *2015 4th International Conference on Advanced Information Technology and Sensor Application (AITS)* (pp. 3-6). IEEE. 10.1109/AITS.2015.8

Xiaojun, C., Xianpeng, L., &Peng, X. (2015, January). IOT-based air pollution monitoring and forecasting system. In *2015 international conference on computer and computational sciences (ICCCS)* (pp. 257-260).IEEE.

Zaidi, A. (2023). Predicting wildfires in Algerian forests using machine learning models. *Heliyon*, *9*(7), e18064. Advance online publication. doi:10.1016/j.heliyon.2023.e18064 PMID:37519679

Zaidi, A. (2023). Smart Implementation of Industrial Internet of Things using Embedded Mechatronic System. *IEEE Embedded Systems Letters*. Advance online publication. doi:10.1109/LES.2023.3309985

Zaidi, A., Ajibade, S.-S. M., Musa, M., & Bekun, F. V. (2023). New Insights into the Research Landscape on the Application of Artificial Intelligence in Sustainable Smart Cities: A Bibliometric Mapping and Network Analysis Approach. *International Journal of Energy Economics and Policy.*, *13*(4), 287–299. doi:10.32479/ijeep.14683

Zaidi, A., & Al-Harbi, O. (2023). Statistical Analysis of a Linear Multi-Step Numerical Treatment. *Journal of Statistics Applications & Probability*, *12*(1), 26. doi:10.18576/jsap/120126

Zaidi, A., & Al Luhayb, A. (2023). Two Statistical Approaches to Justify the Use of the Logistic Function in Binary Logistic Regression. *Mathematical Problems in Engineering*, *2023*, 5525675. doi:10.1155/2023/5525675

# Chapter 15
# Building Information Modeling:
## Harnessing Innovations for Professional Proficiency

**George W. Kennedy**
*Building Technology Unit, Department of Technical Education, Akwa Ibom State College of Education, Nigeria*

**Sam Edidiong Sunday**
*Donfang International Corporation, Calabar, Nigeria*

**Digvijay Pandey**
https://orcid.org/0000-0003-0353-174X
*Department of Technical Education, Government of Uttar Pradesh, India*

**Awi Dennis Kabari**
*Civil Engineering Department, Delta State University of Science and Technology, Ozoro, India*

**Binay Kumar Pandey**
https://orcid.org/0000-0002-4041-1213
*Department of Information Technology, College of Technology, Govind Ballabh Pant University of Agriculture and Technology, Pantnagar, India*

**A. Shaji George**
https://orcid.org/0000-0002-8677-3682
*Business System Department, Almarai Company, TSM, Riyadh, Saudi Arabia*

**Pankaj Dadheech**
https://orcid.org/0000-0001-5783-1989
*Swami Keshvanand Institute of Technology, Management, and Gramothan, India*

## ABSTRACT

*Building information modeling (BIM) has changed architecture, engineering, and construction professionals' built environment conceptualization, design, building, and management. BIM tool evolution, principles, components, modelling methodology and software, user interface and basic functions, construction and post-construction benefits, safety, and intellectual property are covered in this chapter. BIM integrates geometry, spatial relationships, geographical data, and material qualities for interdisciplinary communication and decision-making. This synergy boosts project efficiency by improving design accuracy, timelines, and conflict resolution. Technical knowledge and professional teamwork are needed to use BIM effectively. BIM competency needs a paradigm shift in schooling and professional growth. As AEC digitises, BIM's potential is crucial for professional proficiency.*

DOI: 10.4018/979-8-3693-1335-0.ch015

## 1. INTRODUCTION

Historically, the construction industry has been closely intertwined with processes of industrialization and urbanization (Osuizugbo, 2020). The construction industry encompassing the Architectural, Engineering and Construction sectors, this industry has consistently played a significant role in the socio-economic development of nations (Ullah et al., 2019). The construction industry is saddled with the task of providing infrastructures ranging from social, amenities, and housing to socio-economic development and industrialization (Ruya et al., 2018). Umar (2019) identified two vital roles of the construction project which are the specification of the resulting product (design information) and the initiation and control of the activities required for constructing the facility (management information). Nowadays, building projects (Raghavan, R. et al., 2022) have become increasingly complex and challenging, leading project teams to encounter unprecedented changes (Yusuf, 2020). Planning and constructing built facilities entail the collaboration of diverse stakeholders with various areas of expertise. Throughout this process, the continuous resolution and circulation of construction data among these stakeholders emerge as crucial determinants of success (Bormann et al., 2018).

However, due to disruptions in traditional digital information flows, a loss of pertinent information during data exchange is inevitable. This loss of data (Babu, S. Z. D. et al., 2022) results in increased costs for reproducing data, delays, and wastage, contributing to significant drawbacks in construction projects (Kushwaha, 2016). In response to these challenges, innovative approaches to building design, construction, and management are being adopted globally. While methods such as the design and build approach, project websites, and 3D printing facilitate rapid information exchange, they fail to eliminate issues associated with traditional paper-based communication, including the inability to detect parametric capabilities, conflicts, and constructability issues (Abdullah & Ibrahim, 2016). Presently, information technology applications in the construction industry encompass big data analytics (Aghimien et al., 2021), building information modeling (BIM) (Olanrewaju et al., 2021), digital twins (Opoku et al., 2021) and blockchain (Scott et al., 2021).

Building Information Modeling (BIM) emerged as a solution to the inefficiencies of Computer-Aided Drawing (CAD) by providing a comprehensive digital environment (Swapna, H. R. et al., 2023) that integrates all building information in an electronic file, accessible to various project stakeholders (Abdullah & Ibrahim, 2016). BIM adoption began in the mid-2000s and has gained traction among researchers in both developed and developing countries worldwide, including the Dominican Republic (Silverio & Suresh, 2021), Cambodia (Durdyev et al., 2021), Egypt (Marzouk, Elsaay & Othman, 2022), Pakistan (Akdag & Maqsood, 2019), China (Cui et al., 2021), the United States (Mutis & Mehraj, 2022), the United Kingdom (Dalui et al., 2021). In this chapter, the term 'BIM' refers to an activity, "building information modeling", rather than an object, "building information model". This differentiation underscores that BIM is not a tangible entity or a software category, but rather a human-driven activity that leads to comprehensive process transformations in the construction domain.

According to Al-Ashmori et al. (2020), BIM stands for Building Information Modeling and has emerged as a transformative innovation in the architecture, engineering, and construction (AEC) field, fundamentally altering how professionals conceive, design, construct, and manage built environments. Building Information Modeling (BIM) is a digital representation and collaborative process facilitating the creation, management, and sharing of information about a building project throughout its lifecycle (Hamma-Adama, Kouider & Salman, 2018). Amuda-Yusuf (2018) defines BIM as a methodology utilized in integrating digital descriptions of relationships among building digital objects, enabling stakeholders to query, simulate, and estimate activities and their impacts on the building process as a lifecycle entity. Zigurat (2018) posited that effective BIM utilization in building projects can reduce unbudgeted changes by 40%, shorten project durations by around 7%, save 80% of the time used for project costing valuation, generate nearly a 10% increase in contract value savings, and yield a 3% profit margin for the entire project.

According to Babatunde, Udeaja, and Adekunle (2020), BIM has not been widely adopted by Architectural Engineering and Construction (AEC) organizations in developing nations. BIM has revolutionized the AEC industry by offering a more efficient and integrated approach to design, construction, and maintenance, culminating in a detailed digital representation of a building or infrastructure project that enables professionals to collaborate, analyze, and manage projects effectively (Bryson, 2017). While reports indicate an increase in BIM usage in the construction industry in developed countries, minimal adoption exists in developing countries like Nigeria (Obodoh, 2019; Yusuf, 2020a). In Nigeria, the construction industry is highly fragmented, with various construction professionals typically generating and managing project information individually (Onungwa & Uduma-Olugu, 2017).

Hamma-adama et al. (2017) and Kori et al. (2019) have noted that architectural, mechanical, electrical designs (Pandey, B. K., & Pandey, D., 2023) are predominantly created using 2D CAD platforms in Nigeria, with only a minority, particularly architects, utilizing 3D CAD platforms primarily for visualization or presentation purposes. An examination of the current state of BIM adoption in architectural, engineering, and quantity surveying practices in Nigerian construction industry reveals that while core functions such as drawing, engineering design, and bill of quantity preparation have largely transitioned to computerized methods, professional builders have yet to fully embrace BIM (Yusuf, 2020a). However, effectively leveraging BIM requires a thorough grasp of its technical intricacies and collaborative skills among professionals. Achieving proficiency in BIM entails a paradigm shift in educational curricula and continuous professional development for practitioners. As the AEC sector progresses towards digitalization, embracing the potential of BIM is crucial for attaining and sustaining professional excellence. This chapter therefore present the evolution, components, software (Anand, R. et al., 2022), user interface, and basic functions of BIM tools, along with their benefits in Nigerian construction industries.

## 2. OVERVIEW OF BIM CONCEPT

The advent of technology in the 21st century presents significant potential for revolutionizing traditional processes in the construction industry. The concept of creating and utilizing virtual building models was first discussed in the early 1970s by Professor Charles Eastman at the Georgia Tech School of Architecture (Eastman, 1974). Since its conceptualization, it has undergone several transformations to encompass various aspects beyond design, including estimation, construction processes, life cycle management, performance evaluation, and visualization. The evolution of Building Information Modeling (BIM) signifies a shift from basic Computer-Aided Design (CAD) to a comprehensive digital methodology that reshapes the Architecture, Engineering, and Construction (AEC) industry as follows:

**1. Early Concepts and Origins (1960s-1980s):** The origin of the invention of BIM could be traced back to early computer-aided design (CAD) systems in the 1970s, which laid the foundation for digital design representation at the Georgia Institute of Technology (Yusuf, 2020). However, it was not until the 1980s that pioneering efforts such as "Building Description System" and "Building Modeling" by Charles Eastman began conceptualizing the idea of capturing not only geometric data but also paramet-

*Figure 1. Different BIM dimensions*
*Philip and Mordue (2017)*

ric (Dhanasekar, S. et al., 2023) relationships and attributes within a digital model (JayaLakshmi, G. et al., 2024).

**2. Emergence of Parametric Modeling (1990s):** The 1990s witnessed the emergence of parametric modeling tools, enabling designers to establish relationships between different elements of a design. Software like Autodesk's Revit, introduced in 2000, combined parametric modeling with database-driven information, marking a significant advancement toward true BIM. This period also saw the introduction of Industry Foundation Classes (IFC), an open file format developed to facilitate interoperability between different software applications.

**3. Mainstream Adoption and Collaboration (2000s):** The 2000s saw the widespread acceptance of BIM in the AEC industry. Governments and organizations worldwide began acknowledging the potential benefits of BIM, including improved project coordination, reduced errors, and enhanced lifecycle management. Concepts such as clash detection, virtual walkthroughs, and 4D scheduling became more achievable with the enhanced capabilities of BIM tools. Collaboration and information sharing became central to BIM's evolution, driving the need for standardized processes and open data exchange formats.

**4. Expanding Capabilities (2010s):** The 2010s brought further advancements in BIM technology. Building upon 3D and 4D capabilities, BIM expanded to include dimensions such as 5D (cost estimation) and 6D (facility management and sustainability). Cloud-based collaboration platforms enabled seamless teamwork regardless of geographical locations. Mobile applications facilitated real-time access to BIM models on construction sites. Building owners increasingly recognized the value of BIM for facility management and operation, extending its utility beyond the construction phase. Figures 1 and 2 delineate the dimensions of BIM, ranging from 3D to 7D.

**5. Digital Twins and Smart Cities (2020s and beyond):** As of my last knowledge update in September 2021, the trajectory of BIM's evolution is moving toward even more integrated and data-rich models. The concept of "Digital Twins," where a virtual representation of a physical asset is linked to real-time data (Saxena, A. et al., 2021), is gaining traction. BIM is becoming a foundational technology for the development of smart cities (Pramanik, S. et al., 2023), where data-driven decision-making (Vinodhini, V. et al., 2022) and predictive analysis play a crucial role in urban planning and management (Pandey, B. K. et al., 2023).

The evolution of Building Information Modeling (BIM), as opined by Anifowose et al. (2018) and Amuda-Yusuf (2018), arose from the need to address challenges observed in the 2D system. This ne-

*Figure 2. 6D and 7D model*
*Philip and Mordue (2017)*

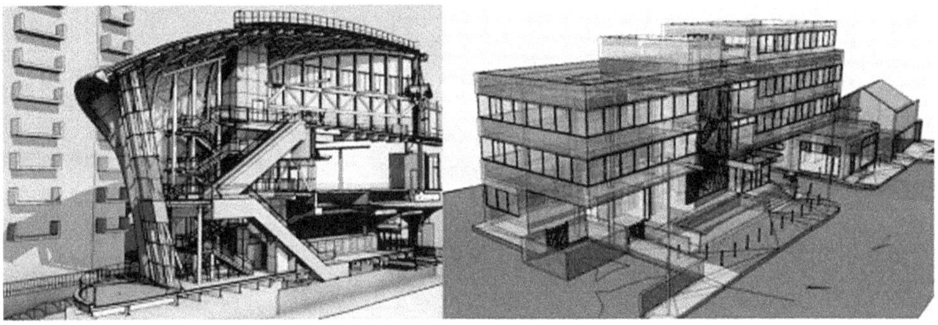

cessitated the introduction of a system that could overcome information sharing and communication (Kirubasri, G. et al., 2021) issues while serving as a repository for storing, organizing, and controlling construction information flow. Various researchers have attempted to contextualize BIM by delineating its functions and capabilities. These capabilities are strategically developed and implemented to manage design and project data, aiming to enhance efficiency (Bessant, Y. A. et al., 2023) across the design, construction, and post-construction phases (Babarinde, 2017; Mohammed & Mohammed, 2018). Figure 3 illustrates the life cycle of BIM in the construction industry (Pandey, B. K. et al., 2024), Figure 4 depicts the sharing of information among project stakeholders within the central model, Figure 5 showcases BIM outside the building: Infrastructure and landscape.

Olanrewaju et al. (2020) assert that the integration of BIM offers both tangible and intangible benefits, ranging from error reduction and enhanced collaboration to heightened client satisfaction and a transition towards more innovative and sustainable practices. According to Baddeley and Chen (2015), Matarneh and Hamed (2017), and Emmanuel-Eze, Ugulu, and Egwunatum (2021), BIM presents a

*Figure 3. Schematic representation of BIM life-cycle in the construction industry*
*Hamma-Adama and Kouider (2018)*

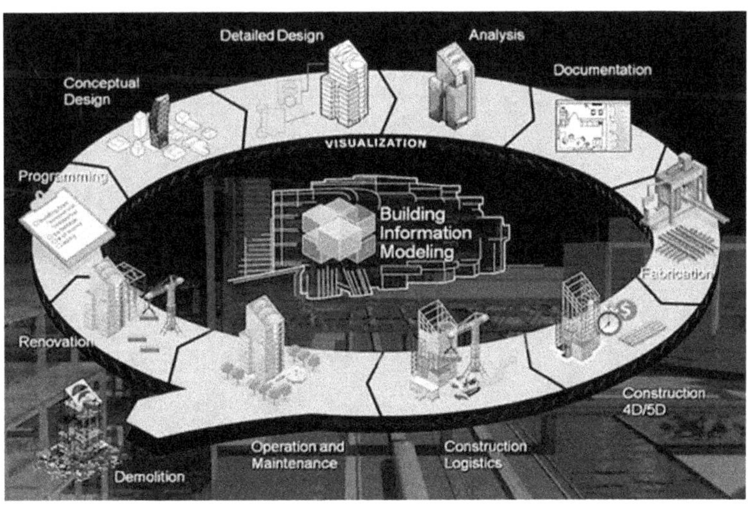

*Figure 4. Schematic representation of data exchange and collaboration*
*Hamma-Adama and Kouider, (2018)*

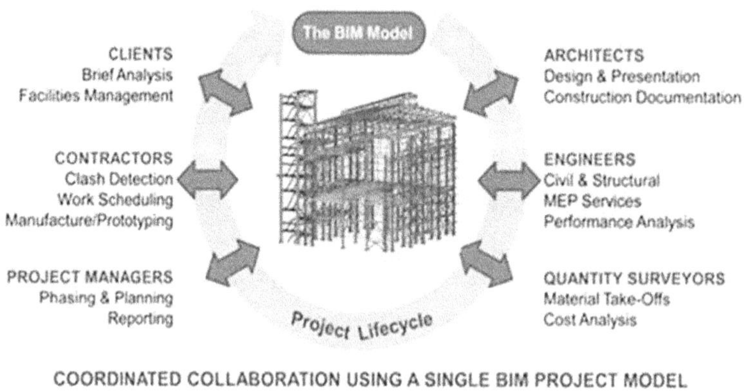


plethora of advantages, including cost savings, accurate visualization, construction sequencing, efficient facility management, regulatory compliance, improved safety, enhanced design quality, client satisfaction, professional development, reduced paperwork, risk mitigation, and enhanced project portfolio. The advantages of employing BIM in construction projects extend beyond mere cost savings and efficient project management. By harnessing BIM's sustainable design tools, comprehensive data becomes readily accessible at every stage of the design process, enabling informed decisions regarding sustainable approaches. This encompasses sustainable design options, critical design analysis, and code compliance. Such a collaborative environment streamlines the decision-making process for the design team, ultimately leading to improved efficiency and cost savings when selecting the most suitable sustainable approach, as shown in the BIM platform in Figure 6 and 7.

*Figure 5. BIM outside the building: Infrastructure and landscape*
*Autodesk (2018)*

*Figure 6. BIM platform for managing sustainable construction*
*Bataw (2015)*

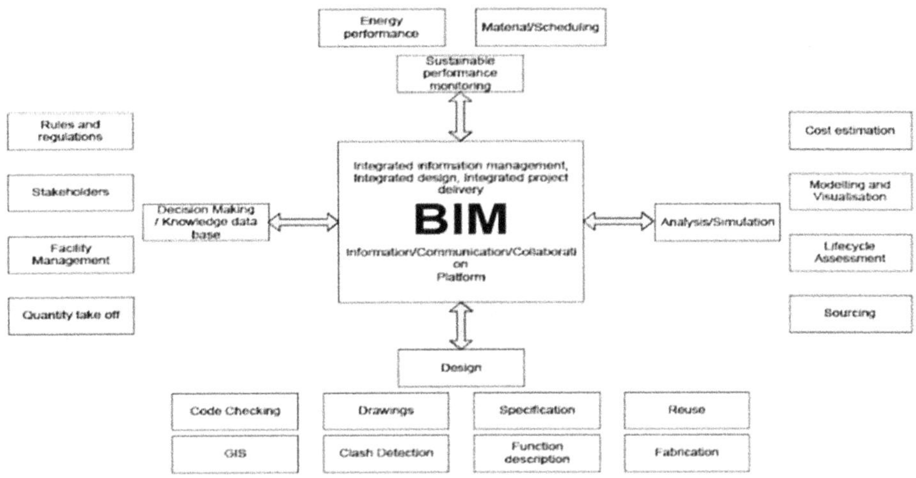


## 3. BUILDING INFORMATION MODELING SOFTWARE

Building Information Modeling (BIM) software has transformed the architecture, engineering, and construction (AEC) industry by facilitating the creation, visualization, analysis, and management of digital models throughout the entire lifecycle of building projects. These software solutions offer a diverse array of functionalities that enhance collaboration, precision, and productivity in the design, construction, and operation of built environments. According to Philip and Mordue (2017), the selection of BIM software depends on various factors including project type, team preferences, collaboration requirements, and

*Figure 7. Benefits of BIM process*
*Goubau (2018)*

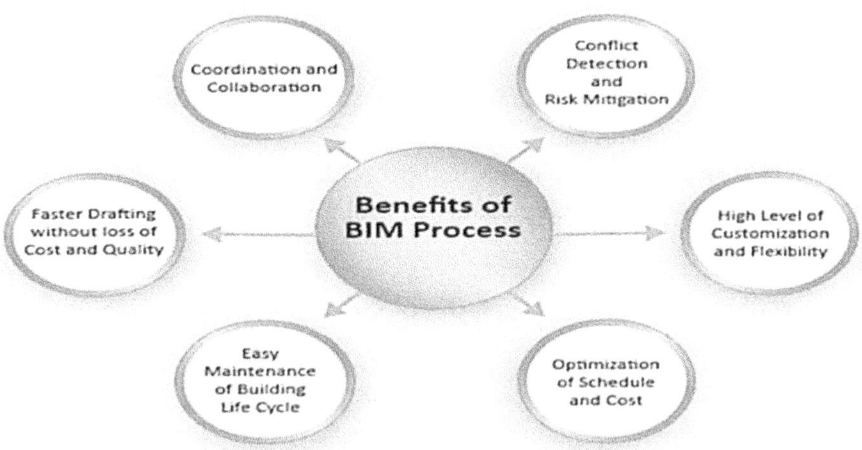

the software's capabilities in terms of modeling, analysis, coordination, and documentation. Below are some examples of BIM software:

**1. Autodesk Revit:** Revit is one of the most widely used BIM software solutions. It offers tools for architectural design, structural engineering, mechanical, electrical and plumbing systems (Autodesk, 2020). Some key features of Autodesk Revit include:

i. Parametric modeling for creating detailed 3D building models.
ii. Family creation for customizing and modeling building components.
iii. Collaboration features for multiple disciplines to work on the same model.
iv. Integration with analysis and visualization tools for simulations and presentations.
v. Clash detection and coordination to identify conflicts between building systems.

**2. Graphisoft ArchiCAD:** ArchiCAD is known for its user-friendly interface and strong focus on architectural design and documentation (Wikipedia, 2020). Some key features of ArchiCAD include:

i. Real-time collaboration through the BIM cloud platform.
ii. Virtual Building™ concept for 3D modeling and coordination.
iii. BIMx for interactive 3D presentations and model exploration.
iv. Energy analysis and sustainability assessment tools.
v. Integration with external rendering engines for high-quality visualizations (Iyyanar, P. et al., 2023).

**3. Bentley Systems AECOsim:** AECOsim Building Designer offers BIM capabilities tailored to infrastructure projects. Some key features of AECOsim include:

i. Comprehensive multidisciplinary modeling.
ii. Parametric modeling for creating complex building elements.
iii. Integration with reality modeling for capturing real-world context.
iv. Clash detection and coordination to prevent clashes and errors.
v. Dynamic views for generating drawings and documents directly from the model.

**4. Vectorworks Architect:** Vectorworks Architect combines BIM features with tools for landscape design and entertainment industry applications (Vectorworks, 2020). Vectorworks Architect features include:

i. Intelligent objects for creating complex building elements.
ii. Landscape design and modeling tools.
iii. Integration with renderers for high-quality visualizations.
iv. Analysis and simulation tools for lighting, energy, and structural analysis.

**5. Solibri Model Checker:** Solibri focuses on model quality assurance and coordination. Solibri features include:

i. Model checking and validation for compliance with standards and requirements.
ii. Clash detection and coordination.
iii. Visualization of issues and discrepancies in the model.
iv. Customizable rulesets for automating quality control processes.

## 4. USER INTERFACE AND BASIC FUNCTIONS OF BIM TOOLS

The interface and fundamental functionalities of BIM tools are designed to simplify complex processes, foster collaboration, and streamline the entire building lifecycle. These tools empower users to efficiently create, visualize, analyze, and manage digital building models, ultimately leading to enhanced project outcomes and increased efficiency in the architecture, engineering, and construction (AEC) industry. BIM serves as a virtual collaboration tool, facilitating the exploration of realistic scenarios that reflect real-world challenges (Olugboyega & Aina, 2018).

BIM tools can be categorized into three primary groups based on their functionalities: visualization, simulation, and computer-aided architectural design (Maina, 2018). These tools allow for the creation of drawings on computers prior to presentation to clients (Dare-abel et al., 2014; Lim et al., 2020; Puteri & Puspitasari, 2021). According to Gohil et al. (2022), examples of computer-aided architectural design (CAAD) software programs include Abis, Allplan, ArchiCAD, AutoCAD, Accurender, Blender, Bricscad, Caddie, Maya, formZ, Spirit, Revit, Lumion, 3Ds Max, CINEMA 4D, Digital Project, SolidWorks, Rhinoceros 3D, Vectorworks, and Google SketchUp. While the user interface may exhibit slight variations across different software, they typically encompass the following elements:

1. **Ribbon/Toolbar:** Contains buttons and icons for accessing various commands and tools.
2. **Project Browser:** Displays a hierarchical list of elements within the model for easy selection and navigation.
3. **Properties Palette:** Provides information and editable parameters for selected elements.
4. **3D Viewport:** Displays the 3D model and allows users to manipulate the view and interact with objects.
5. **Tool Palettes:** House a variety of modeling, editing, and analysis tools.
6. **Drawing/Annotation Area:** Where plans, sections, elevations, and other documentation are created and edited.
7. **Status Bar:** Displays important information about the model's status and any ongoing operations.

BIM tools allow users to create and modify various elements that make up the building model. Common functions include:

1. **Placing Objects:** Users can place architectural components like walls, doors, windows, and structural elements such as columns and beams.
2. **Editing Geometry:** Elements can be adjusted in terms of size, shape, and orientation using tools like Move, Rotate, and Scale.
3. **Parametric Modeling:** Users can define relationships and constraints between elements, ensuring changes are propagated consistently throughout the model.
4. **Copy and Paste:** Allows elements to be duplicated and reused within the model.

Effective navigation and visualization are essential in BIM tools to understand the model's spatial relationships and details (Fargnoli & Lombardi, 2020). This enables the designer to see all of the construction's components while having accurate information about their dimensions, shapes, and locations (Lim et al., 2020). A few visualization tools were named by (Brandt & Timmermans, 2021) as being V-Ray, Artlantis, POV-Ray, YafaRay, Mental Ray, LuxRender, Flamingo, LightWave, RenderZone, RenderMan, Photoshop, Kerkythea, RenderWorks, Maxwell Render, and Adobe After Effects. Basic functions include:

1. **Pan and Zoom:** Users can pan around the model and zoom in or out to view details.
2. **Orbit:** Rotate the view around a point of interest to view the model from different angles.
3. **Walkthroughs:** Create virtual tours through the model to understand its spatial layout.
4. **Sections and Elevations:** Generate views that cut through the model to visualize its internal structure.

BIM tools facilitate the creation of detailed construction documents and annotations. Some common functions include:

1. **Dimensioning:** Add dimensions to convey accurate measurements on drawings.

**2. Text and Labels:** Place annotations, labels, and notes to provide additional information.

**3. Tags and Schedules:** Attach data and annotations to elements, such as door schedules or material lists.

**4. Detail Views:** Create detailed views of specific areas for clear documentation.

According to Matarneh and Hamed (2017), BIM tools offer features to enhance collaboration and coordination among project professionals. Common functions include:

**1. Import and Export:** Import external files and reference them in the model. Export to various formats for sharing.

**2. Clash Detection:** Identify conflicts between different building systems and disciplines.

**3. Collaboration Tools:** Allow multiple users to work on the same model simultaneously, often in a cloud-based environment.

BIM tools may provide analysis and simulation features for better decision-making. Common functions include:

**1. Energy Analysis:** Assess the building's energy performance and sustainability.

**2. Structural Analysis:** Evaluate the structural integrity of elements.

**3. Lighting and Acoustics Analysis:** Simulate lighting conditions and acoustic properties.

## 5. BIM ADOPTION IN GLOBAL CONTEXT AND NIGERIA

Various nations have undertaken initiatives to adopt and implement Building Information Modeling (BIM), which has revolutionized the construction industry. BIM usage has resulted in increased profitability through cost savings and timely project delivery, among other advantages. Liao and Teo (2017) and Darwish et al. (2020) assert that BIM provides a platform for enhancing work processes to drive efficiency and effectiveness in the construction sector by addressing information and communication inefficiencies (Kirubasri, G. et al., 2022), reducing wastage, and promoting collaboration at design and construction stages. Kori and Kivniemi (2015) highlighted benefits ranging from efficient communication and data exchange systems (Devasenapathy, D. et al., 2024) to automated quantification, improved collaboration, coordination of construction documents, enhanced design visualization, clash detection, and cost reduction.

Reports such as the United Kingdom BIM Strategy Report (2012) and BuildSmart (2012) indicate that several developed countries, including the United Kingdom, United States, and Australia, have established strategies for BIM implementation in construction projects, leading to rapid adoption. Matarneh and Hamed (2017) noted the UK government's ambition to become a BIM leader in Europe, with the Efficiency and Reform Group (2011) establishing a roadmap for BIM adoption (Zaidi, A., et al., (2023)). By April 2017, all UK government construction suppliers tendering for centrally procured projects were required to work at BIM Level 2, involving collaborative 3D environment development with attached data, albeit created in separate discipline models (McAuley et al., 2017; Theiler & Smarsly, 2018). The US National BIM standards were introduced in 2015, providing guidelines for BIM standards implementation across planning, design, construction, and operation stages (Cheng & Lu, 2015). BIM usage in the United States saw a significant increase, surpassing that of other developed countries (Eadie et al., 2014), particularly driven by the public sector's championing of implementation across states.

Canada's CanBIM began offering certification for individuals and educational courses and programs (Canada BIM Council, 2018), while Spain mandated BIM use in public sector and infrastructure projects

(Zigurat, 2018). Singh (2017) observed varying BIM adoption levels in developed nations, with Finland mandating BIM implementation since 2002 and by 2007, it was established that all design must be IFC certified. In Hong Kong, Lu et al. (2018) found that BIM maturities of construction-related organizations in Hong Kong vary, with more than half ranging from Stage 0 to 1; and this was attributed to the different developments of their BIM processes and protocols. However, the BIM implementation in Africa differs from one country to another. In Norway, the public sector is serving as the launching pad for adoption, setting out open standards for BIM starting from July 2016 (Liao & Teo, 2017), similarly, Denmark has developed a common language and method for defined information exchange and sharing protocols (Daoud, M. K.,et al.,2024).

The adoption and implementation of new technologies present a viable opportunity for developing countries to address the pressing need for enhancements in built environments and urban infrastructure (Akdag & Maqsood, 2019). The project in Nigeria that has fully implemented Building Information Modeling (BIM) from design to operation is the Eko Atlantic City project (Olanrewaju et al., 2020a). The history of Eko Atlantic City (EAC) dates back to 2003 when the Lagos State Government sought solutions for coastal erosion issues at Bar Beach off Victoria Island. Between 2003 and 2005, efforts focused on assessing the project's feasibility in mitigating coastal erosion problems, leading to the conception of a new city dedicated to shoreline protection and increased housing units (Eko Atlantic, 2016). Eko Atlantic City (EAC) is a distinctive urban development initiative emerging from the Atlantic Ocean, aiming to rejuvenate the shorelines of Victoria Island affected by coastal erosion. The city's construction was undertaken through a public-private partnership between South Energy X Nigeria Limited (SENL) and the Lagos State Government, envisioning Nigeria as a prime tourist destination in Africa, fostering economic growth (Channels Television, 2010; Ajibade, 2017).

Functioning as a metropolis in terms of power administration, the city operates under its governance and is designated as an offshore financial zone, facilitating free capital movement by investors (Olugboyega, 2016). Designed to accommodate a wide array of activities including financial, residential, tourism, commercial, and leisure pursuits, the city upon completion is projected to house 250,000 residents and cater to 150,000 commuters across 10 districts, encompassing Business Districts, Harbor Lights, Marina Districts, Downtown, Eko Island, Avenues, Four Bridges, Eko Drive, East Side Marina, and Ocean Front (South Energy, 2013; Channels Television, 2014; Eko Atlantic, 2016). Noteworthy features of the city's design include independent power supply and water treatment facilities, advanced telecommunications infrastructure, low carbon footprint, absence of on-street parking, an international school, and a hospital, all integrated into the EAC BIM model (South Energy, 2013; Channels Television, 2014; Eko Atlantic, 2016).

Oyedeji (2014) elaborated that most buildings in the EAC project are high-rise structures designed with setbacks to minimize shadow effects on infrastructures. Some prominent buildings include the Eko Corporate Tower, Eko Boulevard, Le Reve Tower, Afren Tower, Eko Pearl Towers (Champagne Pearl, White Pearl, Black Pearl, Indigo Pearl, and Acqua Pearl), Azuri Peninsula, Arep Towers, International School, Atlantic Business and Residence, Eko Mall, Eko Energy Estate, 3-Marinas, and Eko Atlantic Medical Centre (Olugboyega, 2016). Regarding value, Meseko (2014) disclosed waterfront and inner-city plot costs at $2,000/sqm and $1,000/sqm respectively, with the estimated project cost standing at $6 billion.

Olugboyega (2016) recognized the pivotal role BIM played in the EAC project, ensuring comprehensive coverage of all project aspects without oversights. BIM stands as a crucial tool in enhancing lifecycle management and knowledge sharing in construction projects (Olanrewaju et al., 2020). Despite

*Figure 8. Eko Atlantic City (EAC)*
*Eko Brochure (2013); South Energy (2013); Olugboyega (2016)*

its importance, BIM has yet to be fully adopted in most construction projects, with Anifowose et al. (2018) reporting only a 40% level of awareness. The EAC project exemplified the significant benefits achievable through full BIM implementation in the Nigerian construction industry. Figures 8 and 9 depict some of the BIM models developed for the EAC project.

## 6. CHALLENGES FACING BIM ADOPTION IN NIGERIAN CONSTRUCTION PROJECTS

Recently, studies have focused on identifying barriers to BIM adoption in developing countries such as Ghana (Oteng et al., 2018), Pakistan (Akdag & Maqsood, 2019), and Yemen (Gamil and Rahman, 2019), yet research in the Nigerian context remains scarce (Amuda-Yusuf et al., 2017; Anifowose et al., 2018; Hamma-adama and Kouider, 2019; Babatunde et al., 2020a). Despite the strides made by developed nations, inefficient and error-prone practices rooted in traditional 2D and 3D drawing approaches persist as the predominant method in developing countries like Nigeria (Ganiyu et al., 2018, Olugboyega & Aina, 2016). Olanrewaju et al. (2020a) affirm that BIM implementation in Nigeria has not fully leveraged the potential benefits across various construction processes. Onungwa and Uduma-Olugu (2017) contend

*Figure 9. Eko Atlantic City (EAC)*
*Eko Brochure (2013); South Energy (2013); Olugboyega (2016)*

that there is a low level of awareness and technical proficiency in BIM within Nigeria. Kong et al. (2020) identified issues related to time, cost, and design as significant challenges (Ahmad, A. Y. B.,et al.,2024).

While some studies have explored BIM barriers in Nigeria, these have primarily focused on aspects such as incorporating BIM into undergraduate curricula in Nigerian universities from the perspectives of students and academics (Babatunde, Udeaja & Adekunle, 2020); identifying measures to enhance BIM usage in the Nigerian building industry (Onungwa & Uduma-Olugu, 2017); assessing BIM training gaps (Oyewole and Dada, 2019); analyzing the impact of BIM on collaboration (Onungwa & Uduma-Olugu, 2016); BIM awareness (Olapade & Ekemode, 2018); barriers to BIM in SMEs (Saka & Chan, 2020); benefits (Olawumi & Chan, 2019a), as well as drivers and barriers for BIM implementation (Babatunde et al., 2020a). Additionally, higher levels of BIM awareness in Nigeria do not necessarily translate into implementation (Saka & Chan, 2019b). Conversely, other studies, including those by Addy et al. (2018), Gamil & Rahman (2019), and Olawumi & Chan (2019b), underscore the importance of BIM awareness and the need to investigate barriers and antecedents in developing countries, as well as the development of benchmarking models to enhance BIM. However, there remains limited understanding of the barriers to BIM implementation in the Nigerian construction industry, particularly from the perspective of construction professionals (Muniandi, B.,et al.,2019).

This has attracted the attention of numerous researchers that have explored the drivers, obstacles, current status, and competency issues regarding BIM adoption (Babatunde et al., 2020; Fargnoli & Lombardi, 2020; Olanrewaju et al., 2020a; Olorunfemi et al., 2021). Fatima, Saleem, and Lamgir (2015), Jadhav and Ghadge (2016), He et al. (2017), Mohammed, Hasnain, and Quadir (2019), and Olanrewaju et al. (2020) identified several barriers to BIM implementation, including (Daoud, M. K.,et al.,2023):

1. **Limited Awareness and Education:** A significant challenge hindering the adoption of Building Information Modeling (BIM) in Nigeria is the lack of awareness and education among stakeholders. Many professionals in the construction industry are unfamiliar with BIM concepts, methodologies, and potential benefits, leading to reluctance in embracing this transformative technology.
2. **Technological Infrastructure:** Nigeria's technological infrastructure presents a significant hurdle to widespread BIM adoption. Inadequate access to reliable internet connectivity, limited availability of hardware and software resources, and outdated IT infrastructure pose barriers to effectively implementing BIM processes and tools.
3. **Cost Constraints:** The upfront costs associated with implementing BIM can be prohibitive for many construction firms in Nigeria (Bin-Seddeeq et al., 2019; Idrees & Shafiq, 2021). Investing in BIM software, training programs, and hardware upgrades requires financial resources that may not be readily available to all stakeholders.
4. **Resistance to Change:** Resistance to change within the construction industry is a common challenge globally, and Nigeria is no exception. Traditional practices and entrenched workflows often resist the integration of new technologies like BIM. Overcoming resistance to change and fostering a culture of innovation and openness to new methodologies is essential for successful BIM implementation.
5. **Skills Shortages:** A shortage of skilled professionals proficient in BIM methodologies and technologies poses a significant challenge in Nigeria (Pradhananga et al., 2021). The lack of adequately trained personnel capable of designing, implementing, and managing BIM projects hampers the industry's ability to fully leverage the benefits of BIM.

6. **Interoperability Issues**: Incompatibility between different BIM software platforms and file formats presents interoperability challenges in Nigeria. The inability of different software systems to seamlessly exchange information hinders collaboration and data sharing among project stakeholders, undermining the efficiency and effectiveness of BIM workflows.
7. **Regulatory Framework:** The absence of clear regulatory frameworks and standards governing BIM implementation in Nigeria complicates the adoption process. Uncertainty surrounding legal requirements, data ownership, and liability issues may deter stakeholders from fully embracing BIM or lead to inconsistencies in its application across projects.
8. **Cultural Attitudes and Mindsets**: Cultural attitudes towards innovation and technology adoption can influence the uptake of BIM in Nigeria (Fayisa & Wayessa, 2021). Traditional hierarchies, resistance to authority, and aversion to risk-taking may impede the willingness of industry stakeholders to embrace new ways of working and challenge existing norms (Okpala et al., 2021; McNamara & Sepasgozar, 2021).
9. **Data Security and Privacy Concerns**: With the digitization of construction processes comes the risk of data breaches, cyberattacks, and privacy violations (David, S. et al., 2023). Concerns about the security (Khan, B. et al., 2021) and confidentiality of sensitive project information stored in BIM models may deter some stakeholders from fully embracing BIM or sharing data openly.
10. **Infrastructure Challenges:** Beyond technological infrastructure, Nigeria faces broader infrastructure challenges, including unreliable power supply and inadequate transportation networks (Sahani, K. et al., 2023). These infrastructure deficits can disrupt BIM workflows, impede communication (Raja, D. et al., 2024) among project teams, and hinder the seamless exchange of information essential for successful BIM implementation.

Tackling these obstacles necessitates a comprehensive strategy involving educational campaigns, investments in technological infrastructure, skill development, regulatory adjustments, and a cultural shift within Nigeria's construction sector. By surmounting these challenges, Nigeria can fully harness the power of BIM to foster innovation, enhance efficiency, and promote sustainability in construction methodologies nationwide (Ahmad, A. Y. B.,et al.,2023).

## 7. STRATEGIES TO MITIGATE BIM ADOPTION CHALLENGES IN NIGERIAN CONSTRUCTION INDUSTRIES

A well-structured plan for implementing BIM is crucial for organizations seeking to leverage its benefits effectively. According to Matarneh and Hamed (2017), Al-Ashmori, et al. (2020), Makabate, et al. (2021), Olatunji, Dele and Spencer (2022) by aligning objectives, engaging key stakeholders, providing training, managing change, and continuously refining processes, organizations can seamlessly integrate BIM into their operations, thereby enhancing collaboration, efficiency, and project quality. To address challenges in BIM adoption in Nigeria comprehensively, the following strategies are proposed, covering the entire spectrum from conceptualization to construction:

1. **Government Advocacy and Support**: The Nigerian government should take a lead role in advocating for BIM adoption through policy initiatives, incentives, and regulations. This involves mandating BIM usage on public projects, offering tax incentives, and facilitating training programs.

2. **Education and Training Programs**: Establishing comprehensive education and training programs focused on BIM within Nigerian universities, technical colleges, and professional organizations can ensure a skilled workforce proficient in BIM technologies. This should encompass both basic and advanced BIM training for architects, engineers, contractors, and other stakeholders.
3. **Localized BIM Standards**: Developing BIM standards tailored to the Nigerian construction industry, covering areas like data exchange protocols and information management, will ensure alignment with international best practices.
4. **Capacity Building for SMEs**: Small and Medium Enterprises (SMEs) constitute a significant portion of the construction sector in Nigeria. Providing targeted capacity-building initiatives for Small and Medium Enterprises (SMEs) in the construction sector, including subsidized software licenses and technical support, will empower them to adopt BIM effectively.
5. **Public-Private Partnerships (PPP)**: Facilitating partnerships between public institutions, private companies, and international organizations will enable knowledge sharing, technology transfer, and funding opportunities for BIM implementation projects. PPPs can accelerate BIM adoption by pooling resources and expertise from various stakeholders.
6. **Demonstration Projects**: Implementing pilot BIM projects across different sectors and scales, such as infrastructure, commercial buildings, and residential developments, will demonstrate the benefits of BIM in real-world scenarios, encouraging further adoption and investment.
7. **Awareness Campaigns and Stakeholder Engagement**: Launching targeted awareness campaigns and engaging with stakeholders across the construction value chain including government agencies, professional associations, industry bodies (Ekong, M. O. et al., 2023), and academia can raise awareness about the advantages of BIM and foster a collaborative environment for its adoption.
8. **Integration with Digital Infrastructure**: Given Nigeria's increasing digitalization efforts, integrating BIM with other digital infrastructure initiatives, such as Geographic Information Systems (GIS), smart city projects (Pandey, J. K. et al., 2022), and digital twinning, can enhance the interoperability and utility of BIM data beyond the construction phase, creating long-term value.
9. **Access to Reliable Internet and Power Supply**: Addressing infrastructure challenges, such as reliable internet connectivity and stable power supply, is fundamental for successful BIM implementation. Investing in improving infrastructure, particularly in urban and remote areas, ensures seamless access to cloud-based BIM platforms and software tools.
10. **Continuous Evaluation and Improvement**: Establishing mechanisms for ongoing evaluation, feedback, and improvement of BIM processes is essential. Regularly assess the effectiveness of implemented strategies, solicit feedback from users, and adapt policies and interventions accordingly to ensure continuous growth and optimization (Pandey, D. et al., 2021) of BIM adoption in Nigeria.

## 8. CONCLUSION

The evolution of Building Information Modeling (BIM) in Nigeria, from conceptualization to its practical application in construction projects, signifies a paradigm shift in the country's approach to infrastructure development. BIM has emerged as a pivotal tool, transcending conventional methodologies and reshaping the entire project lifecycle, from conception to management. Its adoption in Nigeria heralds promising prospects, promising heightened efficiency, cost-effectiveness, sustainability, and overall project efficacy. By facilitating seamless collaboration among stakeholders, optimizing workflows, and furnishing precise data-driven insights, BIM stands poised to significantly elevate standards within Nigeria's construction sector.

However, harnessing the full potential of BIM demands more than mere technological investment. It necessitates a fundamental shift in mindset, organizational ethos, and collaborative ethos across the industry spectrum. Stakeholders must embrace continual learning rate (Singh, H. et al., 2022) and adaptation to effectively capitalize on BIM's capabilities. Moreover, addressing challenges such as limited technological infrastructure, skill shortages, and regulatory frameworks is paramount to ensuring widespread adoption and successful integration of BIM across Nigeria's construction milieu. Stakeholders must steadfastly commit to nurturing an environment conducive to innovation, cooperation, and excellence. In doing so, they can unlock BIM's transformative prowess, propelling sustainable growth and development within the Nigerian construction landscape.

## REFERENCES

Addy, M., Adinyira, E., & Ayarkwa, J. (2018). Antecedents of building information modelling adoption among quantity surveyors in Ghana: An application of a technology acceptance model. *Journal of Engineering. Design and Technology*, *16*(2), 313–326. doi:10.1108/JEDT-06-2017-0056

Aghimien, D. O., Ikuabe, M., Aigbavboa, C., Oke, A., & Shirinda, W. (2021). Unravelling the Factors Influencing Construction Organizations' Intention to Adopt Big Data Analytics in South Africa. *Construction Economics and Building*, *21*(3), 478–489. doi:10.5130/AJCEB.v21i3.7634

Ahmad, A., Abusaimeh, H., Rababah, A., Alqsass, M., Al-Olima, N., & Hamdan, M. (2024). Assessment of effects in advances of accounting technologies on quality financial reports in Jordanian public sector. *Uncertain Supply Chain Management*, *12*(1), 133–142. doi:10.5267/j.uscm.2023.10.011

Ahmad, A. Y. B. (2024). Ethical implications of artificial intelligence in accounting: A framework for responsible ai adoption in multinational corporations in Jordan. *International Journal of Data and Network Science*, *8*(1), 401–414. doi:10.5267/j.ijdns.2023.9.014

Ahmad, A. Y. B. (2024). Firm Determinants that Influences Implementation of Accounting Technologies in Business Organizations. *WSEAS Transactions on Business and Economics*, *21*, 1–11. doi:10.37394/23207.2024.21.1

Ahmad, A. Y. B., Gongada, T. N., Shrivastava, G., Gabbi, R. S., Islam, S., & Nagaraju, K. (2023). E-Commerce Trend Analysis and Management for Industry 5.0 using User Data Analysis. *International Journal of Intelligent Systems and Applications in Engineering*, *11*(11s), 135–150.

Ahmad, A. Y. B., Hannoon, A., Al-Daoud, K. I., Abu-Alsondos, I. A., & Al-Qaisieh, M. S. (2023). Assessment of Cloud Based Accounting Technology Adoption and Business Performance. *Kurdish Studies, 11*(3).

Ahmad, A. Y. B., Kumari, D. K., Shukla, A., Deepak, A., Chandnani, M., Pundir, S., & Shrivastava, A. (2024). Framework for Cloud Based Document Management System with Institutional Schema of Database. *International Journal of Intelligent Systems and Applications in Engineering*, *12*(3s), 672–678.

Ahmad, A. Y. B., Kumari, D. K., Shukla, A., Deepak, A., Chandnani, M., Pundir, S., & Shrivastava, A. (2024). Framework for Cloud Based Document Management System with Institutional Schema of Database. *International Journal of Intelligent Systems and Applications in Engineering*, *12*(3s), 672–678.

Ahmad, A. Y. B., & Kumari, S. S. (2023). Blockchain Implementation in Financial Sector and Cyber Security System. *2023 International Conference on Artificial Intelligence and Smart Communication (AISC)*, Greater Noida, India. 10.1109/AISC56616.2023.10085045

Ahmad, A. Y. B., Tiwari, A., Nayeem, M. A., Biswal, B. K., Satapathy, D. P., Kulshreshtha, K., & Bordoloi, D. (2024). Artificial Intelligence Perspective Framework of the Smart Finance and Accounting Management Model. *International Journal of Intelligent Systems and Applications in Engineering*, *12*(4s), 586–594.

Ahmad, A. Y. B., William, P., Uike, D., Murgai, A., Bajaj, K. K., Deepak, A., & Shrivastava, A. (2024). Framework for Sustainable Energy Management using Smart Grid Panels Integrated with Machine Learning and IOT based Approach. *International Journal of Intelligent Systems and Applications in Engineering*, *12*(2s), 581–590.

Ajibade, I. (2017). Can a future city enhance urban resilience and sustainability? A political ecology analysis of Eko Atlantic city, Nigeria. *International Journal of Disaster Risk Reduction*, *26*, 85–92. doi:10.1016/j.ijdrr.2017.09.029

Akdag, S. G., & Maqsood, U. (2019). A roadmap for BIM adoption and implementation in developing countries: The Pakistan case. *Archnet-IJAR*, *14*(1), 112–132. doi:10.1108/ARCH-04-2019-0081

Al-Ashmori, Y. Y., Othman, I., Rahmawati, Y., Amran, Y. M., Sabah, S. A., Rafindadi, A. D. U., & Mikić, M. (2020). BIM benefits and its influence on the BIM implementation in Malaysia. *Ain Shams Engineering Journal*, *11*(4), 1013–1019. doi:10.1016/j.asej.2020.02.002

Amuda-Yusuf, G. (2018). Critical success factors for building information modelling implementation. *Construction Economics and Building*, *18*(3), 55–73. doi:10.5130/AJCEB.v18i3.6000

Amuda-Yusuf, G., Adebiyi, R. T., Olowa, T. O. O., & Oladapo, I. B. (2017). Barriers to building information modelling adoption in Nigeria. *Journal of Research Information in Civil Engineering*, *14*(2), 1574–1591.

Anand, R., Singh, J., Pandey, D., Pandey, B. K., Nassa, V. K., & Pramanik, S. (2022). Modern technique for interactive communication in LEACH-based ad hoc wireless sensor network. In *Software Defined Networking for Ad Hoc Networks* (pp. 55–73). Springer International Publishing. doi:10.1007/978-3-030-91149-2_3

Anifowose, M., Babarinde, S. A., & Olanrewaju, O. I. (2018). Adoption Level of Building Information Modelling by selected professionals in Kwara State. *Environmental Technology and Science Journal*, *9*(2), 35–44.

Autodesk. (2018). What is BIM. *BIM and The Future of AEC*. BIM and the future of AEC. https://www.autodesk.com/solutions/bim

Autodesk. (2020). *Building Information Modelling in Practice*. Autodesk whitepaper Report. San Rafael. California.

Babarinde, S. A. (2017). *Assessment of The Adoption Level of Building Information Modelling by Selected Construction Professionals in Kwara-State*. [Unpublished Bachelor Degree Thesis in Department of Quantity Surveying, Federal University of Technology, Minna].

Babatunde, S., Ekundayo, D., Adekunle, A., & Bello, W. (2020a). Comparative analysis of drivers to BIM adoption among AEC firms in developing countries: A case of Nigeria. *Journal of Engineering. Design and Technology*, *18*(6), 1425–1447. doi:10.1108/JEDT-08-2019-0217

Babatunde, S. O., Udeaja, C., & Adekunle, A. O. (2020). Barriers to BIM implementation and ways forward to improve its adoption in the Nigerian AEC firms. *International Journal of Building Pathology and Adaptation*, *39*(1), 48–71. doi:10.1108/IJBPA-05-2019-0047

Babu, S. Z. D., Pandey, D., Naidu, G. T., Sumathi, S., Gupta, A., Bader Alazzam, M., & Pandey, B. K. (2022, July). Analysation of Big Data in Smart Healthcare. In *Artificial Intelligence on Medical Data: Proceedings of International Symposium, ISCMM 2021* (pp. 243-251). Singapore: Springer Nature Singapore.

Baddeley, M., & Chang, C. (2015). Collaborative Building Information Modelling (BIM): Insights from behavioural economics and incentive theory. London: Royal Institution of Chartered Surveyor (RICS).

Bataw, A. (2015). *On the integration of Building Information Modelling in undergraduate Civil engineering programmes in the United Kingdom*. [Thesis, The University of Manchester].

Bessant, Y. A., Jency, J. G., Sagayam, K. M., Jone, A. A. A., Pandey, D., & Pandey, B. K. (2023). Improved parallel matrix multiplication using Strassen and Urdhvatiryagbhyam method. *CCF Transactions on High Performance Computing*, *5*(2), 1–14. doi:10.1007/s42514-023-00149-9

Bin-Seddeeq, A., Assaf, S., Abdallah, A., & Hassanain, M. A. (2019). Time and cost overrun in the Saudi Arabian oil and gas construction industry. *Buildings*, *9*(2), 41. doi:10.3390/buildings9020041

Bormann, A. Konig, M., Koch, C. &Beetz, J. (2008). Building Information Modelling: Why? What? How? Technology Foundations and Industry Practice. *Building Information Modelling*, 1-24. doi:10.1007/978-3-319-92862-3_1

Brandt, P., & Timmermans, S. (2021). Abductive Logic of Inquiry for Quantitative Research in the Digital Age. *Sociological Science*, *8*, 191–210. doi:10.15195/v8.a10

Bryson, B. W. (2017). *Future of Architects: Extinction or Irrelevance*. Design Intelligence. https://www.di.net/articles/future-architects-extinctionirrelevance/

Canada BIM Council (2018). *Building Information Modelling. Industrial Strategy: Government and Industry in Partnership*. Department for Business, Innovation and Skills (BIS).

Cheng, J. C. P., & Lu, Q. (2015). A review of the efforts and roles of the Public Sector for BIM Adoption Worldwide. *Journal of Information Technology in Construction*, *20*, 442–478.

Cui, Q., Hu, X., Liu, X., Zhao, L., & Wang, G. (2021). Understanding Architectural Designers' Continuous Use Intention Regarding BIM Technology: A China Case. *Buildings, 11*(10), 448. doi:10.3390/buildings11100448

Dalui, P., Elghaish, F., Brooks, T., & McIlwaine, S. (2021). Integrated Project Delivery with BIM: A Methodical Approach Within the UK Consulting Sector. *Journal of Information Technology in Construction, 26*, 922–935. doi:10.36680/j.itcon.2021.049

Daoud, M. K., Al-Qeed, M., Ahmad, A. Y. B., & Al-Gasawneh, J. A. (2023). Mobile Marketing: Exploring the Efficacy of User-Centric Strategies for Enhanced Consumer Engagement and Conversion Rates. *International Journal of Membrane Science and Technology, 10*(2), 1252–1262. doi:10.15379/ijmst.vi.1425

Daoud, M. K., Taha, S., Al-Qeed, M., Alsafadi, Y., Ahmad, A., & Allahham, M. (2024). EcoConnect: Guiding environmental awareness via digital marketing approaches. *International Journal of Data and Network Science, 8*(1), 235-242.

Darwish, A. M., Tantawy, M. M., & Elbeltagi, E. (2020). Critical Success Factors for BIM implementation in Construction projects. *Saudi Journal of Civil Engineering, 4*(9), 180–191. doi:10.36348/sjce.2020.v04i09.006

David, S., Duraipandian, K., Chandrasekaran, D., Pandey, D., Sindhwani, N., & Pandey, B. K. (2023). Impact of blockchain in healthcare system. In *Unleashing the Potentials of Blockchain Technology for Healthcare Industries* (pp. 37–57). Academic Press. doi:10.1016/B978-0-323-99481-1.00004-3

Devasenapathy, D., Madhumathy, P., Umamaheshwari, R., Pandey, B. K., & Pandey, D. (2024). Transmission-efficient grid-based synchronized model for routing in wireless sensor networks using Bayesian compressive sensing. *SN Computer Science, 5*(1), 1–11.

Dhanasekar, S., Martin Sagayam, K., Pandey, B. K., & Pandey, D. (2023). Refractive Index Sensing Using Metamaterial Absorbing Augmentation in Elliptical Graphene Arrays. *Plasmonics*, 1–11. doi:10.1007/s11468-023-02152-w

Durdyev, S., Mbachu, J., Thurnell, D., Zhao, L., & Hosseini, M. R. (2021). BIM Adoption in the Cambodian Construction Industry: Key Drivers and Barriers. *ISPRS International Journal of Geo-Information, 10*(4), 1–14. doi:10.3390/ijgi10040215

Eadie, R., Odeyinka, H., Browne, M., Mckeown, C., & Yohanis, M. (2014). Building Information Modelling Adoption: An analysis of the barriers to Implementation. *Journal of Engineering and Architecture, 2*(1), 77–101.

Eastman, C. (1974). The use of computers instead of drawings in building design. *AIA Journal. American Institute of Architects, 63*(3), 46–50.

Eko Atlantic. (2016). *Eko Atlantic Nigeria's New Global City*. EcoAtlantic. https://www.ekoatlantic.com/milestones/Eko-Atlantic-Milestones-Issue-1. pdf

Eko Brochure. (2013). *Eko Pearl Towers*. EkoPearl. www.ekopearl.org

Ekong, M. O., George, W. K., Pandey, B. K., & Pandey, D. (2023). Enhancing the Fundamentals of Industrial Safety Management in TVET for Metaverse Realities. In *Applications of Neuromarketing in the Metaverse* (pp. 19–41). IGI Global. doi:10.4018/978-1-6684-8150-9.ch002

Emmanuel-Eze, C., Ugulu, R. A., & Egwunatum, S. I. (2021). Green Building Materials Products and Service Market in the Construction Industry. *Journal of Engineering. Project & Production Management*, *11*(2), 46–58.

Fargnoli, M., & Lombardi, M. (2020). Building information modelling (BIM) to enhance occupational safety in construction activities: Research trends emerging from one decade of studies. *Buildings*, *10*(6), 98. doi:10.3390/buildings10060098

Fayisa, G. W., & Wayessa, S. G. (2021). Cause of Construction Material Wastage on Public Building Project in Western Oromia. *American Journal of Civil Engineering*, *9*(2), 55–62. doi:10.11648/j.ajce.20210902.14

Gamil, Y., & Rahman, I. A. R. (2019). Awareness and challenges of building information modelling (BIM) implementation in the Yemen construction industry. *Journal of Engineering. Design and Technology*, *17*(5), 1077–1084. doi:10.1108/JEDT-03-2019-0063

Ganiyu, M. (2018). *BIMing the Architectural Curricula – Integrating Building Information Modelling (BIM) in Architectural Education*. Lincoln University.

Gohil, P. P., Malek, M., Bachwani, D., Patel, D., Upadhyaya, D., & Hathiwala, A. (2022). Application of 5D Building Information Modelling for Construction Management. *ECS Transactions*, *107*(1), 26372649. doi:10.1149/10701.2637ecst

Goubau, T. (2018). In *What is BIM? What its Benefits to the Construction Industry?* Aproplan. https://www.aproplan.com/blog/quality-management-plan-construction/what-is-bim-what-are-its-benefits-to-the-construction-industry

Hamma-Adama, M., & Kouider, T. (2018). A Review of Building Information Modelling in Nigeria and Its Potentials. *International Journal of Civil and Environmental Engineering. World Academy of Science and Technology*, *12*(11), 1113–1119.

Hamma-adama, M., & Kouider, T. (2019). What are the barriers and drivers toward BIM adoption in Nigeria? In *Creative Construction Conference*, Budapest University of Technology and Economics. 10.3311/CCC2019-073

Hamma-Adama, M., Kouider, T., & Salman, H. (2018). State of building information modelling (BIM) adoption in Nigeria. *34th Association of Researchers in Construction Management (ARCOM) annual conference: working papers: a productive relationship; balancing fragmentation and integration* (pp. 334-343). Belfast: ARCOM.

Hamma-adama, M., Salman, H. S., & Kouider, T. (2017). *BIM adoption in Nigeria: state of building information modeling adoption in Nigeria*. The ARCOM 2018 Conference, Belfast, UK.

He, Q., Wang, G., Luo, L., Shi, Q., Xie, J., & Meng, X. (2017). Mapping the managerial areas of Building Information Modeling (BIM) using scientometric analysis. *International Journal of Project Management*, *35*(4), 670–685. doi:10.1016/j.ijproman.2016.08.001

Ibrahim, Y. M., & Abdullahi, M. (2016). *Introduction to Building Information Modelling*. Presented at A 3 – Day Workshop/ Annual General Meeting of the Nigerian Institute of Quantity Surveyors, Port-Harcourt.

Idrees, S., & Shafiq, M. T. (2021). Factors for Time and Cost Overrun in Public Projects. *Journal of Engineering. Project & Production Management*, *11*(3), 243–254.

Iyyanar, P., Anand, R., Shanthi, T., Nassa, V. K., Pandey, B. K., George, A. S., & Pandey, D. (2023). A Real-Time Smart Sewage Cleaning UAV Assistance System Using IoT. In *Handbook of Research on Data-Driven Mathematical Modeling in Smart Cities* (pp. 24–39). IGI Global.

Jadhav, S., & Ghadge, A.N. (2016). Application of building information modeling in construction projects: A critical review. *International Journal of Engineering Research*, 26-31.

Jaya, L. G., Pandey, D., Pandey, B. K., Kaur, P., Mahajan, D. A., & Dari, S. S. (2024). Smart Big Data Collection for Intelligent Supply Chain Improvement. In AI and Machine Learning Impacts in Intelligent Supply Chain (pp. 180-195). IGI Global.

Khan, B., Hasan, A., Pandey, D., Ventayen, R. J. M., Pandey, B. K., & Gowwrii, G. (2021). 12 Fusion of Datamining and Artificial Intelligence. *Machine learning and iot for intelligent systems and smart applications*, 201.

Kirubasri, G., Sankar, S., Pandey, D., Pandey, B. K., Nassa, V. K., & Dadheech, P. (2022). Software-defined networking-based Ad hoc networks routing protocols. In *Software defined networking for Ad Hoc networks* (pp. 95–123). Springer International Publishing. doi:10.1007/978-3-030-91149-2_5

Kirubasri, G., Sankar, S., Pandey, D., Pandey, B. K., Singh, H., & Anand, R. (2021, September). A recent survey on 6G vehicular technology, applications and challenges. In *2021 9th International Conference on Reliability, Infocom Technologies and Optimization (Trends and Future Directions)(ICRITO)* (pp. 1-5). IEEE.

Kong, S. W. R., Lau, L. T., Wong, S. Y., & Phan, D. T. (2020). A study on effectiveness of Building Information Modelling (BIM) on the Malaysian construction industry. *2020 IOP Conference Series: Materials Science and Engineering*, *7*(13), 012-035. 10.1088/1757-899X/713/1/012035

Kushwaha, V. (2016). Contribution of Building Information Modelling (BIM) to solve problems in the Architecture, Engineering and Construction (AEC) Industry and Addressing Barriers to Implementation of BIM. *IACSIT International Journal of Engineering and Technology*, *3*(1), 100–105.

Liao, L., & Teo, E. A. L. (2017). Critical Success Factors for enhancing the Building Information Modelling Implementation in building projects in Singapore. *Journal of Civil Engineering and Management*, *23*(8), 1029–1044. doi:10.3846/13923730.2017.1374300

Lim, C. C., Toh, T. C., Lew, Y. L., Lau, C. X., Goh, K. C., Masrom, M. A. N., Zainordin, N., & Loo, S. C. (2020). the Perspective of Students on Drivers and Benefits of Building Information Modelling Incorporation into Quantity Surveying Profession in Klang Valley, Malaysia. *International Journal of Information Management*, *7*(1), 16–29. doi:10.15282/ijim.7.0.2020.5751

Maina, J. J. (2018). Barriers to effective use of CAD and BIM in architecture education in Nigeria. *International Journal of Built Environment and Sustainability*, *5*(3), 175–186. doi:10.11113/ijbes.v5.n3.275

Makabate, C. T., Musonda, I., Okoro, C. S., & Chileshe, N. (2021). Scientometric analysis of BIM adoption by SMEs in the architecture, construction and engineering sector. *Engineering, Construction, and Architectural Management*, *29*(1), 43–58.

Marzouk, M., Elsaay, H., & Othman, A. A. E. (2022). Analyzing BIM implementation in the Egyptian construction industry. *Engineering, Construction, and Architectural Management*, *29*(10), 4177–4190. doi:10.1108/ECAM-07-2020-0523

Matarneh, R. T., & Hamed, S. A. (2017). Exploring the adoption of building information modeling (BIM) in the Jordanian construction industry. *Journal of Architectural Engineering Technology*, *6*(10), 1–7. doi:10.4172/2168-9717.1000189

McAuleyB.HoreA.WestR. (2017). BIM in Ireland 2017, BIM Innovation Capability Programme, CitA Ltd. doi:10.21427/wy14-n123

McNamara, A. J., & Sepasgozar, S. M. (2021). Intelligent contract adoption in the construction industry: Concept development. *Automation in Construction*, *122*, 103–452. doi:10.1016/j.autcon.2020.103452

Meseko, A. A. (2014). Opportunities and Risks for Foreign Real Estate Developers in Nigeria. *International Journal of Economics. Commerce and Management*, *11*(7), 2014.

Mohammed, A., Hasnain, S. A., & Quadir, A. (2019). Implementation of building information modelling (BIM) practices and challenges in construction industry in Qatar. *Journal of Engineering Research and Reports*, *9*(1), 1–9. doi:10.9734/jerr/2019/v9i117008

Mohammed, A. M., & Mohammed, A. H. (2018), Towards Improvement in Facilities Operation and Maintenance through Feedback to the Design Team. *The Built and Human Environment Review, 3*. www.tbher.org/index.php/tbher/article/download/28/29

Muniandi, B., Huang, C., Kuo, C., Yang, T., Chen, K., Lin, Y., Lin, S., & Tsai, T. (2019). A 97% maximum efficiency fully automated control turbo boost topology for battery chargers. *IEEE Transactions on Circuits and Systems. I, Regular Papers*, *66*(11), 4516–4527. doi:10.1109/TCSI.2019.2925374

Mutis, I., & Mehraj, I. (2022). Cloud BIM Governance Framework for Implementation in Construction Firms. *Practice Periodical on Structural Design and Construction*, *27*(1), 04021074. doi:10.1061/(ASCE)SC.1943-5576.0000652

Obodoh, T. (2019). A review on building information modelling in Nigeria and its potentials. *International Journal of Civil and Environmental Engineering*, *12*(11), 1113–1119.

Okpala, I., Nnaji, C., Awolusi, I., & Akanmu, A. (2021). Developing a Success Model for Assessing the Impact of Wearable Sensing Devices in the Construction Industry. *Journal of Construction Engineering and Management*, *147*(7), 04021060. doi:10.1061/(ASCE)CO.1943-7862.0002064

Olanrewaju, O., Ajiboye Babarinde, S., & Salihu, C. (2020). Current State of Building Information Modelling in the Nigerian Construction Industry. *Journal of Sustainable Architecture and Civil Engineering*, *27*(2), 63–77. doi:10.5755/j01.sace.27.2.25142

Olanrewaju, O., Babarinde, S. A., & Salihu, C. (2020a). Current State of Building Information Modelling in the Nigerian Construction Industry. *Journal of Sustainable Architecture and Civil Engineering*, *27*(2), 63–77. doi:10.5755/j01.sace.27.2.25142

Olanrewaju, O. I., Babarinde, S. A., Chileshe, N., & Sandanayake, M. (2021). Drivers for implementation of building information modeling (BIM) within the Nigerian construction industry. *Journal of Financial Management of Property and Construction*, *26*(3), 366–386. doi:10.1108/JFMPC-12-2019-0090

Olapade, D., & Ekemode, B. (2018). Awareness and utilization of building information modelling (BIM) for facility management (FM) in a developing economy: Experience from Lagos, Nigeria. *Journal of Facilities Management*, *16*(4), 387–395. doi:10.1108/JFM-09-2017-0046

Olatunji, J. O., Dele, R. S., & Spencer, O. A. (2022). Building Information Modelling (BIM): Drivers, barriers and socioeconomic benefits. *Covenant Journal of Research in the Built Environment*, *10*(2), 13–23.

Olawumi, T., & Chan, D. (2019a). An empirical survey of the perceived benefits of executing BIM and sustainability practices in the built environment. *Construction Innovation*, *19*(3), 321–342. doi:10.1108/CI-08-2018-0065

Olawumi, T., & Chan, D. (2019b). Development of a benchmarking model for BIM implementation in developing countries. *Benchmarking*, *26*(4), 1210–1232. doi:10.1108/BIJ-05-2018-0138

Olorunfemi, E.T., Oyewobi, L.O., Olanrewaju, O.I. &Olorunfemi, R.T. (2021). *Competencies and the penetration status of building information modelling among built environment professionals in Nigeria.* Proceedings of International Federation of Surveyors, FIG e-Working Week 2021 Smart Surveyors for Land and Water Management - Challenges in a New Reality, Virtual.

Olugboyega, O. (2016). Building Information Modelling-Based Projects in Nigeria Evidences from Eko Atlantic City. *PM World Journal*, (10), 1–14.

Olugboyega, O., & Aina, O. (2016). Analysis of building information modelling usage indices and facilitators in the Nigerian construction industry. *Journal of Logistics. Information and Service Sciences*, *3*(2), 1–36.

Onungwa, I. O., & Uduma-Olugu, N. (2016). *Building Information Modeling in Nigeria and its Impact on Collaboration in Schematic Design Stage and Post Contract Stage of Design. 9th CIDB Post Graduate Conference on Emerging Trends in Construction organizational Practices and Project Management Knowledge Area*, Cape Town, South Africa.

Onungwa, I. O., & Uduma-Olugu, N. (2017). Building Information Modelling and Collaboration in the Nigerian Construction Industry. *9th CIDB Post Graduate Conference on Emerging Trends in Construction organizational Practices and Project Management Knowledge Area,* Cape Town, South Africa. 10.15641/jcbm.1.2.53

Opoku, D. G. J., Perera, S., Osei-Kyei, R., & Rashidi, M. (2021). Digital twin application in the construction industry: A literature review. *Journal of Building Engineering*, *40*, 102726. doi:10.1016/j.jobe.2021.102726

Osuizugbo, I. (2020). Disruptions and Responses within Nigerian Construction Industry amid COVID-19 Threat. *Covenant Journal in Research and Built Environment*, *8*(2), 37–48.

Oteng, T., Ansah, M. K., Kissi, E., & Eshun, B. T. B. (2018), Barriers to the Adoption of Building Information Modeling in Developing Countries: The Case of Ghana. *1st International Conference on Construction Futures*, Wolverhampton., UK.

Oyewole, E., & Dada, J. (2019). Training gaps in the adoption of building information modelling by Nigerian construction professionals. *Built Environment Project and Asset Management*, *9*(3), 399–411. doi:10.1108/BEPAM-10-2017-0090

Pandey, B. K., Nassa, V. K., Pandey, D., George, A. S., Gupta, M., Dadheech, P., & George, A. H. (2023). An Analytical Analysis of the Present-Day Procedures for Disposing of Waste. In *Handbook of Research on Safe Disposal Methods of Municipal Solid Wastes for a Sustainable Environment* (pp. 24–34). IGI Global. doi:10.4018/978-1-6684-8117-2.ch002

Pandey, B. K., & Pandey, D. (2023). Parametric optimization and prediction of enhanced thermoelectric performance in co-doped CaMnO3 using response surface methodology and neural network. *Journal of Materials Science Materials in Electronics*, *34*(21), 1589. doi:10.1007/s10854-023-10954-1

Pandey, B. K., Pandey, D., & Sahani, S. K. (2024). Autopilot control unmanned aerial vehicle system for sewage defect detection using deep learning. *Engineering Reports*, 12852. doi:10.1002/eng2.12852

Pandey, D., Pandey, B. K., & Wairya, S. (2021). Hybrid deep neural network with adaptive galactic swarm optimization for text extraction from scene images. *Soft Computing*, *25*(2), 1563–1580. doi:10.1007/s00500-020-05245-4

Pandey, J. K., Jain, R., Dilip, R., Kumbhkar, M., Jaiswal, S., Pandey, B. K., & Pandey, D. (2022). Investigating Role of IoT in the Development of Smart Application for Security Enhancement. In *IoT Based Smart Applications* (pp. 219–243). Springer International Publishing.

Pessoa, S., Guimarães, A. S., Lucas, S. S., & Simões, N. (2021). 3D printing in the construction industry-A systematic review of the thermal performance in buildings. *Renewable & Sustainable Energy Reviews*, *141*, 110794. doi:10.1016/j.rser.2021.110794

Philip, D., & Mordue, S. (2017). *Bim, Scope and Definition. Common learning outcome for European managers in construction.* Warsaw University of Technology, Civil Engineering Faculty, Department of Construction Engineering and Management. polcen.com.pl

Pradhananga, P., ElZomor, M., & Santi Kasabdji, G. (2021). Identifying the Challenges to Adopting Robotics in the US Construction Industry. *Journal of Construction Engineering and Management*, *147*(5), 05021003. doi:10.1061/(ASCE)CO.1943-7862.0002007

Pramanik, S., Pandey, D., Joardar, S., Niranjanamurthy, M., Pandey, B. K., & Kaur, J. (2023, October). An overview of IoT privacy and security in smart cities. In AIP Conference Proceedings (Vol. 2495, No. 1). AIP Publishing. doi:10.1063/5.0123511

Puteri, S., & Puspitasari, P. (2021). Exploratory study: Conservation practices and the role of architect. *IOP Conference Series. Earth and Environmental Science*, *780*(1), 27–41. doi:10.1088/1755-1315/780/1/012061

Raghavan, R., Verma, D. C., Pandey, D., Anand, R., Pandey, B. K., & Singh, H. (2022). Optimized building extraction from high-resolution satellite imagery using deep learning. *Multimedia Tools and Applications*, *81*(29), 42309–42323. doi:10.1007/s11042-022-13493-9

Raja, D., Kumar, D. R., Santhiyakumari, N., Kumarganesh, S., Sagayam, K. M., Thiyaneswaran, B., Pandey, B. K., & Pandey, D. (2024). A compact dual-feed wide-band slotted antenna for future wireless applications. *Analog Integrated Circuits and Signal Processing*, *118*(2), 1–15. doi:10.1007/s10470-023-02233-0

Ruya, T. F., Chitumu, Z. D., & Kaduma, L. A. (2018). *Challenges of Building Information Modelling Implementation in Africa, A Case of Nigerian Construction Industry. FIG Congress 2018*, Istanbul, Turkey.

Sahani, K., Khadka, S. S., Sahani, S. K., Pandey, B. K., & Pandey, D. (2023). A possible underground roadway for transportation facilities in Kathmandu Valley: A racking deformation of underground rectangular structures. *Engineering Reports*, 12821. doi:10.1002/eng2.12821

Saka, A., & Chan, D. (2020). Profound barriers to building information modelling (BIM) adoption in construction small and medium-sized enterprises (SMEs): An interpretive structural modelling approach. *Construction Innovation*, *20*(2), 261–284. doi:10.1108/CI-09-2019-0087

Saka, A. B., & Chan, D. W. M. (2019b). A global taxonomic review and analysis of the development of BIM research between 2006 and 2017. *Construction Innovation*, *19*(3), 465–490. doi:10.1108/CI-12-2018-0097

Saxena, A., Sharma, N. K., Pandey, D., & Pandey, B. K. (2021). Influence of tourists satisfaction on future behavioral intentions with special reference to desert triangle of Rajasthan. *Augmented Human Research*, *6*(1), 1–9. doi:10.1007/s41133-021-00052-4

Scott, D. J., Broyd, T., & Ma, L. (2021). Exploratory literature review of blockchain in the construction industry. *Automation in Construction*, *132*, 103914. doi:10.1016/j.autcon.2021.103914

Silverio, A. K., Suresh, S., Renukappa, S., & Heesom, D. (2021). Status of BIM implementation in the Dominican Republic construction industry: An empirical study. *Journal of Engineering. Design and Technology*, *21*(2), 417–441. doi:10.1108/JEDT-05-2021-0253

Singh, H., Pandey, B. K., George, S., Pandey, D., Anand, R., Sindhwani, N., & Dadheech, P. (2022, July). Effective Overview of Different ML Models Used for Prediction of COVID-19 Patients. In *Artificial Intelligence on Medical Data: Proceedings of International Symposium, ISCMM 2021* (pp. 185-192). Singapore: Springer Nature Singapore.

Singh, I. (2017). *BIM adoption and implementation around the world: initiatives by major nations.* GeoSpatial World. https://www.geospatialworld.net/blogs/bim-adoption-aroundthe-world/

Swapna, H. R., Bigirimana, E., Madaan, G., Hasan, A., Pandey, B. K., & Pandey, D. (2023). Impact of Neuromarketing on Consumer Psychology in Digitally Connected Networks. In *Applications of Neuromarketing in the Metaverse* (pp. 193–205). IGI Global. doi:10.4018/978-1-6684-8150-9.ch015

Theiler, M., & Smarsly, K. (2018). IFC Monitor: An IFC Schema Extension for Modeling Structural Health Monitoring Systems. *Advanced Engineering Informatics, 37*, 54–65. doi:10.1016/j.aei.2018.04.011

Ullah, K., Lill, I., & Witt, E. (2019). An overview of BIM adoption in the construction industry: Benefits and barriers. *Emerald Reach Proceedings Series, 2*(5), 297–303. doi:10.1108/S2516-285320190000002052

Umar, A. (2019). *An Assessment Of Critical Success Factors For Building Information Modelling Adoption By Clients In The Construction Industry.* [Unpublished Masters Thesis, Ahmadu Bello University, Zaria].

Vectorworks, (2020). *BIM Tools.* Vectorworks. https://www.vectorworks.net/bim-tools

Vinodhini, V., Kumar, M. S., Sankar, S., Pandey, D., Pandey, B. K., & Nassa, V. K. (2022). IoT-based early forest fire detection using MLP and AROC method. *International Journal of Global Warming, 27*(1), 55–70. doi:10.1504/IJGW.2022.122794

Wikipedia. (2020). *Archi CAD.* Wikipedia. https://en.m.wikipedia.autodesk.com/support/revit-2018

Yusuf, A. O. (2020a). Building Information Modeling (BIM): A Necessity for the 21st Century Civil Engineer in Nigeria. *Being a paper presented* at the 2nd International Civil Engineering Conference (ICEC2020). Department of Civil Engineering, Federal University of Technology, Minna, Nigeria.

Yusuf, M. (2020). Assessment of Critical Success Factors for Construction Projects in Pakistan: Advancing and Integrating Construction Education, Research & Practice. *First International Conference on Construction in Developing Countries held on 4-5 August at Karachi, Pakistan.* IEEE.

Zaidi, A. (2023). Predicting wildfires in Algerian forests using machine learning models. *Heliyon, 9*(7), e18064. Advance online publication. doi:10.1016/j.heliyon.2023.e18064 PMID:37519679

Zaidi, A. (2023). Smart Implementation of Industrial Internet of Things using Embedded Mechatronic System. *IEEE Embedded Systems Letters.* Advance online publication. doi:10.1109/LES.2023.3309985

Zaidi, A., Ajibade, S.-S. M., Musa, M., & Bekun, F. V. (2023). New Insights into the Research Landscape on the Application of Artificial Intelligence in Sustainable Smart Cities: A Bibliometric Mapping and Network Analysis Approach. *International Journal of Energy Economics and Policy., 13*(4), 287–299. doi:10.32479/ijeep.14683

Zaidi, A., & Al-Harbi, O. (2023). Statistical Analysis of a Linear Multi-Step Numerical Treatment. *Journal of Statistics Applications & Probability, 12*(1), 26. doi:10.18576/jsap/120126

Zaidi, A., & Al Luhayb, A. (2023). Two Statistical Approaches to Justify the Use of the Logistic Function in Binary Logistic Regression. *Mathematical Problems in Engineering, 2023*, 5525675. doi:10.1155/2023/5525675

Zhao, X. (2017). A scientometric review of global BIM research: Analysis and visualization. *Automation in Construction, 80*, 37–47. doi:10.1016/j.autcon.2017.04.002

Zigurat, U. M. (2018). *BIM Awareness and Level of ICT Penetration in the Nigerian Construction Industry.* [An Unpublished M.Sc. Thesis, School of Construction Management and Engineering, the University of Reading, United Kingdom].

# Chapter 16
# Emerging Engineering Technologies:
## Design, Development, and Industrial Applications of IoT-Based Aquaculture Monitoring System

**P. Selvakumar**
https://orcid.org/0000-0002-3650-4548
*Nehru Institute of Technology, India*

## ABSTRACT

*India is the third-largest fish producer in the world and the second-largest aquaculture country after China. The Indian Blue Revolution highlighted the significance of the aquaculture and fishing industries. This industry is viewed as one that is growing and has the potential to have a significant impact on India. For aquatic life to remain healthy, water quality monitoring is essential. Traditional methods for water quality monitoring involve manual sample collection and complex procedures. They are ineffective, expensive, lack real-time monitoring, and require laboratory testing. To address these issues, the proposed method for monitoring the water quality in aquaculture uses IoT to determine the pH value of the water, which can identify poor water quality that may result in disease or negatively impact living things. Similar to early water quality detection, which prompts the owner to take the proper action, this strategy alerts the owner.*

## INTRODUCTION

Fresh water management is essential due to India's rapidly expanding population, which is driving up agricultural, industrial, and other demands. Fresh water's "chemical, physical, and biological" composition dictates its quality. Monitoring the quality of the water helps identify contaminants, dangerous chemicals, and pollution. The traditional method, which is still used today, entails gathering water samples,

DOI: 10.4018/979-8-3693-1335-0.ch016

analyzing them in a lab, and making suggestions for water treatment, among other things. The current method for monitoring water pollution consists of three basic steps: -

1. Water sampling
2. Sample testing
3. Investigational analysis

All three steps are ineffective, extremely expensive, difficult, and time-consuming. They also demand expert advice. Because of this, rather than relying on a manual process, automation may now be used to monitor water quality and take appropriate action. This has led to some technological advancement in water quality monitoring, which will help to monitor water quality rather than relying on human processes. A small amount of research has been done on the use of machine learning techniques in water standard monitoring in addition to the use of technology to monitor water quality. A subset of artificial intelligence (AI) called machine learning enables machines to learn without having to be logically organized. In one of the studies that used machine learning to track the water quality in rivers, Least Squares Vector Machine was employed. As a result, we have developed an intelligent Internet of Things-based water quality monitoring system that uses a PH sensor and a TDS meter to regularly collect water parameters from different types of water. This is due to the development of machine-to-machine interaction, which leads to the Internet of Things with devices that interact with each other without human intervention.

Through serial communication with the Raspberry Pi 3, these water characteristics are gathered and supplied to the microcontroller. The Pi3 CPU was used to run the K-Means clustering algorithm, which divided the water frameworks into various groups based on pH and TDS. The data set was then trained to determine whether the water quality was good or bad. In order for water authorities to take action, this information is posted on the cloud's website. This was developed as a prototype using the Raspberry Pi 3 and Arduino. It is becoming more and more important to monitor water qualities and characteristics like pH, temperature, and turbidity as the aquaculture industry expands. Water temperature affects when fish eat and how much they grow. In comparison to frozen water, warm water has less dissolved oxygen. Aquatic animals and plants in stationary aquariums require particular care. Their lives could be cut short because there is a smaller amount water in the tank. The right amounts of daylight and carbon dioxide are necessary for the growth of fish and plants. It's essential to constantly monitor the state of the water as a result.

A. The Traditional Approach's Drawbacks
   - Manually collect the sample.
   - Difficult methodology
   - It takes a long time.
   - Measurement precision is poor.
   - Expensive.
   - There is no live tracking.

B. Benefits of the Proposed System
   - There is no requirement for manual collecting.
   - It is simple to use.
   - It is a low-cost and fast process.

## LITERATURE REVIEW

Any research project must include a literature review because it imparts the knowledge and abilities needed to carry out an effective survey. Reviewing previously published materials on the research topic aids in finishing a project. The primary goal of the literature review is to assist researchers in creating a solid foundation and enhancing their project. The following are some of the preceding published papers that are being deliberated for the system's implementation:

In their study titled "Online monitoring of water standard using raspberry Pi3model B," Kalpana claims that the main objective of the planned paper is to without human intervention monitor water quality. The reproduction is used to compute inputs and update values on the server without the need for human intervention. The system collects data from the project's various sensors via the internet using a Raspberry Pi 3 model B. The model is quick, easy, and efficient for water monitoring. Any sensor can be used with the system, and Python is used for the coding. The system is extremely extensible and has a wide range of applications. Monitoring other values, including hydrology, air pollution, commercial and agricultural productivity, and so forth, is possible using it (Kalpana, 2016).

(Nikhil Kedia's, 2006) article "Water Grade Monitoring for Rural Areas-A Sensor Cloud Based Economical Project." Published at the 1st global Conference on Next Generation Computing Technologies in Dehradun, India (NGCT-2015). This study focuses on the complete water value monitoring system, sensor, embedded design, and information display system, as well as the duties of the government, network operators, and villages to ensure the dissemination of the right information. Additionally, the Sensor Cloud is examined. Even though it is at present not possible to robotically improve water quality, careful automation and cost-effective procedures can help improve both the quality of the water and the public's perception of it.

In their study titled "QOI-Aware Energy Management in Internet-of-Things Sensory Environments," Zhanwei Sun, Chi Harold Li, Chatschik Bisdikian, Joel W. Branch, and Bo Yang claim. For delivering a respectable QOI experience in IOT sensory environments, they defined an competent energy administration framework. Unlike past efforts, it is transparent and encourages the use of low-tech methods, preserving long-term energy conservation without caving in to any achievable QOI standards. The innovative notion of "sensor-to-task relevance" with QOI awareness meticulously considers the sensing capabilities offered by a sensor to IOT sensory settings, as well as the QOI requirements of a particular work. The "essential coverage set" of any particular activity is a novel concept for selecting sensors to support a task over time. An energy administration decision is made energetically at runtime under the constraint of service delay, which is the optimal choice for long-term traffic data (Zhanwei Sun et.al 2012).

The study "IoT-based water quality monitoring system" was designed by Vaishnavi V., Daigavane, and M.A. Gaikwad. Real-time and affordable water tracking is made possible by the Internet of Things-based structure. The structure is made up of a number of sensors that are connected to an Arduino board in order to determine the desired value for observing a water property. A wi-fi system that is connected to the internet is used to analyze the sensor data. When the input is ready for visualisation, it is sent to the BLYNK application, where it is displayed using a variety of widgets and the owner is informed.

Anand S., Kaimal A., Jaison R., Santha V., and V.'s "Smart Aquarium" The project entails keeping an eye on the tank's pH, water level, temperature, and fish feeding schedule. Utilizing Programmable Logic Controllers, Supervisory Control, and Data Acquisition, numerous aquarium parameter controls and monitoring procedures are automated. The project could be increased in scope to be used on a larger scale. The processor is secure at home thanks to this clever aquarium defense spell. Due to the fact that PLCs are used as controllers, multiple aquariums can be connected and managed by a single PLC.

"Adaptive Edge Analytics for Distributed Networked Control of Water Systems" is the title of a study created by (Sokratis Kartakis, Weiren Yu, Reza Akhavan, and Julie A, 2016). McCann. This study presents a barrage detection and echolocation method for water distribution networks that combines graph topology analytics, gauzy compression, and incongruity identification. We demonstrate that this approach effectively localizes water barrage occurrences by comparing the reaching times of vibration differences detected at various sensor locations, while also minimizing the amount of communication between sensor equipment and back-end servers. Our consequences can result in communications cost savings of up to 90% when compared to typical periodic reporting scenarios.

(Jayti Bhatt and Jignesh Patoliya's, 2016) "Real Time Water Quality Monitoring System." This study describes how to keep drinking water shielded from contamination by monitoring its quality in real time. We created a brand-new IOT-based water standard tracking method to achieve this goal. Here, we go over the design of an IOT-based system for measuring water quality standards and monitoring it in real time. The sensors in this system gauge the temperature, dissolved oxygen, potential, and other characteristics of the water's quality. The microcontroller processes the sustained standards from the sensors, and the processed data is sent to the core controller, in this case the Raspberry Pi, via the Zigbee protocol. Finally, sensor data can be scanned using cloud computing from an internet browser application.

In their study titled "Smart water standard tracking system with budget-friendly using IoT," (Pasika, Sathish, and Sai Teja Gandla, 2020) reportedly discussed how to design a system that is well-thought-out, affordable, and capable of operating in real-time. The main goal of the system's IOT-based water quality monitoring is to assess if drinking water is suitable for human consumption. The vital sensor for tracking the water grade in this model is coupled to a microcontroller unit (MCU), and further processing is done via a web application. ThingSpeak is utilized as a data analysis and visualization tool. The platform has a number of features for transmitting data to a server and cloud for data storage, as well as for graphically displaying findings using MATLAB programming.

"Water monitoring and analytic based ThingSpeak," according (Miry, Gregor Alexander Aramice, and Abbas Hussien, 2020). The main objective of the suggested structure is to explain why real-time tracking is preferable to traditional water grade monitoring. The IoT-based system uses the internet to check the quality of the water in various locations. The system is composed of a hardware component that utilizes the ThingSpeak platform for data visualization and observation and is written in the Arduino IDE. The ThingSpeak platform will include a ThingSpeak server and cloud that can be used for data storage and analysis. The web and mobile applications for the ThingSpeak platform allow viewing the intended numbers, and an alert notification is sent to the user to alert them to the state of the water. The proposed paradigm may be utilized in other fields, such as healthcare.

## PROPOSED SYSTEM

The suggested paradigm is an embedded system with both hardware and software. Because an embedded structure is based on a microprocessor, it combines software and hardware programming. An inventive and effective system for determining water grade is built using the Internet of Things (IoT). It also has the sensors needed to gather environmental data, move it to a higher layer for data observation, and send it to the ThingSpeak server Patel, Shweta, et al (2020).

The most critical component of the strategy is the embedded structure, which includes NodeMCU as a Wi-Fi component. The central part of the system, which links all of the hardware and software together. To power the Wi-Fi module, a 3.3v/5v power supply is used. To collect the crucial data, we are using a total of three sensors, including a pH sensor, a heat sensor, and an opacity sensor. Each sensor has an input hook and a ground hook that are connected to the ground hook and the analogue hook of the NodeMCU in order for it to operate on a 3.3v/5v power source (Huan, Juan 2020),. After the sensors are connected, data is collected and sent to be visualized, as seen in Figure 1.

The sensors and NodeMCU are programmed in the Arduino IDE using the user-friendly embedded C language after the appropriate board and ports have been configured. The ThingSpeak server is used to create a new channel, configure widgets in accordance with specifications, and obtain an API key to connect the NodeMCU with ThingSpeak in order to visualize the results. The API key is then added to the code before it is uploaded to ThingSpeak. The code for all sensors is then created in the Arduino IDE. ThingSpeak then displays the output, which is the input data gathered from the sensors, as a conclusion. IFTTT Applets are later used to send notifications when the sensor detects changes in the environment (Encinas, Cesar, 2017).

## HARWARE COMPONENT

### pH Sensor

In chemistry, pH refers to the strength or potential of hydrogen in a solution. It is used to determine whether a solution is basic or acidic. At a temperature of 25 °C, a solution is categorized as acidic if its pH is lower than 7, and basic if its pH is higher than 7. The pH of seven is thought to be neutral. Human consumption is both safe and healthy for pH levels between 6.5 and 9.5. To obtain comprehensive information about the quality and characteristics of water, a pH sensor is used to monitor the pH level of the water in the installed tank (as shown in Figure 2). When the readings drop, the water in the tank needs to be refilled (N. Vijayakumar and R. Ramya, 2015).

### NodeMCU

The Node Micro Controller Unit (NodeMCU), an open-source IoT platform, is depicted in Figure 3. The ESP8266 WiFi SoC is used by the NodeMCU. A Wi-Fi module connects each sensor to the Internet of Things. It is a place where software and hardware development takes place. After installing the required boards and ports, the NodeMCU is programmed using the Arduino IDE. The USB cord provides the power. For NodeMCU, a plethora of connectors, features, and specs are accessible (K. Andersson and M. S. Hossain, 2015).

*Figure 1. Algorithm for the proposed system*

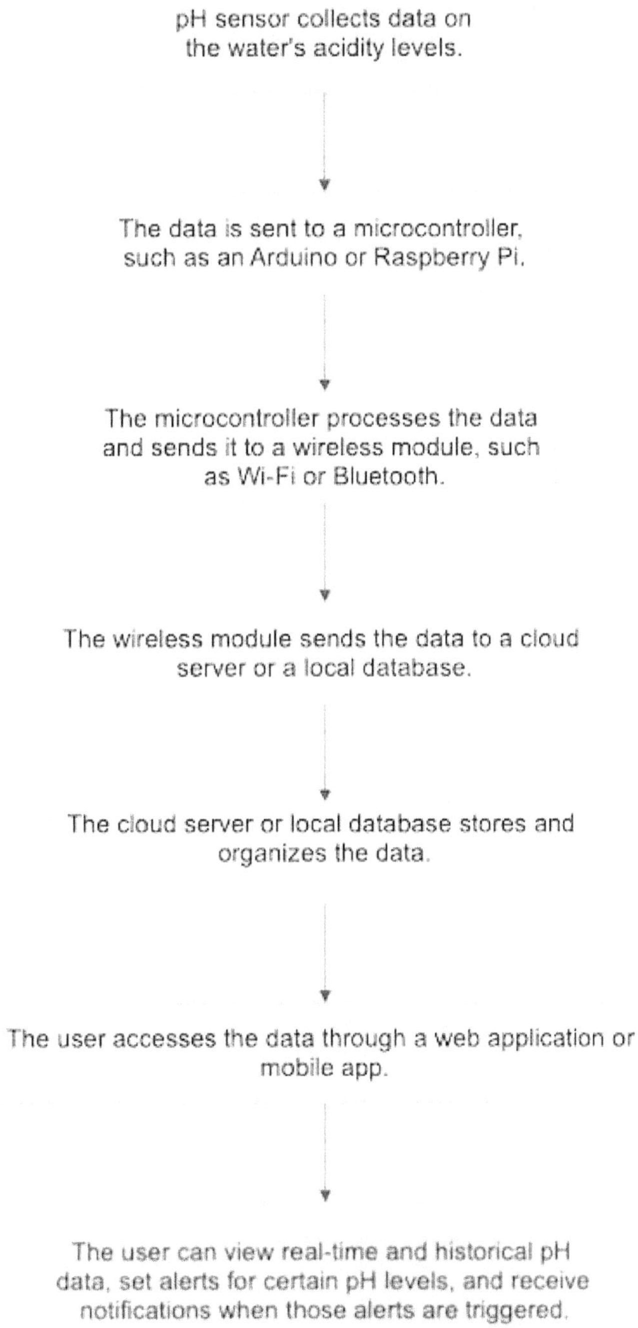

*Figure 2. pH sensor*

## SOFTWARE COMPONENT

### Arduino IDE

The Arduino Integrated Development Environment can be used on computers running Windows, Mac OS, and Linux (IDE). With the help of this IDE, we can program and upload data to all of the project's relevant sensors[16]. This IDE offers Embedded C for coding on all types of hardware and employs

*Figure 3. NodeMCU*

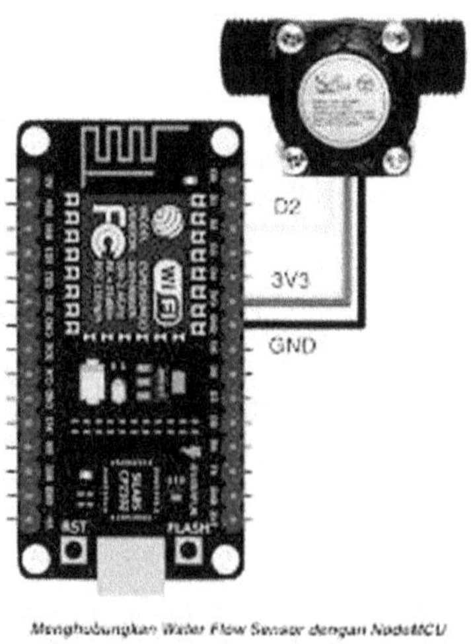

*Figure 4. ThingSpeak server*

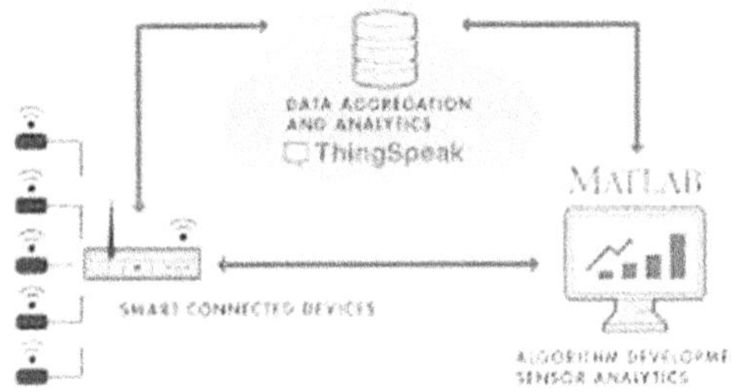


a unique code structuring rule. A software media center for the wiring project, which includes input and output methods, is included in the Arduino IDE in addition to built-in programs for a variety of microcontrollers.

### ThingSpeak Server

ThingSpeak is a free, open-source IOT analytics platform that provides real-time data collection and analysis, cloud data storage, and MATLAB programming for data visualization. The primary purpose of the ThingSpeak server, as depicted in Figure 4, is to securely upload sensor data collected as input to the cloud. Then, using MATLAB, display and analyze your data and graphically present the findings. By adding react and action, we can also add several of our desired gimmicks and carry out numerous operations on them, and the server will produce a response (R. P. N. Budiarti et.al, 2016).

## RESULTS AND DISCUSSION

Figure 5(a) displays the computed values for temperature, turbidity, pH, and ORP. It continuously measures the temperature, turbidity, pH, and ORP. Real-time findings are shown on an LCD, PC, or mobile device. If the obtained result is more than the threshold value, "BAD" will be shown. If the obtained value is less than the threshold value, the comments will be shown as "GOOD". A bar/line graph will be given for your reference. A time series display of sensor data featuring options is shown in Figure 5.

Figure 5. (a) The values of pH, turbidity, temperature, and ORP that were detected as a consequence are displayed in the figure. It continuously checks the temperature, turbidity, pH, and ORP and shows the data in real time on an LCD or mobile device[21]. If the obtained result is more over the threshold value, it will be displayed as "BAD." If the obtained result falls below the threshold value, the remarks will be displayed as "GOOD." A bar/line graph will be supplied for easier comprehension. a time series representation of sensor data for decision-making.

## CONCLUSION

The proposed system's main objective is to use smart water standard tracking in a placed tank. This device will continuously check the water's standards and properties. Monitoring water quality is necessary for

*Figure 5. (a) Computed values for temperature, turbidity, pH, and ORP; (b) Water quality monitor*

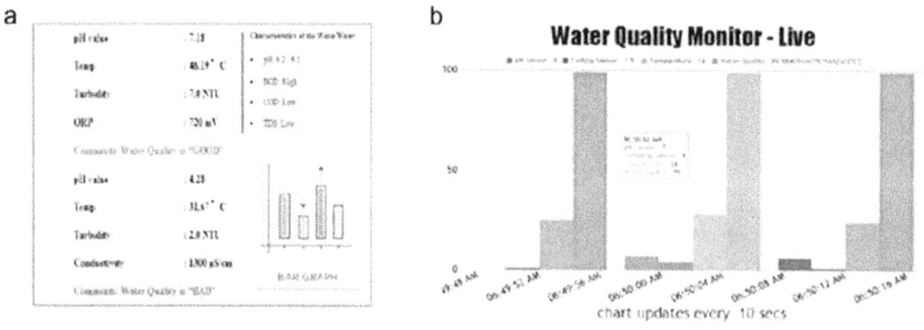


aquatic plants and animals to live in a safe and healthy environment. This article presents a low-cost, highly efficient system for tracking water in real-time. Users can follow and view data gathered from numerous sensors in an organized way using an online tool called ThingSpeak. The software then sends this data to the ThingSpeak server. The ThingSpeak mobile app allows users to follow the evaluation process in addition. Later, IFTTT applets that use ThingSpeak services like ThingHTTP and ThingSpeak Reacts start to appear. The ThingSpeak service will notify the user when the observed values exceed the threshold because it uses IFTTT applet URLs.

## REFERENCES

Andersson, K., & Hossain, M. S. (2015). Heterogeneous Wireless Sensor Networks for Flood Projection Resolution Support Systems. *IEEE Conference on Computer Communications Workshops (INFOCOM WKSHPS),* (pp. 133–137). IEEE.

Budiarti, R. P. N., Widyatmoko, N., & Purnomo, M. H. (2016). Web scraping for automated water standard tracking system: A case study of PDAM Surabaya. International Seminar on Intelligent Technology and Its Applications (ISITIA).

Encinas, C. (2017). Design and implementation of a distributed IoT system for the monitoring of water quality in aquaculture. Wireless Telecommunications Symposium (WTS). IEEE.

Huan, J., Li, H., Wu, F., & Cao, W. (2020). Design of water quality monitoring system for aquaculture ponds based on NB-IoT. *Aquacultural Engineering, 90*, 102088. doi:10.1016/j.aquaeng.2020.102088

Kalpana. (2016). Online monitoring of water standard using raspberry Pi3 model B. *International Journal of Innovative Technology and Research, 4*.

Kartakis, S. & Julie, A. (2016). McCann, Reza Akhavan, IEEE First International Conference on Internet-of-Things Design and Implementation. IEEE.

Kedia, N. (2015). *Water Grade Monitoring for Rural Areas-A Sensor Cloud Based Economical Project.* Dehradun, India.

Miry, A. (2020). Water tracking and analytic based ThingSpeak. *International Journal of Electrical and Computer Engineering 10*(4), 3588.

Vijayakumar, N., & Ramya, R. (2015). The real timemonitoring of water quality in IoT environment. *International Conference on Innovations in Information*. Research Gate.

# Chapter 17
# Deployment of Load Balancing in Cloud Computing:
## Industrial Application and Benefits

**Binay Kumar Pandey**
https://orcid.org/0000-0002-4041-1213
*Department of Information Technology, College of Technology, Govind Ballabh Pant University of Agriculture and Technology, Pantnagar, India*

**A. P. Mukundan**
https://orcid.org/0009-0009-5608-4788
*Champions Group, Singapore & Aligarh Muslim University, India*

**Vinay Kumar Nassa**
https://orcid.org/0000-0002-9606-7570
*Department of Information Communication Technology, Tecnia Institute of Advanced Studies, India*

**Digvijay Pandey**
https://orcid.org/0000-0003-0353-174X
*Department of Technical Education, Government of Uttar Pradesh, India*

**A. Shaji George**
https://orcid.org/0000-0002-8677-3682
*Business System Department, Almarai Company, TSM, Riyadh, Saudi Arabia*

**A. Shahul Hameed**
*Department of Telecommunication, Consolidated Techniques Co. Ltd., Saudi Arabia*

**Pankaj Dadheech**
https://orcid.org/0000-0001-5783-1989
*Swami Keshvanand Institute of Technology, Management, and Gramothan, India*

## ABSTRACT

*Cloud computing involves virtualization, distributed computers, networking, software, and web services. Clouds have clients, datacenters, and servers. It has fault tolerance, high availability, scalability, flexibility, little user overhead, cheap ownership cost, on-demand services, etc. These difficulties require a powerful load balancing algorithm. Memory, CPU, latency, or network-load. Load balancing distributes demand to avoid overloaded distributed system nodes and optimises resource use and job response*

DOI: 10.4018/979-8-3693-1335-0.ch017

*time. Load balancers ensure that processors and network nodes perform similarly. Methods initiated by sender, recipient, or symmetric. Using the divisible load scheduling theorem, a load balancing method will optimise throughput and latency for clouds of different sizes.*

## 1. INTRODUCTION

Cloud computing involves virtualization, distributed computers, networking, software, and web services. Clouds have clients, datacenters, and servers. It has fault tolerance, high availability, scalability, flexibility, little user overhead, cheap ownership cost, on-demand services, etc. These difficulties require a powerful load balancing algorithm. Memory, CPU, latency, or network-load. Load balancing distributes demand to avoid overloaded distributed system nodes and optimizes resource use and job response time. Load balancers ensure that processors and network nodes perform similarly (Pandey, B. K.,et al., 2022). Methods initiated by sender, recipient, or symmetric. Using the divisible load scheduling theorem; a load balancing method will optimize throughput and latency for clouds of different sizes (Bessant, Y. A.,et al.,2023).

The phrase "cloud computing" refers to a collection of practices that includes virtualization, distributed computing, networking, software, and online service provisioning. All of these things are elements that make up cloud computing. Clients, a centralized data storage facility, and a huge number of servers that are spread out around the globe are some of the components that come together to make what is known as a cloud. It has a number of benefits, including on-demand service provisioning, fault tolerance, high availability, scalability, flexibility, decreased overhead for users, and a lower total cost of ownership. In addition to these benefits, it also has a wide range of other benefits (Anthony T.Velte,et al.,2010).

In view of these concerns, it is of the utmost urgency to work towards the construction of a reliable algorithm for load balancing. This must be done immediately. It is likely that the load will have an effect on the capacity of the memory, the degree of delay, or the processing capability of the CPU. The term "load balancing" refers to the process of distributing the workload across the various nodes of a distributed system in order to improve resource utilization and task response time, as well as to avoid a scenario in which some of the nodes are substantially occupied while others are either idle or just doing very little work. Load balancing can also be done in order to avoid a scenario in which some of the nodes are substantially occupied while others are either idle or just doing very little work (George, A. H.,et al.,2021). A situation in which some of the nodes are substantially busy while others are just doing very little work is one that you want to do everything you can to steer clear of using this strategy. Through a process known as load balancing, it is possible to make certain that all of the processors in a system or all of the nodes in a network are working on approximately the same amount of tasks at any particular instant in time. This can be verified by keeping an eye on the total quantity of work being completed by each processor or node. With this strategy, the conversation can be started by either the sender or the recipient, depending on who you want. Alternately, it may take the form of the symmetric type, in which both the sender and the recipient take turns being the one to initiate the conversation (Parthiban, K.,et al.,2021).

The fundamental purpose of this project is to design an efficient method for load balancing by utilizing the divisible load scheduling theorem as a basis. This will allow for the project to meet its primary objective. Because of this, we will have the capability to maximize or decrease a wide variety of per-

*Figure 1. A cloud is used in network diagrams to depict the Internet*

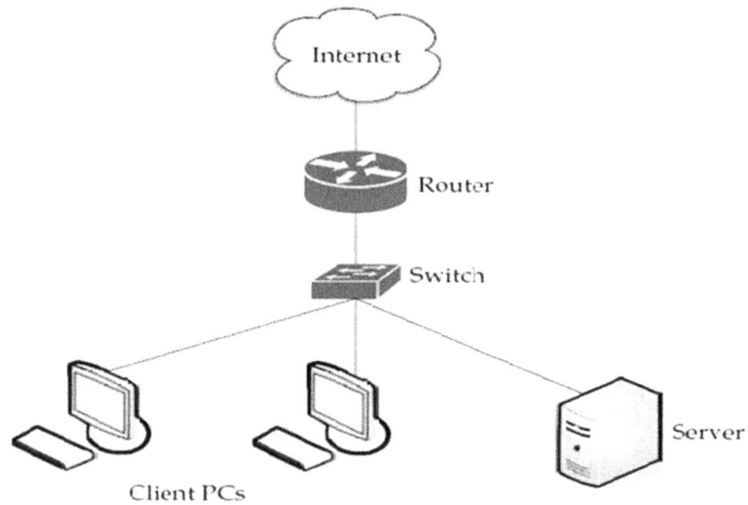


formance factors (for example, throughput and latency) for clouds of varied sizes (the demand imposed by the application will determine the virtual topologies).

The term "cloud computing" refers to a type of on-demand service in which clients are offered shared resources, information, software, and other equipment in accordance with their requirements at a given point in time. This type of service is also known as "utility computing." This category of service can also be referred to as "utility computing." When discussing anything related to the internet, it is likely that one will use this term at some point. The entire internet can be conceptualized as a cloud, which is only one way to think about it. Cloud computing can result in cost savings not only in terms of operational costs but also in terms of capital costs, which can be a significant benefit (Kumar, M. S.,et al.,2021).

At the moment, the practice of load balancing in cloud computing systems can be described as being relatively challenging. Whenever there is a problem, a decentralized approach is always the best method to solve it. primarily because it is not always technically practicable or economically viable to keep one or more idle services running in order to satisfy the necessary needs. This is the primary reason. The architecture of cloud computing is extremely complex, and its components can be found in a wide variety of locations. As a result, it is not possible to individually allocate jobs to the appropriate servers or clients in order to achieve efficient load balancing. There is a degree of unpredictability involved in the process of assigning roles while they are still being filled (Bessant, Y. A.,et al.,2023).

This research looks at a variety of approaches to load balancing in large-scale cloud computing systems and discusses a few of the more promising ones. In order to provide an analysis as well as a comparative research of these methods, as well as to demonstrate various distributed algorithms for load balancing, and to improve a variety of performance parameters for clouds of varying sizes, including throughput, latency, and others, our goal is to demonstrate these distributed algorithms. In addition, we want to demonstrate these distributed algorithms in order to demonstrate various distributed algorithms for load balancing, and to demonstrate various distributed algorithms for load balancing. In (Mequanint Moges,et al.,2005), the idea of load balancing in wireless sensor networks (WSN) was presented; this idea can also be used to cloud computing systems and serve the same purpose. This is due to the fact

*Figure 2. Three components make up a cloud computing solution*

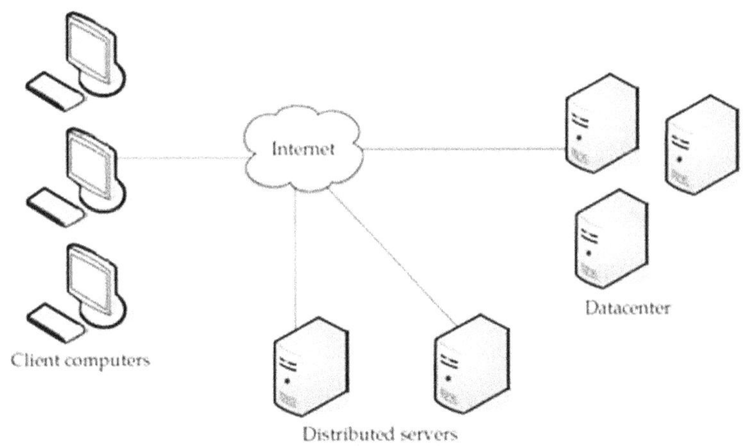


that a WSN is comparable to a cloud in the sense that it has a number of master computers (Servers) and a number of slave computers (Clients) that are coupled together in a convoluted structure. This idea can also be utilized for cloud computing systems given that the Internet as a whole can be conceptualized as a cloud that comprises numerous connection-less and connection-oriented services. An investigation into the relative merits of a number of distinct algorithmic approaches has been carried out by making use of the divisible load scheduling theory described in (Mequanint Moges,et al.,2005).

## 2. THE SERVICES OF CLOUD COMPUTING

When using cloud computing, rather of using locally hosted or remotely situated servers or computers, users can access the services they require from a wide number of sources that are spread out around the globe. The practice of computing in the cloud does not fit neatly into any one particular definition (Zaidi, A.,et al., 2023). In the majority of instances, it is composed of a group of geographically distributed servers that are collectively referred to as masters. Within a network that possesses the scalability and dependability of a datacenter, these masters offer the services and resources that are desired by a variety of clients who are collectively referred to as clients. The scattered computers are responsible for providing the on-demand computing services. Software resources (for example, Software as a Service, or SaaS), physical resources (for example, Platform as a Service, or PaaS), hardware/infrastructure (for example, Hardware as a Service, or HaaS), or infrastructure (for example, Infrastructure as a Service, or IaaS) can all be included in the definition of a service (Babu, S. Z. D., Pandey,et al.,2022). Amazon EC2, sometimes referred to as the Amazon Elastic Compute Cloud is an example of a specific kind of cloud computing service.

## 3. CLOUD COMPONENTS

Clients, a datacenter, and multiple dispersed servers are the three primary elements that make up a cloud computing system. Every component serves a distinct function and contributes in a unique way to the whole.

### 3.1 Clients

When managing information that is associated with the cloud, end users interact with the clients. Customers can be broken down into the following three groups, as shown in (Anthony T.Velte,et al.,2010):

- **Mobile**: Windows Mobile Smartphone, other smartphones, such as a Blackberry or iPhone, and other similar devices.
- They are not involved in the computation process in any way. They don't do anything else but display the information. Servers complete all of the tasks on their behalf. Thin clients do not have any memory built in to their systems.
- **Thick**: These connect to the Internet cloud through a variety of web browsers, including Internet Explorer, Mozilla Firefox, and Google Chrome.

Thin clients are currently more popular than other types of clients due to a number of factors, including their low price, high level of security, low level of power consumption, reduced level of noise, and ability to be quickly replaced and repaired.

### 3.2 Datacenter

A datacenter consists of nothing more than a collection of servers on which various applications are hosted. A connection is made between the end user and the datacenter so that the end user can subscribe to various applications. It's possible for a datacenter to be located in a remote location compared to its customers.

These days, a technique known as virtualization is employed in order to install a piece of software that enables numerous instances of applications that run on virtual servers.

### 3.3 Distributed Servers

Distributed servers are the components of a cloud that are dispersed across the Internet and are in charge of hosting a range of applications. They are responsible for making the cloud work. Despite this, the user will have the impression that he is accessing the application through his local system even though he is actually accessing the application over the cloud.

### 3.4 Type of Clouds

Based on the domain or environment in which clouds are used, clouds can be divided into 3 categories:

- **Public Clouds**
- **Private Clouds**
- **Hybrid Clouds (combination of both private and public clouds)**

*Figure 3. Full virtualization*

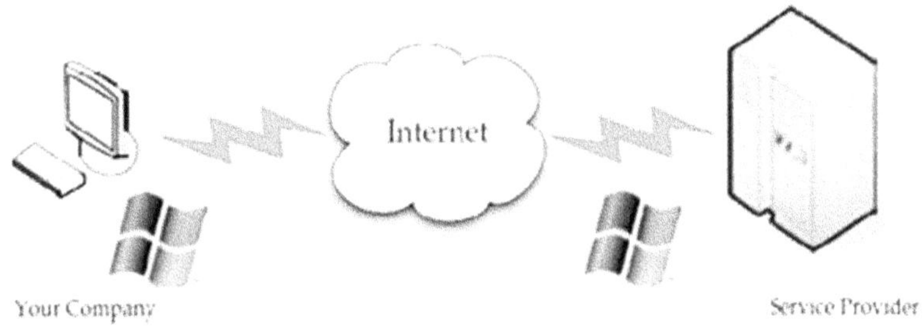


## 3.5 Virtualization

• In the context of cloud computing, this is a really helpful concept to keep in mind. The term "virtualization" refers to "something that isn't real," but it provides all of the features of a real object. It is the software equivalent of a computer that can run a variety of different programmers in the same manner as a physical device.

• Cloud computing is related to virtualization since, with virtualization, an end user is able to utilise a variety of cloud computing services. Depending on the level of virtualization implemented, the distant datacenter will be responsible for delivering a variety of services.

• In the case of clouds, there are two distinct varieties of virtualization, as shown in (Anthony T.Velte,et al.,2010):

• Full virtualization
• Para virtualization

## 3.6 Full Virtualization

In the case of complete virtualization, the installation of the operating system of one machine onto another computer is carried out in its entirety. This is done in order to simulate the full functionality of both machines.

• The services are provided in a completely virtualized fashion by the remote datacenter in this case. According to the information presented in (Anthony T.Velte,et al.,2010), full virtualization has been successful for a number of different reasons:

• • Isolating users from one another and the control program
• • Emulating hardware on another machine Sharing a computer system with numerous users
• • Isolating users from one another and the control programmers

## 3.7 Para Virtualization

Through the use of system resources such as memory and CPU in an effective manner, hardware virtualization makes it possible for many operating systems to coexist on a single computer. software like VMware, for example. In this location, not all of the services are completely available; rather, only some of the services are supplied. para virtualization, the hardware allows multiple operating systems to run

*Figure 4. Para virtualization*

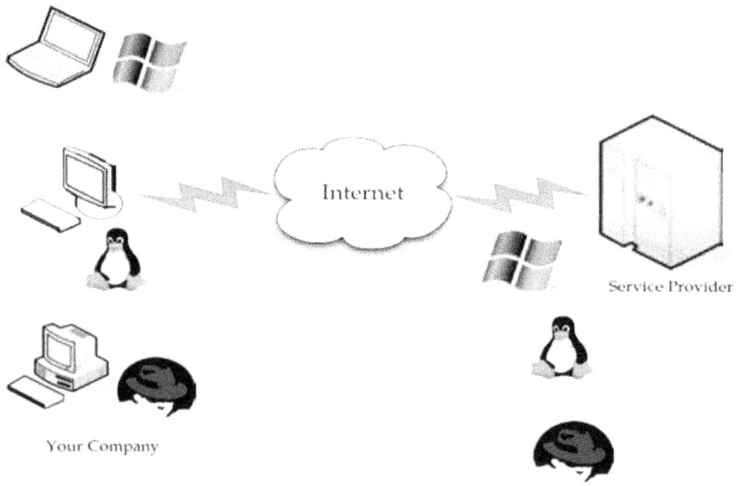


on single machine by efficient use of system resources such as memory and processor. e.g. VMware software. Here all the services are not fully available, rather the services are provided partially.

Para virtualization has the following advantages as given in (Anthony T.Velte,et al.,2010):

- **Disaster recovery**: If there is a problem with the system, the guest instances will be relocated to a different piece of hardware until the original machine can be fixed or replaced.
- **Migration: Because** the hardware can be quickly replaced, migrating, or relocating the various pieces of a new machine, is faster and easier. This is because the hardware can be simply replaced.
- **Capacity management**: When operating in a virtualized environment, it is much easier and faster to add additional processing power to hard drives, and it is also simpler and quicker to expand the capacity of existing hard drives. Because system components or hardwires may easily be moved, replaced, or repaired, capacity management is simplified and made even simpler. This makes it much easier to manage available storage space.

## 4. SERVICES PROVIDED BY CLOUD COMPUTING

Service means different types of applications provided by different servers across the cloud. It is generally given as "as a service". Services in a cloud are of 3 types as given in (Anthony T.Velte,et al.,2010):

- Software as a Service (SaaS)
- Platform as a Service (PaaS)
- Hardware as a Service (HaaS) or Infrastructure as a Service (IaaS)

### 4.1 Software as a Service (SaaS)

The user accesses many software applications hosted on various servers across the internet when utilizing a SaaS platform. The user operates the software in its current state without making any modifications, and they are not required to make a large number of modifications or integrate it with any other

*Figure 5. Software as a service (SaaS)*
*(adopted from Anthony T. Velte, et al., 2010)*

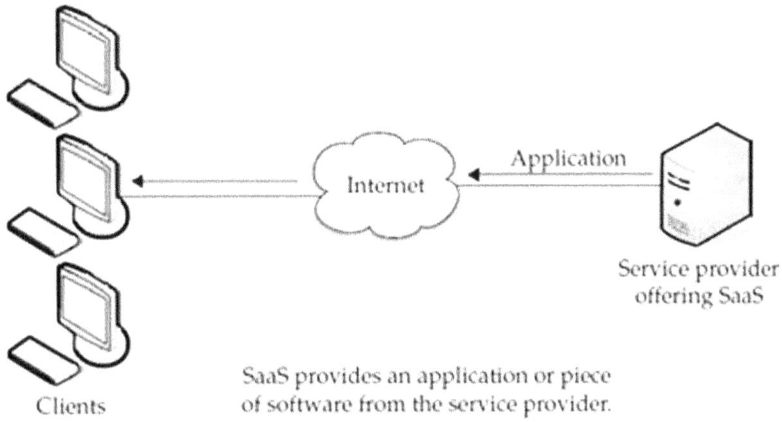


systems. The supplier is responsible for all necessary upgrades and patches while also ensuring that the infrastructure continues to function [2].

The customer will be responsible for paying for the amount of time that he makes use of the product (Muniandi, B.,et al.,2019). Software that performs a straightforward job and does not require communication with any other systems is an excellent choice for deployment as software (Pandey, B. K.,et al.,2011) that is provided as a service (Zaidi, A., et al., (2023)). Customers who do not have an interest in performing software development but who require powerful applications can also profit from software as a service (SaaS).

Some of these applications include:

- Customer resource management (CRM)
- Video conferencing
- IT service management
- Accounting
- Web analytics
- Web content management

**Benefits**: The most significant advantage of using SaaS is that it is more cost-effective than purchasing the application in its entirety (Daoud, M. K.,et al., 2023). When compared to the organization, the service provider typically provides applications that are both more affordable and more dependable (Anthony T.Velte,et al.,2010).The following are some additional advantages (mentioned in (Anthony T.Velte,et al.,2010)): A working knowledge of the internet, improved marketing, and less employees, more Internet dependability and data security, increased bandwidth, and so forth(Pramanik, S.,et al.,2021).

Obstacles:

- SaaS isn't of any help when the organization has a very specific computational need that doesn't match to the SaaS services
- While making the contract with a new vendor, there may be a problem. Because the old vendor may charge the moving fee. Thus it will increase the unnecessary costs.
- SaaS faces challenges from the availability of cheaper hardwires and open source applications.

*Figure 6. Platform as a service (PaaS)*
*(adopted from Anthony T. Velte et al., 2010)*

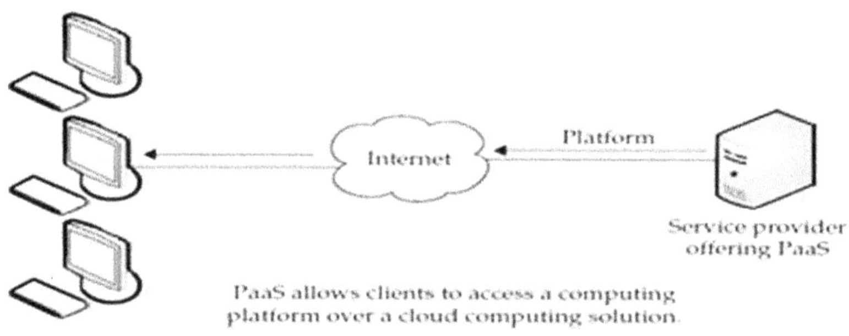


## 4.2 Platform as a Service (PaaS)

PaaS makes available over the internet all of the resources necessary for the development of software applications and web services (Pandey, B. K.,et al.,2022), and this may be done without the need to download or install any additional software (Ahmad, A. Y. B.,et al.,2023).

The design and development of software, as well as its testing, deployment, and hosting, are all included in PaaS services. The collaboration of teams, the integration of databases and web services, the storing and versioning of data, and other services are all examples of additional possibilities ((Ahmad, A. Y. B.,et al.,2024)).

## 4.3 Downfall

• Lack of portability among different providers.

• If the service provider is out of business, the user's applications, data will be lost.

## 4.10 Hardware as a Service (HaaS)

• IaaS is an abbreviation that stands for "Infrastructure as a Service." It provides the hardware to an organisation as a service so that the organisation can put anything it wants into the hardware in accordance with its own preferences (Anthony T.Velte,et al.,2010).The user is able to "rent" resources (taken from (Anthony T.Velte,et al.,2010)) through the use of HaaS.

*Figure 7. Hardware as a service*

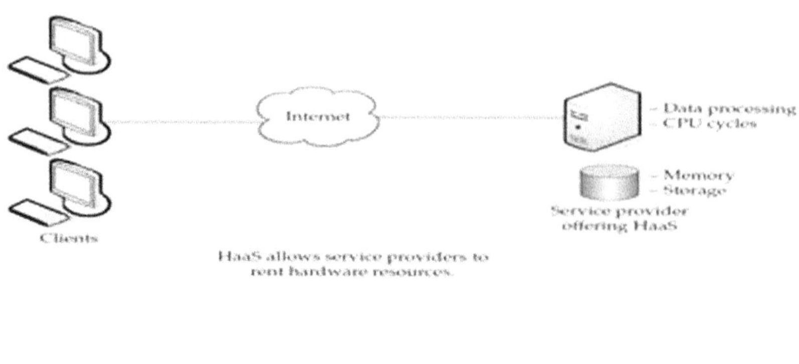

- Server space
- Network equipment
- Memory
- CPU cycles
- Storage space

Cloud computing offers a variety of applications of the Service Oriented Architecture (SOA) and Internet of Services (IoS) types, including fault tolerance, high scalability, availability, flexibility, decreased information technology overhead for the user, reduced cost of owner-ship, on demand services, and so on. The development of a reliable algorithm for load balancing is of the utmost importance with regard to these challenges (Daoud, M. K.,et al.,2024).

## 5. CONCLUSION

This chapter presents an overview of the underlying ideas that drive cloud computing, as well as information on the services that may be given by systems that make use of cloud computing. Additionally, the chapter discusses the services that may be offered by systems that make use of cloud computing. In addition, the chapter provides information on the various services that can be made available through the use of cloud computing. In addition to this, it gives information on cloud computing services and studies the use of load balancing in cloud computing for a range of different commercial applications. Moreover, it examines the use of load balancing in cloud computing.

## REFERENCES

Ahmad, A., Abusaimeh, H., Rababah, A., Alqsass, M., Al-Olima, N., & Hamdan, M. (2024). Assessment of effects in advances of accounting technologies on quality financial reports in Jordanian public sector. *Uncertain Supply Chain Management*, *12*(1), 133–142. doi:10.5267/j.uscm.2023.10.011

Ahmad, A. Y. B. (2024). Ethical implications of artificial intelligence in accounting: A framework for responsible ai adoption in multinational corporations in Jordan. *International Journal of Data and Network Science*, *8*(1), 401–414. doi:10.5267/j.ijdns.2023.9.014

Ahmad, A. Y. B. (2024). Firm Determinants that Influences Implementation of Accounting Technologies in Business Organizations. *WSEAS Transactions on Business and Economics*, *21*, 1–11. doi:10.37394/23207.2024.21.1

Ahmad, A. Y. B., Gongada, T. N., Shrivastava, G., Gabbi, R. S., Islam, S., & Nagaraju, K. (2023). E-Commerce Trend Analysis and Management for Industry 5.0 using User Data Analysis. *International Journal of Intelligent Systems and Applications in Engineering*, *11*(11s), 135–150.

Ahmad, A. Y. B., Hannoon, A., Al-Daoud, K. I., Abu-Alsondos, I. A., & Al-Qaisieh, M. S. (2023). Assessment of Cloud Based Accounting Technology Adoption and Business Performance. Kurdish Studies, 11(3).

Ahmad, A. Y. B., Kumari, D. K., Shukla, A., Deepak, A., Chandnani, M., Pundir, S., & Shrivastava, A. (2024). Framework for Cloud Based Document Management System with Institutional Schema of Database. *International Journal of Intelligent Systems and Applications in Engineering*, *12*(3s), 672–678.

Ahmad, A. Y. B., Kumari, D. K., Shukla, A., Deepak, A., Chandnani, M., Pundir, S., & Shrivastava, A. (2024). Framework for Cloud Based Document Management System with Institutional Schema of Database. *International Journal of Intelligent Systems and Applications in Engineering*, *12*(3s), 672–678.

Ahmad, A. Y. B., & Kumari, S. S. M. S, S. K. Guha, A. Gehlot and B. Pant, "Blockchain Implementation in Financial Sector and Cyber Security System," *2023 International Conference on Artificial Intelligence and Smart Communication (AISC)*, Greater Noida, India, 2023, pp. 586-590, 10.1109/AISC56616.2023.10085045

Ahmad, A. Y. B., Tiwari, A., Nayeem, M. A., Biswal, B. K., Satapathy, D. P., Kulshreshtha, K., & Bordoloi, D. (2024). Artificial Intelligence Perspective Framework of the Smart Finance and Accounting Management Model. *International Journal of Intelligent Systems and Applications in Engineering*, *12*(4s), 586–594.

Ahmad, A. Y. B., William, P., Uike, D., Murgai, A., Bajaj, K. K., Deepak, A., & Shrivastava, A. (2024). Framework for Sustainable Energy Management using Smart Grid Panels Integrated with Machine Learning and IOT based Approach. *International Journal of Intelligent Systems and Applications in Engineering*, *12*(2s), 581–590.

Anthony, T. Velte, Toby J.Velte, Robert Elsenpeter, Cloud Computing A Practical Approach, TATA McGRAW-HILL Edition 2010.

Babu, S. Z. D., Pandey, D., Naidu, G. T., Sumathi, S., Gupta, A., Bader Alazzam, M., & Pandey, B. K. (2022, July). Analysation of Big Data in Smart Healthcare. In Artificial Intelligence on Medical Data: Proceedings of International Symposium, ISCMM 2021 (pp. 243-251). Singapore: Springer Nature Singapore.

Bessant, Y. A., Jency, J. G., Sagayam, K. M., Jone, A. A. A., Pandey, D., & Pandey, B. K. (2023). Improved parallel matrix multiplication using Strassen and Urdhvatiryagbhyam method. *CCF Transactions on High Performance Computing*, *5*(2), 1–14. doi:10.1007/s42514-023-00149-9

Daoud, M. K. & Alqudah, M. . (2023). The Relationship Between Mobile Marketing and Customer Perceptions in Jordanian Commercial Banks: The Electronic Quality as A Mediator Variable. *IJMST*, *10*(2), 1360-1371.

Daoud, M. K., Al-Qeed, M., Ahmad, A. Y. B., & Al-Gasawneh, J. A. (2023). Mobile Marketing: Exploring the Efficacy of User-Centric Strategies for Enhanced Consumer Engagement and Conversion Rates. *International Journal of Membrane Science and Technology*, *10*(2), 1252–1262. doi:10.15379/ijmst.vi.1425

Daoud, M. K., Taha, S., Al-Qeed, M., Alsafadi, Y., Ahmad, A., & Allahham, M. (2024). EcoConnect: Guiding environmental awareness via digital marketing approaches. International Journal of Data and Network Science, 8(1), 235-242.

George, A. H., Fernando, M., George, A. S., Baskar, T., & Pandey, D. (2021). Metaverse: The next stage of human culture and the internet. [IJARTET]. *International Journal of Advanced Research Trends in Engineering and Technology*, *8*(12), 1–10.

Kumar, M. S., Sankar, S., Nassa, V. K., Pandey, D., Pandey, B. K., & Enbeyle, W. (2021). Innovation and creativity for data mining using computational statistics. In *Methodologies and Applications of Computational Statistics for Machine Intelligence* (pp. 223–240). IGI Global. doi:10.4018/978-1-7998-7701-1.ch012

Mequanint Moges, Thomas G.Robertazzi, "Wireless Sensor Networks: Scheduling for Measurement and Data Reporting", August 31, 2005.

Muniandi, B., Huang, C., Kuo, C., Yang, T., Chen, K., Lin, Y., Lin, S., & Tsai, T. (2019). A 97% maximum efficiency fully automated control turbo boost topology for battery chargers. *IEEE Transactions on Circuits and Systems. I, Regular Papers*, *66*(11), 4516–4527. doi:10.1109/TCSI.2019.2925374

Pandey, B. K., Pandey, D., Nassa, V. K., George, S., Aremu, B., Dadeech, P., & Gupta, A. (2022, July). Effective and secure transmission of health information using advanced morphological component analysis and image hiding. In *Artificial Intelligence on Medical Data: Proceedings of International Symposium, ISCMM 2021* (pp. 223-230). Singapore, Springer Nature Singapore.

Pandey, B. K., Pandey, D., Wairya, S., Agarwal, G., Dadeech, P., Dogiwal, S. R., & Pramanik, S. (2022). Application of integrated steganography and image compressing techniques for confidential information transmission. *Cyber Security and Network Security*, 169-191.

Pandey, B. K., Pandey, S. K., & Pandey, D. (2011). A survey of bioinformatics applications on parallel architectures. *International Journal of Computer Applications*, *23*(4), 21–25. doi:10.5120/2877-3744

Parthiban, K., Pandey, D., & Pandey, B. K. (2021). Impact of SARS-CoV-2 in online education, predicting and contrasting mental stress of young students: A machine learning approach. *Augmented Human Research*, *6*(1), 10. doi:10.1007/s41133-021-00048-0

Pramanik, S., Ghosh, R., Pandey, D., Samanta, D., Dutta, S., & Dutta, S. (2021). Techniques of Steganography and Cryptography in Digital Transformation. In Emerging Challenges, Solutions, and Best Practices for Digital Enterprise Transformation (pp. 24-44). IGI Global. doi:10.4018/978-1-7998-8587-0.ch002

Zaidi, A. (2023). Predicting wildfires in Algerian forests using machine learning models. *Heliyon*, *9*(7), e18064. doi:10.1016/j.heliyon.2023.e18064 PMID:37519679

Zaidi, A. (2023). Smart Implementation of Industrial Internet of Things using Embedded Mechatronic System. *IEEE Embedded Systems Letters*. Advance online publication. doi:10.1109/LES.2023.3309985

Zaidi, A., Ajibade, S.-S. M., Musa, M., & Bekun, F. V. (2023). New Insights into the Research Landscape on the Application of Artificial Intelligence in Sustainable Smart Cities: A Bibliometric Mapping and Network Analysis Approach. *International Journal of Energy Economics and Policy.*, *13*(4), 287–299. doi:10.32479/ijeep.14683

Zaidi, A., & Al-Harbi, O. (2023). Statistical Analysis of a Linear Multi-Step Numerical Treatment. *Journal of Statistics Applications & Probability*, *12*(1), 26. Advance online publication. doi:10.18576/jsap/120126

Zaidi, A., & Al Luhayb, A. (2023). Two Statistical Approaches to Justify the Use of the Logistic Function in Binary Logistic Regression. *Mathematical Problems in Engineering*, *2023*, 5525675. doi:10.1155/2023/5525675

# Chapter 18
# Optimized Throughput-Based Broadcasting for Next Generation Wireless Networks

**Digvijay Pandey**
https://orcid.org/0000-0003-0353-174X
*Department of Technical Education, Government of Uttar Pradesh, India*

**Binay Kumar Pandey**
https://orcid.org/0000-0002-4041-1213
*Department of Information Technology, College of Technology, Govind Ballabh Pant University of Agriculture and Technology, Pantnagar, India*

**Vinay Kumar Nassa**
https://orcid.org/0000-0002-9606-7570
*Department of Information Communication Technology, Tecnia Institute of Advanced Studies, India*

**Darshan A. Mahajan**
https://orcid.org/0000-0002-1239-6343
*NICMAR University, India*

**A. Shaji George**
https://orcid.org/0000-0002-8677-3682
*Business System Department, Almarai Company, TSM, Riyadh, Saudi Arabia*

**Pankaj Dadheech**
https://orcid.org/0000-0001-5783-1989
*Swami Keshvanand Institute of Technology, Management, and Gramothan, India*

## ABSTRACT

*Remote areas benefit from lower costs for transporting enormous amounts of data between a source node and multiple networking devices. Noise and constraints necessitated a 64 kb/s broadcaster. Bit defect rates, abnormalities, and frequent retransmission prevented data movement.This could greatly minimize channel usage. The maximum transmission unit distance was 64B. Data packet size limits exacerbated issues. Massive data traffic, connection imprecision, and changing topology affect network structure, making data dissemination from source nodes to all devices difficult.This book chapter suggests using an advanced throughput optimal broadcast in point-to-multipoint wireless networks to improve wire-less link precision by using the Mayfly optimization method, a recent swarm intelligence soft computing technique, to improve geometrical configuration of interfering terminals and forecasted per-flow throughput*

DOI: 10.4018/979-8-3693-1335-0.ch018

## INTRODUCTION

According to one definition of the term "broadcast challenge," this refers to the task of effectively dispersing the packet transmission from origin nodes to the various other nodes depicted in figure 1. Broadcast media is a term that refers to the wireless infrastructure that is extensively utilized in a number of applications. These applications include the dissemination of information, battle communication, catastrophe risk management, network function estimation, and effective data propagation, particularly in wireless mobile networks.

Because of the fundamental nature of its structure, the issue of broadcasting over wireless connections has been subjected to an extensive amount of scrutiny and investigation in a variety of published publications. As a direct consequence of this, various algorithms for improving performance measurements have been proposed. Broadcasting that uses little energy, broadcasting that is optimal throughout, broadcasting that has low latency, and broadcasting that places restrictions on retransmissions are only a few examples. In addition to the aforementioned attributes, the point-to-point aspect of the wireless channel has been an important distinguishing feature (Pandey, D.,et al.,2021).The benefit of wireless broadcast would be useful across the board for network-wide broadcast application sectors, with the aim of effectively dispersing packets across all of the devices that are part of the system. In addition, synchronous transmitting units on wireless networks have had their availability severely restricted to non-interfering schedules because of interference from other nodes (Pandey, B. K.,et al.,2022).

This research describes the effective packet propagation difficulties in wireless networks that have point-to-multipoint wireless broadcast stations, as well as dynamic policy to assist in the expansion of broadcast network potential and the Mayfly optimization method, a recent swarm intelligence soft computing technique, to improve the geometrical configuration of an activated non-interfering linkage. Other topics covered include dynamic policy to aid in the expansion of broadcast network potential, as well as dynamic policy to aid in the expansion of broadcast network potential.

## RELATED WORK

According to (Sinha, A., Tassiulas, L., & Modiano, E. (2019)), the difficulty of establishing concentrated point-to-point connections in wireless networks should have been investigated. This resulted in the development of a variety of broadcasting strategies. As the authors studied the rigorous framework, which presupposes that judgements are made either by a centralized unit or by an individual, the full challenge of designing the broadcast policy must have been considered. People were able to obtain a randomized packet system with the help of the aforementioned things, which requires information to be shared between neighbouring nodes. The methodology would be perfect for meeting strict time constraints like deadlines. The entirety of the study focused on investigating both combined optimal scheduling as well as the issues of propagation in wireless networks that had connections.

A technique for resolving flow problems with links that makes use of a recently created virtualization component as part of the solution (Ahmad, A. Y. B.,et al., 2023). This is said to be the very first known approach in wireless networks to have broadcast utility, and it has been recognized as such. A system model that includes attempting to verify a finite-horizon toughness disadvantage variation and extracting an optimal policy for quite a comfy wireless network variant will be investigated, as will the fundamental problem of throughput-optimized broadcasting via wireless connections. This policy was

*Figure 1. Layout of packet transmission from source node to other Nodes*

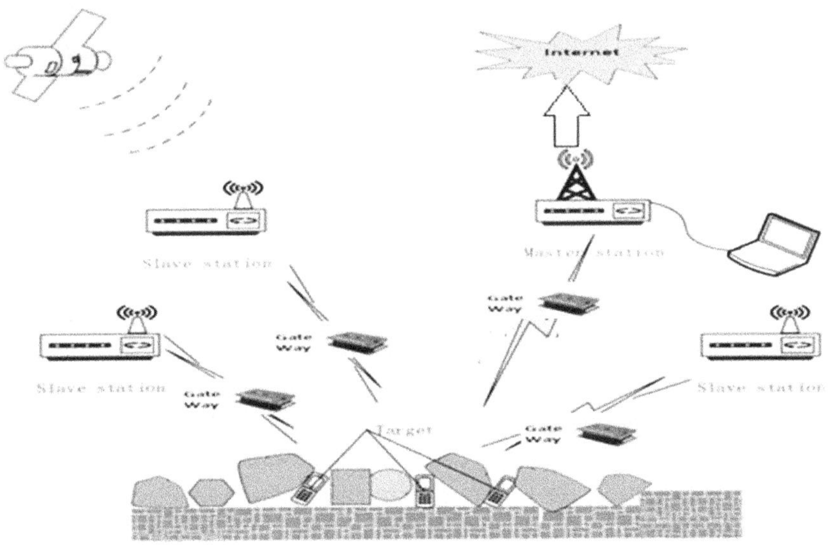


implemented in a network that did not allow for any level of relaxation, and when combined with a priority-based packet strategy, it produces a method that maximizes throughput .This research (Anand, R.,et al., 2022) examines the feasibility of broadcast data transmission by employing opportunity route discovery for point-to-multipoint communication. The evaluation is conceptual. It is proposed that a receiver negotiating process chance broadcast (RNOB) technique be implemented. This procedure would incorporate a decentralized co-operation system in addition to the priority scheduling scheme. The simulations demonstrate that RNOB performs better than standard protocols in terms of the veracity of the data being transmitted and the efficiency of doing so for WSNs throughout the IoT network (Ahmad, A. Y. B.,et al., 2024).

According to (Biason, A., &Zorzi, M. (2019), the effect of multicast mixed with mm Waves creates a problem, the root of which is the manner in which beams have to be constructed in order to communicate directly. This article explores a few essential aspects that lie at the junction of the two technological options, notably how it may be inefficient to limit wireless networks to unicast traffic alone (Daoud, M. K.,et al.,2023). The genetic algorithm uses a real picture encoding, and the positioning method would be constructed by assessing the connectivity limits with both unidentified nodes as well as a small number of anchor nodes. Finally, the model would be completed in order to solve the problem (Pandey, B. K., et al., 2022).

His article (Abadal Cavallé, S. (2016)) provides information about the most important components of the internet's layout, as well as the challenges of implementing broadcast-based wireless networks on chips (Daoud, M. K.,et al.,2024). These challenges have essentially been reduced to an improvement in the medium access control techniques that are able to manage vast numbers of processing units. It is essential to evaluate the broadcasting performance and scalability of a variety of MAC designs prior to taking into account the impact that the proposed framework might have on the efficacy, flexibility, and fully programmability of upcoming multi-core frameworks, programming environments, and concurrent methodologies. This analysis (Boopathi, S., Pandey, B. K., & Pandey, D. (2023)) offers a comprehensive conceptual model for optimizing mobile networks and recommends two effective choice strategies that had previously gotten little attention in recent findings. These models and recommendations were made possible by the findings of previous research. The first would be a supervised machine learning modeling as well as optimizing decision-making processes through enhancement, while the other one appears to be a straightforward and uncomplicated reinforcement training strategy. The significance and veracity of these investigations have been demonstrated by the fact that the suggested solutions have been validated through the use of genuine experimental studies and computer models (Muniandi, B.,et al.,2019). According to what is mentioned in (Kirubasri, G.,et al.,2021), in order for wireless communication systems to fully utilize the benefits they have over cabled communication technology, the difficulty of intervention needs to be addressed in the appropriate manner. The results of the investigation indicate that interference from a wire-free voice sender has a considerable, and most likely negative, effect on the amount of available bandwidth in a wire-free 802.11 b/g/n system while the system is operating in a laboratory environment. The findings highlight the importance of further investigation into the interruption as well as network management, and they endorse a comprehensive methodology in which systems have always been optimized for specific commercial processes premised not just on transmission coverage but also on network system performance. In addition, the findings highlight the importance of additional investigation into the interruption as well as network management. In order to fully understand what participants have an effect on the throughput of the network, it is necessary to conduct a comprehensive assessment of these interventions. This analysis must take into account not only the medium access layer but also the physical layer. Identifying a specific radio frequency that is starting to produce intervention is the first step. In this paper (George, A. S.,et al.,2022), we challenge the conventional notion that "too much information requires so much strength as well as radio signal emissions," and we argue that future cellular connections require an intelligent radio surroundings. An impactful wireless premise is described in which environmental artefacts have been encased in synthetic thin films of electro-magnetic as well as adaptable substances (keep referring to them as reprogrammable intelligent meta-surfaces), which seem to be able to change their properties in response to external stimuli. Intelligent radio environments have always had the potential to provide modern wireless systems with continuous wire-less networking in addition to the requirement to transmit data without producing new impulses but instead reusing previously generated radio waves and concentrating on the two primary types of reprogrammable smart meta-surfaces that are utilized during wireless connections.

According to, the idea of disturbance matching has recently come to light as a revolutionary method for boosting performance in multi-hop wireless systems. (Guo, W.,et al.,2011) explains that this method was developed. In order to keep a transmitter "aligned" with the vacant dimension of many other transmitters, there must always be some level of disturbance. This ensures that there is no unwanted disturbance. In this study, an approach to improving the throughput performance of multi-hop wireless systems by utilizing noise alignments is explained. Nonlinear programming, also known as NLP, was applied in order to provide a solution to the problem of throughput optimization for a multi-hop wireless network that made use of interfering alignments. A branch-and-bound system is essential for overcoming such a barrier because it is claimed to ensure a theoretically possible efficiency limit and also employs numerical results to test the computing performance and suggest ideas for improvements in throughput.

The position of the gateways needs to be optimized in (Sharma, M.,et al.,2021) Discuss Methodology in order to maximize connection speeds. This will allow for reduced implementation costs while also enhancing throughput performance. Additionally, a large number of gateways are deployed in order to alleviate overpopulation from within the network infrastructure. This causes disruption and adds to the expense of the project. Within Wireless Mesh Networks, a number of specialists concentrated solely on the installation of network interfaces or mesh routers. The difficulty of gate-way positioning needs to be centred on networks that make use of a wide variety of wireless devices to make it possible for machines to communicate with one another. It might be challenging to keep up with all of the different wireless approaches as well as the gate-way positioning for the highest possible utilization of bandwidth. In this paper, a "Novel Gateway Select Algorithm" was suggested. Its purpose was to locate the most promising prospective gateway in order to maximize performance under conditions of high traffic across coordination networks.

According to (Arani, A. H.,et a;.,2016), heterogeneous wireless systems, also known as Het-Nets, would make an excellent option for the development of future wireless links. This would be done with the intention of providing speedier data solutions in a manner that is both customized and efficient financially. Despite this, fluctuations in traffic load appear to be more noticeable in HetNets as a result of the coexistence of multiple components, which necessitates extra flexible resource utilization. On the other hand, efficient resource planning will have a significant impact on the experiences of users' apparatus as well as the efficiency of the network. It will also address the combined user apparatus association and downstream resource allocation algorithm for HetNets that is being investigated in this study. A user apparatus association approach that is dependent on the predicted workload as well as the signal-to-interference-and-noise ratio on base stations at the user apparatus location has therefore been presented as a solution to this problem (Pandey, B. K., et al., 2021).

According to (Oshima, K.,et al., 2022), highly heterogeneous topologies will become the dominating forms for next space-based networks. This is due to the fact that the actualization of a large number of derivative functionality will be predominantly dependent upon inter-satellite connections. Nodes reflecting a wide variety of satellites would come together to construct a networking structure capable of dynamic topologies. Regrettably, only a small number of investigations on these linkages have been carried out. The purpose of this study is to investigate a broadcasting strategy that may be utilized in space-based dynamic systems in order to establish a paradigm that incorporates both centralized and decentralized channels. In addition, it was suggested that monitoring performance measures may be used to evaluate the interconnectivity in complex networks in addition to the effectiveness of multipath routing. We also analyze the influence of the information survival rate, which has a significant influence on the packet delivery and, unfortunately, can waste communication links, in order to evaluate the logic of the intergenerational effects and investigate the performance of multi-group connections (M. K. Daoud,et al., 2023).

In the end, the goal is to explore the performance of multi-group connections and verify the logic of associated indications. Additionally, the goal is to investigate the effects of data life expectancy, which has a direct effect on packet delivery but, unfortunately, can waste transmission costs. The results related survival rates were reported in the final section, and it is believed that performance measurements, including those recommended in this research will be of major aid for the building of future space-based systems and broadcasting mechanisms. According to the data, whenever the ranging error is 50%, the positioning error only increases by roughly 15% in comparison to the error that would have occurred if ranging had not been used. If the ranging error is similar to a node's wireless range inside of extremely harsh circumstances, and the positioning error and positioning ratio have not been significantly affected despite this, then the ranging error has been successfully compensated for. The framework of this algorithm might be suitable for approval, and it even has a perfect limit.

It would appear that a fundamental quality of wireless channels is their inherent point-to-multipoint existence. This means that a packet that is delivered by a node can be listened to by all of the nodes that are in the immediate vicinity of that node. The entire feature, which is also famous for being that of the obvious benefit of wireless broadcasting, would be especially helpful in network-wide broadcast applications, the purpose of which is to distribute equality-based packets amongst all of the nodes in a network. Furthermore, the number of instantaneous data transfers in a wireless network would be determined by the number of non-interfering feasible schedules. This is because there is interference between the nodes in the network. Constructing a broadcast algorithm that makes effective use of the broadcast advantage while taking into account the interference constraints is a difficult task to solve.

## PROBLEM STATEMENT

Broadcasting seems to be the basic network capability of data delivery from an origin node to most other nodes in a system. There are many difficulties in designing effectual wireless broadcast methodologies. A broadcast policy must always use non-interfering links.

Only optimized geometrical configurations of interfering channels are conceivable; this enables precise computation of contingent packet loss likelihood and busy schedule on the geometrical configuration of the attempting to interfere stations and anticipated per-flowthroughput. The Mayfly optimization algorithm (MOA), a novel swarm intelligence optimization algorithm, and an approach for finding non-interfering links at each time slot are presented.

## PROPOSED METHODOLOGY

In the beginning, the throughput-optimal broadcasting policy on wireless networks using point-to-multipoint connections and flexible scheduling restrictions has been achieved. This plan has been developed using strong precedence-relaxed virtual network methodology that has previously been utilized to develop throughput-optimal guidelines in networking comprising point-to-point links. Packet routing and scheduling decisions were made on those networks by addressing typical optimization issues weighed by virtual queues (George, D. A. S.,et Al.., 2022).

This approach is proven throughput optimum, employing a hybrid of the Lyapunov approach and an adversarial queueing theory sampling path argument. Simulation results show that the suggested strategy in is effective, including both stationary and non-stationary network conditions (Zaidi, A., et al., (2023)). The existing streamlined method given in assumes that an error-free wireless network seems to be likewise interference-free. Therefore With the use of the Mayfly Optimization Algorithm and geometrical arrangement, a much more realistic wireless link design will integrate the probability of packet losses with specific reception nodes due to both fading and receiver noise (Zipeng Ye,et al.,2021).

Mayfly is an optimization algorithm. Crossover technologies and local search were employed depending on social conduct. The toughest mayfly could survive if it was assumed that after birth, the mayfly would always be an adolescent. Every mayfly's location in the search area symbolizes a viable solution. The algorithm's operating concept seems to be as follows: As a suggested solution, two categories of mayflies, male and female, were generated at random in the problem space and denoted by d-dimensional vectors.

$A= (A_{1,} A_{2,} A_{3,\ldots\ldots} A_{d})$

Its efficiency is determined against a predetermined goal function*(A)*

A speed vector has been specifically defined:

$K = (K_1, K_2, \ldots\ldots\ldots\ldots K_d)$

to depict the mayfly's shift in location. Each mayfly's flying orientation is the consequence of a dynamical interplay between individualistic and social flying time. Each time, the mayfly would change its path in order to achieve its individual perfect location, which would be the best position achieved by any mayfly in a group (Kirubasri, G.,et al.,2022).

The mayfly optimization algorithm's construction consists primarily of the steps outlined:

1. Configure the speed parameters as well as initialize both female and male mayfly populations.
2. Considering the $m_i^t$ is the present location of male mayfly i in the solution space at timestep t. $f$ present location of a female mayfly within the search area $f_i^t$ at time step t.
3. 3. The male mayfly travels at the pace given below.

$$s_{ij}^{t+1} = s_{ij}^t + c_1 e^{-\beta r_m^2}(mbest_{ij} - m_i^t) + c_2 e^{-\beta r_g^2}(gbest_{ij} - m_i^t) \tag{1}$$

The method, whereby the female mayfly attracts the male mayfly being represented and the equation to decide the female mayfly's acceleration is as follows:

$$s_{ij}^{t+1} = \begin{cases} s_{ij}^t + c_2 e^{-\beta r_{lt}^t\left(m_{ij}^t - f_{ij}^t\right)}, iff\left(f_i\right) > f\left(m_i\right) \\ s_{ij}^t + fl * r,, iff\left(f_i\right) < f\left(m_i\right) \end{cases} \tag{2}$$

where $s_{ij}^t$ is the change of speed of mayfly *i (i = 1, 2, ..., n)* in dimension *j (j = 1, 2, ..., d)* at time stept, $m_{ij}^t$ and $f_{ij}^t$ would be the positioning of male and female mayfly *i* in dimension *j* at time step *t*, $c_1$ and $c_2$ are positive attraction constant, $\beta$ denotes acoefficient of fixed visibility, $r_{lt}$ depicts the cartesian distance exists between male and female mayfly. Ultimately, *fl* represents as random walking coefficient, that is used, whenever female is not particularly interested in the male, thus it flies randomly, and *r* is a random value in the range [−1 1]

Determine and classify the fitness value to even get $pbest_i$ and is the optimal position that mayfly *i* has never been to. Considering the minimization problem under consideration, the particular optimal position $pbest_i$ at the next time step *t+1* is computed by applying equation (3).

$$pbest_i = \begin{cases} A_i^{t+1}, iff\left(A_i^t\right) < f\left(pbest_i\right) \\ iskeptthesamevalue, otherwise \end{cases} \tag{3}$$

Wherein *f* would be the objective function that describes the solution's efficiency, and gbest, perhaps the global optimum solution, is denoted as:

$$gbest \in \{pbest_1, pbest_{12}, \ldots pbest_n \mid f(cbest)\} = \min\{f(pbest_1, pbest_2, \ldots pbest_n)\} \quad (4)$$

The mating procedure between two mayflies is represented by the crossover operator. Parents are chosen in much the same manner that females are tempted to become males. The choice, for instance, might be random or dependent upon their fitness function. As a result, the best female finds the best man, the second-best female finds the second-best male, and so on. The cross generates two descendants, the outcomes of which are as described in the following:

$$Offspring_1 = H^* \text{ male} + (1 - H)^* \text{emale} \quad (5)$$

$$Offspring_{2=} H^* \text{ female} + (1 - H) * \text{Male} \quad (6)$$

The male parent usually referred to as the father, while the female parent is referred to as the mother. H is a random value within a certain range, and also the offspring's starting speed is set to 0.

Male and female mayflies change their locations in succession, and they mate. Applying a velocity to the previous location yields the following location.

$s_{ij}^t$ it can be expressed as:

$$m_i^{t+1} = m_i^t + s_{ij}^t \quad (7)$$

$$f_i^{t+1} = f_i^t + s_{ij}^t \quad (8)$$

Modifypbest and gbest after calculating fitness;

If the stop criteria is satisfied, quit and displays the results; alternatively, continue Steps 3–5.

The mayfly optimization method is depicted in Algorithm 1 as pseudo code. The updating procedure of the female and male mayfly communities is combined with a pseudo code.

The detailed Mayfly Optimization Technique is given below

Input consists of objective function *f(A)*, size of male population *m*, size of female population size *f*,visibility- coefficient denoted by β, initial velocity for male mayfly represented by $s_m$, initial velocity for female mayfly given by$s_n$, total number ofiterationdenoted by *T*, symbol $C_1$ and $C_2$ are the affirmative attraction factors, dimensionsolution given by *d*, size of population size represented by *L*. the output solution of this algorithms is dented by optimal gbest.

```
1. Start
2. Assess each solution within the perspective of the objective function.f(A)
```

```
3. Determine the optimum value gbest, among all feasible solutions.
4: do
{
5: Using equation (3), determine the pbest value, which is the optimum solu-
tion for every male mayfly.
6: for (i=1: i<=f; i++)
{
7. for (j=1: j<=d; j++)
{
8: Equation 2 is used to determine a female's fly speed.
9:          Equation number 8 is used to determine female fly position.
10:    }
11 :}
12: for (i=1: i<=m; i++)
13:  {
 14.      for (j=1: j<=d; j++)
             {
15.     Equation number 1 is used to determine male velocity
16              Equation number 7 is used to determine male position
16:}
17:              upgrade the value of pbest
18:    }
19: To calculate the ranking of male mayflies
20: To calculate the ranking of male mayfliesfemale mayflies
21: Mayfly mating takes place.
22: Assess their offspring.
23: Distribute offspring into male as well as female groupings at random.
24: Substitute the ineffective solution with one of the most effectual alter-
natives.
25: pbest and gbest must be updated.
26: while (t ≤ T)
27:  Stop
```

The process of optimizing the layout of the nodes is illustrated in Figure 2. First and foremost, the aim functions need to be stated. In general, the dimensions of the target variable are proportional to both the number of base stations and the dimensions of the base station coordinates (Pandey, B. K.,et al.,2022). Elements such as the transmission range of the positioning system determine the boundaries of the goal region. The total number of mayflies, including female and male, is represented by the population (B. Muniandi et al.,2019). During the starting stages of population development, it is necessary to ascertain the beginning location, initial velocity, visibility coefficient, and positive attractiveness coefficients (also known as C1, C2). It is necessary to include details regarding the maximum number of iterations, denoted by the letter T, as this is the variable that determines whether the optimization method is finished. The one-of-a-kind optimization approach is shown in Figure 2.

*Figure 2. Optimization process stages for graphical configuration of wireless network*

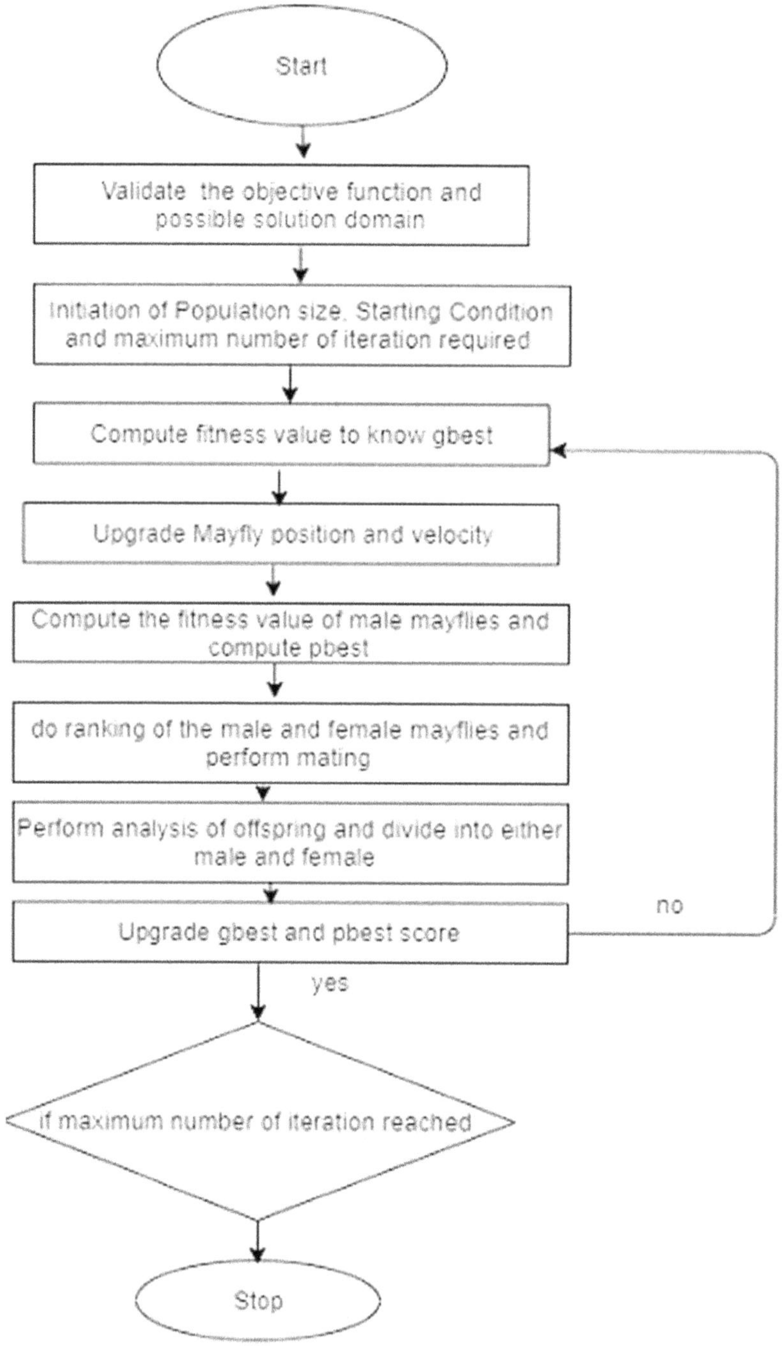


## CONCLUSION

The problem of effective packet dispersion in wireless networks that have many hops of wireless broadcast stations is investigated in this research. The strategy is the result of carefully crafting an architecture that allows for relaxed precedence virtualization. This is employed in the process of policy formulation, particularly in situations in which networks are linked. After that, decisions are made on the resolution of network-related difficulties, scheduling, and the packet routing tasks. The Mayfly optimization technique, which is a modern swarm intelligence computational intelligence technology, has been used to improve the geometrical configuration of an enabled non-interfering connection of a wireless network. This was accomplished by integrating the Lyapunov approach with sample path arguments, which caused the policy to become throughput optimum.

In order to optimize the geometric arrangement of transmitting base stations and other nodes, the objective function Mayfly optimization has been utilized. Using particle swarm optimization and genetic algorithms, the mathematical model of mayfly optimization of such a range of geometric configurations for the base stations would be evaluated for certain other geometric configurations of nodes used in wireless networks. This evaluation would take place for certain other geometric configurations of nodes used in wireless networks.

## REFERENCES

Anand, R., Singh, J., Pandey, D., Pandey, B. K., Nassa, V. K., & Pramanik, S. (2022). Modern technique for interactive communication in LEACH-based ad hoc wireless sensor network. In *Software Defined Networking for Ad Hoc Networks* (pp. 55–73). Springer International Publishing. doi:10.1007/978-3-030-91149-2_3

Biason, A., & Zorzi, M. (2019). Multicast via point to multipoint transmissions in directional 5G mmWave communications. *IEEE Communications Magazine*, *57*(2), 88–94. doi:10.1109/MCOM.2019.1700679

Pandey, B. K., Pandey, D., Nassa, V. K., George, S., Aremu, B., Dadeech, P., & Gupta, A. (2022, July). Effective and secure transmission of health information using advanced morphological component analysis and image hiding. In *Artificial Intelligence on Medical Data: Proceedings of International Symposium, ISCMM 2021* (pp. 223-230). Singapore: Springer Nature Singapore.

Pandey, D., Wairya, S., Al Mahdawi, R. S., Najim, S. A. D. M., Khalaf, H. A., Al Barzinji, S. M., & Obaid, A. J. (2021). Secret data transmission using advanced steganography and image compression. *International Journal of Nonlinear Analysis and Applications*, *12*(Special Issue), 1243–1257.

Pandey, B. K., Pandey, D., Wairya, S., Agarwal, G., Dadeech, P., Dogiwal, S. R., & Pramanik, S. (2022). Application of integrated steganography and image compressing techniques for confidential information transmission. *Cyber Security and Network Security*, 169-191.

Ahmad, A., Abusaimeh, H., Rababah, A., Alqsass, M., Al-Olima, N., & Hamdan, M. (2024). Assessment of effects in advances of accounting technologies on quality financial reports in Jordanian public sector. *Uncertain Supply Chain Management*, *12*(1), 133–142. doi:10.5267/j.uscm.2023.10.011

Ahmad, A. Y. B. (2024). Ethical implications of artificial intelligence in accounting: A framework for responsible ai adoption in multinational corporations in Jordan. *International Journal of Data and Network Science*, *8*(1), 401–414. doi:10.5267/j.ijdns.2023.9.014

Ahmad, A. Y. B. (2024). Firm Determinants that Influences Implementation of Accounting Technologies in Business Organizations. *WSEAS Transactions on Business and Economics, 21*, 1–11. doi:10.37394/23207.2024.21.1

Ahmad, A. Y. B., Gongada, T. N., Shrivastava, G., Gabbi, R. S., Islam, S., & Nagaraju, K. (2023). E-Commerce Trend Analysis and Management for Industry 5.0 using User Data Analysis. *International Journal of Intelligent Systems and Applications in Engineering, 11*(11s), 135–150.

Ahmad, A. Y. B., Hannoon, A., Al-Daoud, K. I., Abu-Alsondos, I. A., & Al-Qaisieh, M. S. (2023). Assessment of Cloud Based Accounting Technology Adoption and Business Performance. *Kurdish Studies, 11*(3).

Ahmad, A. Y. B., Kumari, D. K., Shukla, A., Deepak, A., Chandnani, M., Pundir, S., & Shrivastava, A. (2024). Framework for Cloud Based Document Management System with Institutional Schema of Database. *International Journal of Intelligent Systems and Applications in Engineering, 12*(3s), 672–678.

Ahmad, A. Y. B., Kumari, D. K., Shukla, A., Deepak, A., Chandnani, M., Pundir, S., & Shrivastava, A. (2024). Framework for Cloud Based Document Management System with Institutional Schema of Database. *International Journal of Intelligent Systems and Applications in Engineering, 12*(3s), 672–678.

Ahmad, A. Y. B., & Kumari, S. S. (2023). Blockchain Implementation in Financial Sector and Cyber Security System. *2023 International Conference on Artificial Intelligence and Smart Communication (AISC)*, Greater Noida, India. 10.1109/AISC56616.2023.10085045

Ahmad, A. Y. B., Tiwari, A., Nayeem, M. A., Biswal, B. K., Satapathy, D. P., Kulshreshtha, K., & Bordoloi, D. (2024). Artificial Intelligence Perspective Framework of the Smart Finance and Accounting Management Model. *International Journal of Intelligent Systems and Applications in Engineering, 12*(4s), 586–594.

Ahmad, A. Y. B., William, P., Uike, D., Murgai, A., Bajaj, K. K., Deepak, A., & Shrivastava, A. (2024). Framework for Sustainable Energy Management using Smart Grid Panels Integrated with Machine Learning and IOT based Approach. *International Journal of Intelligent Systems and Applications in Engineering, 12*(2s), 581–590.

Arani, A. H., Mehbodniya, A., Omidi, M. J., & Adachi, F. (2016, September). Distributed load balancing user association and self- anizing resource allocation in HetNets. In *2016 IEEE 84th Vehicular Technology Conference (VTC-Fall)* (pp. 1-5). IEEE.

Daoud, M. K. (2023). The Relationship Between Mobile Marketing and Customer Perceptions in Jordanian Commercial Banks: The Electronic Quality as A Mediator Variable. *IJMST, 10*(2).

Daoud, M. K., Al-Qeed, M., Ahmad, A. Y. B., & Al-Gasawneh, J. A. (2023). Mobile Marketing: Exploring the Efficacy of User-Centric Strategies for Enhanced Consumer Engagement and Conversion Rates. *International Journal of Membrane Science and Technology, 10*(2), 1252–1262. doi:10.15379/ijmst.vi.1425

Daoud, M. K., Taha, S., Al-Qeed, M., Alsafadi, Y., Ahmad, A., & Allahham, M. (2024). EcoConnect: Guiding environmental awareness via digital marketing approaches. *International Journal of Data and Network Science, 8*(1), 235-242.

George, D. A. S., George, A. H., Baskar, T., & Pandey, D. (2021). XDR: The Evolution of Endpoint Security Solutions-Superior Extensibility and Analytics to Satisfy the Organizational Needs of the Future. International Journal of Advanced Research in Science [IJARSCT]. *Tongxin Jishu*, *8*(1), 493–501.

George, A. S., George, A. H., Baskar, T., & Pandey, D. (2022). The Transformation of the workspace using Multigigabit Ethernet. *Partners Universal International Research Journal*, *1*(3), 34–43.

Guo, W., & Huang, X. (2011, April). On throughput enhancement of multi-hop wireless networks using interference alignment. In *2011 20th Annual Wireless and Optical Communications Conference (WOCC)* (pp. 1-6). IEEE. 10.1109/WOCC.2011.5872297

Kirubasri, G., Sankar, S., Pandey, D., Pandey, B. K., Nassa, V. K., & Dadheech, P. (2022). Software-defined networking-based Ad hoc networks routing protocols. In *Software defined networking for Ad Hoc networks* (pp. 95–123). Springer International Publishing. doi:10.1007/978-3-030-91149-2_5

Kirubasri, G., Sankar, S., Pandey, D., Pandey, B. K., Singh, H., & Anand, R. (2021, September). A recent survey on 6G vehicular technology, applications and challenges. In *2021 9th International Conference on Reliability, Infocom Technologies and Optimization (Trends and Future Directions)(ICRITO)* (pp. 1-5). IEEE.

Muniandi, B., Huang, C., Kuo, C., Yang, T., Chen, K., Lin, Y., Lin, S., & Tsai, T. (2019). A 97% maximum efficiency fully automated control turbo boost topology for battery chargers. *IEEE Transactions on Circuits and Systems. I, Regular Papers*, *66*(11), 4516–4527. doi:10.1109/TCSI.2019.2925374

Muniandi, B., Huang, C.-J., Kuo, C.-C., Yang, T.-F., Chen, K.-H., Lin, Y.-H., Lin, S.-R., & Tsai, T.-Y. (2019, November). A 97% Maximum Efficiency Fully Automated Control Turbo Boost Topology for Battery Chargers. *IEEE Transactions on Circuits and Systems. I, Regular Papers*, *66*(11), 4516–4527. doi:10.1109/TCSI.2019.2925374

Oshima, K., Yamamoto, D., Yumoto, A., Kim, S. J., Ito, Y., Hasegawa, M., & Initiative, I. D. (2022). Online machine learning algorithms to optimize performances of complex wireless communication systems. *Mathematical Biosciences and Engineering*, *19*(2), 2056–2094. doi:10.3934/mbe.2022097 PMID:35135242

Pandey, B. K., Pandey, D., & Agarwal, A. (2022). Encrypted information transmission by enhanced steganography and image transformation. [IJDAI]. *International Journal of Distributed Artificial Intelligence*, *14*(1), 1–14. doi:10.4018/IJDAI.297110

Pandey, B. K., Pandey, D., Wairya, S., & Agarwal, G. (2021). An advanced morphological component analysis, steganography, and deep learning-based system to transmit secure textual data. [IJDAI]. *International Journal of Distributed Artificial Intelligence*, *13*(2), 40–62. doi:10.4018/IJDAI.2021070104

Sharma, M., Sharma, B., Gupta, A. K., Khosla, D., Goyal, S., & Pandey, D. (2021). A study and novel AI/ML-based framework to detect COVID-19 virus using smartphone embedded sensors. In *Sustainability Measures for COVID-19 Pandemic* (pp. 59–74). Springer Nature Singapore. doi:10.1007/978-981-16-3227-3_4

Sinha, A., Tassiulas, L., & Modiano, E. (2019). Throughput-Optimal Broadcast in Wireless Networks with Dynamic Topology. *IEEE Transactions on Mobile Computing*, *18*(5), 1203–1216. doi:10.1109/TMC.2018.2846246

Ye, Z., & Zhou, Q. (2021). Performance Evaluation Indicators of Space Dynamic Networks under Broadcast Mechanism. Space: Science & Technology. doi:10.34133/2021/9826517

Zaidi, A. (2023). Predicting wildfires in Algerian forests using machine learning models. *Heliyon*, *9*(7), e18064. doi:10.1016/j.heliyon.2023.e18064 PMID:37519679

Zaidi, A. (2023). Smart Implementation of Industrial Internet of Things using Embedded Mechatronic System. *IEEE Embedded Systems Letters*. Advance online publication. doi:10.1109/LES.2023.3309985

Zaidi, A., Ajibade, S.-S. M., Musa, M., & Bekun, F. V. (2023). New Insights into the Research Landscape on the Application of Artificial Intelligence in Sustainable Smart Cities: A Bibliometric Mapping and Network Analysis Approach. *International Journal of Energy Economics and Policy.*, *13*(4), 287–299. doi:10.32479/ijeep.14683

Zaidi, A., & Al-Harbi, O. (2023). Statistical Analysis of a Linear Multi-Step Numerical Treatment. *Journal of Statistics Applications & Probability*, *12*(1), 26. doi:10.18576/jsap/120126

Zaidi, A., & Al Luhayb, A. (2023). Two Statistical Approaches to Justify the Use of the Logistic Function in Binary Logistic Regression. *Mathematical Problems in Engineering*, *2023*, 5525675. doi:10.1155/2023/5525675

# Compilation of References


Aazam, M., & Huh, E. N. (2014). Fog computing and smart gateway based communication for cloud of things. *Proceedings of the 2nd IEEE International Conference on Future Internet of Things and Cloud (FiCloud' 14)*. IEEE. 10.1109/FiCloud.2014.83

Abarrategi, A., Gutiérrez, M. C., Moreno-Vicente, C., Hortigüela, M. J., Ramos, V., López-Lacomba, J. L., Ferrer, M. L., & del Monte, F. (2008, January). Multiwall carbon nanotube scaffolds for tissue engineering purposes. *Biomaterials*, *29*(1), 94–102. doi:10.1016/j.biomaterials.2007.09.021 PMID:17928048

Abdulkarim, Y. I., Awl, H. N., Muhammadsharif, F. F., Saeed, S. R., Sidiq, K. R., Khasraw, S. S., ... Pandey, D. (2023). Metamaterial-based sensors loaded corona-shaped resonator for COVID-19 detection by using microwave techniques. *Plasmonics*, 1–16.

Abioye, S. O., Oyedele, L. O., Akanbi, L., Ajayi, A., Delgado, J. M. D., Bilal, M., & Ahmed, A. (2021). Artificial intelligence in the construction industry: A review of present status, opportunities and future challenges. *Journal of Building Engineering*, *44*, 103299. doi:10.1016/j.jobe.2021.103299

Abson, D. J., & Rothwell, J. S. (2013). Review of type IV cracking of weldments in 9–12%Cr creep strength enhanced ferritic steels. *International Materials Reviews*, *58*(8), 437–473. doi:10.1179/1743280412Y.0000000016

Addy, M., Adinyira, E., & Ayarkwa, J. (2018). Antecedents of building information modelling adoption among quantity surveyors in Ghana: An application of a technology acceptance model. *Journal of Engineering. Design and Technology*, *16*(2), 313–326. doi:10.1108/JEDT-06-2017-0056

Afrizal, S. H., Handayani, P. W., Hidayanto, A. N., Eryando, T., Budiharsana, M., & Martha, E. (2019). Barriers and challenges to Primary Health Care Information System (PHCIS) adoption from health management perspective: A qualitative study. *Informatics in Medicine Unlocked*, *17*, 100198. doi:10.1016/j.imu.2019.100198

Aghimien, D. O., Ikuabe, M., Aigbavboa, C., Oke, A., & Shirinda, W. (2021). Unravelling the Factors Influencing Construction Organizations' Intention to Adopt Big Data Analytics in South Africa. *Construction Economics and Building*, *21*(3), 478–489. doi:10.5130/AJCEB.v21i3.7634

Ahmad, A. Y. A. B. (2019). Empirical Analysis on Accounting Information System Usage in Banking Sector in Jordan. *Academy of Accounting and Financial Studies Journal*, *23*(5), 1–9.

Ahmad, A. Y. A. B., Gongada, T. N., Shrivastava, G., Gabbi, R. S., Islam, S., & Nagaraju, K. (2023). E-Commerce Trend Analysis and Management for Industry 5.0 using User Data Analysis. *International Journal of Intelligent Systems and Applications in Engineering*, *11*(11s), 135–150.

Ahmad, A. Y. A. B., & Kumari, S. S. (2023). Blockchain Implementation in Financial Sector and Cyber Security System. *2023 International Conference on Artificial Intelligence and Smart Communication (AISC)*, Greater Noida, India. 10.1109/AISC56616.2023.10085045

Ahmad, A. Y. B. (2024). Ethical implications of artificial intelligence in accounting: A framework for responsible ai adoption in multinational corporations in Jordan. *International Journal of Data and Network Science*, *8*(1), 401–414. doi:10.5267/j.ijdns.2023.9.014

Ahmad, A. Y. B. (2024). Firm Determinants that Influences Implementation of Accounting Technologies in Business Organizations. *WSEAS Transactions on Business and Economics*, *21*, 1–11. doi:10.37394/23207.2024.21.1

Ahmad, A. Y. B., Kumari, D. K., Shukla, A., Deepak, A., Chandnani, M., Pundir, S., & Shrivastava, A. (2024). Framework for Cloud Based Document Management System with Institutional Schema of Database. *International Journal of Intelligent Systems and Applications in Engineering*, *12*(3s), 672–678.

Ahmad, A. Y. B., Tiwari, A., Nayeem, M. A., Biswal, B. K., Satapathy, D. P., Kulshreshtha, K., & Bordoloi, D. (2024). Artificial Intelligence Perspective Framework of the Smart Finance and Accounting Management Model. *International Journal of Intelligent Systems and Applications in Engineering*, *12*(4s), 586–594.

Ahmad, A. Y. B., William, P., Uike, D., Murgai, A., Bajaj, K. K., Deepak, A., & Shrivastava, A. (2024). Framework for Sustainable Energy Management using Smart Grid Panels Integrated with Machine Learning and IOT based Approach. *International Journal of Intelligent Systems and Applications in Engineering*, *12*(2s), 581–590.

Ahmad, A., Abusaimeh, H., Rababah, A., Alqsass, M., Al-Olima, N., & Hamdan, M. (2024). Assessment of effects in advances of accounting technologies on quality financial reports in Jordanian public sector. *Uncertain Supply Chain Management*, *12*(1), 133–142. doi:10.5267/j.uscm.2023.10.011

Ahn, D. G. (2021). *Directed Energy Deposition (DED) Process: State of the Art, International Journal of Precision Engineering and Manufacturing - Green Technology*. Korean Society for Precision Engineering., doi:10.1007/s40684-020-00302-7

Ajibade, I. (2017). Can a future city enhance urban resilience and sustainability? A political ecology analysis of Eko Atlantic city, Nigeria. *International Journal of Disaster Risk Reduction*, *26*, 85–92. doi:10.1016/j.ijdrr.2017.09.029

Akbar, S., Coiera, E., & Magrabi, F. (2020). Safety concerns with consumer-facing mobile health applications and their consequences: A scoping review. *Journal of the American Medical Informatics Association : JAMIA*, *27*(2), 330–340. doi:10.1093/jamia/ocz175 PMID:31599936

Akdag, S. G., & Maqsood, U. (2019). A roadmap for BIM adoption and implementation in developing countries: The Pakistan case. *Archnet-IJAR*, *14*(1), 112–132. doi:10.1108/ARCH-04-2019-0081

Al-Ashmori, Y. Y., Othman, I., Rahmawati, Y., Amran, Y. M., Sabah, S. A., Rafindadi, A. D. U., & Mikić, M. (2020). BIM benefits and its influence on the BIM implementation in Malaysia. *Ain Shams Engineering Journal*, *11*(4), 1013–1019. doi:10.1016/j.asej.2020.02.002

Albermany S, Baqer FM. EEG authentication system using fuzzy vault scheme. *Journal of Discrete Mathematical Sciences and Cryptography*. 2021:1-6.

Albermany, S. A., & Safdar, G. A. (2014). Keyless security in wireless networks. *Wireless Personal Communications*, *79*(3), 1713–1731. doi:10.1007/s11277-014-1954-1

Alberti, E. A., Bueno, B. M. P., & D'Oliveira, A. S. C. M. (2016). Additive manufacturing using plasma transferred arc. *International Journal of Advanced Manufacturing Technology*, *83*(9), 1861–1871. doi:10.1007/s00170-015-7697-7

Alexey, A. (2003, August 1). Tomchenko, Gregory P. Harmer, Brent T. Marquis, John W. Allen. Semiconducting metal oxide sensor array for the selective detection of combustion gases. *Sensors and Actuators. B, Chemical*, *93*(1–3), 126–134. doi:10.1016/S0925-4005(03)00240-5

Alim, S., Vejayan, J., Yusoff, M. M., & Kafi, A. K. M. (2018). Recent uses of carbon nanotubes & gold nanoparticles in electrochemistry with application in biosensing: A review. *Biosensors & Bioelectronics*, *121*, 121–136. doi:10.1016/j.bios.2018.08.051 PMID:30205246

Ali, S. (2019). Rehman,S. A. U. Luan, H.-Y. Farid M. U. and Huang, H. *The Science of the Total Environment*, *646*, 1126–1139. doi:10.1016/j.scitotenv.2018.07.348 PMID:30235599

Alsagabi, S. (2016). High Temperature Deformation Behavior of P92 Steel. *Transactions of the Indian Institute of Metals*, *69*(8), 1513–1518. doi:10.1007/s12666-015-0725-3

Alvarez, M. G., Prucca, C., Milanesio, M. E., Durantini, E. N., & Rivarola, V. (2006). Photodynamic activity of a new sensitizer derived from porphyrin-C60 dyad and its biological consequences in a human carcinoma cell line. *The International Journal of Biochemistry & Cell Biology*, *38*(12), 2092–2101. doi:10.1016/j.biocel.2006.05.019 PMID:16899389

Amuda-Yusuf, G. (2018). Critical success factors for building information modelling implementation. *Construction Economics and Building*, *18*(3), 55–73. doi:10.5130/AJCEB.v18i3.6000

Amuda-Yusuf, G., Adebiyi, R. T., Olowa, T. O. O., & Oladapo, I. B. (2017). Barriers to building information modelling adoption in Nigeria. *Journal of Research Information in Civil Engineering*, *14*(2), 1574–1591.

Anand, R., Khan, B., Nassa, V. K., Pandey, D., Dhabliya, D., Pandey, B. K., & Dadheech, P. (2023). Hybrid convolutional neural network (CNN) for Kennedy Space Center hyperspectral image. *Aerospace Systems*, *6*(1), 71–78. doi:10.1007/s42401-022-00168-4

Anand, R., Lakshmi, S. V., Pandey, D., & Pandey, B. K. (2023). An enhanced ResNet-50 deep learning model for arrhythmia detection using electrocardiogram biomedical indicators. *Evolving Systems*, 1–15.

Anand, R., Singh, J., Pandey, D., Pandey, B. K., Nassa, V. K., & Pramanik, S. (2022). Modern technique for interactive communication in LEACH-based ad hoc wireless sensor network. In *Software Defined Networking for Ad Hoc Networks* (pp. 55–73). Springer International Publishing. doi:10.1007/978-3-030-91149-2_3

Andersson, K., & Hossain, M. S. (2015). Heterogeneous Wireless Sensor Networks for Flood Projection Resolution Support Systems. *IEEE Conference on Computer Communications Workshops (INFOCOM WKSHPS)*, (pp. 133–137). IEEE.

Andersson, V., Sawatzky, R., & Öhlén, J. (2022). Relating person-centredness to quality-of-life assessments and patient-reported outcomes in healthcare: A critical theoretical discussion. *Nursing Philosophy*, *23*(3), e12391. doi:10.1111/nup.12391 PMID:35502530

Anifowose, M., Babarinde, S. A., & Olanrewaju, O. I. (2018). Adoption Level of Building Information Modelling by selected professionals in Kwara State. *Environmental Technology and Science Journal*, *9*(2), 35–44.

Anna, A. (2012, June 1). Shvedova, Antonio Pietroiusti, Bengt Fadeel, Valerian E. Kagan. Mechanisms of carbon nanotube-induced toxicity: Focus on oxidative stress. *Toxicology and Applied Pharmacology*, *261*(2), 121–133. doi:10.1016/j.taap.2012.03.023 PMID:22513272

Ansari, R. (2013). Mirnezhad, M. and Sahmani, S. "An accurate molecular mechanics model for computation of sizedependent elastic properties of armchair and zigzag singlewalled carbon nanotubes,". *Meccanica*, *48*(6), 1355–1367. doi:10.1007/s11012-012-9671-x

Anthony, T. Velte, Toby J.Velte, Robert Elsenpeter, Cloud Computing A Practical Approach, TATA McGRAW-HILL Edition 2010.

Arani, A. H., Mehbodniya, A., Omidi, M. J., & Adachi, F. (2016, September). Distributed load balancing user association and self- anizing resource allocation in HetNets. In *2016 IEEE 84th Vehicular Technology Conference (VTC-Fall)* (pp. 1-5). IEEE.

Arbogast, J.W, Darmanyan, A.P, Foote, C.S, Rubin, Y, Diederich, F.N, & Alvarez, M.M. (1991). Photophysical properties of C60. *Journal of Physical Chemistry*, *95*(1), 11–12. doi:10.1021/j100154a006

Artaza, T., Suárez, A., Veiga, F., Braceras, I., Tabernero, I., Larrañaga, O., & Lamikiz, A. (2020). Wire arc additive manufacturing Ti6Al4V aeronautical parts using plasma arc welding: Analysis of heat-treatment processes in different atmospheres. *Journal of Materials Research and Technology*, *9*(6), 15454–15466. doi:10.1016/j.jmrt.2020.11.012

Artyukhov, V. I., Penev, E. S., & Yakobson, B. I. (2014). Why Nanotubes Grow Chiral. *Nature Communications*, *5*(1), 4892–4892. doi:10.1038/ncomms5892 PMID:25224858

Aryanto, K. Y. E., Abyong, I. W., Seputra, K. A., Pradnyana, G. A., & Paramartha, A. Y. (2020, January). Design of the Integrated Health Information System in the District of Bangli. In *3rd International Conference on Innovative Research Across Disciplines* (ICIRAD 2019) (pp. 240-245). Atlantis Press. https://doi.org/10.2991/assehr.k.200115.039

Astefanei, A. (2015). Nu'n˜ez, O. Galceran, M.T. *Analytica Chimica Acta*, *882*, 1–21. doi:10.1016/j.aca.2015.03.025 PMID:26043086

Atzori, L., Iera, A., & Morabito, G. (2011). IoT: Giving a social structure to the internet of things. *IEEE Communications Letters*, *15*(11), 1193–1195. doi:10.1109/LCOMM.2011.090911.111340

Augustine, S., Singh, J., Srivastava, M., Sharma, M., Das, A., & Malhotra, B. D. (2017). Recent advances in carbon based nanosystems for cancer theranostics. *Biomaterials Science*, *5*(5), 901–952. doi:10.1039/C7BM00008A PMID:28401206

Autodesk. (2018). What is BIM. *BIM and The Future of AEC*. BIM and the future of AEC. https://www.autodesk.com/solutions/bim

Autodesk. (2020). *Building Information Modelling in Practice*. Autodesk whitepaper Report. San Rafael. California.

Azarbad, M., Ebrahimzadeh, A., & Babajani-Feremi, A. (2010). Brain tissue segmentation using an unsupervised clustering technique based on PSO algorithm. *17th Iranian Conference of Biomedical Engineering (ICBME)*. IEEE.

Azzam, T., & Domb, A. J. (2004). Current developments in gene transfection agents. *Current Drug Delivery*, *1*(2), 165–193. doi:10.2174/1567201043479902 PMID:16305382

Baader, F., Calvanese, D., McGuinness, D., Nardi, D., & Patel-Schneider, P. (Eds.). (2002). *The Description Logic Handbook*. Cambridge University Press.

Babarinde, S. A. (2017). *Assessment of The Adoption Level of Building Information Modelling by Selected Construction Professionals in Kwara-State*. [Unpublished Bachelor Degree Thesis in Department of Quantity Surveying, Federal University of Technology, Minna].

Babatunde, S. O., Udeaja, C., & Adekunle, A. O. (2020). Barriers to BIM implementation and ways forward to improve its adoption in the Nigerian AEC firms. *International Journal of Building Pathology and Adaptation*, *39*(1), 48–71. doi:10.1108/IJBPA-05-2019-0047

Babatunde, S., Ekundayo, D., Adekunle, A., & Bello, W. (2020a). Comparative analysis of drivers to BIM adoption among AEC firms in developing countries: A case of Nigeria. *Journal of Engineering. Design and Technology*, *18*(6), 1425–1447. doi:10.1108/JEDT-08-2019-0217

Babu, S. Z. D., & Pandey, D. (2022). Analysation of Big Data in Smart Healthcare. In *Artificial Intelligence on Medical Data: Proceedings of International Symposium, ISCMM 2021*, (pp. 243-251). Singapore: Springer Nature Singapore.

Babu, S. Z. D., Pandey, D., Naidu, G. T., Sumathi, S., Gupta, A., Bader Alazzam, M., & Pandey, B. K. (2022, July). Analysation of Big Data in Smart Healthcare. In *Artificial Intelligence on Medical Data: Proceedings of International Symposium, ISCMM 2021* (pp. 243-251). Singapore: Springer Nature Singapore.

Babu., . . .. SZD. Analysation of Big Data in Smart Healthcare. n: Gupta, M, Ghatak, S, Gupta, A, Mukherjee, AL (eds) Artificial Intelligence on Medical Data Lecture Notes in Computational Vision and Biomechanics. 2023;37. https://doi.org/ doi:10.1007/978-981-19-0151-5_21

Baddeley, M., & Chang, C. (2015). Collaborative Building Information Modelling (BIM): Insights from behavioural economics and incentive theory. London: Royal Institution of Chartered Surveyor (RICS).

Baduge, S. K., Thilakarathna, S., Perera, J. S., Arashpour, M., Sharafi, P., Teodosio, B., Shringi, A., & Mendis, P. (2022). Artificial intelligence and smart vision for building and construction 4.0: Machine and deep learning methods and applications. *Automation in Construction*, *141*, 104440. doi:10.1016/j.autcon.2022.104440

Baghchi, M., Moriyama, H., & Shahidi, F. (2013). *Bio-Nanotechnology: A Revolution in Food, Biomedical and Health Sciences*. Wiley-Blackwell. doi:10.1002/9781118451915

Bagheri, M., Borhani, T. N. G., & Zahedi, G. (2012). Estimation of flash point and autoignition temperature of organic sulfur chemicals. *Energy Conversion and Management*, *58*, 185–196. doi:10.1016/j.enconman.2012.01.014

Bambach, M., Sizova, I., Sydow, B., Hemes, S., & Meiners, F. (2020). Hybrid manufacturing of components from Ti-6Al-4V by metal forming and wire-arc additive manufacturing. *Journal of Materials Processing Technology*, *282*, 116689. Advance online publication. doi:10.1016/j.jmatprotec.2020.116689

Bambach, M., Sizova, I., Sydow, B., Hemes, S., & Meiners, F. (2021). Perspectives on additive manufacturing in Industry 4.0. In M. Manjaiah & ... (Eds.), *Woodhead Publishing Reviews: Mechanical Engineering Series* (pp. 127–150). Woodhead Publishing. https://doi.org/https://doi.org/10.1016/B978-0-12-822056-6.00001-1

Banerjee, S., Ghosh, S., Bhowmick, G. D., Pandey, D., Pandey, B. K., & Dadheech, P. (2023). Biofuels and Health Hazards: An Overview. Handbook of Research on Safe Disposal Methods of Municipal Solid Wastes for a Sustainable Environment, 324-344.

Banerjee, S., Kaur, D., Pandey, D., Pal, S. R., Ahuchaogu, A. A., Iqbal, M. O., & Pandey, B. K. (2020). Industry 4.0 in the Nutraceutical Industry and Public Health Nutrition. In Robotics and Automation in Industry 4.0 (pp. 396-422). CRC Press.

Barbadikar, R., Deshmukh, S., Maddi, L., Laha, K., Parameswaran, P., Ballal, R., Peshwe, R., Paretkar, K., Nandagopal, M., & Mathew, D. (2015). Effect of normalizing and tempering temperatures on microstructure and mechanical properties of P92 steel. *International Journal of Pressure Vessels and Piping*, *132–133*, 97–105. doi:10.1016/j.ijpvp.2015.07.001

Barnaghi, P., Presser, M., & Moessner, K. (2010). Publishing Linked Sensor Data. *Proceedings of the 3rd International Workshop on Semantic Sensor Networks*. IEEE.

Basri, M. H., Farmaciawaty, D. A., Rachmania, I. N., Widjaja, F. B., & Wiyana, A. F. (2021, May). Healthcare Ecosystem Mapping in Primary Care: A Case Study in West Java, Indonesia. In *International Conference on Business and Engineering Management* (ICONBEM 2021) (pp. 184-190). Atlantis Press. https://doi.org/10.2991/aebmr.k.210522.025

Bataw, A. (2015). *On the integration of Building Information Modelling in undergraduate Civil engineering programmes in the United Kingdom.* [Thesis, The University of Manchester].

Beatriz, R. C. M., & Karla, F. R. (2019). Recent advances on the use of carbon nanotubes as smart biomaterials. *Journal of Materials Chemistry. B, 1*(2), 1–20.

Behera, A. (2020). Processes and Application in Additive Manufacturing: Practices in Aerospace, Automobile, Medical, and Electronic Industries. In *Additive Manufacturing applications for metals and composites* (pp. 25–47). IGI Global. doi:10.4018/978-1-7998-4054-1.ch002

Berners-Lee, T. (2000). *Weaving the Web: The Original Design and Ultimate Design of the World Wide Web by its inventor.* Harper Business.

Bessant, Y. A., Jency, J. G., Sagayam, K. M., Jone, A. A. A., Pandey, D., & Pandey, B. K. (2023). Improved parallel matrix multiplication using Strassen and Urdhvatiryagbhyam method. *CCF Transactions on High Performance Computing, 5*(2), 1–14. doi:10.1007/s42514-023-00149-9

Bhakta, P., & Barthunia, B. (2020, April–June). Fullerene and its applications: A review. *Journal of Indian Academy of Oral Medicine and Radiology, 32*(2), 159–163. doi:10.4103/jiaomr.jiaomr_191_19

Bhat, P. (2013). Mulgund, S. *Indo Am. J. Pharm. Res., 3*(8), 6549–6554.

Bian, Z., Ru, J. W., Wei, H. W., & Zhang, T. (2004, January). andAkihisa Inoue.Carbon-Nanotube-Reinforced Zr-Based Bulk Metallic GlassComposites and Their Properties. *Advanced Functional Materials, 14*(1), 55–63. doi:10.1002/adfm.200304422

Biason, A., & Zorzi, M. (2019). Multicast via point to multipoint transmissions in directional 5G mmWave communications. *IEEE Communications Magazine, 57*(2), 88–94. doi:10.1109/MCOM.2019.1700679

Biju, V. (2014). Chemical modifications and bioconjugate reactions of nanomaterials for sensing, imaging, drug delivery and therapy. *Chemical Society Reviews, 43*(3), 744–764. doi:10.1039/C3CS60273G PMID:24220322

Bin-Seddeeq, A., Assaf, S., Abdallah, A., & Hassanain, M. A. (2019). Time and cost overrun in the Saudi Arabian oil and gas construction industry. *Buildings, 9*(2), 41. doi:10.3390/buildings9020041

Bjerrum, E. J., & Threlfall, R. (2017). *Molecular generation with recurrent neural networks (RNNs).* arXiv preprint arXiv:1705.04612.

Bolskar, R. D., Benedetto, A. F., Husebo, L. O., Price, R. E., Jackson, E. F., Wallace, S., Wilson, L. J., & Alford, J. M. (2003). First Soluble M@C $_{60}$ Derivatives Provide Enhanced Access to Metallofullerenes and Permit in Vivo Evaluation of Gd@C $_{60}$ [C(COOH) $_{2}$ ] $_{10}$ as a MRI Contrast Agent. *Journal of the American Chemical Society, 125*(18), 5471–5478. doi:10.1021/ja0340984 PMID:12720461

Bonomi, F., Milito, R., Natarajan, P., & Zhu, J. (2014). *"Fog computing: a platform for internet of things and analytics," in Big Data and Internet ofThings: A Road Map for Smart Environments.* Springer.

Bonomi, F., Milito, R., Zhu, J., & Addepalli, S. (2012). Fog computing and its role in the internet of things. *Proceedings of the 1st ACM MCC Workshop on Mobile Cloud Computing*, (pp. 13–16). 10.1145/2342509.2342513

Boonsiritomachai, W., & Pitchayadejanant, K. (2019). Determinants affecting mobile banking adoption by generation Y based on the Unified Theory of Acceptance and Use of Technology Model modified by the Technology Acceptance Model concept. *Kasetsart Journal of Social Sciences, 40*(2), 349–358.

Boopathi, S., Pandey, B. K., & Pandey, D. (2023). Advances in artificial intelligence for image processing: techniques, applications, and optimization. In *Handbook of research on thrust technologies' effect on image processing* (pp. 73–95). IGI Global. doi:10.4018/978-1-6684-8618-4.ch006

Bordel Sánchez, B., Alcarria, R., Sánchez de Rivera, D., & Robles, T. (2018). Process execution in CyberPhysical Systems using cloud and Cyber-Physical Internet services. *The Journal of Supercomputing, 74*(8), 4127–4169. doi:10.1007/s11227-018-2416-4

Bormann, A. Konig, M., Koch, C. &Beetz, J. (2008). Building Information Modelling: Why? What? How? Technology Foundations and Industry Practice. *Building Information Modelling*, 1-24. doi:10.1007/978-3-319-92862-3_1

Bosi, S., Da Ros, T., Spalluto, G., Balzarini, J., & Prato, M. (2003). Synthesis and Anti-HIV properties of new water-soluble bis-functionalized[60]fullerene derivatives. *Bioorganic & Medicinal Chemistry Letters, 13*(24), 4437–4440. doi:10.1016/j.bmcl.2003.09.016 PMID:14643341

Boumia, L. (2014). Zidour, M. Benzair, A. and Tounsi, A. "A Timoshenko beam model for vibration analysis of chiral singlewalled carbon nanotubes,". *Physica E, Low-Dimensional Systems and Nanostructures, 59*, 186–191. doi:10.1016/j.physe.2014.01.020

Brandt, P., & Timmermans, S. (2021). Abductive Logic of Inquiry for Quantitative Research in the Digital Age. *Sociological Science, 8*, 191–210. doi:10.15195/v8.a10

Broilo, M. & Natale, F.G.B (2010). A Stochastic Approach to Image Retrieval Using Relevance Feedback and Particle Swarm Optimization. IEEE Transactions on Multimedia, 12(4).

Broy, M., Cengarle, M. V., & Geisberger, E. (2012). Cyber-Physical Systems: Imminent Challenges. In *Large-Scale Complex IT Syst. Dev., Operat. and Manag. - 17th Monterey Workshop*. IEEE.

Bryson, B. W. (2017). *Future of Architects: Extinction or Irrelevance*. Design Intelligence. https://www.di.net/articles/future-architects-extinctionirrelevance/

Budiarti, R. P. N., Widyatmoko, N., & Purnomo, M. H. (2016). Web scraping for automated water standard tracking system: A case study of PDAM Surabaya. International Seminar on Intelligent Technology and Its Applications (ISITIA).

Bujnowska-Fedak, M. M., Waligóra, J., & Mastalerz-Migas, A. (2019). The internet as a source of health information and services. *Advancements and Innovations in Health Sciences*, 1-16. https://doi.org/ doi:10.1007/5584_2019_396

Canada BIM Council (2018). *Building Information Modelling. Industrial Strategy: Government and Industry in Partnership*. Department for Business, Innovation and Skills (BIS).

Cardenas-Benitez, B., Djordjevic, I., Hosseini, S., Madou, M. J., & Martinez-Chapa, S. O. (2018). Covalent functionalization of carbon nanomaterials for biosensor applications: An update. *Journal of the Electrochemical Society, 165*(3), B103–B117. doi:10.1149/2.0381803jes

Cardoso, J., Sheth, A., Miller, J., Arnold, J., & Kochut, K. (2004). Quality of service for workflows and web service processes. *Journal of Web Semantics, 1*(3), 281–308. doi:10.1016/j.websem.2004.03.001

Carleo, G., & Troyer, M. (2017). Solving the quantum many-body problem with artificial neural networks. *Science, 355*(6325), 602–606. doi:10.1126/science.aag2302 PMID:28183973

Carter, E. A. (2008). Challenges in modeling materials properties without experimental input. *Science*, *321*(5890), 800–803. doi:10.1126/science.1158009 PMID:18687955

Çetin, M., Aytekin, E., Yavuz, B., & Bozdağ-Pehlivan, S. (2017). Chapter 7 - Nanoscience in Targeted Brain Drug Delivery. Nanotechnology Methods for Neurological Diseases and Brain Tumors Drug Delivery across the Blood-Brain Barrier. Academic Press. doi:10.1016/B978-0-12-803796-6.00007-1

Chahal, A., & Gulia, P. (2019). Machine learning and deep learning. *International Journal of Innovative Technology and Exploring Engineering*, *8*(12), 4910–4914. doi:10.35940/ijitee.L3550.1081219

Chatterjee, A., & Deopura, B. L. (2002). Carbon nanotubes and nanofibre: An overview. *Fibers and Polymers*, *3*(4), 134–139. doi:10.1007/BF02912657

Chen, M., Guan, R., & Yang, S. (2019). Hybrids of Fullerenes and 2D Nanomaterials. doi:10.1002/advs.201800941

Cheng, H., Xue, L., Wang, P., Zeng, P., & Yu, H. (2017). Ontology-based web service integration for flexible manufacturing systems. In *15th Int. Conf. on Ind. Inf.*, (pp. 351–356). IEEE. 10.1109/INDIN.2017.8104797

Cheng, J. C. P., & Lu, Q. (2015). A review of the efforts and roles of the Public Sector for BIM Adoption Worldwide. *Journal of Information Technology in Construction*, *20*, 442–478.

Chen, Z., Ma, L., Liu, Y., & Chen, C. (2012). Applications of Functionalized Fullerenes in Tumor Theranostics. *Theranostics*, *2*(3), 238–250. doi:10.7150/thno.3509 PMID:22509193

Chik, M. W., Hussain, Z., Zulkefeli, M., Tripathy, M., Kumar, S., Majeed, A. B. A., & Byrappa, K. (2019). Polymer-wrapped single-walled carbon nanotubes: A transformation toward better applications in healthcare. *Drug Delivery and Translational Research*, *9*(2), 578–594. doi:10.1007/s13346-018-0505-9 PMID:29594914

Chi, Y., Bhonsle, J. B., Canteenwala, T., Huang, J.-P., Shiea, J., Chen, B.-J., & Chiang, L. Y. (1998). Novel Water-soluble Hexa(sulfobutyl)fullerenes as Potent Free Radical Scavengers. *Chemistry Letters*, *27*(5), 465–466. doi:10.1246/cl.1998.465

Chowdhury, D. F. (2011). *5.49 - Carbon Nanotube for Drug Delivery and Controlled Release* (2nd ed., Vol. 5). Reference Module in Biomedical Sciences Comprehensive Biotechnology., doi:10.1016/B978-0-08-088504-9.00518-3

Ciortea, A., Mayer, S., & Michahelles, F. (2018). Repurposing Manufacturing Lines on the Fly with Multi-agent Systems for the Web of Things. In *Proc. of the 17th Int. Conf. on Autonomous Agents and Multi-Agent Systems*, (pp. 813–822). Int. Found. for Autonomous Agents and Multiagent Systems / ACM.

Colaço, M. J., Orlande, H. R. B., & Dulikravich, G. S. (2006). Inverse and optimization problems in heat transfer. *Journal of the Brazilian Society of Mechanical Sciences and Engineering*, *28*(1), 1–24. doi:10.1590/S1678-58782006000100001

Colitti, W., Steenhaut, K., Caro, N. De., (2011). Integrating Wireless Sensor Networks with the Web", Extending the Internet to Low power and Lossy Networks (IP+SN).

Colucci, S., Di Noia, T., Pinto, A., Ruta, M., Ragone, A., & Tinelli, E. (2007). A Nonmonotonic Approach to Semantic Matchmaking and Request Refinement in E-Marketplaces. *International Journal of Electronic Commerce*, *12*(2), 127–154. doi:10.2753/JEC1086-4415120205

Cory, A., Henson, J., Pschorr, K., Sheth, A. P., & Thirunarayan, K. (2009). SemSOS: Semantic sensor Observation Service. *Proceedings of the International Symposium on Collaborative Technologies and Systems*. IEEE.

Courbis, C., & Finkelstein, A. (2005). Weaving Aspects into Web Service Orchestrations. *Proceeding of International Conference of Web Services (ICWS '05)*, IEEE. 10.1109/ICWS.2005.129

Cuan-Baltazar, J. Y., Muñoz-Perez, M. J., Robledo-Vega, C., Pérez-Zepeda, M. F., & Soto-Vega, E. (2020). Misinformation of COVID-19 on the internet: Infodemiology study. *JMIR Public Health and Surveillance*, *6*(2), e18444. doi:10.2196/18444 PMID:32250960

Cui, Q., Hu, X., Liu, X., Zhao, L., & Wang, G. (2021). Understanding Architectural Designers' Continuous Use Intention Regarding BIM Technology: A China Case. *Buildings*, *11*(10), 448. doi:10.3390/buildings11100448

Czyrska-filemonowicz, A., Zielińska-lipiec, A., & Ennis, P. J. (2006). Modified 9% Cr Steels for Advanced Power Generation : Microstructure and Properties. *Journal of Achievements in Materials and Manufacturing Engineering*, *19*(2), 43–48.

Dai, S., You, R., Lu, Z., Huang, X., Mamitsuka, H., & Zhu, S. (2020). FullMeSH: Improving large-scale MeSH indexing with full text. *Bioinformatics (Oxford, England)*, *36*(5), 1533–1541. doi:10.1093/bioinformatics/btz756 PMID:31596475

Dalui, P., Elghaish, F., Brooks, T., & McIlwaine, S. (2021). Integrated Project Delivery with BIM: A Methodical Approach Within the UK Consulting Sector. *Journal of Information Technology in Construction*, *26*, 922–935. doi:10.36680/j.itcon.2021.049

Daoud, M. K. & Alqudah, M. . (2023). The Relationship Between Mobile Marketing and Customer Perceptions in Jordanian Commercial Banks: The Electronic Quality as A Mediator Variable. *IJMST*, *10*(2), 1360-1371.

Daoud, M. K., . Alqudah, M. . Al-Qeed, B. A. . Al Qaied, and A. Y. A. B. . Ahmad, "The Relationship Between Mobile Marketing and Customer Perceptions in Jordanian Commercial Banks: The Electronic Quality as A Mediator Variable", ijmst, vol. 10, no. 2, pp. 1360-1371, Jun. 2023.

Daoud, M. K., Taha, S., Al-Qeed, M., Alsafadi, Y., Ahmad, A., & Allahham, M. (2024). EcoConnect: Guiding environmental awareness via digital marketing approaches. *International Journal of Data and Network Science*, *8*(1), 235-242.

Daoud, M. K., Al-Qeed, M., Ahmad, A. Y. B., & Al-Gasawneh, J. A. (2023). Mobile Marketing: Exploring the Efficacy of User-Centric Strategies for Enhanced Consumer Engagement and Conversion Rates. *International Journal of Membrane Science and Technology*, *10*(2), 1252–1262. doi:10.15379/ijmst.vi.1425

Daraz, L., Morrow, A. S., Ponce, O. J., Beuschel, B., Farah, M. H., Katabi, A., Alsawas, M., Majzoub, A. M., Benkhadra, R., Seisa, M. O., Ding, J. F., Prokop, L., & Murad, M. H. (2019). Can patients trust online health information? A meta-narrative systematic review addressing the quality of health information on the internet. *Journal of General Internal Medicine*, *34*(9), 1884–1891. doi:10.1007/s11606-019-05109-0 PMID:31228051

Darwish, A. M., Tantawy, M. M., & Elbeltagi, E. (2020). Critical Success Factors for BIM implementation in Construction projects. *Saudi Journal of Civil Engineering*, *4*(9), 180–191. doi:10.36348/sjce.2020.v04i09.006

Dass, A., & Moridi, A. (2019). State of the Art in Directed Energy Deposition: From Additive Manufacturing to Materials Design. *Coatings*, *9*(7), 418. doi:10.3390/coatings9070418

Dave, P., Chandarana, A., Goel, P., & Ganatra, A. (2021). An amalgamation of YOLOv4 and XGBoost for next-gen smart traffic management system. *PeerJ. Computer Science*, *7*, e586. doi:10.7717/peerj-cs.586 PMID:34239973

David, S., Duraipandian, K., Chandrasekaran, D., Pandey, D., Sindhwani, N., & Pandey, B. K. (2023). Impact of blockchain in healthcare system. In *Unleashing the Potentials of Blockchain Technology for Healthcare Industries* (pp. 37–57). Academic Press. doi:10.1016/B978-0-323-99481-1.00004-3

Davis, E. L., Kelly, P. J., Deane, F. P., Baker, A. L., Buckingham, M., Degan, T., & Adams, S. (2020). The relationship between patient-centered care and outcomes in specialist drug and alcohol treatment: A systematic literature review. *Substance Abuse*, *41*(2), 216–231. doi:10.1080/08897077.2019.1671940 PMID:31638870

De Virgilio, R., Di Sciascio, E., Ruta, M., Scioscia, F., & Torlone, R. (2011). Semantic-based rfid data management. In *Unique Radio Innovation for the 21st Century* (pp. 111–141). Springer. doi:10.1007/978-3-642-03462-6_6

Deepa, R., Anand, R., Pandey, D., Pandey, B. K., & Karki, B. (2022). Comprehensive performance analysis of classifiers in diagnosis of epilepsy. *Mathematical Problems in Engineering*, *2022*, 2022. doi:10.1155/2022/1559312

Demirer, G. S., Zhang, H., Matos, J. L., Goh, N. S., Cunningham, F. J., Sung, Y., Chang, R., Aditham, A. J., Chio, L., Cho, M.-J., Staskawicz, B., & Landry, M. P. (2019). High aspect ratio nanomaterials enable delivery of functional genetic material without DNA integration in mature plants. *Nature Nanotechnology*, *14*(5), 456–464. doi:10.1038/s41565-019-0382-5 PMID:30804481

Deng, S., & Berry, V. (2016). Wrinkled, rippled and crumpled graphene: An overview of formation mechanism, electronic properties, and applications. *Materials Today*, *19*(4), 197–212. doi:10.1016/j.mattod.2015.10.002

Deng-xu, H. E., Hua-zheng, Z. H. U., & Gui-qing, L. I. U. (2011). Glowworm swarm optimization algorithm for solving multi-constrained QoS multicast routing problem. *Seventh International Conference on Computational Intelligence and Security*, (pp. 66-70). IEEE.

Devasenapathy, D., Madhumathy, P., Umamaheshwari, R., Pandey, B. K., & Pandey, D. (2024). Transmission-efficient grid-based synchronized model for routing in wireless sensor networks using Bayesian compressive sensing. *SN Computer Science*, *5*(1), 1–11.

Dhanasekar, S., Martin Sagayam, K., Pandey, B. K., & Pandey, D. (2023). Refractive Index Sensing Using Metamaterial Absorbing Augmentation in Elliptical Graphene Arrays. *Plasmonics*, 1–11. doi:10.1007/s11468-023-02152-w

Dibike, Y. B., Velickov, S., & Solomatine, D. (2000, March). Support vector machines: Review and applications in civil engineering. In *Proceedings of the 2nd Joint Workshop on Application of AI in Civil Engineering* (pp. 215-218). Research Gate.

Diliman, Q. C. (2005). Traffic impact assessment for sustainable traffic management and transportation planning in urban areas. *Proceedings of the Eastern Asia Society for Transportation Studies*, *5*, 2342–2351.

Dilla, S. M., Tasnim, T., Amraeni, Y., & Selviana, A. (2022). Factors Relating to the Success of the Health Information System at the Buton District Health Center. *Waluya The International Science of Health Journal*, *1*(2), 29–46. doi:10.54883/wish.v1i2.8

Dimiduk, D. M., Holm, E. A., & Niezgoda, S. R. (2018). Perspectives on the impact of machine learning, deep learning, and artificial intelligence on materials, processes, and structures engineering. *Integrating Materials and Manufacturing Innovation*, *7*(3), 157–172. doi:10.1007/s40192-018-0117-8

Ding, C., Tong, L., Feng, J., & Fu, J. (2016). Recent Advances in Stimuli-Responsive Release Function Drug Delivery Systems for Tumor Treatment. *Molecules (Basel, Switzerland)*, *21*(12), 1715. doi:10.3390/molecules21121715 PMID:27999414

Ding, D., Pan, Z., Cuiuri, D., & Li, H. (2015). Wire-feed additive manufacturing of metal components: Technologies, developments and future interests. *International Journal of Advanced Manufacturing Technology*, *81*(1), 465–481. doi:10.1007/s00170-015-7077-3

Ding, R., Lu, G., Yan, Z., & Wilson, M. (2001). Recent advances in the preparation and utilization of carbon nanotubes for hydrogen storage. *Journal of Nanoscience and Nanotechnology*, *1*(1), 7–29. doi:10.1166/jnn.2001.012 PMID:12914026

Dinovitzer, M., Chen, X., Laliberte, J., Huang, X., & Frei, H. (2019). Effect of wire and arc additive manufacturing (WAAM) process parameters on bead geometry and microstructure. *Additive Manufacturing*, *26*, 138 146. doi:10.1016/j.addma.2018.12.013

Divya, A., Kavithanjali, T., & Dharshini, P. (2019, March). IoT Enabled Forest Fire Detection and Early Warning System. In *2019 IEEE International Conference on System, Computation, Automation and Networking (ICSCAN)* (pp. 1-5). IEEE. 10.1109/ICSCAN.2019.8878808

Dokas, I. M., Feehan, J., & Imran, S. (2013). EWaSAP: An early warning sign identification approach based on a systemic hazard analysis. *Safety Science*, *58*, 11–26. doi:10.1016/j.ssci.2013.03.013

Dolez Patricia, I. (2015). *Nanomaterials Definitions, Classifications, and Applications*. Nanoengineering Global Approaches to Health and Safety Issues., doi:10.1016/B978-0-444-62747-6.00001-4

Dominey-Howes, D. (2002). Documentary and geological records of tsunamis in the Aegean Sea region of Greece and their potential value to risk assessment and disaster management. *Natural Hazards*, *25*(3), 195–224. doi:10.1023/A:1014808804611

Dubey, V., Kumar, P., & Chauhan, N. (2019). Forest fire detection system using IoT and artificial neural network. In *International Conference on Innovative Computing and Communications* (pp. 323-337).Springer, Singapore. 10.1007/978-981-13-2324-9_33

Ducrot, P., Montagni, I., Nguyen Thanh, V., Serry, A. J., & Richard, J. B. (2021). Evolution of online health-related information seeking in France from 2010 to 2017: Results from nationally representative surveys. *Journal of Medical Internet Research*, *23*(4), e18799. doi:10.2196/18799 PMID:33851927

Dugan, L. L., Gabrielsen, J. K., Yu, S. P., Lin, T.-S., & Choi, D. W. (1996). Buckminsterfullerenol Free Radical Scavengers Reduce Excitotoxic and Apoptotic Death of Cultured Cortical Neurons. *Neurobiology of Disease*, *3*(2), 129–135. doi:10.1006/nbdi.1996.0013 PMID:9173920

Dugan, L. L., Turetsky, D. M., Du, C., Lobner, D., Wheeler, M., Almli, C. R., Shen, C. K.-F., Luh, T.-Y., Choi, D. W., & Lin, T.-S. (1997). Carboxyfullerenes as neuroprotective agents. *Proceedings of the National Academy of Sciences of the United States of America*, *94*(17), 9434–9439. doi:10.1073/pnas.94.17.9434 PMID:9256500

Durdyev, S., Mbachu, J., Thurnell, D., Zhao, L., & Hosseini, M. R. (2021). BIM Adoption in the Cambodian Construction Industry: Key Drivers and Barriers. *ISPRS International Journal of Geo-Information*, *10*(4), 1–14. doi:10.3390/ijgi10040215

Dutta, S., Mukherjee, U., & Pandey, D. (2023). Recent Trends of Addressing COVID-19 Disease by AI/ML. In Handbook of Research on Thrust Technologies' Effect on Image Processing (pp. 395-419). IGI Global.

Dwivedi, Y. K., Rana, N. P., Jeyaraj, A., Clement, M., & Williams, M. D. (2019). Re-examining the unified theory of acceptance and use of technology (UTAUT): Towards a revised theoretical model. *Information Systems Frontiers*, *21*(3), 719–734. doi:10.1007/s10796-017-9774-y

Eadie, R., Odeyinka, H., Browne, M., Mckeown, C., & Yohanis, M. (2014). Building Information Modelling Adoption: An analysis of the barriers to Implementation. *Journal of Engineering and Architecture*, *2*(1), 77–101.

Eastman, C. (1974). The use of computers instead of drawings in building design. *AIA Journal. American Institute of Architects*, *63*(3), 46–50.

Eko Atlantic. (2016). *Eko Atlantic Nigeria's New Global City*. EcoAtlantic. https://www.ekoatlantic. com/milestones/Eko-Atlantic-Milestones-Issue-1. pdf

Eko Brochure. (2013). *Eko Pearl Towers*. EkoPearl. www.ekopearl.org

Ekong, M. O., George, W. K., Pandey, B. K., & Pandey, D. (2023). Enhancing the Fundamentals of Industrial Safety Management in TVET for Metaverse Realities. In *Applications of Neuromarketing in the Metaverse* (pp. 19–41). IGI Global. doi:10.4018/978-1-6684-8150-9.ch002

Emmanuel-Eze, C., Ugulu, R. A., & Egwunatum, S. I. (2021). Green Building Materials Products and Service Market in the Construction Industry. *Journal of Engineering. Project & Production Management*, *11*(2), 46–58.

Encinas, C. (2017). Design and implementation of a distributed IoT system for the monitoring of water quality in aquaculture. Wireless Telecommunications Symposium (WTS). IEEE.

Enes, R. F., Tomé, A. C., Cavaleiro, J. A. S., Amorati, R., Fumo, M. G., Pedulli, G. F., & Valgimigli, L. (2006). Synthesis and Antioxidant Activity of [60]Fullerene–BHT Conjugates. *Chemistry (Weinheim an der Bergstrasse, Germany)*, *12*(17), 4646–4653. doi:10.1002/chem.200501495 PMID:16534828

Ennaceur, S., Bugis, B., & Al-Mohaithef, M. (2023). e-Health information literacy of health science university students: Actual situation and Improvement strategies. *Journal of Electronic Resources Librarianship*, *35*(2), 91–105. doi:10.1080/1941126X.2023.2197749

Ennis, P.J. & Czyrska-Filemonowicz, A. (2003). Recent advances in creep-resistant steels for power plant applications', *Sadhana - Academy Proceedings in Engineering Sciences*, *28*(3–4), 709–730. . doi:10.1007/BF02706455

Esmaeilzadeh, P., Dharanikota, S., & Mirzaei, T. (2021). The role of patient engagement in patient-centric health information exchange (HIE) initiatives: An empirical study in the United States. *Information Technology & People*. Advance online publication. doi:10.1108/ITP-05-2020-0316

Faber, F. A., Hutchison, L., Huang, B., Gilmer, J., Schoenholz, S. S., Dahl, G. E., & von Lilienfeld, O. A. (2017). *Fast machine learning models of electronic and energetic properties consistently reach approximation errors better than DFT accuracy*. arXiv preprint arXiv:1702.05532.

Fahmi, A. S., & Ayuningtyas, D. (2020, November). Readiness Analysis of Electronic Information System in Public Health Center as Supporting Tools in Implementation of Fraud Prevention System on UHC Program: Literature Review. In *International Conference of Health Development. Covid-19 and the Role of Healthcare Workers in the Industrial Era* (ICHD 2020) (pp. 100-109). Atlantis Press. https://doi.org/10.2991/ahsr.k.201125.019

Farghali, A. A. (2017). Abdel Tawab, H. A. Abdel Moaty S. A. and Khaled, R. *Journal of Nanostructure in Chemistry*, *7*, 101–111. doi:10.1007/s40097-017-0227-4

Fargnoli, M., & Lombardi, M. (2020). Building information modelling (BIM) to enhance occupational safety in construction activities: Research trends emerging from one decade of studies. *Buildings*, *10*(6), 98. doi:10.3390/buildings10060098

Fauzan, A., & Noviandi, N. (2020). Evaluation of Optima Regional Health Information System with HOT-Fit on Technology Aspects Approach in Johar Baru Health Center Jakarta. [JICHI]. *Journal of Intelligent Computing and Health Informatics*, *1*(1), 9–14. doi:10.26714/jichi.v1i1.5397

Fayisa, G. W., & Wayessa, S. G. (2021). Cause of Construction Material Wastage on Public Building Project in Western Oromia. *American Journal of Civil Engineering*, *9*(2), 55–62. doi:10.11648/j.ajce.20210902.14

Fiorito, S., Serafino, A., Andreola, F., Togna, A., & Togna, G. (2006). Toxicity and biocompatibility of carbon nanoparticles. *Journal of Nanoscience and Nanotechnology*, *6*(3), 591–599. doi:10.1166/jnn.2006.125 PMID:16573109

Ford, S. (2016). Additive manufacturing and sustainability : an exploratory study of the advantages and challenges. *Journal of Cleaner Production*. . doi:10.1016/j.jclepro.2016.04.150

Fraihat, B. A. M., Ahmad, A. Y. B., Alaa, A. A., Alhawamdeh, A. M., Soumadi, M. M., Aln'emi, E. A. S., & Alkhawaldeh, B. Y. S. (2023). Evaluating Technology Improvement in Sustainable Development Goals by Analysing Financial Development and Energy Consumption in Jordan. *International Journal of Energy Economics and Policy*, *13*(4), 348–355. doi:10.32479/ijeep.14438

Freitas, R. A. Jr. (2000). Nanodentistry. *The Journal of the American Dental Association*, *131*(11), 1559–1565. doi:10.14219/jada.archive.2000.0084 PMID:11103574

Friel, R. J. (2015) 'Power ultrasonics for additive manufacturing and consolidating of materials', Power Ultrasonics: Applications of High-Intensity Ultrasound, (pp. 313–335). Springer. doi:10.1016/B978-1-78242-028-6.00013-2

Galluzzi, L., Vitale, I., Abrams, J., Alnemri, E. S., Baehrecke, E. H., Blagosklonny, M. V., Dawson, T. M., Dawson, V. L., El-Deiry, W. S., Fulda, S., Gottlieb, E., Green, D. R., Hengartner, M. O., Kepp, O., Knight, R. A., Kumar, S., Lipton, S. A., Lu, X., Madeo, F., & Kroemer, G. (2012). Molecular definitions of cell death subroutines: Recommendations of the Nomenclature Committee on Cell Death 2012. *Cell Death and Differentiation*, *19*(1), 107–120. doi:10.1038/cdd.2011.96 PMID:21760595

Galvão, R. K. H., Araújo, M. C. U., Fragoso, W. D., Silva, E. C., José, G. E., Soares, S. F. C., & Paiva, H. M. (2008). A variableelimination method to improve the parsimony of MLR models using the successive projections algorithm. *Chemometrics and Intelligent Laboratory Systems*, *92*(1), 83–91. doi:10.1016/j.chemolab.2007.12.004

Gamil, Y., & Rahman, I. A. R. (2019). Awareness and challenges of building information modelling (BIM) implementation in the Yemen construction industry. *Journal of Engineering. Design and Technology*, *17*(5), 1077–1084. doi:10.1108/JEDT-03-2019-0063

Ganiyu, M. (2018). *BIMing the Architectural Curricula – Integrating Building Information Modelling (BIM) in Architectural Education*. Lincoln University.

Gao, H., Xu, W., Sun, J., & Tang, Y. (2010). Multilevel Thresholding for Image Segmentation Through an Improved Quantum-Behaved Particle Swarm Algorithm. *IEEE Transactions on Instrumentation and Measurement, 59*(4).

Gaurav, K., & Bansal, H. (2013), Particle Swarm Optimization (PSO) Technique for Image Enhancement. *IJECT, 4*.

Gel'atko, M., Hatala, M., Botko, F., Vandžura, R., Hajnyš, J., Šajgalík, M., & Török, J. (2023). Stress Relieving Heat Treatment of 316L Stainless Steel Made by Additive Manufacturing Process. *Materials (Basel)*, *16*(19), 6461. doi:10.3390/ma16196461 PMID:37834599

Georgakilas, V., Perman, J. A., Tucek, J., & Zboril, R. (2015). Broad Family of Carbon Nanoallotropes: Classification, Chemistry, and Applications of Fullerenes, Carbon Dots, Nanotubes, Graphene, Nanodiamonds, and Combined Superstructures. *Chemical Reviews*, *115*(11), 4744–4822. doi:10.1021/cr500304f PMID:26012488

George, D. A. S., George, A. H., Baskar, T., & Pandey, D. (2021). XDR: The Evolution of Endpoint Security Solutions-Superior Extensibility and Analytics to Satisfy the Organizational Needs of the Future. International Journal of Advanced Research in Science [IJARSCT]. *Tongxin Jishu*, *8*(1), 493–501.

George, A. H., Fernando, M., George, A. S., Baskar, T., & Pandey, D. (2021). Metaverse: The next stage of human culture and the internet. [IJARTET]. *International Journal of Advanced Research Trends in Engineering and Technology*, *8*(12), 1–10.

George, A. S., Baskar, T., George, A. H., Pandey, D., & Martin, A. G. (2022). A Review of 6G: Towards The Future. *Partners Universal International Research Journal*, *1*(4), 1–12.

George, A. S., George, A. H., Baskar, T., & Pandey, D. (2022). The Transformation of the workspace using Multigigabit Ethernet. *Partners Universal International Research Journal, 1*(3), 34–43.

Ghazal, M., Albashaireh, Z. S., & Kern, M. (2008). Wear resistance of nanofilled composite resin and feldspathic ceramic artificial teeth. *The Journal of Prosthetic Dentistry, 100*(6), 441–448. doi:10.1016/S0022-3913(08)60262-0 PMID:19033028

Gholamzadeh, A., & Karimi Taheri, A. (2009). The prediction of hot flow behavior of Al-6%Mg alloy. *Mechanics Research Communications, 36*(2), 252–259. doi:10.1016/j.mechrescom.2008.06.011

Gibson, I., Rosen, D., & Stucker, B. (2015). *3D Printing*. Rapid Prototyping, and Direct Digital Manufacturing, Additive Manufacturing Technologies.

Gille, S., Griese, L., & Schaeffer, D. (2021). Preferences and experiences of people with chronic illness in using different sources of health information: Results of a mixed-methods study. *International Journal of Environmental Research and Public Health, 18*(24), 13185. doi:10.3390/ijerph182413185 PMID:34948792

Goh, G. B., Hodas, N. O., & Vishnu, A. (2017). Deep learning for computational chemistry. *Journal of Computational Chemistry, 38*(16), 1291–1307. doi:10.1002/jcc.24764 PMID:28272810

Gohil, P. P., Malek, M., Bachwani, D., Patel, D., Upadhyaya, D., & Hathiwala, A. (2022). Application of 5D Building Information Modelling for Construction Management. *ECS Transactions, 107*(1), 26372649. doi:10.1149/10701.2637ecst

Gokhale, M. M., & Somani, R. R. (2015). Fullerenes: Chemistry and its applications. *Mini-Reviews in Organic Chemistry, 12*(4), 355–366. doi:10.2174/1570193X12666150930224428

Gomez-Chabla, R., Real-Aviles, K., Moran, C., Grijalva, P., & Recalde, T. (2019, January). IoT applications in agriculture: A systematic literature review. In *2nd International Conference on ICTs in Agronomy and Environment* (pp. 68-76). Springer, Cham. 10.1007/978-3-030-10728-4_8

Gonzalez, K. A. (2002). Synthesis and In vitro characterization of a tissue-Selective fullerene: Vectoring C60(OH)16AMBP to mineralized bone. *Bioorganic & Medicinal Chemistry, 10*(6), 1991–1997. doi:10.1016/S0968-0896(02)00049-4 PMID:11937359

Gornet, T. (2017). History of Additive Manufacturing. Research Gate. doi:10.4018/978-1-5225-2289-8.ch001

Goubau, T. (2018). In *What is BIM? What its Benefits to the Construction Industry?* Aproplan. https://www.aproplan.com/blog/quality-management-plan-construction/what-is-bim-what-are-its-benefits-to-the-construction-industry

Grushevskaya, H. V., & Krylova, N. G. (2018). Carbon nanotubes as a high-performance platform for target delivery of anticancer quinones. *Current Pharmaceutical Design, 24*(43), 5207–5218. doi:10.2174/1381612825666190117095132 PMID:30652640

Guan, L., Suenaga, K., & Iijima, S. (2008). Smallest Carbon Nanotube Assigned with Atomic Resolution Accuracy. *Nano Letters, 8*(2), 459–462. doi:10.1021/nl072396j PMID:18186659

Guan, M., Ge, J., Wu, J., Zhang, G., Chen, D., Zhang, W., Zhang, Y., Zou, T., Zhen, M., Wang, C., Chu, T., Hao, X., & Shu, C. (2016). Fullerene/photosensitizer nanovesicles as highly efficient and clearable phototheranostics with enhanced tumour accumulation for cancer therapy. *Biomaterials, 103*, 75–85. doi:10.1016/j.biomaterials.2016.06.023 PMID:27376559

Gubbi, J., Buyya, R., Marusic, S., & Palaniswami, M. (2013). Internet of Things (IoT): A vision, architectural elements, and future directions. *Future Generation Computer Systems, 29*(7), 1645–1660. doi:10.1016/j.future.2013.01.010

Gul, R. Zainab, N. Asad, M. Salma, B. (2019). An Overview of the Recent Progress in the Synthesis and Applications of Carbon Nanotubes. *J. carbo Res. 5*(3), pp. 2-31.

Guo, W., & Huang, X. (2011, April). On throughput enhancement of multi-hop wireless networks using interference alignment. In *2011 20th Annual Wireless and Optical Communications Conference (WOCC)* (pp. 1-6). IEEE. 10.1109/WOCC.2011.5872297

Gupta, A. K., Sharma, R., Pandey, D., Nassa, V. K., Pandey, B. K., George, A. S., & Dadheech, P. (2023). Performance Analysis of Eight-Channel WDM Optical Network with Different Optical Amplifiers for Industry 4.0. In *Innovation and Competitiveness in Industry 4.0 Based on Intelligent Systems* (pp. 197–212). Springer International Publishing. doi:10.1007/978-3-031-29775-5_9

Gupta, A., Anand, R., Pandey, D., Sindhwani, N., Wairya, S., Pandey, B. K., & Sharma, M. (2021). Prediction of breast cancer using extremely randomized clustering forests (ERCF) technique: Prediction of breast cancer. [IJDST]. *International Journal of Distributed Systems and Technologies*, *12*(4), 1–15. doi:10.4018/IJDST.287859

Gupta, R., Mane, M., Bhardwaj, S., Nandekar, U., Afaq, A., Dhabliya, D., & Pandey, B. K. (2023). Use of Artificial Intelligence for Image Processing to Aid Digital Forensics: Legislative Challenges. In *Handbook of Research on Thrust Technologies' Effect on Image Processing* (pp. 433–447). IGI Global. doi:10.4018/978-1-6684-8618-4.ch026

Hamma-Adama, M., Kouider, T., & Salman, H. (2018). State of building information modelling (BIM) adoption in Nigeria. *34th Association of Researchers in Construction Management (ARCOM) annual conference: working papers: a productive relationship; balancing fragmentation and integration* (pp. 334-343). Belfast: ARCOM.

Hamma-adama, M., Salman, H. S., & Kouider, T. (2017). *BIM adoption in Nigeria: state of building information modeling adoption in Nigeria.* The ARCOM 2018 Conference, Belfast, UK.

Hamma-Adama, M., & Kouider, T. (2018). A Review of Building Information Modelling in Nigeria and Its Potentials. *International Journal of Civil and Environmental Engineering. World Academy of Science and Technology*, *12*(11), 1113–1119.

Hamma-adama, M., & Kouider, T. (2019). What are the barriers and drivers toward BIM adoption in Nigeria? In *Creative Construction Conference*, Budapest University of Technology and Economics. 10.3311/CCC2019-073

Han, M., & Yao, W. (2010). Remote sensing image fusion using particle swam optimization. *International Conference on Intelligent Control and Information Processing, 241*, 13-15

Han, M., Dong, T. & Hou, D. (2020). Carbon nanotube based Janus composite membrane of oil fouling resistance for direct contact membrane distillation. *Journal of Membrane Science.*

Han, Y. (2020). *A simple and effective flexible electrochemiluminescence sensor for lidocaine detection.*

Hao, Y., Sun, G., Fan, T., Sun, X., Liu, Y., Zhang, N., Zhao, L., Zhong, R., & Peng, Y. (2019). Prediction on the mutagenicity of nitroaromatic compounds using quantum chemistry descriptors based QSAR and machine learning derived classification methods. *Ecotoxicology and Environmental Safety*, *186*, 109822. doi:10.1016/j.ecoenv.2019.109822 PMID:31634658

Harahap, N. C., Handayani, P. W., & Hidayanto, A. N. (2021). Barriers in health information systems and technologies to support maternal and neonatal referrals at primary health centers. *Healthcare Informatics Research*, *27*(2), 153–161. doi:10.4258/hir.2021.27.2.153 PMID:34015881

Hasanah, A., Rangkuti, D. N., Pinem, G. Z. D., & Nasution, F. (2021). Evaluation of the Health Information System in Teladan Public Health Center in 2020 Using the Health Metrics Network Instrument Approach. *International Archives of Medical Sciences and Public Health*, 2(1).

Hasnain, M. S., & Nayak, A. K. (2019). *Carbon Nanotubes for Targeted Drug Delivery.* Springer Briefs in Applied Sciences and Technology.

Hazra, A., Adhikari, M., Amgoth, T., & Srirama, S. N. (2021, November). A Comprehensive Survey on Interoperability for IIoT: Taxonomy, Standards, and Future Directions. *ACM Computing Surveys*, *55*(1), 1–35. doi:10.1145/3485130

He, H. (2013). Carbon Nanotubes: Applications in Pharmacy and Medicine. BioMed Research International.

He, A., Xie, G., Yang, X., Wang, X., & Zhang, H. (2015). A physically-based constitutive model for a nitrogen alloyed ultralow carbon stainless steel. *Computational Materials Science*, *98*, 64–69. doi:10.1016/j.commatsci.2014.10.044

He, A., Xie, G., Zhang, H., & Wang, X. (2013). A comparative study on Johnson-Cook, modified Johnson-Cook and Arrhenius-type constitutive models to predict the high temperature flow stress in 20CrMo alloy steel. *Materials & Design*, *52*, 677–685. doi:10.1016/j.matdes.2013.06.010

He, J., Zhang, D., Zhang, W., Qiu, C., & Zhang, W. (2017). Constitutive equation and hot compression deformation behavior of homogenized Al-7.5Zn-1.5Mg-0.2Cu-0.2Zr alloy. *Materials (Basel)*, *10*(10), 1–12. doi:10.3390/ma10101193 PMID:29057825

Hemamalini, V., Rajarajeswari, S., Nachiyappan, S., Sambath, M., Devi, T., Singh, B. K., & Raghuvanshi, A. (2022). Food quality inspection and grading using efficient image segmentation and machine learning-based system. *Journal of Food Quality*, *2022*, 1–6. doi:10.1155/2022/5262294

He, Q., Wang, G., Luo, L., Shi, Q., Xie, J., & Meng, X. (2017). Mapping the managerial areas of Building Information Modeling (BIM) using scientometric analysis. *International Journal of Project Management*, *35*(4), 670–685. doi:10.1016/j.ijproman.2016.08.001

Hidayat, I. F., & Hermanto, B. R. (2020, October). A preliminary implementation of hl7 fhir to achieve interoperability in indonesia's local ehr. In 2020 *27th International Conference on Telecommunications* (ICT) (pp. 1-5). IEEE. 10.1109/ICT49546.2020.9239534

Hills, G., Lau, C., Wright, A., Fuller, S., Bishop, M. D., Srimani, T., Kanhaiya, P., Ho, R., Amer, A., Stein, Y., Murphy, D., Arvind, Chandrakasan, A., & Shulaker, M. M. (2019). Modern microprocessor built from complementary carbon nanotube transistors. *Nature*, *572*(7771), 595–602. doi:10.1038/s41586-019-1493-8 PMID:31462796

Hirlekar, R., Yamagar, M., Garse, H., Vij, M., & Kadam, V. (2009). Carbon nanotubes and its applications: A review. *Asian Journal of Pharmaceutical and Clinical Research*, *2*, 17–27.

Howard, J. B. (1992). *Symposium (International) on Combustion*. Elsevier.

Hsu, H.-C., Chiang, L.-Y., Chen, W.-J., & Lee, Y.-T. (2000). Water-Soluble Hexasulfobutyl [60] Fullerene Inhibits Plasma Lipid Peroxidation by Direct Association with Lipoproteins. *Journal of Cardiovascular Pharmacology*, *36*(4), 423–427. doi:10.1097/00005344-200010000-00002 PMID:11026641

Hsu, S., & Luo, P. (2019). *From Nanoarchitectonics to Tissue Architectonics: Nanomaterials for Tissue Engineering. Advanced Supramolecular Nanoarchitectonics*. Elsevier Inc., doi:10.1016/B978-0-12-813341-5.00012-7

Huang, Y., Wang, S., Xiao, Z., & Liu, H. (2017). Critical Condition of Dynamic Recrystallization in 35CrMo Steel. *Metals*, *7*(5), 161. doi:10.3390/met7050161

Huan, J., Li, H., Wu, F., & Cao, W. (2020). Design of water quality monitoring system for aquaculture ponds based on NB-IoT. *Aquacultural Engineering*, *90*, 102088. doi:10.1016/j.aquaeng.2020.102088

Hush, M. R. (2017). Machine learning for quantum physics. *Science*, *355*(6325), 580–580. doi:10.1126/science.aam6564 PMID:28183936

Hu, X., Yu, S., Yang, G., Long, W., Guo, T., Tian, J., Liu, M., Li, X., Zhang, X., & Wei, Y. (2022). Facile synthesis of inorganic–organic hybrid fluorescent nanoparticles with AIE feature using hexachlorocyclotriphosphazene as the bridge. *Journal of Molecular Liquids*, *345*, 117693. doi:10.1016/j.molliq.2021.117693

Hu, Z., Guan, W., Wang, W., Huang, L., Xing, H., & Zhu, Z. (2007). Protective effect of a novel cystine C60 derivative on hydrogen peroxide-induced apoptosis in rat pheochromocytoma PC12 cells. *Chemico-Biological Interactions*, *167*(2), 135–144. doi:10.1016/j.cbi.2007.02.009 PMID:17353010

Hu, Z., Zhao, F., Wang, Y., Huang, Y., Chen, L., Li, N., Li, J., Li, Z., & Yi, G. (2014). Facile fabrication of a C60-polydopamine-graphene nanohybrid for single light induced photothermal and photodynamic therapy. *Chemical Communications (Cambridge)*, *50*(74), 10815–10818. doi:10.1039/C4CC04416A PMID:25089303

Ibrahim, Y. M., & Abdullahi, M. (2016). *Introduction to Building Information Modelling*. Presented at A 3 – Day Workshop/ Annual General Meeting of the Nigerian Institute of Quantity Surveyors, Port-Harcourt.

Idrees, S., & Shafiq, M. T. (2021). Factors for Time and Cost Overrun in Public Projects. *Journal of Engineering. Project & Production Management*, *11*(3), 243–254.

Iijima, S. (1991). Helical microtubules of graphitic carbon. *Nature*, *354*(6348), 56–58. doi:10.1038/354056a0

Iijima, S. (2002). Carbon nanotubes: Past, present and future. Phys. B. *Physica B, Condensed Matter*, *323*(1-4), 1–5. doi:10.1016/S0921-4526(02)00869-4

Iijima, S., Ajayan, P. M., & Ichihashi, T. (1992). Growth model for carbon nanotubes. *Physical Review Letters*, *69*(21), 3100–3103. doi:10.1103/PhysRevLett.69.3100 PMID:10046725

Iijima, S., & Ichihashi, T. (1993). Single-shell carbon nanotubes of 1-nm diameter. *Nature*, *363*(6430), 603–605. doi:10.1038/363603a0

Indarto, A., Ikhsan, N. A., & Wibowo, I. (2020). *Applications of carbon nanotubes for controlling waterborne pathogens*. Waterborne Pathogens Detection and Treatment., doi:10.1016/B978-0-12-818783-8.00020-7

Islam, M. R., Mondal, C., Azam, M. K., & Islam, A. S. M. J. (2016). Text detection and recognition using enhanced MSER detection and a novel OCR technique. In *2016 5th International Conference on Informatics, Electronics and Vision (ICIEV)*, (pp. 15-20). IEEE.

Isobe, H., Nakanishi, W., Tomita, N., Jinno, S., Okayama, H., & Nakamura, E. (2006). Nonviral Gene Delivery by Tetraamino Fullerene. *Molecular Pharmaceutics*, *3*(2), 124–134. doi:10.1021/mp050068r PMID:16579641

Isobe, H., Tomita, N., Jinno, S., Okayama, H., & Nakamura, E. (2001). Synthesis and transfection capability of multifunctionalized fullerene polyamine. *Chemistry Letters*, *12*(12), 1214–1215. doi:10.1246/cl.2001.1214

Ivanova, O., Williams, C., & Campbell, T. (2013). Additive manufacturing (AM) and nanotechnology: Promises and challenges. *Rapid Prototyping Journal*, *19*(5), 353–364. doi:10.1108/RPJ-12-2011-0127

Ivántabernero, P. A., Álvarez, P. & Suárez, A. (2018). Study on Arc Welding Processes for High Deposition Rate Additive Manufacturing. Procedia CIRP, 68, 358–362. doi:10.1016/j.procir.2017.12.095

Iyyanar, P., Anand, R., Shanthi, T., Nassa, V. K., Pandey, B. K., George, A. S., & Pandey, D. (2023). A Real-Time Smart Sewage Cleaning UAV Assistance System Using IoT. In *Handbook of Research on Data-Driven Mathematical Modeling in Smart Cities* (pp. 24–39). IGI Global.

Jadhav, S., & Ghadge, A.N. (2016). Application of building information modeling in construction projects: A critical review. *International Journal of Engineering Research*, 26-31.

Jadon, A., Omama, M., Varshney, A., Ansari, M. S., & Sharma, R. (2019). Firenet: A specialized lightweight fire & smoke detection model for real-time iot applications. arXiv preprint arXiv:1905.11922.

Jakl, A., Schoffer, L., Husinsky, M., & Wagner, M. (2018), Augmented Reality for Industry 4.0: Architecture and User Experience. In *Proceeding of the 11th Forum Media Technology, CER-WS,* 38-42.

Jandl, C., Nurgazina, J., Schoffer, L., Reichl, C., Wagner, M., & Moser, T. (2019). SensiTrack – A Privacy by Design Concept for Industrial IoT Applications, In *Proceeding of the 24th IEEE International Conference on Emerging Technologies and Factory Automation*, (pp. 1782-1789). IEEE. 10.1109/ETFA.2019.8869186

Jariwala, D., Sangwan, V. K., Lauhon, L. J., Marks, T. J., & Hersam, M. C. (2013). Carbon nanomaterials for electronics, optoelectronics, photovoltaics, and sensing. *Chemical Society Reviews*, *42*(7), 2824–2860. doi:10.1039/C2CS35335K PMID:23124307

Jaya Lakshmi. G., Pandey, D., Pandey, B. K., Kaur, P., Mahajan, D. A., & Dari, S. S. (2024). Smart Big Data Collection for Intelligent Supply Chain Improvement. In AI and Machine Learning Impacts in Intelligent Supply Chain (pp. 180-195). IGI Global.

Jaya, L. G., Pandey, D., Pandey, B. K., Kaur, P., Mahajan, D. A., & Dari, S. S. (2024). Smart Big Data Collection for Intelligent Supply Chain Improvement. In AI and Machine Learning Impacts in Intelligent Supply Chain (pp. 180-195). IGI Global.

Jayakumar, D. N., & Venkatesh, P. (2014). Glowworm swarm optimization algorithm with topsis for solving multiple objective environmental economic dispatch problem. *Applied Soft Computing*, *23*, 375–386.

JayaLakshmi. G., Pandey, D., Pandey, B. K., Kaur, P., Mahajan, D. A., & Dari, S. S. (2024). Smart Big Data Collection for Intelligent Supply Chain Improvement. In AI and Machine Learning Impacts in Intelligent Supply Chain (pp. 180-195). IGI Global.

Jayaram, K., Janani, K., Jeyaguru, R., Kumaresh, R., & Muralidharan, N. (2019, March). Forest Fire Alerting System With GPS Co-ordinates Using IoT. In 2019, *5th International Conference on Advanced Computing & Communication Systems (ICACCS)* (pp. 488-491). IEEE. 10.1109/ICACCS.2019.8728383

Jayatilake, S. M. D. A. C., & Ganegoda, G. U. (2021). Involvement of machine learning tools in healthcare decision making. *Journal of Healthcare Engineering*, *2021*, 1–20. Advance online publication. doi:10.1155/2021/6679512 PMID:33575021

Jeschke, S., Brecher, C., Meisen, T., Ozdemir, D., & Eschert, T. (2017), Industrial Internet of Things and Cyber Manufacturing Systems. In Industrial Internet of Things. Springer.

Jha, S., Tewari, A., Mishra, S., & Toppo, S. (2017). Constitutive Relations for Ti-6Al-4V Hot Working. *Procedia Engineering*, *173*, 755–762. doi:10.1016/j.proeng.2016.12.089

Jiang, G. (2006). Li, G. J. *Photochem. Photobiol. B: Biol.*, *85*(3), 223–227. doi:10.1016/j.jphotobiol.2006.08.001

Jiang, J. L., & Wong, W. K. (2018). Fundamentals of common computer vision techniques for textile quality control. In *Applications of computer vision in fashion and textiles* (pp. 3–15). Woodhead Publishing. doi:10.1016/B978-0-08-101217-8.00001-4

Jiao, F., Liu, Y., Qu, Y., Li, W., Zhou, G., Ge, C., Li, Y., Sun, B., & Chen, C. (2010). Studies on anti-tumor and antimetastatic activities of fullerenol in a mouse breast cancer model. *Carbon, 48*(8), 2231–2243. doi:10.1016/j.carbon.2010.02.032

Jiao, Z., Hu, P., Xu, H., & Wang, Q. (2020a). Machine learning and deep learning in chemical health and safety: A systematic review of techniques and applications. *Journal of Chemical Health and Safety, 27*(6), 316–334. doi:10.1021/acs.chas.0c00075

Jiao, Z., Ji, C., Yuan, S., Zhang, Z., & Wang, Q. (2020b). Development of machine learning based prediction models for hazardous properties of chemical mixtures. *Journal of Loss Prevention in the Process Industries, 67*, 104226. doi:10.1016/j.jlp.2020.104226

Jiao, Z., Yuan, S., Ji, C., Mannan, M. S., & Wang, Q. (2019). Optimization of dilution ventilation layout design in confined environments using Computational Fluid Dynamics (CFD). *Journal of Loss Prevention in the Process Industries, 60*, 195–202. doi:10.1016/j.jlp.2019.05.002

Ji, Z. Q., Sun, H., Wang, H., Xie, Q., Liu, Y., & Wang, Z. (2006). Biodistribution and tumour uptake of C60(OH)x in mice. *Journal of Nanoparticle Research, 8*(1), 53–63. doi:10.1007/s11051-005-9001-5

Jonas, J., Quelennec, X., Jiang, L., & Martin, É. (2009). The Avrami kinetics of dynamic recrystallization. *Acta Materialia, 57*(9), 2748–2756. doi:10.1016/j.actamat.2009.02.033

Kalpana. (2016). Online monitoring of water standard using raspberry Pi3 model B. *International Journal of Innovative Technology and Research, 4*.

Kamboj, M., Arora, R., & Gupta, H. (2016). Comparative evaluation of the efficacy of synthetic nanocrystalline hydroxyapatite bone graft (Ostim ®) and synthetic microcrystalline hydroxyapatite bone graft (Osteogen ®) in the treatment of human periodontal intrabony defects: A clinical and denta scan study. *Journal of Indian Society of Periodontology, 20*(4), 423–428. doi:10.4103/0972-124X.184036 PMID:28298825

Kameswari, G., Himaja, B., & Tirumala, N. (2018). Wild Fire Detection using. *Internet of Things : Engineering Cyber Physical Human Systems.*

Karakostas, B. (2013). A DNS architecture for the Internet of things: A case study in transport logistics. *Procedia Computer Science, 19*, 594–601. doi:10.1016/j.procs.2013.06.079

Karousis, N., Martinez, I. S., Ewels, C. P., & Tagmatarchis, N. (2016). Structure, properties, functionalization, and applications of carbon nanohorns. *Chemical Reviews, 116*(8), 4850–4883. doi:10.1021/acs.chemrev.5b00611 PMID:27074223

Kartakis, S. & Julie, A. (2016). McCann, Reza Akhavan, IEEE First International Conference on Internet-of-Things Design and Implementation. IEEE.

Käsermann, F., & Kempf, C. (1997). Photodynamic inactivation of enveloped viruses by buckminsterfullerene. *Antiviral Research, 34*(1), 65–70. doi:10.1016/S0166-3542(96)01207-7 PMID:9107386

Katam, R., Kalapatapu, P., & Pasupuleti, V. D. K. (2022, June). A review on technological advancements in the field of data driven structural health monitoring. In *European workshop on structural health monitoring* (pp. 371-380). Cham: Springer International Publishing.

Katasonov, A., Kaykova, O., Khriyenko, O., Nikitin, S., & Terziyan, V. (2008). Smart Semantic Middleware for the Internet of Things. *Proceedings of the 5th International Conference of Informatics in Control, Automation and Robotics*, (pp. 11-15). IEEE.

Kazeem, K. O., Olawumi, T. O., & Osunsanmi, T. (2023). Roles of Artificial Intelligence and Machine Learning in Enhancing Construction Processes and Sustainable Communities. *Buildings, 13*(8), 2061. doi:10.3390/buildings13082061

Kedia, N. (2015). *Water Grade Monitoring for Rural Areas-A Sensor Cloud Based Economical Project*. Dehradun, India.

Keskinock, P., & Tayur, S. (2001). Quantitive analysis of Internet-enabled supply chain. *Interfaces*, *31*(2), 70–89. doi:10.1287/inte.31.2.70.10626

Khajeh, A., & Modarress, H. (2010). QSPR prediction of flash point of esters by means of GFA and ANFIS. *Journal of Hazardous Materials*, *179*(1-3), 715–720. doi:10.1016/j.jhazmat.2010.03.060 PMID:20381958

Khan, A. U., Oriol, M., Kiran, M., Jiang, M., & Djemame, K. (2012). Security risks and their management in cloud computing. *4th IEEE Int. Conf. Cloud Comput. Technol. Sci. Proc.* (pp. 121–128). IEEE. 10.1109/CloudCom.2012.6427574

Khan, B., Hasan, A., Pandey, D., Ventayen, R. J. M., Pandey, B. K., & Gowwrii, G. (2021). 12 Fusion of Datamining and Artificial Intelligence. *Machine learning and iot for intelligent systems and smart applications*, 201.

Khan, R., Khan, S. U., Zaheer, R., & Khan, S. (2012). Future internet: the internet of things architecture, possible applications and key challenges. *Proceedings of the 10th International Conference on Frontiers of Information Technology (FIT '12)*. IEEE. 10.1109/FIT.2012.53

Khedkar, S. P., Canessane, R. A., & Najafi, M. L. (2021). Prediction of traffic generated by IoT devices using statistical learning time series algorithms. *Wireless Communications and Mobile Computing*, *2021*, 1–12. doi:10.1155/2021/5366222

Khin Mya Mya, A. (2012). Sreekumaran Nair, V. Jagadeesh Babu, Rajendiran Murugana and Seeram Ramakrishna.A review on nanomaterials for environmental remediation. *Energy & Environmental Science*, *5*, 8075.

Khurshed, A. S., & Bilal, A. T. (2016). Synthesis of carbon nanotubes by catalytic chemical vapour deposition: A review on carbon sources, catalysts and substrates. *Materials Science in Semiconductor Processing*, *41*(5), 67–82.

Kimera, R., Kaggwa, F., Mwavu, R., Mugonza, R., Tumuhimbise, W., Munguci, G., & Kamuganga, F. (2020). *An Overview of Data Science Innovations*. Challenges and Limitations Towards Real-World Implementations in Global Health.

Kim, S., Park, J., Youn, Y. S., Oh, K. T., Bae, J. H., & Lee, E. S. (2015). Hoechst 33258–conjugated hyaluronated fullerene for efficient photodynamic tumor therapy and necrotic tumor targeting. *Journal of Bioactive and Compatible Polymers*, *30*(3), 275–288. doi:10.1177/0883911515574136

Kirubasri, G., Sankar, S., Pandey, D., Pandey, B. K., Singh, H., & Anand, R. (2021, September). A recent survey on 6G vehicular technology, applications and challenges. In *2021 9th International Conference on Reliability, Infocom Technologies and Optimization (Trends and Future Directions)(ICRITO)* (pp. 1-5). IEEE.

Kirubasri, G., Sankar, S., Pandey, D., Pandey, B. K., Singh, H., & Anand, R. (2021, September). A Recent Survey on 6G Vehicular Technology, Applications and Challenges. In *2021 9th International Conference on Reliability, Infocom Technologies and Optimization (Trends and Future Directions)(ICRITO)* (pp. 1-5). IEEE.

Kirubasri, G., Sankar, S., Pandey, D., Pandey, B. K., Nassa, V. K., & Dadheech, P. (2022). Software-defined networking-based Ad hoc networks routing protocols. In *Software defined networking for Ad Hoc networks* (pp. 95–123). Springer International Publishing. doi:10.1007/978-3-030-91149-2_5

Koch, F. A., & Oldham, W. K. (1985). Oxidation-reduction potential–a tool for monitoring, control and optimization of biological nutrient removal systems. *Water Science and Technology*, *17*(11-12), 259–281. doi:10.2166/wst.1985.0237

Kong, S. W. R., Lau, L. T., Wong, S. Y., & Phan, D. T. (2020). A study on effectiveness of Building Information Modelling (BIM) on the Malaysian construction industry. *2020 IOP Conference Series: Materials Science and Engineering*, *7*(13), 012-035. 10.1088/1757-899X/713/1/012035

Kour, R., Arya, S., Young, S.-J., Gupta, V., Bandhoria, P., & Khosla, A. (2020). Recent advances in carbon nanomaterials as electrochemical sensors. *Journal of the Electrochemical Society*, *167*(3), 37555. doi:10.1149/1945-7111/ab6bc4

Kratschmer, W., Lamb, L. D., Fostiropoulos, K., & Huffman, D. R. (1990). Solid C60: A new form of carbon. *Nature*, *347*(6291), 27. doi:10.1038/347354a0

Krishna, R., Manjaiah, M., & Mohan, C. B. (2021). Developments in additive manufacturing. In M. Manjaiah, (Eds.), *Woodhead Publishing Reviews: Mechanical Engineering Series* (pp. 37–62). Woodhead Publishing.

Krishna, V., Singh, A. P., Sharma, P., Iwakuma, N., Wang, Q., Zhang, Q., Knapik, J., Jiang, H., Grobmyer, S. R., Koopman, B., & Moudgil, B. (2010). Polyhydroxy fullerenes for non-invasive cancer imaging and therapy. *Small*, *6*(20), 2236–2241. doi:10.1002/smll.201000847 PMID:20818623

Kroto, H. W., Heath, J. R., O'Brien, S. C., Curl, R. F., & Smalley, R. E. (1985). C60: Buckminsterfullerene. *Nature*, *318*(6042), 162–163. doi:10.1038/318162a0

Krusic P.J, Wasserman E, Keizer P.N, Morton J.R, Preston K.F. Radical reactions of C60 Science. 1991;254:1183–5

Kuila, T., Bose, S., Mishra, A., Khanra, P., Kim, N., & Lee, J. (2012). Chemical functionalization of graphene and its applications. *Progress in Materials Science*, *57*(7), 1061–1105. doi:10.1016/j.pmatsci.2012.03.002

Kujala, S., Hörhammer, I., Väyrynen, A., Holmroos, M., Nättiaho-Rönnholm, M., Hägglund, M., & Johansen, M. A. (2022). Patients' Experiences of Web-Based Access to Electronic Health Records in Finland: Cross-sectional Survey. *Journal of Medical Internet Research*, *24*(6), e37438. doi:10.2196/37438 PMID:35666563

Kumar Pandey, B., Pandey, D., Nassa, V. K., Ahmad, T., Singh, C., George, A. S., & Wakchaure, M. A. (2021). Encryption and steganography-based text extraction in IoT using the EWCTS optimizer. *Imaging Science Journal*, *69*(1-4), 38–56. doi:10.1080/13682199.2022.2146885

Kumar, M. S., Sankar, S., Nassa, V. K., Pandey, D., Pandey, B. K., & Enbeyle, W. (2021). Innovation and creativity for data mining using computational statistics. In *Methodologies and Applications of Computational Statistics for Machine Intelligence* (pp. 223–240). IGI Global. doi:10.4018/978-1-7998-7701-1.ch012

Kushwaha, V. (2016). Contribution of Building Information Modelling (BIM) to solve problems in the Architecture, Engineering and Construction (AEC) Industry and Addressing Barriers to Implementation of BIM. *IACSIT International Journal of Engineering and Technology*, *3*(1), 100–105.

Kwag, D. S., Park, K., Oh, K. T., & Lee, E. S. (2013). Hyaluronated fullerenes with photoluminescent and antitumoral activity. *Chemical Communications (Cambridge)*, *49*(3), 282–284. doi:10.1039/C2CC36596K PMID:23174913

Kyzyma, E. (2015). J. Surf. Investig. *X-ray Synchrotron Neutron Tech.*, *9*(1), 1–5. doi:10.1134/S1027451015010127

Laasraoui, A., & Jonas, J. J. (1991). Prediction of steel flow stresses at high temperatures and strain rates. *Metallurgical Transactions. A, Physical Metallurgy and Materials Science*, *22*(7), 1545–1558. doi:10.1007/BF02667368

Lacerda, L., Bianco, A., Prato, M., & Kostarelos, K. (2006). Carbon nanotubes as nanomedicines: From toxicology to pharmacology. *Advanced Drug Delivery Reviews*, *58*(14), 1460–1470. doi:10.1016/j.addr.2006.09.015 PMID:17113677

Lai, Y. L. (2003). Murugan, P. Hwang, K. *Life Sciences*, *72*(11), 1271–1278. doi:10.1016/S0024-3205(02)02374-3 PMID:12570927

Lasi, H., Fettke, P., Kemper, H.-G., Feld, T., & Hoffmann, M. (2014). Industry 4.0. *BISE*, *6*(4), 239–242.

LeCun, Y., Bengio, Y., & Hinton, G. (2015). *Deep learning. nature*, *521*(7553), 436-444.

Lee, D. J., Ahn, Y. S., Youn, Y. S., & Lee, E. S. (2012). Poly(ethylene glycol)-crosslinked fullerenes for high efficient phototherapy. *Polymers for Advanced Technologies*, *24*(2), 220–227. doi:10.1002/pat.3074

Lee, D., & Yoon, S. N. (2021). Application of artificial intelligence-based technologies in the healthcare industry: Opportunities and challenges. *International Journal of Environmental Research and Public Health*, *18*(1), 271. doi:10.3390/ijerph18010271 PMID:33401373

Lee, H. L., & Billington, C. (1992). Managing supply chain inventories: Pitfalls and opportunities. *Sloan Management Review*, *33*(3), 65–77.

Lee, H. Y., Jin, S. W., Henning-Smith, C., Lee, J., & Lee, J. (2021). Role of health literacy in health-related information-seeking behavior online: Cross-sectional study. *Journal of Medical Internet Research*, *23*(1), e14088. doi:10.2196/14088 PMID:33502332

Lee, J., Kao, H.-A., & Yang, S. (2014). Service Innovation and Smart Analytics for Industry 4.0 and Big Data Environment. *Procedia CIRP*, *16*, 3–8. doi:10.1016/j.procir.2014.02.001

Lee, Y., Buchanan, B. G., Mattison, D. M., Klopman, G., & Rosenkranz, H. S. (1995). Learning rules to predict rodent carcinogenicity of non-genotoxic chemicals. *Mutation Research. Fundamental and Molecular Mechanisms of Mutagenesis*, *328*(2), 127–149. doi:10.1016/0027-5107(94)00202-G PMID:7739598

Lefort, L., Henson, C., Taylor, K., Barnaghi, P., Compton, M., Corcho, O., Garcia-Castro, R., Graybeal, J., Herzog, A. & Janowicz, K. (2005). *Semantic Sensor Network XG Final Report*. W3. https://www.w3.org/2005/Incubator/ssn/XGR-ssn/

Lelisho, M. E., Pandey, D., Alemu, B. D., Pandey, B. K., & Tareke, S. A. (2022). The Negative Impact of Social Media during COVID-19 Pandemic. *Trends in Psychology*, *31*(1), 1–20. doi:10.1007/s43076-022-00192-5

Liao, L., & Teo, E. A. L. (2017). Critical Success Factors for enhancing the Building Information Modelling Implementation in building projects in Singapore. *Journal of Civil Engineering and Management*, *23*(8), 1029–1044. doi:10.3846/13923730.2017.1374300

Liao, W. H., Kao, Y., & Li, Y.-S. (2011). A sensor deployment approach using glowworm swarm optimization algorithm in wireless sensor networks. *Expert Systems with Applications*, *38*(10), 12180–12188.

Li, F., Chen, S., Shi, J., Tian, H., & Zhao, Y. (2017). Evaluation and Optimization of a Hybrid Manufacturing Process Combining Wire Arc Additive Manufacturing with Milling for the Fabrication of Stiffened Panels. *Applied Sciences (Basel, Switzerland)*, *7*(12), 1233. doi:10.3390/app7121233

Li, L., & Horrocks, I. (2004). A software framework for matchmaking based on semantic web technology. *International Journal of Electronic Commerce*, *8*(4), 39–60. doi:10.1080/10864415.2004.11044307

Li, L., & Li, D. (2008). Image texture classification based on immune particle swarm optimization. *Acta Geodaetica et Cartographica Sinica.*, *37*(2), 185–189.

Li, L., Zhen, M., Wang, H., Sun, Z., Jia, W., Zhao, Z., Zhou, C., Liu, S., Wang, C., & Bai, C. (2020). Functional Gadofullerene Nanoparticles Trigger Robust Cancer Immunotherapy Based on Rebuilding an Immunosuppressive Tumor Microenvironment. *Nano Letters*, *20*(6), 4487–4496. doi:10.1021/acs.nanolett.0c01287 PMID:32407113

Lim, C. C., Toh, T. C., Lew, Y. L., Lau, C. X., Goh, K. C., Masrom, M. A. N., Zainordin, N., & Loo, S. C. (2020). the Perspective of Students on Drivers and Benefits of Building Information Modelling Incorporation into Quantity Surveying Profession in Klang Valley, Malaysia. *International Journal of Information Management*, *7*(1), 16–29. doi:10.15282/ijim.7.0.2020.5751

Lim, E.-K., Kim, T., Paik, S., Haam, S., Huh, Y.-M., & Lee, K. (2015). Nanomaterials for theranostics: Recent advances and future challenges. *Chemical Reviews*, *115*(1), 327–394. doi:10.1021/cr300213b PMID:25423180

Lin, A. M., Chyi, B. Y., Wang, S. D., Yu, H.-H., Kanakamma, P. P., Luh, T.-Y., Chou, C. K., & Ho, L. T. (1999). Carboxyfullerene Prevents Iron-Induced Oxidative Stress in Rat Brain. *Journal of Neurochemistry*, *72*(4), 1634–1640. doi:10.1046/j.1471-4159.1999.721634.x PMID:10098871

Lin, C., Xia, C., Chen, M., & Chen, S. (2010). Constitutive descriptions for hot compressed 2124-T851 aluminum alloy over a wide range of temperature and strain rate. *Computational Materials Science*, *50*(1), 227–233. doi:10.1016/j.commatsci.2010.08.003

Lin, P. A., Gomez-Ballesteros, J. L., Burgos, J. C., Balbuena, P. B., Natarajan, B., & Sharma, R. (2017). Direct Evidence of Atomic-Scale Structural Fluctuations in Catalyst Nanoparticles. *Journal of Catalysis*, *349*, 149–155. doi:10.1016/j.jcat.2017.03.009 PMID:28740274

Lin, Y. C., Chen, M. S., & Zhang, J. (2009). Modeling of flow stress of 42CrMo steel under hot compression. *Materials Science and Engineering A*, *499*(1–2), 88–92. doi:10.1016/j.msea.2007.11.119

Lin, Y. C., & Chen, X. M. (2011). A critical review of experimental results and constitutive descriptions for metals and alloys in hot working. *Materials & Design*, *32*(4), 1733–1759. doi:10.1016/j.matdes.2010.11.048

Lin, Y. X., Wang, Y., Blake, S., Yu, M., Mei, L., Wang, H., & Shi, J. (2020). RNA nanotechnology-mediated cancer immunotherapy. *Theranostics*, *10*(1), 281–299. doi:10.7150/thno.35568 PMID:31903120

Li, Q., Hong, L., Li, H., & Liu, C. (2017). Graphene oxide-fullerene C60 (GO-C60) hybrid for photodynamic and photothermal therapy triggered by near-infrared light. *Biosensors & Bioelectronics*, *89*, 477–482. doi:10.1016/j.bios.2016.03.072 PMID:27055602

Li, Q., Huang, C., Liu, L., Hu, R., & Qu, J. (2018). Enhancing type I photochemistry in photodynamic therapy under near infrared light by using antennae–fullerene complexes. *Cytometry. Part A*, *93*(10), 997–1003. doi:10.1002/cyto.a.23596 PMID:30230226

Li, R., Zhen, M., Guan, M., Chen, D., Zhang, G., Ge, J., Gong, P., Wang, C., & Shu, C. (2013). A novel glucose colorimetric sensor based on intrinsic peroxidase-like activity of C60-carboxyfullerenes. *Biosensors & Bioelectronics*, *47*, 502–507. doi:10.1016/j.bios.2013.03.057 PMID:23628844

Liu, C. Y., Zhang, R. J., & Yan, Y. N. (2011). Hot deformation behaviour and constitutive modelling of P92 heat resistant steel. *Materials Science and Technology*, *27*(8), 1281–1286. doi:10.1179/026708310X12683158443323

Liu, F., Ma, J., Peng, L., Huang, G., Zhang, W., Xie, H., & Mi, X. (2020). Hot deformation behavior and microstructure evolution of Cu-Ni-Co-Si alloys. *Materials (Basel)*, *13*(9), 2042. doi:10.3390/ma13092042 PMID:32349437

Liu, J., Ohta, S., Sonoda, A., Yamada, M., Yamamoto, M., Nitta, N., Murata, K., & Tabata, Y. (2007). Preparation of PEG-conjugated fullerene containing $Gd^{3+}$ ions for photodynamic therapy. *Journal of Controlled Release*, *117*(1), 104–110. doi:10.1016/j.jconrel.2006.10.008 PMID:17156882

Liu, J., & Tabata, Y. (2010). Photodynamic therapy of fullerene modified with pullulan on hepatoma cells. *Journal of Drug Targeting*, *18*(8), 602–610. doi:10.3109/10611861003599479 PMID:20180750

Liu, Q., Xu, L., Zhang, X., Li, N., Zheng, J., Guan, M., Fang, X., Wang, C., & Shu, C. (2013). Enhanced photodynamic efficiency of an aptamer-guided fullerene photosensitizer toward tumour cells. *Chemistry, an Asian Journal*, *8*(10), 2370–2376. doi:10.1002/asia.201300039 PMID:23907978

Liu, R., Ward, L., Wolverton, C., Agrawal, A., Liao, W. K., & Choudhary, A. (2016, August). Deep learning for chemical compound stability prediction. In *Proceedings of ACM SIGKDD workshop on large-scale deep learning for data mining (DL-KDD)* (pp. 1-7). ACM.

Liu, Y., Zhang, C., Wang, Y., Xu, X., Zhu, H., & Jiang, J. (2022). Reveal the hot deformation behaviour and microstructure evolution in additively manufactured 316L stainless steel. *Materials Science and Engineering A, 861*(August), 144290. doi:10.1016/j.msea.2022.144290

Liu, Z., Pagani, M., Zinniker, D., DeConto, R., Huber, M., Brinkhuis, H., Shah, S. R., Leckie, R. M., & Pearson, A. (2009, February 27). Global Cooling During the Eocene-Oligocene Climate Transition. March (2009). *Science, 323*(5918), 1187–1190. doi:10.1126/science.1166368 PMID:19251622

Li, Y., Ji, H., Li, W., Li, Y., Pei, W., & Liu, J. (2018). Hot deformation characteristics-Constitutive equation and processing maps-of 21-4N Heat- resistant steel. *Materials (Basel), 12*(1), 89. doi:10.3390/ma12010089 PMID:30591698

Li, Z., Zhang, F., Pan, L., Zhu, X., & Zhang, Z. (2015). Preparation and characterization of injectable Mitoxantrone poly (lactic acid)/fullerene implants for in vivo chemo-photodynamic therapy. *Journal of Photochemistry and Photobiology. B, Biology, 149*, 51–57. doi:10.1016/j.jphotobiol.2015.05.018 PMID:26046749

Lobov, A., Lopez, F. U., Herrera, V. V., Puttonen, J., & Lastra, J. L. M. (2008). Semantic Web Services framework for manufacturing industries. In *Int. Conf. on Rob. and Biomim.*, (pp. 2104–2108). IEEE.

Long, D. D. (2018). Practice-informed research: Contemporary challenges and ethical decision-making. *The Right for the Elderly to Commit Suicide, 15*(2), 15.

Lorenz, A., Jones, B., Wimpenny, I., & Jackson, R. (2020). A review of hybrid manufacturing. *Proceedings - 26th Annual International Solid Freeform Fabrication Symposium - An Additive Manufacturing Conference,* (pp. 96–108).

Lounasvuori, M. M., Kelly, D., & Foord, J. S. (2018). Carbon black as lowcost alternative for electrochemical sensing of phenolic compounds. *Carbon, 129*, 252–257. doi:10.1016/j.carbon.2017.12.020

Luan, J., Sun, C., Li, X., & Zhang, Q. (2014). Constitutive model for AZ31 magnesium alloy based on isothermal compression test. *Materials Science and Technology, 30*(2), 211–219. doi:10.1179/1743284713Y.0000000341

Luechtefeld, T., Rowlands, C., & Hartung, T. (2018). Big-data and machine learning to revamp computational toxicology and its use in risk assessment. *Toxicology Research, 7*(5), 732–744. doi:10.1039/c8tx00051d PMID:30310652

Lu, P., Chen, S., & Zheng, Y. (2012). Artificial intelligence in civil engineering. *Mathematical Problems in Engineering*, 2012.

Maass, W., Filler, A. (2006). Towards an infrastructure for semantically annotated physical products. INFORMATIK 2006–Informatik für Menschen–Band 2, Beiträge der 36. *Jahrestagung der Gesellschaft für Informatik eV* (GI).

Ma, D., Gao, J., Zhang, Z., Zhao, H., & Wang, Q. (2020). Locating the gas leakage source in the atmosphere using the dispersion wave method. *Journal of Loss Prevention in the Process Industries, 63*, 104031. doi:10.1016/j.jlp.2019.104031

Ma, D., & Zhang, Z. (2016). Contaminant dispersion prediction and source estimation with integrated Gaussian-machine learning network model for point source emission in atmosphere. *Journal of Hazardous Materials, 311*, 237–245. doi:10.1016/j.jhazmat.2016.03.022 PMID:27035273

Madhumathy, P., & Pandey, D. (2022). Deep learning based photo acoustic imaging for non-invasive imaging. *Multimedia Tools and Applications, 81*(5), 7501–7518. doi:10.1007/s11042-022-11903-6

Madjido, M., Espressivo, A., Maula, A. W., Fuad, A., & Hasanbasri, M. (2019). Health information system research situation in Indonesia: A bibliometric analysis. *Procedia Computer Science, 161*, 781–787. doi:10.1016/j.procs.2019.11.183

Madkour, L. H. (2020). *Nanoparticles Induce Oxidativeand Endoplasmic Reticulum Antioxidant TherapeuticDefenses.* Springer International Publishing. doi:10.1007/978-3-030-37297-2

Madkour, L. H. (2020). Nucleic Acids as Gene Anticancer Drug Delivery Therapy. Elsevier.

Madkour, L.H. (2020). *Reactive Oxygen Species (ROS), Nanoparticles, and Endoplasmic Reticulum (ER) Stress-Induced Cell Death Mechanisms.* Imprint: Academic Press.

Madkour, L.H. (2023). Graphitic Carbon Nitride Quantum Dots (g-C3N4): Fundamentals and Applications.International Research Journal of Basic and Clinical Studies, 8(4), 1-14.

Madkour, L. H. (2022). *Book: Nanoparticle-Based Drug Delivery in Cancer Treatment.First Published.* Imprint CRC Press., doi:10.1201/9781003229674

Madkour, L. H. (2023). Carbon Nanotubes—Based Intelligent Platform for Cancer Vaccine Co-delivery Nanocarriers Immunotherapy Achievements: Challenges In vitro and In vivo. *International Journal of Nursing Care and Research, 1*(2), 1–55.

Mahon, B. (2004). *The Man Who Changed Everything: The Life of James Clerk Maxwell.* John Wiley & Sons Ltd.

Maina, J. J. (2018). Barriers to effective use of CAD and BIM in architecture education in Nigeria. *International Journal of Built Environment and Sustainability, 5*(3), 175–186. doi:10.11113/ijbes.v5.n3.275

Makabate, C. T., Musonda, I., Okoro, C. S., & Chileshe, N. (2021). Scientometric analysis of BIM adoption by SMEs in the architecture, construction and engineering sector. *Engineering, Construction, and Architectural Management, 29*(1), 43–58.

Malhotra, P., Pandey, D., Pandey, B. K., & Patra, P. M. (2021). Managing agricultural supply chains in COVID-19 lockdown. *International Journal of Quality and Innovation, 5*(2), 109–118. doi:10.1504/IJQI.2021.117181

Manju, K., Anand, R., Pandey, B. K., Nassa, V. K., Shahul, A., George, A. H., & Dogiwal, S. R. (2023). Determination of Early Onset Glaucoma Using OCT Image. In *Handbook of Research on Thrust Technologies' Effect on Image Processing* (pp. 244–255). IGI Global. doi:10.4018/978-1-6684-8618-4.ch015

Mansurov, Z. (2012). Producing nanomaterials in combustion. *Combustion, Explosion, and Shock Waves, 48*(5), 561–569. doi:10.1134/S0010508212050073

Marchesan, S., Da Ros, T., Spalluto, G., Balzarini, J., & Prato, M. (2005). Anti-HIV properties of cationic fullerene derivatives. *Bioorganic & Medicinal Chemistry Letters, 15*(15), 3615–3618. doi:10.1016/j.bmcl.2005.05.069 PMID:15978810

Marinakis, Y., Iordanidou, G. R., & Marinaki, M. (2013). Particle swarm optimization for the vehicle routing problem with stochastic demands. *Applied Soft Computing, 13*(4), 1693–1704. doi:10.1016/j.asoc.2013.01.007

Marinelli, G., Martina, F., Ganguly, S., Williams, S., Lewtas, H., Hancock, D., Mehraban, S., & Lavery, N. (2019). Microstructure and thermal properties of unalloyed tungsten deposited by Wire + Arc Additive Manufacture. *Journal of Nuclear Materials, 522*, 45–53. doi:10.1016/j.jnucmat.2019.04.049

Marquez, F. (2023).*Non-Destructive Testing and Condition Monitoring Techniques in Wind Energy.A volume in Wind Energy Engineering. Copyright.* Elsevier Inc., doi:10.1016/C2021-0-02588-8

Marrella, A. (2018). *Automated Planning for Business Process Management.*

Marzouk, M., Elsaay, H., & Othman, A. A. E. (2022). Analyzing BIM implementation in the Egyptian construction industry. *Engineering, Construction, and Architectural Management*, *29*(10), 4177–4190. doi:10.1108/ECAM-07-2020-0523

Mashal, I., Alsaryrah, O., Chung, T. Y., Yang, C. Z., Kuo, W. H., & Agrawal, D. P. (2015). Choices for interaction with things on Internet and underlying issues. *Ad Hoc Networks*, *28*, 68–90. doi:10.1016/j.adhoc.2014.12.006

Mashino, T., Okuda, K., Hirota, T., Hirobe, M., Nagano, T., & Mochizuki, M. (1999). Inhibition of E-coli growth by fullerene derivatives and inhibition mechanism. *Bioorganic & Medicinal Chemistry Letters*, *9*(20), 2959–2962. doi:10.1016/S0960-894X(99)00515-6 PMID:10571155

Mashino, T., Shimotohno, K., Ikegami, N., Nishikawa, D., Okuda, K., Takahashi, K., Nakamura, S., & Mochizuki, M. (2005). Humanimmunodeficiency virus-reverse transcriptase inhibition and hepatitis C virus RNA-dependent RNA polymerase inhibition activities of fullerene derivatives. *Bioorganic & Medicinal Chemistry Letters*, *15*(4), 1107–1109. doi:10.1016/j.bmcl.2004.12.030 PMID:15686922

Matarneh, R. T., & Hamed, S. A. (2017). Exploring the adoption of building information modeling (BIM) in the Jordanian construction industry. *Journal of Architectural Engineering Technology*, *6*(10), 1–7. doi:10.4172/2168-9717.1000189

Mather, T., Kumaraswamy, S., & Latif, S. (2009). Cloud. *Security and Privacy*, 299.

Mathur, R., Alver, B., San Roman, A., Wilson, B. G., Wang, X., Agoston, A. T., Park, P. J., Shivdasani, R. A., & Roberts, C. W. M. (2017). ARID1.A loss impairs enhancer-mediated gene regulation and drives colon cancer in mice. *Nature Genetics*, *49*(2), 296–302. doi:10.1038/ng.3744 PMID:27941798

Matis, J. H., Birkett, T., & Boudreaux, D. (1989). An Application of the Markov Chain Approach to Forecasting Cotton Yields from Surveys. *Agricultural Systems*, *29*(4), 357–370. doi:10.1016/0308-521X(89)90097-8

Maurya, K., Yeom, T., Kang, W., Park, H., Hong, K., & Reddy, S. (2022). Optimization of hybrid manufacturing process combining forging and wire-arc additive manufactured Ti-6Al-4V through hot deformation characterization. *Journal of Alloys and Compounds*, *894*, 162453. doi:10.1016/j.jallcom.2021.162453

McAuleyB.HoreA.WestR. (2017). BIM in Ireland 2017, BIM Innovation Capability Programme, CitA Ltd. doi:10.21427/wy14-n123

McNamara, A. J., & Sepasgozar, S. M. (2021). Intelligent contract adoption in the construction industry: Concept development. *Automation in Construction*, *122*, 103–452. doi:10.1016/j.autcon.2020.103452

McQueen, H. J., & Ryan, N. D. (2002). Constitutive analysis in hot working. *Materials Science and Engineering A*, *322*(1–2), 43–63. doi:10.1016/S0921-5093(01)01117-0

Mehdi, H. (2012). Removal of noise patterns in handwritten images using expectation maximization and fuzzy inference systems. *Elsevier journal of pattern recognition*, pp. 4237-4249.

Meiryani, M., Aliffiyah, L., Lusianah, L., Lidiyawati, L., Lesmana, T., & Kriswanto, K. (2022, June). Benefits of Information Technology in Improving the Quality of Health Services. In *Proceedings of the 2022 6th International Conference on E-Education, E-Business and E-Technology* (pp. 97-101). ACM.10.1145/3549843.3549858

Menear, M., Girard, A., Dugas, M., Gervais, M., Gilbert, M., & Gagnon, M. P. (2022). Personalized care planning and shared decision making in collaborative care programs for depression and anxiety disorders: A systematic review. *PLoS One*, *17*(6), e0268649. doi:10.1371/journal.pone.0268649 PMID:35687610

Meneses, A. A. M., Machado, M. D., & Schirru, R. (2008). Particle swarm optimization applied to the nuclear reload problem of a pressurized water reactor. *Progress in Nuclear Energy*, *51*(2), 319–326. doi:10.1016/j.pnucene.2008.07.002

Menon, V., Pandey, D., Khosla, D., Kaur, M., Vashishtha, H. K., George, A. S., & Pandey, B. K. (2022). A Study on COVID–19, Its Origin, Phenomenon, Variants, and IoT-Based Framework to Detect the Presence of Coronavirus. In *IoT Based Smart Applications* (pp. 1–13). Springer International Publishing.

Mequanint Moges, Thomas G.Robertazzi, "Wireless Sensor Networks: Scheduling for Measurement and Data Reporting", August 31, 2005.

Meseko, A. A. (2014). Opportunities and Risks for Foreign Real Estate Developers in Nigeria. *International Journal of Economics. Commerce and Management, 11*(7), 2014.

Meslie, Y., Enbeyle, W., Pandey, B. K., Pramanik, S., Pandey, D., Dadeech, P., & Saini, A. (2021). Machine intelligence-based trend analysis of COVID-19 for total daily confirmed cases in Asia and Africa. In *Methodologies and Applications of Computational Statistics for Machine Intelligence* (pp. 164–185). IGI Global. doi:10.4018/978-1-7998-7701-1.ch009

Mhapsekar, K., McConaha, M., & Anand, S. (2018). Additive Manufacturing Constraints in Topology Optimization for Improved Manufacturability. *Journal of Manufacturing Science and Engineering, 140*(5), 051017. doi:10.1115/1.4039198

Mi, H., Liu, Y., Jiao, Z., Wang, W., & Wang, Q. (2020). A numerical study on the optimization of ventilation mode during emergency of cable fire in utility tunnel. *Tunnelling and Underground Space Technology, 100*, 103403. doi:10.1016/j.tust.2020.103403

Minor, M., Montani, S., & Recio-García, J. A. (2014). Process-oriented Case-based Reasoning. *Information Systems, 40*, 103–105. doi:10.1016/j.is.2013.06.004

Miry, A. (2020). Water tracking and analytic based ThingSpeak. *International Journal of Electrical and Computer Engineering 10*(4), 3588.

Mirzadeh, H., Cabrera, J. M., & Najafizadeh, A. (2011). Constitutive relationships for hot deformation of austenite. *Acta Materialia, 59*(16), 6441–6448. doi:10.1016/j.actamat.2011.07.008

Mirzadeh, H., Cabrera, M., Prado, M., & Najafizadeh, A. (2011). Hot deformation behavior of a medium carbon microalloyed steel. *Materials Science and Engineering A, 528*(10–11), 3876–3882. doi:10.1016/j.msea.2011.01.098

Modi, C., Patel, D., Borisaniya, B., Patel, A., & Rajarajan, M. (2013). A survey on security issues and solutions at different layers of Cloud computing. *The Journal of Supercomputing, 63*(2), 561–592. doi:10.1007/s11227-012-0831-5

Mohammed, A. M., & Mohammed, A. H. (2018), Towards Improvement in Facilities Operation and Maintenance through Feedback to the Design Team. *The Built and Human Environment Review, 3*. www.tbher.org/index.php/tbher/article/download/28/29

Mohammed, A., Hasnain, S. A., & Quadir, A. (2019). Implementation of building information modelling (BIM) practices and challenges in construction industry in Qatar. *Journal of Engineering Research and Reports, 9*(1), 1–9. doi:10.9734/jerr/2019/v9i117008

Monostori, L. (2014). Cyber-physical Production Systems: Roots, Expectations and R&D Challenges. *Procedia CIRP, 17*, 9–13. doi:10.1016/j.procir.2014.03.115

Montenegro, G., Kushalnagar, N., Hui, J., & Culler, D. (2007). *Transmission of IPv6 packets over IEEE 802.15.4 networks.*

Mori, T., Takada, H., Ito, S., Matsubayashi, K., Miwa, N., & Sawaguchi, T. (2006). Preclinical studies on the safety of fullerene upon acute oral administration and evaluation for no mutagenesis. *Toxicology, 225*(1), 48–54. doi:10.1016/j.tox.2006.05.001 PMID:16782258

Mostafaei, M. A., & Kazeminezhad, M. (2012). The prediction of hot flow behavior of Al-6%Mg alloy. *Materials Science and Engineering A*, *535*, 216–221. doi:10.1016/j.msea.2011.12.067

Moussa, F. (1997). *Fullerenes*, *97*(42), 332–336.

Mroz, P., Pawlak, A., Satti, M., Lee, H., Wharton, T., Gali, H., Sarna, T., & Hamblin, M. R. (2007). Functionalized fullerenes mediate photodynamic killing of cancer cells: Type I versus Type II photochemical mechanism. *Free Radical Biology & Medicine*, *43*(5), 711–719. doi:10.1016/j.freeradbiomed.2007.05.005 PMID:17664135

Mroz, P., Xia, Y., Asanuma, D., Konopko, A., Zhiyentayev, T., Huang, Y.-Y., Sharma, S. K., Dai, T., Khan, U. J., Wharton, T., & Hamblin, M. R. (2011). Intraperitoneal photodynamic therapy mediated by a fullerene in a mouse model of abdominal dissemination of colon adenocarcinoma. *Nanomedicine (London)*, *7*(6), 965–974. doi:10.1016/j.nano.2011.04.007 PMID:21645643

Müller, G. (2018). Workflow Modeling Assistance by Casebased Reasoning. Springer.

Muniandi, B., Huang, C., Kuo, C., Yang, T., Chen, K., Lin, Y., Lin, S., & Tsai, T. (2019). A 97% maximum efficiency fully automated control turbo boost topology for battery chargers. *IEEE Transactions on Circuits and Systems. I, Regular Papers*, *66*(11), 4516–4527. doi:10.1109/TCSI.2019.2925374

Musyarofah, S., Muliawati, R., & Mushidah, M. (2020, July). Health Service Information System in Kendal District toward Kendal Smart City. In Proceedings of the Third Andalas International Public Health Conference, AIPHC 2019, Padang, West Sumatera, Indonesia. 10.4108/eai.9-10-2019.2297244

Mutis, I., & Mehraj, I. (2022). Cloud BIM Governance Framework for Implementation in Construction Firms. *Practice Periodical on Structural Design and Construction*, *27*(1), 04021074. doi:10.1061/(ASCE)SC.1943-5576.0000652

Nabwire, S., Suh, H. K., Kim, M. S., Baek, I., & Cho, B. K. (2021). Application of artificial intelligence in phenomics. *Sensors (Basel)*, *21*(13), 4363. doi:10.3390/s21134363 PMID:34202291

Nagarajan, S. M., Muthukumaran, V., Vinoth Kumar, V., Beschi, I. S., & Magesh, S. (2021). *Fine Tuning Smart Manufacturing Enterprise Systems: A Perspective of Internet of Things-Based Service-Oriented Architecture, Handbook of Research on Innovations and Applications of AI, IoT, and Cognitive Technologies*. IGI Global Publication. doi:10.4018/978-1-7998-6870-5.ch006

Nagode, A., Kosec, L., Ule, B., & Kosec, G. (2011). Review of creep resistant alloys for power plant applications. *Metalurgija*, *50*(1), 45–48.

Najmon, J. C., Raeisi, S., & Tovar, A. (2019). *Review of additive manufacturing technologies and applications in the aerospace industry*. Additive Manufacturing for the Aerospace Industry., doi:10.1016/B978-0-12-814062-8.00002-9

Nakashima, S., Yoshie, M., Sano, H., & Bahar, A. (2009). Effect of a test dentifrice containing nano-sized calcium carbonate on remineralization of enamel lesions in vitro. *Journal of Oral Science*, *51*(1), 69–77. doi:10.2334/josnusd.51.69 PMID:19325202

Nancy, V. (2008, February 28). Integrated microfluidic systems with an immunosensor modified with carbon nanotubes for detection of prostate specific antigen (PSA) in human serum samples. *Biosensors & Bioelectronics*, *23*(7), 1145–1151. doi:10.1016/j.bios.2007.11.003 PMID:18162392

National Science and Technology Council Committee on Technology Subcommittee on the Materials Genome Initiative. (2011). *Materials Genome Initiative for global competitiveness*. National Science and Technology Council.

National Science and Technology Council Committee on Technology Subcommittee on the Materials Genome Initiative. (2014). *Materials Genome Initiative strategic plan*. National Science and Technology Council.

Navas, H., Picher, M., Andrieux-Ledier, A., Fossard, F., Michel, T., Kozawa, A., Maruyama, T., Anglaret, E., Loiseau, A., & Jourdain, V. (2017). Unveiling the Evolutions of Nanotube Diameter Distribution During the Growth of Single-Walled Carbon Nanotubes. *ACS Nano*, *11*(3), 3081–3088. doi:10.1021/acsnano.7b00077 PMID:28285520

Nayan, N., Singh, G., Murty, N., Jha, K., Pant, G., Koshy, M., & Ramamurty, U. (2014). Hot deformation behaviour and microstructure control in AlCrCuNiFeCo high entropy alloy. *Intermetallics*, *55*, 145–153. doi:10.1016/j.intermet.2014.07.019

Negi, S., Nambolan, A., Kapil, S., Joshi, P. R. M., Karunakaran, P., & Bhargava, P. (2020). Review on electron beam based additive manufacturing. *Rapid Prototyping Journal*, *26*(3), 485–498. doi:10.1108/RPJ-07-2019-0182

Negri, V., Pacheco-Torres, J., Calle, D., & López-Larrubia, P. (2020). Carbon Nanotubes in Biomedicine. *Topics in Current Chemistry (Cham)*, *378*(1), 15. doi:10.1007/s41061-019-0278-8 PMID:31938922

Ni, J., Yang, H., Yao, J., Li, Z., & Qin, P. (2019). Toxic gas dispersion prediction for point source emission using deep learning method. *Human and Ecological Risk Assessment*.

Ni, L., Ahmad, S. F., Alshammari, T. O., Liang, H., Alsanie, G., Irshad, M., Alyafi-AlZahri, R., BinSaeed, R. H., Al-Abyadh, M. H. A., Abu Bakir, S. M. M., & Ayassrah, A. Y. B. A. (2023). The role of environmental regulation and green human capital towards sustainable development: The mediating role of green innovation and industry upgradation. *Journal of Cleaner Production*, *421*, 138497. doi:10.1016/j.jclepro.2023.138497

Nimibofa, A., Newton, E. A., Cyprain, A. Y., & Donbebe, W. (2018). Fullerenes:Synthesis and Applications. *J. Mater. Sci. Res.*, *7*, 22.

Ning, H., & Wang, Z. (2011). Future internet of things architecture: Like mankind neural system or social organization framework? *IEEE Communications Letters*, *15*(4), 461–463. doi:10.1109/LCOMM.2011.022411.110120

Nishitha, P., & Pandey, D. (2021). A study on student perception towards online education during covid-19 crisis. *Augmented Human Research*, *6*(1), 16. doi:10.1007/s41133-021-00055-1

Nobusawa, K., Akiyama, M., Ikeda, A., & Naito, M. (2012). pH responsive smart carrier of [60] fullerene with 6-amino-cyclodextrin inclusion complex for photodynamic therapy. *Journal of Materials Chemistry*, *22*(42), 22610–22613. doi:10.1039/c2jm34791a

Obaid, A. J., & Sharma, S. (2020). Recent trends and development of heuristic artificial intelligence approach in mechanical system and engineering product design. *Saudi Journal of Engineering and Technology*, *5*(2), 86–93. doi:10.36348/sjet.2020.v05i02.008

Obodoh, T. (2019). A review on building information modelling in Nigeria and its potentials. *International Journal of Civil and Environmental Engineering*, *12*(11), 1113–1119.

Ocviyanti, D., & Sari, J. M. (2021). Analysis of antenatal care quality in cases of referred pregnant women in emergency rooms based on MCH book records. *Indonesian Journal of Obstetrics and Gynecology*, *121-125*, 121–125. doi:10.32771/inajog.v9i3.1339

Okpala, I., Nnaji, C., Awolusi, I., & Akanmu, A. (2021). Developing a Success Model for Assessing the Impact of Wearable Sensing Devices in the Construction Industry. *Journal of Construction Engineering and Management*, *147*(7), 04021060. doi:10.1061/(ASCE)CO.1943-7862.0002064

Olanrewaju, O. I., Babarinde, S. A., Chileshe, N., & Sandanayake, M. (2021). Drivers for implementation of building information modeling (BIM) within the Nigerian construction industry. *Journal of Financial Management of Property and Construction*, *26*(3), 366–386. doi:10.1108/JFMPC-12-2019-0090

Olanrewaju, O., Ajiboye Babarinde, S., & Salihu, C. (2020). Current State of Building Information Modelling in the Nigerian Construction Industry. *Journal of Sustainable Architecture and Civil Engineering*, *27*(2), 63–77. doi:10.5755/j01.sace.27.2.25142

Olapade, D., & Ekemode, B. (2018). Awareness and utilization of building information modelling (BIM) for facility management (FM) in a developing economy: Experience from Lagos, Nigeria. *Journal of Facilities Management*, *16*(4), 387–395. doi:10.1108/JFM-09-2017-0046

Olatunji, J. O., Dele, R. S., & Spencer, O. A. (2022). Building Information Modelling (BIM): Drivers, barriers and socioeconomic benefits. *Covenant Journal of Research in the Built Environment*, *10*(2), 13–23.

Olawumi, T., & Chan, D. (2019a). An empirical survey of the perceived benefits of executing BIM and sustainability practices in the built environment. *Construction Innovation*, *19*(3), 321–342. doi:10.1108/CI-08-2018-0065

Olawumi, T., & Chan, D. (2019b). Development of a benchmarking model for BIM implementation in developing countries. *Benchmarking*, *26*(4), 1210–1232. doi:10.1108/BIJ-05-2018-0138

Olorunfemi, E.T., Oyewobi, L.O., Olanrewaju, O.I. &Olorunfemi, R.T. (2021). *Competencies and the penetration status of building information modelling among built environment professionals in Nigeria.* Proceedings of International Federation of Surveyors, FIG e-Working Week 2021 Smart Surveyors for Land and Water Management - Challenges in a New Reality, Virtual.

Olugboyega, O. (2016). Building Information Modelling-Based Projects in Nigeria Evidences from Eko Atlantic City. *PM World Journal*, (10), 1–14.

Olugboyega, O., & Aina, O. (2016). Analysis of building information modelling usage indices and facilitators in the Nigerian construction industry. *Journal of Logistics. Information and Service Sciences*, *3*(2), 1–36.

Onungwa, I. O., & Uduma-Olugu, N. (2016). *Building Information Modeling in Nigeria and its Impact on Collaboration in Schematic Design Stage and Post Contract Stage of Design. 9th CIDB Post Graduate Conference on Emerging Trends in Construction organizational Practices and Project Management Knowledge Area*, Cape Town, South Africa.

Onungwa, I. O., & Uduma-Olugu, N. (2017). Building Information Modelling and Collaboration in the Nigerian Construction Industry. *9th CIDB Post Graduate Conference on Emerging Trends in Construction organizational Practices and Project Management Knowledge Area,* Cape Town, South Africa. 10.15641/jcbm.1.2.53

Onyancha, R. B., Aigbe, U. O., Ukhurebor, K. E., & Muchiri, P. W. (2021). A facile review on the synthesis and applications of carbon nanotubes in heavy-metal remediation and biomedical fields. *Journal of Molecular Structure*, *1238*, 130462. doi:10.1016/j.molstruc.2021.130462

Opoku, D. G. J., Perera, S., Osei-Kyei, R., & Rashidi, M. (2021). Digital twin application in the construction industry: A literature review. *Journal of Building Engineering*, *40*, 102726. doi:10.1016/j.jobe.2021.102726

Oshima, K., Yamamoto, D., Yumoto, A., Kim, S. J., Ito, Y., Hasegawa, M., & Initiative, I. D. (2022). Online machine learning algorithms to optimize performances of complex wireless communication systems. *Mathematical Biosciences and Engineering*, *19*(2), 2056–2094. doi:10.3934/mbe.2022097 PMID:35135242

Osuizugbo, I. (2020). Disruptions and Responses within Nigerian Construction Industry amid COVID-19 Threat. *Covenant Journal in Research and Built Environment*, *8*(2), 37–48.

Oteng, T., Ansah, M. K., Kissi, E., & Eshun, B. T. B. (2018), Barriers to the Adoption of Building Information Modeling in Developing Countries: The Case of Ghana. *1st International Conference on Construction Futures*, Wolverhampton., UK.

Ottosen, T., Mani, N. S., & Fratta, M. N. (2019). Health information literacy awareness and capacity building: Present and future. *IFLA Journal, 45*(3), 207–215. doi:10.1177/0340035219857441

Oyewole, E., & Dada, J. (2019). Training gaps in the adoption of building information modelling by Nigerian construction professionals. *Built Environment Project and Asset Management, 9*(3), 399–411. doi:10.1108/BEPAM-10-2017-0090

Pai, P., Nair, K., Jamade, S., Shah, R., Ekshinge, V., & Jadhav, N. (2006). Pharmaceutical applications of carbon tubes and nanohorns Curr. *Pharmaceutical Research, 1*, 11–15.

Pal, K. (2020). Information Sharing for Manufacturing Supply Chain Management Based on Blockchain Technology. Cross-Industry Use of Blockchain Technology and Opportunities for the Future. IGI Global Publication. doi:10.4018/978-1-7998-3632-2.ch001

Pal, K., & Ul-Haque, A. (2000). *Internet of Things and Blockchain Technology in Apparel Manufacturing Supply Chain Data Management.* In the proceeding of 11th International Conference on Ambient Systems, Networks and Technologies (ANT-2020), Procedia Computer Science, Warsaw, Poland.

Pal, K. (2017). Supply Chain Coordination Based on Web Services. In H. K. Chan, N. Subramanian, & M. D. Abdulrahman (Eds.), *Supply Chain Management in the Big Data Era* (pp. 137–171). IGI Global Publication. doi:10.4018/978-1-5225-0956-1.ch009

Pal, K. (2018). A Big Data Framework for Decision Making in Supply Chain. IGI Publication.

Pal, K., & Karakostas, B. (2014). A Multi Agent-Based Service Framework for Supply Chain Management, In the proceeding of International Conference on Ambient Systems, Networks and Technology. *Procedia Computer Science, 32*, 53–60. doi:10.1016/j.procs.2014.05.397

Pandey, B. K., Pandey, D., Nassa, V. K., George, S., Aremu, B., Dadeech, P., & Gupta, A. (2022, July). Effective and secure transmission of health information using advanced morphological component analysis and image hiding. In *Artificial Intelligence on Medical Data: Proceedings of International Symposium, ISCMM 2021* (pp. 223-230). Singapore, Springer Nature Singapore.

Pandey, B. K., Pandey, D., Nassa, V. K., George, S., Aremu, B., Dadeech, P., & Gupta, A. (2022, July). Effective and secure transmission of health information using advanced morphological component analysis and image hiding. In *Artificial Intelligence on Medical Data: Proceedings of International Symposium, ISCMM 2021* (pp. 223-230). Singapore: Springer Nature Singapore.

Pandey, B. K., Pandey, D., Wairya, S., Agarwal, G., Dadeech, P., Dogiwal, S. R., & Pramanik, S. (2022). Application of integrated steganography and image compressing techniques for confidential information transmission. *Cyber Security and Network Security,* 169-191.

Pandey, C., Mohan, M., Kumar, P., Thakre, G. and Saini, N. (2019) 'Role of evolving microstructure on the mechanical behaviour of P92 steel welded joint in as-welded and post weld heat treated state', *Journal of Materials Processing Technology, 263*(June 2018), 241–255. . doi:10.1016/j.jmatprotec.2018.08.032

Pandey, D., Hasan, A., Pandey, B. K., Lelisho, M. E., George, A. H., & Shahul, A. (2023). COVID-19 epidemic anxiety, mental stress, and sleep disorders in developing country university students. *CSI Transactions on ICT*, 1-9.

Pandey, D., Ogunmola, G. A., Enbeyle, W., Abdullahi, M., Pandey, B. K., & Pramanik, S. (2021). COVID-19: A framework for effective delivering of online classes during lockdown. *Human Arenas,* 1-15.

Pandey, B. K., Devi, P., George, A. S., Nassa, V. K., Dadheech, P., Thankachan, B., & Dogiwal, S. R. (2023). Review on Different Types of Disturbing Noise for Degraded Complex Image. In *Handbook of Research on Thrust Technologies' Effect on Image Processing* (pp. 15–24). IGI Global. doi:10.4018/978-1-6684-8618-4.ch002

Pandey, B. K., Nassa, V. K., Pandey, D., George, A. S., Gupta, M., Dadheech, P., & George, A. H. (2023). An Analytical Analysis of the Present-Day Procedures for Disposing of Waste. In *Handbook of Research on Safe Disposal Methods of Municipal Solid Wastes for a Sustainable Environment* (pp. 24–34). IGI Global. doi:10.4018/978-1-6684-8117-2.ch002

Pandey, B. K., & Pandey, D. (2023). Parametric optimization and prediction of enhanced thermoelectric performance in co-doped CaMnO3 using response surface methodology and neural network. *Journal of Materials Science Materials in Electronics*, *34*(21), 1589. doi:10.1007/s10854-023-10954-1

Pandey, B. K., Pandey, D., & Agarwal, A. (2022). Encrypted information transmission by enhanced steganography and image transformation. [IJDAI]. *International Journal of Distributed Artificial Intelligence*, *14*(1), 1–14. doi:10.4018/IJDAI.297110

Pandey, B. K., Pandey, D., Alkhafaji, M. A., Güneşer, M. T., & Şeker, C. (2023). A Reliable Transmission and Extraction of Textual Information Using Keyless Encryption, Steganography, and Deep Algorithm with Cuckoo Optimization. In *Micro-Electronics and Telecommunication Engineering: Proceedings of 6th ICMETE 2022* (pp. 629–636). Springer Nature Singapore. doi:10.1007/978-981-19-9512-5_57

Pandey, B. K., Pandey, D., Dadheech, P., Mahajan, D. A., George, A. S., & Hameed, A. S. (2023). Review on Smart Sewage Cleaning UAV Assistance for Sustainable Development. In *Handbook of Research on Safe Disposal Methods of Municipal Solid Wastes for a Sustainable Environment* (pp. 69–79). IGI Global. doi:10.4018/978-1-6684-8117-2.ch005

Pandey, B. K., Pandey, D., Gupta, A., Nassa, V. K., Dadheech, P., & George, A. S. (2023). Secret data transmission using advanced morphological component analysis and steganography. In *Role of Data-Intensive Distributed Computing Systems in Designing Data Solutions* (pp. 21–44). Springer International Publishing. doi:10.1007/978-3-031-15542-0_2

Pandey, B. K., Pandey, D., & Sahani, S. K. (2024). Autopilot control unmanned aerial vehicle system for sewage defect detection using deep learning. *Engineering Reports*, 12852. doi:10.1002/eng2.12852

Pandey, B. K., Pandey, D., Wairya, S., & Agarwal, G. (2021). An advanced morphological component analysis, steganography, and deep learning-based system to transmit secure textual data. [IJDAI]. *International Journal of Distributed Artificial Intelligence*, *13*(2), 40–62. doi:10.4018/IJDAI.2021070104

Pandey, B. K., Pandey, D., Wariya, S., & Agarwal, G. (2021). A deep neural network-based approach for extracting textual images from deteriorate images. *EAI Endorsed Transactions on Industrial Networks and Intelligent Systems*, *8*(28), e3–e3. doi:10.4108/eai.17-9-2021.170961

Pandey, B. K., Pandey, S. K., & Pandey, D. (2011). A survey of bioinformatics applications on parallel architectures. *International Journal of Computer Applications*, *23*(4), 21–25. doi:10.5120/2877-3744

Pandey, C., Mahapatra, M., Kumar, P., & Saini, N. (2018). 'Some studies on P91 steel and their weldments', *Journal of Alloys and Compounds. Journal of Alloys and Compounds*, *V*, 332–364. doi:10.1016/j.jallcom.2018.01.120

Pandey, D., Nassa, V. K., Jhamb, A., Mahto, D., Pandey, B. K., George, A. H., & Bandyopadhyay, S. K. (2021). An integration of keyless encryption, steganography, and artificial intelligence for the secure transmission of stego images. In *Multidisciplinary approach to modern digital steganography* (pp. 211–234). IGI Global. doi:10.4018/978-1-7998-7160-6.ch010

Pandey, D., & Pandey, B. K. (2022). An efficient deep neural network with adaptive galactic swarm optimization for complex image text extraction. In *Process Mining Techniques for Pattern Recognition* (pp. 121–137). CRC Press. doi:10.1201/9781003169550-10

Pandey, D., Pandey, B. K., & Wairya, S. (2021). Hybrid deep neural network with adaptive galactic swarm optimization for text extraction from scene images. *Soft Computing*, *25*(2), 1563–1580. doi:10.1007/s00500-020-05245-4

Pandey, D., Pandey, B. K., & Wariya, S. (2019). Study of various techniques used for video retrieval. *Journal of Emerging Technologies and Innovative Research*, *6*(6), 850–853.

Pandey, D., Wairya, S., Al Mahdawi, R. S., Najim, S. A. D. M., Khalaf, H. A., Al Barzinji, S. M., & Obaid, A. J. (2021). Secret data transmission using advanced steganography and image compression. *International Journal of Nonlinear Analysis and Applications*, *12*(Special Issue), 1243–1257.

Pandey, D., Wairya, S., Sharma, M., Gupta, A. K., Kakkar, R., & Pandey, B. K. (2022). An approach for object tracking, categorization, and autopilot guidance for passive homing missiles. *Aerospace Systems*, *5*(4), 553–566. doi:10.1007/s42401-022-00150-0

Pandey, J. K., Jain, R., Dilip, R., Kumbhkar, M., Jaiswal, S., Pandey, B. K., & Pandey, D. (2022). Investigating Role of IoT in the Development of Smart Application for Security Enhancement. In *IoT Based Smart Applications* (pp. 219–243). Springer International Publishing.

Pannu, A. (2015). Artificial intelligence and its application in different areas. *Artificial Intelligence*, *4*(10), 79–84.

Pan, Y., Jiang, J., Wang, R., & Cao, H. (2008). Advantages of support vector machine in QSPR studies for predicting auto-ignition temperatures of organic compounds. *Chemometrics and Intelligent Laboratory Systems*, *92*(2), 169–178. doi:10.1016/j.chemolab.2008.03.002

Pan, Y., Jiang, J., Wang, R., Cao, H., & Cui, Y. (2009). A novel QSPR model for prediction of lower flammability limits of organic compounds based on support vector machine. *Journal of Hazardous Materials*, *168*(2-3), 962–969. doi:10.1016/j.jhazmat.2009.02.122 PMID:19329246

Parthiban, K., Pandey, D., & Pandey, B. K. (2021). Impact of SARS-CoV-2 in online education, predicting and contrasting mental stress of young students: A machine learning approach. *Augmented Human Research*, *6*(1), 10. doi:10.1007/s41133-021-00048-0

Patel, A. R., Ramaiya, K. K., Bhatia, C. V., Shah, H. N., & Bhavsar, S. N. (2021). Artificial intelligence: Prospect in mechanical engineering field—a review. Data Science and Intelligent Applications. *Proceedings of ICDSIA*, *2020*, 267–282.

Patel, D., Gosh, A., Banerjee, S. J., & Jana, D. (2016). GSM Based Fire Sensor Alarm Using Arduino. *International Journal of Scientific and Engineering Research*, *7*(4).

Pellarini, F., Pantarotto, D., Da Ros, T., Giangaspero, A., Tossi, A., & Prato, M. (2001). A Novel [60]Fullerene Amino Acid for Use in Solid-Phase Peptide Synthesis. *Organic Letters*, *3*(12), 1845–1848. doi:10.1021/ol015934m PMID:11405726

Pena-opez, I. (2005). Internet Report 2005. *Internet of Things : Engineering Cyber Physical Human Systems*, 2005.

Penev, E. S., Bets, K. V., Gupta, N., & Yakobson, B. I. (2018). Transient Kinetic Selectivity in Nanotubes Growth on Solid Co–W Catalyst. *Nano Letters*, *18*(8), 5288–5293. doi:10.1021/acs.nanolett.8b02283 PMID:29979600

Penkova, A. V., Acquah, S. F. A., Piotrovskiy, L. B., Markelov, D. A., Semisalova, A. S., & Kroto, H. W. (2017). Fullerene derivatives as nano-additives in polymer composites. *Russian Chemical Reviews*, *86*(6), 530–566. doi:10.1070/RCR4712

Pennetta, C. (2020). Floresta, G. Graziano, A. C. E. Cardile, V. Rubino, L. Galimberti, M. Rescifina A. and Barbera, V. *Nanomaterials (Basel, Switzerland), 10*, 1073. doi:10.3390/nano10061073 PMID:32486371

Pertiwi, J., & Nugrahaeni, A. (2020). Examination of Technology Acceptance Model on Attitude Toward SIHA in Wonosobo Indonesia. *International Proceedings The 2nd ISMoHIM* 2020. IEEE. doi:10.32585/jikemb.v2i1.810

Pessoa, S., Guimarães, A. S., Lucas, S. S., & Simões, N. (2021). 3D printing in the construction industry-A systematic review of the thermal performance in buildings. *Renewable & Sustainable Energy Reviews, 141*, 110794. doi:10.1016/j.rser.2021.110794

Pham, V. P., Jang, S.-H., Whang, D., & Choi, J.-Y. (2017). Direct growth of graphene on rigid and flexible substrates: Progress, applications, and challenges. *Chemical Society Reviews, 46*(20), 6276–6300. doi:10.1039/C7CS00224F PMID:28857098

Philip, D., & Mordue, S. (2017). *Bim, Scope and Definition. Common learning outcome for European managers in construction.* Warsaw University of Technology, Civil Engineering Faculty, Department of Construction Engineering and Management. polcen.com.pl

Pierson, H. A., & Gashler, M. S. (2017). Deep learning in robotics: A review of recent research. *Advanced Robotics, 31*(16), 821–835. doi:10.1080/01691864.2017.1365009

Pooja, A.P., Jayanth, J. and Koliwad, S.(2011). Classification of RS data using Decision Tree Approach. *International Journal of Computer Applications, 23*(3).

Pradhananga, P., ElZomor, M., & Santi Kasabdji, G. (2021). Identifying the Challenges to Adopting Robotics in the US Construction Industry. *Journal of Construction Engineering and Management, 147*(5), 05021003. doi:10.1061/(ASCE)CO.1943-7862.0002007

Pramanik, S., Ghosh, R., Pandey, D., Samanta, D., Dutta, S., & Dutta, S. (2021). Techniques of Steganography and Cryptography in Digital Transformation. In Emerging Challenges, Solutions, and Best Practices for Digital Enterprise Transformation (pp. 24-44). IGI Global. doi:10.4018/978-1-7998-8587-0.ch002

Pramanik, S., Pandey, D., Joardar, S., Niranjanamurthy, M., Pandey, B. K., & Kaur, J. (2023, October). An overview of IoT privacy and security in smart cities. In AIP Conference Proceedings (Vol. 2495, No. 1). AIP Publishing.Albermany S.). doi:10.1063/5.0123511

Pramanik, R., & Bag, S. (2018). Shape decomposition-based handwritten compound character recognition for Bangla OCR. *Journal of Visual Communication and Image Representation, 50*, 123–134. doi:10.1016/j.jvcir.2017.11.016

Puteri, S., & Puspitasari, P. (2021). Exploratory study: Conservation practices and the role of architect. *IOP Conference Series. Earth and Environmental Science, 780*(1), 27–41. doi:10.1088/1755-1315/780/1/012061

Puttonen, J., Lobov, A., & Lastra, J. L. M. (2013). Semantics-Based Composition of Factory Automation Processes Encapsulated by Web Services. *IEEE Transactions on Industrial Informatics, 9*(4), 2349–2359. doi:10.1109/TII.2012.2220554

Puttonen, J., Lobov, A., Soto, M. A. C., & Lastra, J. L. M. (2010). A Semantic Web Services-based approach for production systems control. *Advanced Engineering Informatics, 24*(3), 285–299. doi:10.1016/j.aei.2010.05.012

Qian, F., Chen, L., Li, J., Ding, C., Chen, X., & Wang, J. (2019). Direct prediction of the toxic gas diffusion rule in a real environment based on LSTM. *International Journal of Environmental Research and Public Health, 16*(12), 2133. doi:10.3390/ijerph16122133 PMID:31212880

Qian, L., Fang, G., Zeng, P., & Wang, X. (2015). Correction of flow stress and determination of constitutive constants for hot working of API X100 pipeline steel. *International Journal of Pressure Vessels and Piping, 132–133*(November), 43–51. doi:10.1016/j.ijpvp.2015.05.008

Quintero, F. A., Patel, S. J., Munoz, F., & Sam Mannan, M. (2012). Review of existing QSAR/QSPR models developed for properties used in hazardous chemicals classification system. *Industrial & Engineering Chemistry Research, 51*(49), 16101–16115. doi:10.1021/ie301079r

Raghavan, R., Verma, D. C., Pandey, D., Anand, R., Pandey, B. K., & Singh, H. (2022). Optimized building extraction from high-resolution satellite imagery using deep learning. *Multimedia Tools and Applications, 81*(29), 42309–42323. doi:10.1007/s11042-022-13493-9

Raghuvanshi, A., Singh, U. K., & Joshi, C. (2022). A review of various security and privacy innovations for IoT applications in healthcare. *Advanced Healthcare Systems, 4*, 43–58. doi:10.1002/9781119769293.ch4

Rahmani Didar, B., & Balbuena, P. B. (2017). Growth of Carbon Nanostructures on Cu Nanocatalysts. *The Journal of Physical Chemistry. C, Nanomaterials and Interfaces, 121*(13), 7232–7239. doi:10.1021/acs.jpcc.6b12816

Raja, D., Kumar, D. R., Santhiyakumari, N., Kumarganesh, S., Sagayam, K. M., Thiyaneswaran, B., Pandey, B. K., & Pandey, D. (2024). A compact dual-feed wide-band slotted antenna for future wireless applications. *Analog Integrated Circuits and Signal Processing, 118*(2), 1–15. doi:10.1007/s10470-023-02233-0

Ramadani, L. (2019). *Health Information Systems Enabled Transformation of Service Ecosystems: The Case of Indonesian Healthcare.*

Rana, M. M., Ibrahim, D. S., Mohd Asyraf, M. R., Jarin, S., & Tomal, A. (2017). A review on recent advances of CNTs as gas sensors. *Sensor Review, 37*(2), 127–136. doi:10.1108/SR-10-2016-0230

Rancan, F., Rosan, S., Boehm, F., Cantrell, A., Brellreich, M., Schoenberger, H., Hirsch, A., & Moussa, F. (2002). Cytotoxicity and photocytotoxicity of a dendritic C60 mono-adduct and a malonic acid C60 tris-adduct on Jurkat cells. *Journal of Photochemistry and Photobiology. B, Biology, 67*(3), 157–162. doi:10.1016/S1011-1344(02)00320-2 PMID:12167314

Raoof, M., Mackeyev, Y., Cheney, M. A., Wilson, L. J., & Curley, S. A. (2012). Internalization of C60 fullerenes into cancer cells with accumulation in the nucleus via the nuclear pore complex. *Biomaterials, 33*(10), 2952–2960. doi:10.1016/j.biomaterials.2011.12.043 PMID:22245558

Rawat, W., & Wang, Z. (2017). Deep Convolutional Neural Networks for Image Classification: A Comprehensive Review. *Neural Computation, 29*(9), 1–98. doi:10.1162/neco_a_00990 PMID:28599112

Ren, J., Li, L., Chen, C., Chen, X., Cai, Z., Qiu, L., Wang, Y., Zhu, X., & Peng, H. (2013). Twisting Carbon Nanotube Fibers for Both Wire-Shaped Micro-Supercapacitor and Micro-Battery. *Advanced Materials, 25*(8), 1155–1159. doi:10.1002/adma.201203445 PMID:23172722

Revathi, T. K., Sathiyabhama, B., Sankar, S., Pandey, D., Pandey, B. K., & Dadeech, P. (2022). An intelligent model for coronary heart disease diagnosis. *Networking Technologies in Smart Healthcare: Innovations and Analytical Approaches*, 234.

Rezayat, S., Boushehri, S. V. S., Salmanian, B., Omidvari, A. H., Tarighat, S., Esmaeili, S., Sarkar, S., Amirshahi, N., Alyautdin, R. N., & Orlova, M. A. (2009). The porphyrin–fullerene nanoparticles to promote the ATP overproduction in myocardium: 25Mg2+-magnetic isotope effect. *European Journal of Medicinal Chemistry, 44*(4), 1554–1569. doi:10.1016/j.ejmech.2008.07.030 PMID:18782645

Rima, K., & Singh, D. P. (2021). *Nano-Biotechnological Approach of Plant Genetic Engineering*. Policy Issues in Genetically Modified Crops.A Global Perspective. doi:10.1016/B978-0-12-820780-2.00021-2

Ríos, S., Colegrove, P. A., & Williams, S. W. (2019). Metal transfer modes in plasma Wire + Arc additive manufacture. *Journal of Materials Processing Technology*, *264*, 45–54. doi:10.1016/j.jmatprotec.2018.08.043

Robert, G., Donetto, S., & Williams, O. (2021). Co-designing healthcare services with patients. The Palgrave Handbook of Co-Production of Public Services and Outcomes, 313-333. https://doi.org/ doi:10.1007/978-3-030-53705-0_16

Rosli, A., Alkahari, A., Mohd, M., Shajahan, R., Faiz, H., & Safarudin, G. (2021). Review on effect of heat input for wire arc additive manufacturing process. *Journal of Materials Research and Technology*, *11*, 2127–2145. doi:10.1016/j.jmrt.2021.02.002

Rouse, J. G., Yang, J., Ryman-Rasmussen, J. P., Barron, A. R., & Monteiro-Riviere, N. A. (2007). Effects of mechanical flexion on the penetration of fullerene amino acid-derivatized peptide nanoparticles through the skin. *Nano Letters*, *7*(1), 155–160. doi:10.1021/nl062464m PMID:17212456

Rumman, R., Lewis, A., Hascoet, Y., & Quinton, S. (2019). Laser metal deposition and wire arc additive manufacturing of materials: An overview. *Archives of Metallurgy and Materials*, *64*(2), 467–473. doi:10.24425/amm.2019.127561

Russomanno, D. J., Kothari, C. R., & Thomas, O. A. (2005). Building a Sensor Ontology: A Practical Approach Leveraging ISO and OGC Models. *2005 International Conference on Artificial Intelligence*, (pp. 637-643). IEEE.

Ruta, M., Colucci, S., Scioscia, F., Di Sciascio, E., & Donini, F. M. (2011). Finding commonalities in RFID semantic streams. *Procedia Computer Science*, *5*, 857–864. doi:10.1016/j.procs.2011.07.118

Ruya, T. F., Chitumu, Z. D., & Kaduma, L. A. (2018). *Challenges of Building Information Modelling Implementation in Africa, A Case of Nigerian Construction Industry. FIG Congress 2018*, Istanbul, Turkey.

Ryan, E. M. (2018) *On Wire and Arc Additive Manufacture of Aluminium*. [Thesis\.

Saeedfar, K., Heng, L., Ling, T., & Rezayi, M. (2013). Potentiometric Urea Biosensor Based on an Immobilised Fullerene-Urease Bio-Conjugate. *Sensors (Basel)*, *13*(12), 16851–16866. doi:10.3390/s131216851 PMID:24322561

Sahani, K., Khadka, S. S., Sahani, S. K., Pandey, B. K., & Pandey, D. (2023). A possible underground roadway for transportation facilities in Kathmandu Valley: A racking deformation of underground rectangular structures. *Engineering Reports*, 12821. doi:10.1002/eng2.12821

Sahu, A., Gupta, R., Sagar, S., Kumar, M., & Sagar, R. (2017, July-December). A study of psychiatric comorbidity after traumatic limb amputation: A neglected entity. *Industrial Psychiatry Journal*, *26*(2), 228–232. doi:10.4103/ipj.ipj_80_16 PMID:30089974

Sahub, C., Tuntulani, T., Nhujak, T., & Tomapatanaget, B. (2018). Effective biosensor based on graphene quantum dots via enzymatic reaction for directly photoluminescence detection of organophosphate pesticide. *Sensors and Actuators. B, Chemical*, *258*, 88–97. doi:10.1016/j.snb.2017.11.072

Said, O., & Masud, M. (2013). Towards internet of things: Survey and future vision. *International Journal of Computer Networks*, *5*(1), 1–17.

Saifuddin, N. Raziah, A. Z. & Junizah, A. R. (2013). Carbon Nanotubes: A Review on Structure and Their Interaction with Proteins. Hindawi Publishing Corporation.

Saikumar, T., & Sriramya, P. (2019). *IoT Enabled Forest Fire Detection and Altering the Authorities.International Journal of Recent Technology and Engineering*. IJRTE.

Saka, A. B., & Chan, D. W. M. (2019b). A global taxonomic review and analysis of the development of BIM research between 2006 and 2017. *Construction Innovation, 19*(3), 465–490. doi:10.1108/CI-12-2018-0097

Saka, A., & Chan, D. (2020). Profound barriers to building information modelling (BIM) adoption in construction small and medium-sized enterprises (SMEs): An interpretive structural modelling approach. *Construction Innovation, 20*(2), 261–284. doi:10.1108/CI-09-2019-0087

Salah, L. S., Ouslimani, N., Bousba, D., Huynen, I., Danlée, Y., & Aksas, H. (2021). Carbon Nanotubes (CNTs) from Synthesis to Functionalized (CNTs) Using Conventional and New Chemical Approaches. Journal of Nanomaterials.

Salas-Trevino, D., Saucedo-Cardenas, O., & de Jesus Loera-Arias, M. (2018). Carbon nanotubes: An alternative for platinum-based drugs delivery systems. *JBUON, 23*(3), 541–549. PMID:30003717

Samantaray, D., Mandal, S., & Bhaduri, A. K. (2010). Constitutive analysis to predict high-temperature flow stress in modified 9Cr-1Mo (P91) steel. *Materials & Design, 31*(2), 981–984. doi:10.1016/j.matdes.2009.08.012

Same, S., & Samee, G. (2018). Carbon nanotube biosensor for diabetes disease. *Crescent Journal of Medical and Biological Sciences, 5*(1), 1–6.

Sames, J., List, A., Pannala, S., Dehoff, R., & Babu, S. (2016). The metallurgy and processing science of metal additive manufacturing. *International Materials Reviews, 61*(5), 315–360. doi:10.1080/09506608.2015.1116649

Samuel, A. L. (1959). Some studies in machine learning using the game of checkers. *IBM Journal of Research and Development, 3*(3), 210–229. doi:10.1147/rd.33.0210

Sanchez-Valencia, J. R., Dienel, T., Groning, O., Shorubalko, I., Mueller, A., Jansen, M., Amsharov, K., Ruffieux, P., & Fasel, R. (2014). Controlled Synthesis of Single-Chirality Carbon Nanotubes. *Nature, 512*(7512), 61–64. doi:10.1038/nature13607 PMID:25100481

Saravana, K. R., & Vijayalakshmi, R. (2006). Nanotechnology in dentistry. *Indian Journal of Dental Research, 17*(2), 62–65. doi:10.4103/0970-9290.29890 PMID:17051869

Sarkar, I., Adhikari, M., Kumar, N., & Kumar, S. (2021). Dynamic task placement for deadline-aware IoT applications in federated for networks. *IEEE Internet of Things Journal, 1*, 2021.

Sarkar, T. K., Mailloux, R. J., Oliner, A. A., Salazar-Palma, M., & Sengupta, D. L. (2006). *History of Wireless*. John Wiley & Sons Inc. doi:10.1002/0471783021

Savitha, N., & Malathi, S. (2018, October). A survey on fire safety measures for industry safety using IOT. In *2018 3rd International Conference on Communication and Electronics Systems (ICCES)* (pp. 1199-1205).IEEE. 10.1109/CESYS.2018.8723930

Saxena, A., Agarwal, A., Pandey, B. K., & Pandey, D. (2024). Examination of the Criticality of Customer Segmentation Using Unsupervised Learning Methods. *Circular Economy and Sustainability*, 1–14. doi:10.1007/s43615-023-00336-4

Saxena, A., Sharma, N. K., Pandey, D., & Pandey, B. K. (2021). Influence of tourists satisfaction on future behavioral intentions with special reference to desert triangle of Rajasthan. *Augmented Human Research, 6*(1), 1–9. doi:10.1007/s41133-021-00052-4

Schneider, M., Hippchen, B., Abeck, S., Jacoby, M., & Herzog, R. (2018). Enabling IoT platform interoperability using a systematic development approach by example. In *Proceedings of the Global Internet of Things Summit (GIoTS'18).* IEEE. 10.1109/GIOTS.2018.8534549

Schütt, K., Kindermans, P. J., Sauceda Felix, H. E., Chmiela, S., Tkatchenko, A., & Müller, K. R. (2017). Schnet: A continuous-filter convolutional neural network for modeling quantum interactions. *Advances in Neural Information Processing Systems*, •••, 30.

Schwerdtfeger, P., Wirz, L., & Avery, J. (2013). Program fullerene: A software package for constructing and analyzing structures of regular fullerenes. *Journal of Computational Chemistry*, *34*(17), 1508–1526. doi:10.1002/jcc.23278 PMID:23559399

Scott, D. J., Broyd, T., & Ma, L. (2021). Exploratory literature review of blockchain in the construction industry. *Automation in Construction*, *132*, 103914. doi:10.1016/j.autcon.2021.103914

Sebastien, W., Giorgia, W., Monica, P., Cedric, B., Jean-Paul, K., & Renato, B. (2005). Targeted delivery of amphotericin B to cells by using functionalized carbon nanotubes. *Angewandte Chemie*, *117*(39), 6516–6520. doi:10.1002/ange.200501613

Segler, M. H., Preuss, M., & Waller, M. P. (2017). Learning to plan chemical syntheses. arXiv preprint arXiv:1708.04202.

Segler, M. H., Kogej, T., Tyrchan, C., & Waller, M. P. (2018). Generating focused molecule libraries for drug discovery with recurrent neural networks. *ACS Central Science*, *4*(1), 120–131. doi:10.1021/acscentsci.7b00512 PMID:29392184

Seiger, R., Huber, S., & Schlegel, T. (2018). Toward an execution system for self-healing workflows in cyber-physical systems. *Software & Systems Modeling*, *17*(2), 551–572. doi:10.1007/s10270-016-0551-z

Seixas Gomes de Almeida, B., & Coppo Leite, V. (2019, December 4). Victor CoppoLeite (2019),Particle Swarm Optimization: A Powerful Technique for Solving Engineering Problems. *Swarm Intelligence*. Advance online publication. doi:10.5772/intechopen.89633

Sellars, C. M., & McTegart, W. J. (1966). On the mechanism of hot deformation. *Acta Metallurgica*, *14*(9), 1136–1138. doi:10.1016/0001-6160(66)90207-0

Semeniuk, M., Yi, Z., Poursorkhabi, V., Tjong, J., Jaffer, S., Lu, Z. H., & Sain, M. (2019). Future perspectives and review on organic carbon dots in electronic applications. *ACS Nano*, *13*(6), 6224–6255. doi:10.1021/acsnano.9b00688 PMID:31145587

Sengupta, R., Sengupta, D., Kamra, A. K., & Pandey, D. (2020). Artificial Intelligence and Quantum Computing for a Smarter Wireless Network. *Artificial Intelligence*, *7*(19), 2020.

Sennan, S., Pandey, D., Alotaibi, Y., & Alghamdi, S. (2022). A Novel Convolutional Neural Networks Based Spinach Classification and Recognition System. *Computers, Materials & Continua*, *73*(1), 343–361. doi:10.32604/cmc.2022.028334

Senthilnath, J., Omkar, S. N., Mani, V., Tejovanth, N., Diwakar, P. G., & Archana Shenoy, B. (2011). Multi-spectral satellite image classification using glowworm swarm optimization. *Geoscience and Remote Sensing Symposium*, (pp. 47-50). IEEE.

Senyoni, W. F. (2021). *Institutional Work in Strengthening Health Information Systems in Fragmented Settings: An Action Research Study of Information Dashboards in East African Community and Indonesia.*

Serda, M., Ware, M. J., Newton, J. M., Sachdeva, S., Krzykawska-Serda, M., Nguyen, L., Law, J., Anderson, A. O., Curley, S. A., Wilson, L. J., & Corr, S. J. (2018). Development of photoactive Sweet-C60 for pancreatic cancer stellate cell therapy. *Nanomedicine (London)*, *13*(23), 2981–2993. doi:10.2217/nnm-2018-0239 PMID:30501557

Shaikh, T. N., & Mukane, S. M. (2017). Segmentation of Pomegranate Leaf for Detection of Disease Using Image Processing. *International Journal of Innovative Research in Computer and Communication Engineering*, *5*(4), 378–384.

Sharma, N., Kumar, B., & Singh, V. (2014). Recognition of Off-line Hand printed English Characters, Numerals, and Special Symbols. *IEEE International Conference- Confluence The Next Generation Information Technology Summit.*

Sharma, S., Pandey, B. K., Pandey, D., Anand, R., Sharma, A., & Saini, S. (2023, March). Character Recognition Technique Implementation for Complicated Deteriorated Scene. In *2023 6th International Conference on Information Systems and Computer Networks (ISCON)* (pp. 1-4). IEEE. 10.1109/ISCON57294.2023.10112185

Sharma, A. K., Ansari, M. F. R., Siddiqui, M. F., & Baig, M. A. (2017). IOT enabled forest fire detection and online monitoring system. [IJCTER]. *Int. J. Curr. Trends Eng. Res., 3*, 50–54.

Sharma, M., Pandey, D., Khosla, D., Goyal, S., Pandey, B. K., & Gupta, A. K. (2022). Design of a GaN-based Flip Chip Light Emitting Diode (FC-LED) with au Bumps & Thermal Analysis with different sizes and adhesive materials for performance considerations. *Silicon, 14*(12), 7109–7120. doi:10.1007/s12633-021-01457-x

Sharma, M., Saripalli, S. R., Gupta, A. K., Palta, P., & Pandey, D. (2023). Image Processing-Based Method of Evaluation of Stress from Grain Structures of Through Silicon Via (TSV). *International Journal of Image and Graphics*, 2550008. doi:10.1142/S0219467825500081

Sharma, M., Sharma, B., Gupta, A. K., Khosla, D., Goyal, S., & Pandey, D. (2021). A study and novel AI/ML-based framework to detect COVID-19 virus using smartphone embedded sensors. In *Sustainability Measures for COVID-19 Pandemic* (pp. 59–74). Springer Nature Singapore. doi:10.1007/978-981-16-3227-3_4

Sharma, M., Sharma, B., Gupta, A. K., & Pandey, D. (2023). Recent developments of image processing to improve explosive detection methodologies and spectroscopic imaging techniques for explosive and drug detection. *Multimedia Tools and Applications, 82*(5), 6849–6865. doi:10.1007/s11042-022-13578-5

Sharma, M., Talwar, R., Pandey, D., Nassa, V. K., Pandey, B. K., & Dadheech, P. (2024). A Review of Dielectric Resonator Antennas (DRA)-Based RFID Technology for Industry 4.0. *Robotics and Automation in Industry, 4*(0), 303–324.

Sharma, R., Ghosh, A., & Joshi, P. K. (2013). Decision tree approach for classification of remotely sensed satellite data using open source support. Indian Academy of Sciences. *Journal of Earth System Science, 122*(5), 1237–1247. doi:10.1007/s12040-013-0339-2

Shen, B., Lu, S., Sun, C., Song, Z., Zhang, F., Kang, J., Cao, Y., & Xiang, M. (2022, June). Effects of Amino Hyperbranched Polymer-Modified Carbon Nanotubes on the Crystallization Behavior of Poly (L-Lactic Acid) (PLLA). *Polymers, 14*(11), 2188. doi:10.3390/polym14112188 PMID:35683866

Shen, D., Wu, G., & Suk, H.-I. (2017b). Deep Learning in Medical Image Analysis. *Annual Review of Biomedical Engineering, 19*(1), 221–248. doi:10.1146/annurev-bioeng-071516-044442 PMID:28301734

ShenL.YueZ.FengF. (2017). MSR-net: Low-light Image Enhancement Using Deep Convolutional Network. arXiv:1711.02488v1

Shen, R., Jiao, Z., Parker, T., Sun, Y., & Wang, Q. (2020). Recent application of Computational Fluid Dynamics (CFD) in process safety and loss prevention: A review. *Journal of Loss Prevention in the Process Industries, 67*, 104252. doi:10.1016/j.jlp.2020.104252

Shetty, N. J., Swati, P., & David, K. (2013). Nanorobots: Future in dentistry. *The Saudi Dental Journal, 25*(2), 49–52. doi:10.1016/j.sdentj.2012.12.002 PMID:23960556

Shi, H., Gu, R., Xu, W., Huang, H., Xue, L., Wang, W., Zhang, Y., Si, W., & Dong, X. (2019). Near-Infrared Light-Harvesting Fullerene-Based Nanoparticles for Promoted Synergetic Tumor Phototheranostics. *ACS Applied Materials & Interfaces, 11*(48), 44970–44977. doi:10.1021/acsami.9b17716 PMID:31702130

Shi, J. Yu., & Wang, X. (2013). *PEGylated fullerene/iron oxide nanocomposites for photodynamic therapy, targeted drug delivery and MR imaging* (Vol. 34). Biomaterials. doi:10.1016/j.biomaterials.2013.08.049

Shi, J., Liu, Y., Wang, L., Gao, J., Zhang, J., Yu, X., Ma, R., Liu, R., & Zhang, Z. (2014). A tumoural acidic pH-responsive drug delivery system based on a novel photosensitizer (fullerene) for in vitro and in vivo chemo-photodynamic therapy. *Acta Biomaterialia, 10*(3), 1280–1291. doi:10.1016/j.actbio.2013.10.037 PMID:24211343

Shi, J., Li, X., Khan, F., Chang, Y., Zhu, Y., & Chen, G. (2019). Artificial bee colony Based Bayesian Regularization Artificial Neural Network approach to model transient flammable cloud dispersion in congested area. *Process Safety and Environmental Protection, 128*, 121–127. doi:10.1016/j.psep.2019.05.046

Shi, J., Wang, Z., & Wang, L. (2013). Photodynamic therapy of a 2-methoxyestradiol tumour-targeting drug delivery system mediated by Asn-Gly-Arg in breast cancer. *International Journal of Nanomedicine, 8*, 1551. doi:10.2147/IJN.S40011 PMID:23637528

Shi, J., Zhang, H., Wang, L., Li, L., Wang, H., Wang, Z., Li, Z., Chen, C., Hou, L., Zhang, C., & Zhang, Z. (2013). PEI-derivatized fullerene drug delivery using folate as a homing device targeting to tumor. *Biomaterials, 34*(1), 251–261. doi:10.1016/j.biomaterials.2012.09.039 PMID:23069706

Shinde, S. R., Karode, A. H., & Suralkar, D. S. (2017). Review on IOT Based Environment Monitoring System. *International Journal of Electronics and Communication Engineering and Technology, 8*(2).

Shinohara, H. (2000). Endohedral metallofullerenes. *Reports on Progress in Physics, 63*(6), 843–892. doi:10.1088/0034-4885/63/6/201

Shi, R. X., & Liu, Z. D. (2011). Hot deformation behavior of P92 steel used for ultra-super-critical power plants. *Journal of Iron and Steel Research International, 18*(7), 53–58. doi:10.1016/S1006-706X(11)60090-3

Shukla, R. (2021). The SCONUL Seven Pillars of Information Literacy: Core Model to Test the Skills of LIS Students. Library Philosophy and Practice, 5851, University of Jammu, India..

Shulaker, M. M., Hills, G., Patil, N., Wei, H., Chen, H.-Y., Wong, H.-S. P., & Mitra, S. (2013). Carbon Nanotube Computer. *Nature, 501*(7468), 526–530. doi:10.1038/nature12502 PMID:24067711

Sidi Salah, L., Chouai, M., Danlée, Y., Huynen, I., & Ouslimani, N. (2020). Simulation and optimization of electromagnetic absorption of polycarbonate/CNT composites using machine learning. *Micromachines, 11*(8), 778. doi:10.3390/mi11080778 PMID:32824164

Silverio, A. K., Suresh, S., Renukappa, S., & Heesom, D. (2021). Status of BIM implementation in the Dominican Republic construction industry: An empirical study. *Journal of Engineering. Design and Technology, 21*(2), 417–441. doi:10.1108/JEDT-05-2021-0253

Singh, H., Pandey, B. K., George, S., Pandey, D., Anand, R., Sindhwani, N., & Dadheech, P. (2022, July). Effective Overview of Different ML Models Used for Prediction of COVID-19 Patients. In *Artificial Intelligence on Medical Data: Proceedings of International Symposium, ISCMM 2021* (pp. 185-192). Singapore: Springer Nature Singapore.

Singh, I. (2017). *BIM adoption and implementation around the world: initiatives by major nations.* GeoSpatial World. https://www.geospatialworld.net/blogs/bim-adoption-aroundthe-world/

Singh, S., Madaan, G., Kaur, J., Swapna, H. R., Pandey, D., Singh, A., & Pandey, B. K. (2023). Bibliometric Review on Healthcare Sustainability. Handbook of Research on Safe Disposal Methods of Municipal Solid Wastes for a Sustainable Environment, (pp. 142-161). IEEE.

Singh, S., Madaan, G., Kaur, J., Swapna, H. R., Pandey, D., Singh, A., & Pandey, B. K. (2023). Bibliometric Review on Healthcare Sustainability. Handbook of Research on Safe Disposal Methods of Municipal Solid Wastes for a Sustainable Environment, 142-161.

Singh, A., Kapil, S., & Das, M. (2020). A comprehensive review of the methods and mechanisms for powder feedstock handling in directed energy deposition. *Additive Manufacturing*, *35*, 101388. doi:10.1016/j.addma.2020.101388

Singh, D., Sharma, N., Gupta, M., & Sharma, S. (2017). Development of system for early fire detection using ArduinoUNO. *International Journal of Engineering Science*, 10857.

Singh, J., Pandey, D., & Singh, A. K. (2023). Event detection from real-time twitter streaming data using community detection algorithm. *Multimedia Tools and Applications*, *83*(8), 1–28. doi:10.1007/s11042-023-16263-3

Singh, S., Patel, H., Goyani, M., & Patel, P. (2021, January). *Multidisciplinary International Research Journal of Gujarat Technological University. VOLUME*, *3*(1), 97–120.

Sinha, A., Jain, R., Zhao, H., Karolia, P., & Jadon, N. (2018, January 8). Voltammetric sensing based on the use of advanced carbonaceous nanomaterials: A review. *Mikrochimica Acta*, *185*(2), 89. doi:10.1007/s00604-017-2626-0 PMID:29594390

Sinha, A., Tassiulas, L., & Modiano, E. (2019). Throughput-Optimal Broadcast in Wireless Networks with Dynamic Topology. *IEEE Transactions on Mobile Computing*, *18*(5), 1203–1216. doi:10.1109/TMC.2018.2846246

Siste, K., Hanafi, E., Sen, L. T., Christian, H., Adrian, Siswidiani, L. P., Limawan, A. P., Murtani, B. J., & Suwartono, C. (2020). The impact of physical distancing and associated factors towards internet addiction among adults in Indonesia during COVID-19 pandemic: A nationwide web-based study. *Frontiers in Psychiatry*, *11*, 580977. doi:10.3389/fpsyt.2020.580977 PMID:33101092

Smola, A. (1999). *Regression Estimation with Support Vector Learning Machines* [Master's Thesis, The Technical University of Munich].

Song, T., Liu, F., Deng, N., Qian, S., Cui, T., Guan, Y., Arnolda, L., Zhang, Z., & Yu, P. (2021). A comprehensive 6A framework for improving patient self-management of hypertension using mHealth services: Qualitative thematic analysis. *Journal of Medical Internet Research*, *23*(6), e25522. doi:10.2196/25522 PMID:34152272

Sosnowska, M., Kutwin, M., Jaworski, S., Strojny, B., Wierzbicki, M., Szczepaniak, J., Łojkowski, M., Święszkowski, W., Bałaban, J., Chwalibog, A., & Sawosz, E. (2019). Mechano-signalling, induced by fullerene C60 nanofilms, arrests the cell cycle in the G2/M phase and decreases proliferation of liver cancer cells. *International Journal of Nanomedicine*, *14*, 6197–6215. doi:10.2147/IJN.S206934 PMID:31496681

Sotiropoulou, S., Gavalas, V., Vamvakaki, V., & Chaniotakis, N. (2003). Novel carbon materials in biosensor systems. *Biosensors & Bioelectronics*, *18*(2-3), 211–215. doi:10.1016/S0956-5663(02)00183-5 PMID:12485767

Soung, Y. H., Nguyen, T., Cao, H., Lee, J., & Chung, J. (2016). Emerging roles of exosomes in cancer invasion and metastasis. *BMB Reports*, *49*(1), 18–25. doi:10.5483/BMBRep.2016.49.1.239 PMID:26592936

Sridar, S., Klecka, M. A., & Xiong, W. (2022). Interfacial characteristics of P91 steel - Inconel 740H bimetallic structure fabricated using wire-arc additive manufacturing. *Journal of Materials Processing Technology*, *300*, 117396. doi:10.1016/j.jmatprotec.2021.117396

Srivastava, M., & Rathee, S. (2022). Additive manufacturing: Recent trends, applications and future outlooks. *Progress in Additive Manufacturing*, *7*(2), 261–287. doi:10.1007/s40964-021-00229-8

Stojmenovic, I., & Wen, S. (2014). "The fog computing paradigm: scenarios and security issues," in Proceedings of the Federated Conference on Computer Science and Information Systems (FedCSIS' 14), 1–8, IEEE, Warsaw, Poland, September 2014. 10.15439/2014F503

Stulov, A., Tikhonov, A., & Snitko, I. (2020, September). Fundamentals of artificial intelligence in power transformers smart design. In *2020 International Ural Conference on Electrical Power Engineering (UralCon)* (pp. 34-38). IEEE. 10.1109/UralCon49858.2020.9216245

Subashini, S., & Kavitha, V. (2011, January). A survey on security issues in service delivery models of cloud computing. *Journal of Network and Computer Applications*, *34*(1), 1–11. doi:10.1016/j.jnca.2010.07.006

Sudha, P. N., Sangeetha, K., Vijayalakshmi, K., & Barhoum, A. (2018). *Nanomaterials history, classification, unique properties, production and market.Emerging Applications of Nanoparticles and Architectural Nanostructures.Copyright.* Elsevier Inc., doi:10.1016/B978-0-323-51254-1.00012-9

Sulistyawati, S., Wibowo, T. A., Rokhmayanti, R., Nugroho, A. S. D., Sukesi, T. W., Hastuti, S. K. W., Mulasari, S. A., & Feletto, M. (2023). Introduction and implementation of an immunization information system in the Indonesian province of Daerah Istimewa Yogyakarta: Lessons for scaling-up. *BMC Health Services Research*, *23*(1), 12. doi:10.1186/s12913-022-08910-6 PMID:36604659

Sun, C., Zuo, X., Xiang, Y., & Yang, J. (2016). Investigation on Hot Deformation Behavior and Hot Processing Map of BSTMUF601 Super-Alloy. *Metals*, *6*(3), 70. doi:10.3390/met6030070

Suykens, J. A. K., Van Gestel, T., De Brabanter, J., De Moor, B., & Vandewalle, J. (2002). *Least Squares Support VectorMachines.* World Scientific. doi:10.1142/5089

Svetlizky, D., Das, M., Zheng, B., Vyatskikh, L., Bose, S., Bandyopadhyay, A., Schoenung, M., Lavernia, J., & Eliaz, N. (2021). Directed energy deposition (DED) additive manufacturing: Physical characteristics, defects, challenges and applications. *Materials Today*, *49*, 271–295. doi:10.1016/j.mattod.2021.03.020

Swaminathan, J. M. (2000). *Supply chain management.* International Encyclopedia of the Social and Behavioural Sciences, Elsevier Sciences.

Swapna, H. R., Bigirimana, E., Madaan, G., Hasan, A., Pandey, B. K., & Pandey, D. (2023). Impact of Neuromarketing on Consumer Psychology in Digitally Connected Networks. In *Applications of Neuromarketing in the Metaverse* (pp. 193–205). IGI Global. doi:10.4018/978-1-6684-8150-9.ch015

Swire-Thompson, B., & Lazer, D. (2020). Public health and online misinformation: Challenges and recommendations. *Annual Review of Public Health*, *41*(1), 433–451. doi:10.1146/annurev-publhealth-040119-094127 PMID:31874069

Tabata, Y. (1997). Photodynamic effect of polyethylene glycol–modified fullerene on tumour. *Japanese Journal of Cancer Research*, *88*, 1108. doi:10.1111/j.1349-7006.1997.tb00336.x PMID:9439687

Tabian, I., Fu, H., & Sharif Khodaei, Z. (2019). A convolutional neural network for impact detection and characterization of complex composite structures. *Sensors (Basel)*, *19*(22), 4933. doi:10.3390/s19224933 PMID:31726762

Tai-Seale, M., Downing, N. L., Jones, V. G., Milani, R. V., Zhao, B., Clay, B., Sharp, C. D., Chan, A. S., & Longhurst, C. A. (2019). Technology-enabled consumer engagement: Promising practices at four health care delivery organizations. *Health Affairs*, *38*(3), 383–390. doi:10.1377/hlthaff.2018.05027 PMID:30830826

Tajbakhsh, N., Jae, Y., & Shin, Y. (2016). Convolutional Neural Networks for Medical Image Analysis: Full Training or Fine Tuning? *IEEE Transactions on Medical Imaging*, *35*(5), 1299–1312. doi:10.1109/TMI.2016.2535302 PMID:26978662

Tamilmani, K., Rana, N. P., Wamba, S. F., & Dwivedi, R. (2021). The extended Unified Theory of Acceptance and Use of Technology (UTAUT2): A systematic literature review and theory evaluation. *International Journal of Information Management*, *57*, 102269. doi:10.1016/j.ijinfomgt.2020.102269

Tang, L., Li, J., Pan, T., Yin, Y., Mei, Y., Xiao, Q., Wang, R., Yan, Z., & Wang, W. (2022). Versatile carbon nanoplatforms for cancer treatment and diagnosis: Strategies, applications and future perspectives. *Theranostics*, *12*(5), 2290–2321. doi:10.7150/thno.69628 PMID:35265211

Tan, N. X., Rao, H. B., Li, Z. R., & Li, X. Y. (2009). Prediction of chemical carcinogenicity by machine learning approaches. *SAR and QSAR in Environmental Research*, *20*(1-2), 27–75. doi:10.1080/10629360902724085 PMID:19343583

Taylor, R., Hare, J. P., Abdul-Sada, A. K., & Kroto, H. W. (1990). Isolation, separation and characterization of the fullerenes C60 and C70: The third form of carbon. *Journal of the Chemical Society. Chemical Communications*, (20), 1423–1425. doi:10.1039/c39900001423

Tegos, G. P., Demidova, T. N., Arcila-Lopez, D., Lee, H., Wharton, T., Gali, H., & Hamblin, M. R. (2005). Cationic fullerenes are effective and selective antimicrobial photosensitizers. *Chemistry & Biology*, *12*(10), 1127–1135. doi:10.1016/j.chembiol.2005.08.014 PMID:16242655

Theiler, M., & Smarsly, K. (2018). IFC Monitor: An IFC Schema Extension for Modeling Structural Health Monitoring Systems. *Advanced Engineering Informatics*, *37*, 54–65. doi:10.1016/j.aei.2018.04.011

Thrash, T. P., Cagle, D. W., Alford, J. M., Wright, K., Ehrhardt, G. J., Mirzadeh, S., & Wilson, L. J. (1999). Toward fullerene-based radiopharmaceuticals: High-yield neutron activation of endohedral 165Ho metallofullerenes. *Chemical Physics Letters*, *308*(3-4), 329–336. doi:10.1016/S0009-2614(99)00581-3

Tian, P., Tang, L., Teng, K. S., & Lau, S. P. (2018). Graphene quantum dots from chemistry to applications. *Materials Today. Chemistry*, *10*, 221–258. doi:10.1016/j.mtchem.2018.09.007

Tobias, G., Mendoza, E., & Ballesteros, B. (2016). Functionalization of Carbon Nanotubes. In B. Bhushan (Ed.), *Encyclopedia of Nanotechnology* (pp. 1281–1291). Springer. doi:10.1007/978-94-017-9780-1_48

Tokuyama, H., Yamago, S., Nakamura, E., Shiraki, T., & Sugiura, Y. (1993). Photoinduced biochemical activity of fullerene carboxylic acid. *Journal of the American Chemical Society*, *115*(17), 7918–7919. doi:10.1021/ja00070a064

Tripathi, R. P., Sharma, M., Gupta, A. K., Pandey, D., Pandey, B. K., Shahul, A., & George, A. H. (2023). Timely prediction of diabetes by means of machine learning practices. *Augmented Human Research*, *8*(1), 1. doi:10.1007/s41133-023-00062-4

Tsao, N., Luh, T.-Y., Chou, C.-K., Chang, T.-Y., Wu, J.-J., Liu, C.-C., & Lei, H.-Y. (2002). In vitro action of carboxyfullerene. *The Journal of Antimicrobial Chemotherapy*, *49*(4), 641–649. doi:10.1093/jac/49.4.641 PMID:11909838

Tüzün, B., Çağlar Yavuz, S., & Sarıpınar, E. (2017). 4D-QSAR analysis and pharmacophore modeling: Propoxy methylphenyl oxasiazole derivatives by electron conformatitional-genetic algorithm method. *Journal of Physical & Theoretical Chemistry*, *14*(2), 149–164.

Tüzün, B., & Saripinar, E. (2020). Molecular docking and 4D-QSAR model of methanone derivatives by electron conformational-genetic algorithm method. *Journal of the Indian Chemical Society*, *17*, 985–1000.

Ullah, K., Lill, I., & Witt, E. (2019). An overview of BIM adoption in the construction industry: Benefits and barriers. *Emerald Reach Proceedings Series*, *2*(5), 297–303. doi:10.1108/S2516-285320190000002052

Umar, A. (2019). *An Assessment Of Critical Success Factors For Building Information Modelling Adoption By Clients In The Construction Industry*. [Unpublished Masters Thesis, Ahmadu Bello University, Zaria].

Utari, S. S., Madhumala, R. B., Haryanti, S. C., Rachmawati, U. A., & Atmoko, S. P. U. (2022, December). Enhancing Security on Indonesian E-health Cloud. In *Second International Conference on Advanced Technologies in Intelligent Control, Environment, Computing & Communication Engineering* (ICATIECE) (pp. 1-3). IEEE. 10.1109/ICATIECE56365.2022.10046767

van Schalkwyk, M. C., Bourek, A., Kringos, D. S., Siciliani, L., Barry, M. M., De Maeseneer, J., & McKee, M. (2020). The best person (or machine) for the job: Rethinking task shifting in healthcare. *Health Policy (Amsterdam), 124*(12), 1379–1386. doi:10.1016/j.healthpol.2020.08.008 PMID:32900551

Varelas, G., Voutsakist, E., Raftopoulout, P., Petrakis, E. G. M., & Milios, E. (2005). Semantic Similarity methods in WordNet and their application to information retrieval on the Web. *Proceedings of the 7th annual ACM international workshop on web information and data management.* ACM. 10.1145/1097047.1097051

Vashist, A., Kaushik, A., Vashist, A., Sagar, V., Ghosal, A., Gupta, Y. K., Ahmad, S., & Nair, M. (2018). Advances in carbon nanotubes-hydrogel hybrids in nanomedicine for therapeutics. *Advanced Healthcare Materials, 7*(9), 170–187. doi:10.1002/adhm.201701213 PMID:29388356

Vassilakaki, E., & Moniarou-Papaconstaninou, V. (2023). Librarians' support in improving health literacy: A systematic literature review. *Journal of Librarianship and Information Science, 55*(2), 500–514. doi:10.1177/09610006221093794

Vectorworks, (2020). *BIM Tools*. Vectorworks. https://www.vectorworks.net/bim-tools

Venkatasubramanian, V. (2019). The promise of artificial intelligence in chemical engineering: Is it here, finally? *AIChE Journal. American Institute of Chemical Engineers, 65*(2), 466–478. doi:10.1002/aic.16489

Verawati, F., Abidin, Z., Renaldi, R., Yanthi, D., & Widodo, M. D. (2022). Analysis of the Implementation of the Generic Regional Health Information System (SIKDA) at Simpang Tiga Health Center Pekanbaru City: Analisis Pelaksanaan Sistem Informasi Kesehatan Daerah (SIKDA) Generik Di Puskesmas Simpang Tiga Kota Pekanbaru. *Jurnal Olahraga dan Kesehatan* (ORKES), *1*(2), 229-240. doi:10.56466/orkes/Vol1.Iss2.7

Verma, J., Khedkar, V. M., & Coutinho, E. C. (2010). 3D-QSAR in drug design-a review. *Current Topics in Medicinal Chemistry, 10*(1), 95–115. doi:10.2174/156802610790232260 PMID:19929826

Vermesan, O., Friess, P., & Guillemin, P. (2011). "Internet of things strategic research roadmap," in Internet of Things. *Global Technological and Societal Trends, 1*, 9–52.

Vijayakumar, N., & Ramya, R. (2015). The real timemonitoring of water quality in IoT environment. *International Conference on Innovations in Information*. Research Gate.

Vijayalakshmi, S. R., & Muruganand, S. (2017, February). A survey of Internet of Things in fire detection and fire industries. In *2017 International Conference on I-SMAC (IoT in Social, Mobile, Analytics and Cloud)(I-SMAC)* (pp. 703-707). IEEE. 10.1109/I-SMAC.2017.8058270

Vinodhini, V., Kumar, M. S., Sankar, S., Pandey, D., Pandey, B. K., & Nassa, V. K. (2022). IoT-based early forest fire detection using MLP and AROC method. *International Journal of Global Warming, 27*(1), 55–70. doi:10.1504/IJGW.2022.122794

Vora, J., Parmar, H., Chaudhari, R., Khanna, S., Doshi, M., & Patel, V. (2022). Experimental investigations on mechanical properties of multi-layered structure fabricated by GMAW-based WAAM of SS316L. *Journal of Materials Research and Technology, 20*, 2748–2757. doi:10.1016/j.jmrt.2022.08.074

Vouk, M. A. (2008). Cloud computing - Issues, research and implementations. *Proc. Int. Conf. Inf. Technol. Interfaces*. ITI. 10.2498/cit.1001391

Wang, M., Liu, H., Chen, F., & Liu, J. (2015, August). Forest fire warning system based on GIS and WSNs. In *2015 4th International Conference on Advanced Information Technology and Sensor Application (AITS)* (pp. 3-6). IEEE. 10.1109/AITS.2015.8

Wang, B., Chen, B., & Zhao, J. (2015). The real-time estimation of hazardous gas dispersion by the integration of gas detectors, neural network and gas dispersion models. *Journal of Hazardous Materials*, *300*, 433–442. doi:10.1016/j.jhazmat.2015.07.028 PMID:26223017

Wang, B., Fu, W., Lv, Z., Jiang, P., Zhang, W., & Tian, Y. (2008). Study on hot deformation behavior of 12%Cr ultra-super-critical rotor steel. *Materials Science and Engineering A*, *487*(1–2), 108–113. doi:10.1016/j.msea.2007.10.007

Wang, B., Qi, X., Liu, J., Zhou, R., Lin, C., Shangguan, J., Zhang, Z., Zhao, L., & Li, G. (2019). MYH9 Promotes Growth and Metastasis via Activation of MAPK/AKT Signaling in Colorectal Cancer. *Journal of Cancer*, *10*(4), 874–884. doi:10.7150/jca.27635 PMID:30854093

Wang, B., Xu, K., & Wang, Q. (2019). Prediction of upper flammability limits for fuel mixtures using quantitative structure–property relationship models. *Chemical Engineering Communications*, *206*(2), 247–253. doi:10.1080/00986445.2018.1483350

Wang, B., Yi, H., Xu, K., & Wang, Q. (2017). Prediction of the self-accelerating decomposition temperature of organic peroxides using QSPR models. *Journal of Thermal Analysis and Calorimetry*, *128*(1), 399–406. doi:10.1007/s10973-016-5922-8

Wang, J., Xie, L., Wang, T., Wu, F., Meng, J., Liu, J., & Xu, H. (2017). Visible light-switched cytosol release of siRNA by amphiphilic fullerene derivative to enhance RNAi efficacy in vitro and in vivo. *Acta Biomaterialia*, *59*, 158–169. doi:10.1016/j.actbio.2017.05.031 PMID:28511875

Wang, L., Liu, F., Zuo, Q., & Chen, F. (2013). Prediction of flow stress for N08028 alloy under hot working conditions. *Materials & Design*, *47*, 737–745. doi:10.1016/j.matdes.2012.12.074

Wang, M., Jin, P., & Wang, J. (2014). Hot deformation and processing maps of 7005 aluminum alloy. *High-Temperature Materials and Processes*, *33*(4), 369–375. doi:10.1515/htmp-2013-0083

Wang, R., Chen, B., Qiu, S., Zhu, Z., Wang, Y., Wang, Y., & Qiu, X. (2018). Comparison of machine learning models for hazardous gas dispersion prediction in field cases. *International Journal of Environmental Research and Public Health*, *15*(7), 1450. doi:10.3390/ijerph15071450 PMID:29996467

Wang, S., Huang, Y., Xiao, Z., Liu, Y., & Liu, H. (2017). A Modified Johnson-Cook Model for Hot Deformation Behavior of 35CrMo Steel. *Metals*, *7*(9), 337. doi:10.3390/met7090337

Wang, T., Upponi, J. R., & Torchilin, V. P. (2012). Design of multifunctional non-viral gene vectors to overcome physiological barriers: Dilemmas and strategies. *International Journal of Pharmaceutics*, *427*(1), 3–20. doi:10.1016/j.ijpharm.2011.07.013 PMID:21798324

Wang, T., & Wang, C. (2019). Functional Metallofullerene Materials and Their Applications in Nanomedicine, Magnetics, and Electronics. *Small*, *15*(48), e1901522. doi:10.1002/smll.201901522 PMID:31131986

Wang, Y., Meng, Y., Wang, S., Li, C., Shi, W., Chen, J., Wang, J., & Huang, R. (2015). Direct solvent-derived polymer-coated nitrogen-doped carbon nanodots with high water solubility for targeted fluorescence imaging of glioma. *Small*, *11*(29), 3575–3581. doi:10.1002/smll.201403718 PMID:25808813

Wang, Z., Fu, W., Wang, B., Zhang, W., Lv, Z., & Jiang, P. (2010). Study on hot deformation characteristics of 12%Cr ultra-super-critical rotor steel using processing maps and Zener-Hollomon parameter. *Materials Characterization*, *61*(1), 25–30. doi:10.1016/j.matchar.2009.10.001

Wattimena, E. S. N., Razak, A., Pallutturi, S., Syafar, D. M., Saleh, L. M., & Rahmadani, S. (2020). Analysis Health Information Systems Quality on the Effectiveness of Public Health Center Officers in Ambon City. *Medico-Legal Update*, *20*(3).

Wei, J. N., Duvenaud, D., & Aspuru-Guzik, A. (2016). Neural networks for the prediction of organic chemistry reactions. *ACS Central Science*, *2*(10), 725–732. doi:10.1021/acscentsci.6b00219 PMID:27800555

Wei, L.-F. (2001). Shih, J.-S. *Analytica Chimica Acta*, *437*(1), 77–85. doi:10.1016/S0003-2670(01)00941-2

Wei, P., Zhang, L., Lu, Y., Man, N., & Wen, L. (2010). C60(Nd) nanoparticles enhance chemotherapeutic susceptibility of cancer cells by modulation of autophagy. *Nanotechnology*, *21*(49), 495101. doi:10.1088/0957-4484/21/49/495101 PMID:21071824

Weyrich, M., & Ebert, C. (2016). Reference architectures for the internet of things. *IEEE Software*, *33*(1), 112–116. doi:10.1109/MS.2016.20

Wikipedia. (2020). *Archi CAD*. Wikipedia. https://en.m.wikipedia.autodesk.com/support/revit-2018

Winkler, V. J. (2011). *"Securing the Cloud," Cloud Comput. Secur. Tech. tactics*. Elsevier.

Withorn, T., Eslami, J., Lee, H., Clarke, M., Caffrey, C., Springfield, C., Ospina, D., Andora, A., Castañeda, A., Mitchell, A., Kimmitt, J. M., Vermeer, W., & Haas, A. (2021). Library instruction and information literacy 2020. *RSR. Reference Services Review*, *49*(3/4), 329–418. doi:10.1108/RSR-07-2021-0046

Witte, P., Beuerle, F., Hartnagel, U., Lebovitz, R., Savouchkina, A., Sali, S., Guldi, D., Chronakis, N., & Hirsch, A. (2007). Water solubility, antioxidant activity and cytochrome C binding of four families of exohedral adducts of C60 and C70. *Organic & Biomolecular Chemistry*, *5*(22), 3599–3613. doi:10.1039/b711912g PMID:17971989

Wooley, P. S. (2011). Identifying Cloud Computing Security Risks. *Continuity in Education*, *1277*(February).

Wu, M., Lu, T. J., Ling, F. Y., Sun, J., & Du, H. Y. (2010). Research on the architecture of internet of things. *Proceedings of the 3rd International Conference on Advanced Computer Theory and Engineering, 10*(5). IEEE.

Wu, B., Qian, C., Ni, W., & Fan, S. (2012). The improvement of glowworm swarm optimization for continuous optimization problems. *Expert Systems with Applications*, *39*(7), 6335–6342.

Wu, G., Kim, M., Wang, Q., Munsell, B. C., & Shen, D. (2016). Scalable High Performance Image Registration Framework by Unsupervised Deep Feature Representations Learning. *IEEE Transactions on Biomedical Engineering*, *63*(7), 1505–1516. doi:10.1109/TBME.2015.2496253 PMID:26552069

Wu, H., Lorenson, A., Anderson, B., Witteman, L., Wu, H., Meredig, B., & Morgan, D. (2017). Robust FCC solute diffusion predictions from ab-initio machine learning methods. *Computational Materials Science*, *134*, 160–165. doi:10.1016/j.commatsci.2017.03.052

Wu, S., Zheng, Q., Xing, X., Dong, Y., Wang, Y., You, Y., Chen, R., Hu, C., Chen, J., Gao, D., Zhao, Y., Wang, Z., Xue, T., Ren, Z., & Cui, J. (2018). Matrix stiffness-upregulated LOXL2 promotes fibronectin production, MMP9 and CXCL12 expression and BMDCs recruitment to assist pre-metastatic niche formation. *Journal of Experimental & Clinical Cancer Research*, *37*(1), 99. doi:10.1186/s13046-018-0761-z PMID:29728125

Wu, T., Zhang, X., Yuan, Q., Xue, J., Lu, G., Liu, Z., Wang, H., Wang, H., Ding, F., Yu, Q., Xie, X., & Jiang, M. (2016). Fast Growth of Inch-Sized Single-Crystalline Graphene from a Controlled Single Nucleus on Cu–Ni Alloys. *Nature Materials*, *15*(1), 43–47. doi:10.1038/nmat4477 PMID:26595118

Wu, Y., & Wang, G. (2018). Machine learning based toxicity prediction: From chemical structural description to transcriptome analysis. *International Journal of Molecular Sciences*, *19*(8), 2358. doi:10.3390/ijms19082358 PMID:30103448

Xiaojun, C., Xianpeng, L., &Peng, X. (2015, January). IOT-based air pollution monitoring and forecasting system. In *2015 international conference on computer and computational sciences (ICCCS)* (pp. 257-260).IEEE.

Xin, H., Tarus, I., Cheng, L., Veljkovic, M., Persem, N., & Lorich, L. (2021). Experiments and numerical simulation of wire and arc additive manufactured steel materials. *Structures*, *34*, 1393–1402. doi:10.1016/j.istruc.2021.08.055

Xu, X., Zhang, Z., Dong, J., Yi, D., Niu, J., Wu, M., Lin, L., Yin, R., Li, M., Zhou, J., Wang, S., Sun, J., Duan, X., Gao, P., Jiang, Y., Wu, X., Peng, H., Ruoff, R. S., Liu, Z., & Liu, K. (2017). Ultrafast Epitaxial Growth of Metre-Sized Single-Crystal Graphene on Industrial Cu Foil. *Science Bulletin*, *62*(15), 1074–1080. doi:10.1016/j.scib.2017.07.005 PMID:36659334

Xu, Z. P., Zeng, Q. H., Lu, G. Q., & Yu, A. B. (2006). Inorganic nanoparticles as carriers for efficient cellular delivery. *Chemical Engineering Science*, *61*(3), 1027–1040. doi:10.1016/j.ces.2005.06.019

Yadav, B. C., & Kumar, R. (2008). Structure, properties and applications of fullerenes. *International Journal of Nanotechnology and Applications*, *2*(1), 15–24.

Yahya, F., Chang, V., Walters, J., & Wills, B. (2014). *Security Challenges in Cloud Storage*. doi:10.1109/CloudCom.2014.171

Yamakoshi, Y., Umezawa, N., Ryu, A., Arakane, K., Miyata, N., Goda, Y., Masumizu, T., & Nagano, T. (2003). Active oxygen species generated from photoexcited fullerene (C 60) as potential medicines: O 2-• versus 1 O 2. *Journal of the American Chemical Society*, *125*(42), 12803–12809. doi:10.1021/ja0355574 PMID:14558828

Yang, H., Li, Z.H. and Zhang, Z.L. (2007) 'Investigation on Zener-Hollomon parameter in the warm-hot deformation behavior of CF53', *Journal of Shanghai Jiaotong University (Science)*, *12 E*(3), 352–358.. doi:10.1631/jzus.2006.A1453

Yang, B., Chen, Y., & Shi, J. (2019). Reactive oxygen species (ROS) based Nanomedicine. *Chemical Reviews*, *119*(8), 4881–4985. doi:10.1021/acs.chemrev.8b00626 PMID:30973011

Yang, F., Wang, X., Si, J., Zhao, X., Qi, K., Jin, C., Zhang, Z., Li, M., Zhang, D., Yang, J., Zhang, Z., Xu, Z., Peng, L.-M., Bai, X., & Li, Y. (2017). Water-Assisted Preparation of High-Purity Semiconducting (14,4) Carbon Nanotubes. *ACS Nano*, *11*(1), 186–193. doi:10.1021/acsnano.6b06890 PMID:28114760

Yang, F., Wang, X., Zhang, D., Yang, J., Luo, D., Xu, Z., Wei, J., Wang, J.-Q., Xu, Z., Peng, F., Li, X., Li, R., Li, Y., Li, M., Bai, X., Ding, F., & Li, Y. (2014). Chirality-Specific Growth of Single-Walled Carbon Nanotubes on Solid Alloy Catalysts. *Nature*, *510*(7506), 522–524. doi:10.1038/nature13434 PMID:24965654

Yang, N., Chen, X., Ren, T., Zhang, P., & Yang, D. (2015). Carbon nanotube based biosensors.Sensors and Actuators B: Chemical Volume 207. *Sensors and Actuators. B, Chemical*, *207*(February), 690–715. doi:10.1016/j.snb.2014.10.040

Yang, N., Dai, Q., Chu, H., Zhang, C., Wang, W., & Xiao, P. (2017). Effect of Aluminum Alloying on the Hot Deformation Behavior of Nano-bainite Bearing Steel. *Journal of Materials Engineering and Performance*, *26*(12), 5954–5962. doi:10.1007/s11665-017-3018-7

Yang, X., Qin, L., Wang, Y., Zhang, B., Liu, Z., Ma, H., Lu, J., Huang, X., Shi, D., & Hu, Z. (2014). Detection of Mycobacterium tuberculosis based on H37Rv binding peptides using surface functionalized magnetic microspheres coupled with quantum dots – a nano detection method for Mycobacterium tuberculosis. *International Journal of Nanomedicine*, *9*, 77. doi:10.2147/IJN.S71700 PMID:24379667

Yan, P., Liu, Z., Liu, W., Bao, H., & Weng, Y. (2013). Hot deformation behavior of a new 9%Cr heat resistant steel G115. *Journal of Iron and Steel Research International*, *20*(9), 73–79. doi:10.1016/S1006-706X(13)60159-4

Ye, Z., & Zhou, Q. (2021). Performance Evaluation Indicators of Space Dynamic Networks under Broadcast Mechanism. Space: Science & Technology. doi:10.34133/2021/9826517

Ye, L., Kollie, L., Liu, X., Guo, W., Ying, X., Zhu, J., Yang, S., & Yu, M. (2021). Antitumor Activity and Potential Mechanism of Novel Fullerene Derivative Nanoparticles. *Molecules (Basel, Switzerland)*, *26*(11), 3252. doi:10.3390/molecules26113252 PMID:34071369

Yin, F., Hua, L., Mao, H., & Han, X. (2013). Constitutive modeling for flow behavior of GCr15 steel under hot compression experiments. *Materials & Design*, *43*, 393–401. doi:10.1016/j.matdes.2012.07.009

Yin, L. J., Bai, K. K., Wang, W. X., Li, S. Y., Zhang, Y., & He, L. (2017). Landau quantization of dirac fermions in graphene and its multilayers. *Frontiers of Physics*, *12*(4), 127208. doi:10.1007/s11467-017-0655-0

Yin, R., Wang, M., Huang, Y.-Y., Landi, G., Vecchio, D., Chiang, L. Y., & Hamblin, M. R. (2015). Antimicrobial photodynamic inactivation with decacationic functionalized fullerenes: Oxygen-independent photokilling in presence of azide and new mechanistic insights. *Free Radical Biology & Medicine*, *79*, 14–27. doi:10.1016/j.freeradbiomed.2014.10.514 PMID:25451642

Yoo, S. W., Lee, C. M., & Kim, D. H. (2023). Effect of Functionally Graded Material (FGM) Interlayer in Metal Additive Manufacturing of Inconel-Stainless Bimetallic Structure by Laser Melting Deposition (LMD) and Wire Arc Additive Manufacturing (WAAM). *Materials (Basel)*, *16*(2), 535. Advance online publication. doi:10.3390/ma16020535 PMID:36676271

Yoshioka, N., Honiden, S., & Fnkelstein, A. (2004). Security Patterns: A Method for Constructing Secure and Efficient Inter-Company Coordination Systems. *Proceedings of Enterprise Distributed Object Computing Conference 2004 (EDOC'04)*. IEEE.. 10.1109/EDOC.2004.1342507

Youle, R. J. (2005). Karbowski, M. *Nature Reviews. Molecular Cell Biology*, *6*(8), 657–663. doi:10.1038/nrm1697 PMID:16025099

Yovadiani, A., Ramadani, L., & Dewi, F. (2021). Toward National One Health Data: Reconfiguring Local Health Information Systems through Enterprise Architecture. Ultima InfoSys. *Jurnal Ilmu Sistem Informasi*, *12*(2), 101–107. doi:10.31937/si.v12i2.2274

Yuli, Z. H. A. N. G., Xiaoping, M. A., & Yanzi, M. I. A. O. (2011). Localization of Multiple Odor Sources Using Modified Glowworm Swarm Optimization with Collective Robots. *Proceedings of the 30th Chinese Control*, (pp. 1899-1904). IEEE.

Yu, M. F., Files, B. S., Arepalli, S., & Ruoff, R. S. (2000). Tensile loading of ropes of single wall carbon nanotubes and their mechanical properties. *Physical Review Letters*, *84*(24), 5552–5555. doi:10.1103/PhysRevLett.84.5552 PMID:10990992

Yumna, Q. (2023). *Mostapha Tarfaoui, Khalid Lafdi. 13 - Background, advancement, and applications of in situ structural health monitoring based on different modes of failure detection in composites: a review*. Non-Destructive Testing and Condition Monitoring Techniques in Wind Energy Wind Energy Engineering.

Yusuf, A. O. (2020a). Building Information Modeling (BIM): A Necessity for the 21st Century Civil Engineer in Nigeria. *Being a paper presented* at the 2nd International Civil Engineering Conference (ICEC2020). Department of Civil Engineering, Federal University of Technology, Minna, Nigeria.

Yusuf, M. (2020). Assessment of Critical Success Factors for Construction Projects in Pakistan: Advancing and Integrating Construction Education, Research & Practice. *First International Conference on Construction in Developing Countries held on 4-5 August at Karachi, Pakistan*. IEEE.

Yusuf, S. M., Cutler, S., & Gao, N. (2019). Review: The Impact of Metal Additive Manufacturing on the Aerospace Industry. *Metals*, *9*(12), 1286. doi:10.3390/met9121286

Zaabab, A. H., Zhang, Q. J., & Nakhla, M. (1995). A neural network modeling approach to circuit optimization and statistical design. *IEEE Transactions on Microwave Theory and Techniques*, *43*(6), 1349–1358. doi:10.1109/22.390193

Zaidi, A. (2023). Predicting wildfires in Algerian forests using machine learning models. *Heliyon*, *9*(7), e18064. doi:10.1016/j.heliyon.2023.e18064 PMID:37519679

Zaidi, A. (2023). Smart Implementation of Industrial Internet of Things using Embedded Mechatronic System. *IEEE Embedded Systems Letters*. doi:10.1109/LES.2023.3309985

Zaidi, A., Ajibade, S.-S. M., Musa, M., & Bekun, F. V. (2023). New Insights into the Research Landscape on the Application of Artificial Intelligence in Sustainable Smart Cities: A Bibliometric Mapping and Network Analysis Approach. *International Journal of Energy Economics and Policy.*, *13*(4), 287–299. doi:10.32479/ijeep.14683

Zaidi, A., & Al Luhayb, A. (2023). Two Statistical Approaches to Justify the Use of the Logistic Function in Binary Logistic Regression. *Mathematical Problems in Engineering*, *2023*, 5525675. doi:10.1155/2023/5525675

Zaidi, A., & Al-Harbi, O. (2023). Statistical Analysis of a Linear Multi-Step Numerical Treatment. *Journal of Statistics Applications & Probability*, *12*(1), 26. doi:10.18576/jsap/120126

Zakaria, M. R., Akil, H. M., Muhammad, H. A. K., Ullah, F., Javed, F., & Nosbi, N. (2019, November). Hybrid carbon fiber-carbon nanotubes reinforced polymer composites: A review. *Composites. Part B, Engineering*, *176*(1), 107313. doi:10.1016/j.compositesb.2019.107313

Zaytseva, O., & Neumann, G. (2016). Carbon nanomaterials: Production, impact on plant development, agricultural and environmental applications. *Chemical and Biological Technologies in Agriculture*, *3*(1), 17. doi:10.1186/s40538-016-0070-8

Zhang, D., Wen, L., Huang, R., Wang, H., Hu, X., & Xing, D. (2018). Mitochondrial specific photodynamic therapy by rare-earth nanoparticles mediated near-infrared graphene quantum dots. *Biomaterials*, *153*, 14–26. doi:10.1016/j.biomaterials.2017.10.034 PMID:29096398

Zhang, F., Peng, W., Yang, Y., Dai, W., & Song, J. (2019). A Novel Method for Identifying Essential Genes by Fusing Dynamic Protein–Protein Interactive Networks. *Genes*, *10*(1), 31. doi:10.3390/genes10010031 PMID:30626157

Zhang, H., Hong, M., Chen, P., Xie, A., & Shen, Y. (2016, April). 3D and ternary rGO/MCNTs/$Fe_3O_4$ composite hydrogels: Synthesis, characterization and their electromagnetic wave absorption properties. *Journal of Alloys and Compounds*, *665*(25), 381–387. doi:10.1016/j.jallcom.2016.01.091

Zhang, H., Hou, L., Jiao, X., Ji, Y., Zhu, X., & Zhang, Z. (2015). Transferrin-mediated fullerenes nanoparticles as Fe(2+)-dependent drug vehicles for synergistic anti-tumor efficacy. *Biomaterials*, *37*, 353–366. doi:10.1016/j.biomaterials.2014.10.031 PMID:25453964

Zhang, M., Li, J., Xing, G., He, R., Li, W., Song, Y., & Guo, H. (2011). Variation in the internalization of differently sized nanoparticles induces different DNA-damaging effects on a macrophage cell line. *Archives of Toxicology, 85*(12), 1575–1588. doi:10.1007/s00204-011-0725-y PMID:21881955

Zhang, P., Yi, C., Chen, G., Qin, H., & Wang, C. (2016). Constitutive Model Based on Dynamic Recrystallization Behavior during Thermal Deformation of a Nickel-Based Superalloy. *Metals, 6*(7), 161. doi:10.3390/met6070161

Zhang, X. (2022). *Application of Artificial Intelligence Recognition Technology in Digital Image Processing*. Wireless Communications and Mobile Computing- Hindawi., doi:10.1155/2022/7442639

Zhao, J., & Kumar, V. V. (2021). *Handbook of Research on Innovations and Applications of AI, IoT, and Cognitive Technologies*. IGI Global Publication. doi:10.4018/978-1-7998-6870-5

Zhao, S., Blaabjerg, F., & Wang, H. (2020). An overview of artificial intelligence applications for power electronics. *IEEE Transactions on Power Electronics, 36*(4), 4633–4658. doi:10.1109/TPEL.2020.3024914

Zhao, X. (2017). A scientometric review of global BIM research: Analysis and visualization. *Automation in Construction, 80*, 37–47. doi:10.1016/j.autcon.2017.04.002

Zhao, Y.-L., & Fraser Stoddart, J. (2009). Noncovalent Functionalization of Single-Walled Carbon Nanotubes. *Accounts of Chemical Research, 42*(8), 1161–1171. doi:10.1021/ar900056z PMID:19462997

Zhiwei, Y., Zhengbing, H., Xin, Z., & Bin, X. (2010). Image Segmentation Based on 2-D Fisher and Chaos Particle Swarm Optimization Algorithm. *2nd International Symposium on Information Engineering and Electronic Commerce (IEEC)*. IEEE.

Zhong, X. (2012). Yuan, X. Chai, Y. *Chemical Communications (Cambridge), 48*(4), 597–599. doi:10.1039/C1CC16081H

Zhou, L., Wang, B., Jiang, J., Pan, Y., & Wang, Q. (2017). Quantitative structure-property relationship (QSPR) study for predicting gas-liquid critical temperatures of organic compounds. *Thermochimica Acta, 655*, 112–116. doi:10.1016/j.tca.2017.06.021

Zhou, W., Huo, J., Yang, Y., Zhang, X., Li, S., Zhao, C., Ma, H., Liu, Y., Liu, J., Li, J., Zhen, M. M., Li, J., Fang, X., & Wang, C. (2020). Aminated Fullerene Abrogates Cancer Cell Migration by Directly Targeting Myosin Heavy Chain 9. *ACS Applied Materials & Interfaces, 12*(51), 56862–56873. doi:10.1021/acsami.0c18785 PMID:33305958

Zhu, B., Hou, X., Liu, S., Ma, W., Dong, M., Wen, H., Wei, Q., Du, S., & Zhang, Y. (2021). Iot equipment monitoring system based on c5. 0 decision tree and time-series analysis. *IEEE Access : Practical Innovations, Open Solutions, 10*, 36637–36648. doi:10.1109/ACCESS.2021.3054044

Zhu, J., Ji, Z., Wang, J., Sun, R., Zhang, X., Gao, Y., Sun, H., Liu, Y., Wang, Z., Li, A., Ma, J., Wang, T., Jia, G., & Gu, Y. (2008). Tumor-Inhibitory Effect and Immunomodulatory Activity of Fullerol $C_{60}(OH)_x$. *Small, 4*(8), 1168–1175. doi:10.1002/smll.200701219 PMID:18574800

Zhu, L., He, J., & Zhang, Y. (2018). A two-stage constitutive model of X12CrMoWVNbN10-1-1 steel during elevated temperature. *Materials Research Express, 5*(2), 1–11. doi:10.1088/2053-1591/aaa911

Zigurat, U. M. (2018). *BIM Awareness and Level of ICT Penetration in the Nigerian Construction Industry.* [An Unpublished M.Sc. Thesis, School of Construction Management and Engineering, the University of Reading, United Kingdom].

# About the Contributors

**Khaoula Alaoui** received his B.Sc. in Chemistry (2013) and she was awarded his M.Sc. degree in Organic chemistry and Environnemental Sciences (2015) from Ibn Tofail University. Then she started to PhD thesis. She had published more than 20 papers in international journals. His work has been cited 300 total citations by 204 documents with h-index=10 (since 2016 in SCOPUS) and presented several communications in symposia and national/international meetings. She has supervised more than 30 Master theses in corrosion sciences. She is a reviewer at several national and international journals. She is the author of a 4 book schapters. Her main research interests are in the field of corrosion sciences and molecular modeling simulation using new tools and approaches such as Artificial intelligence, Machine and Deep Learning.

**Younes El Kacimi** is currently working as Quality Director in Automotive Industry. He has more than 15 years of industry core experience in Quality and Project Management and academic levels. Younes has built an international profile over the years through research, teaching, and professional experience. He holds expertise in corrosion and corrosion prevention by studying and developing advanced engineering tools and technology. Younes has authored 1 Patent, 8 Books and Books chapters and many other publications in corrosion sciences, nanomaterials and analytical chemistry. He is also an editorial board member and reviewer in different well-renowned international journals.

***

**Pankaj Dadheech** is currently working as a Professor & Deputy Head in the Department of Computer Science & Engineering (NBA Accredited), Swami Keshvanand Institute of Technology, Management & Gramothan (SKIT), Jaipur, Rajasthan, India (Accredited by NAAC A++ Grade). He has more than 18 years of experience in teaching. He is currently working a Professor & Dy. HOD in the Department of Computer Science & Engineering (NBA Accredited), Swami Keshvanand Institute of Technology, Management & Gramothan (SKIT), Jaipur, Rajasthan, India. He has published 25 Patents at Intellectual Property India, Office of the Controller General of Patents, Design and Trade Marks, Department of Industrial Policy and Promotion, Ministry of Commerce and Industry, Government of India. He has published 8 International Patents (USA, South African, Australian, Germany) & 2 Copyrights. He has 73 publications in various International & National Journals, 63 papers in various National & International conferences. He has published 9 Books & 38 Book Chapters.

**A. Shaji George** is a highly respected and accomplished figure in the fields of education and the ICT industry. With over 30 years of experience in designing and deploying large-scale projects, Dr. George has made significant contributions to industry. He has a passion for working with clients to help them overcome complex business challenges, drive innovation, and deliver results. Dr. George is recognized as an expert in the field of ICT and has published over 250 research papers in national and international journals and conferences. He is also the author of 30 books that have been published by reputed international publishers. His extensive contributions to the field have earned him 20 awards, recognizing his outstanding achievements and contributions to the industry. Dr. George is the founder of PU Publications, PUIRJ, PUIIJ and PUIRP International Journal. He has a keen interest in wireless networking, cloud computing, big data analytics, the Internet of Things, and industrial automation systems. He works with clients from a wide range of industries to develop strategies and implement next-generation solutions that transform their businesses. Dr. George's wealth of knowledg

**George Williams Kennedy** is a scholar that has published several papers in international journal and book Chapters. He is married to Barr. Esther.

**Rita Komalasari** is a lecturer at YARSI University. Her current work is focused on Empowering Informed Healthcare Choices in Rural Areas: SIKDA: A Novel Model for Enhancing Patient Access to Reliable Consumer Health Information in Indonesia.

**Loutfy H. Madkour** is a professor of physical chemistry and nanoscience at Tanta University (Egypt). He earned his BSc, MSc, and PhD in physical chemistry from the Cairo, Minia, and Tanta universities in Egypt, respectively. He has published 200 peer-reviewed original research articles, 20 review articles, and 8 books. He is an editorial board member of several international journals, including International Journal of Industrial Chemistry (IJIC—Springer); International

**Darshan Mahajan** is working as an associate professor and having 18 years of experience in education.

**Nicholus Malatji** is a material and metallurgy researcher and academician.

**Arijit Mondal** is pursuing BTech in Computer Science & Engineering (AIML) at Brainware University, West Bengal, India.

**Balakumar Muniandi** is an Associate Professor of Practice, Electrical and Computer science Engineering department

**Ramaingam Nagarajan** received his B.E. in Electrical and Electronics Engineering from Madurai Kamarajar University, Madurai, India, in 1997. He received his M.E. in Power Electronics and Drives from Anna University, Chennai, India, in 2008. He received his Ph.D in Electrical Engineering from Anna University, Chennai, India, in 2014. He has worked in the industry as an Electrical Engineer. He is currently working as Professor of Electrical and Electronics Engineering at Gnanamani College of Technology, Namakkal, Tamilnadu, India. His current research interest includes Power Electronics, Power System, Soft Computing Techniques, Cloud Computing, Network Security, Wireless Communication

System and Renewable Energy Sources. He is published more than 115 research papers and 50 Books in various referred international Publications.

**Japheth Obiko** is a material and metallurgy researcher and academician

**P. Selvakumar**, Department of Science and Humanitites at Nehru Institute of Technology, Coimbatore, Tamilnadu, India-641105, where he has been since 2023. His area of interest is the field of phytochemistry. His Credit includes 30 Journal publications, 12 paptents and 20 conference papers, conducted 5 seminars and conferences and honoured with best teacher Awards.

**Kamalendu Pal** is currently associated with the Department of Computer Science, School of Science and Technology at City, University of London. He has a diverse academic background, having received a BSc (Hons) degree in Physics from Calcutta University, India, followed by a Postgraduate Diploma in Computer Science from Pune, India. He has also worked as a research assistant for the government of India's two research and development organizations, under the Council of Scientific and Industrial Research (CSIR). Later, Kamalendu graduated with his MSc in Software Systems Technology from the University of Sheffield, a Postgraduate Diploma in Artificial Intelligence from Kingston University, an MPhil in Computer Science from University College London, an MBA from the University of Hull, and a diploma in Higher Education teaching practice from City, University of London, United Kingdom. Kamalendu is an accomplished researcher, having published over a hundred international research articles (including book chapters) in the scientific community. His research papers have been published in prestigious journals such as the ACM SIGMIS Database, Expert Systems with Applications, Decision Support Systems, and international conferences. His research interests include various fields of computer science - such as knowledge-based systems, decision support systems, teaching and learning practices, blockchain technology, software engineering, service-oriented computing, sensor network simulation, and supply chain management. Kamalendu is also on the editorial board of an international computer science journal. He is a member of the British Computer Society, the Institution of Engineering and Technology, and the IEEE Computer Society.

**Binay Kumar Pandey** currently working as an Assistant Professor in Department of Information Technology of Govind Ballabh Pant University of Agriculture and Technology Pantnagar Uttrakhand, India. He obtained his M. Tech with Specialization in Bioinformatics from Maulana Azad National Institute of Technology Bhopal M. P. India, in 2008 . He obtained his First Degree B. Tech at the IET Lucknow (Uttar Pradesh Technical University, Uttar Pradesh and Lucknow) India, in 2005. In 2010, he joined Department of Information Technology of College of Technology in Govind Ballabh Pant University of Agriculture and Technology Pantnagar as an Assistant Professor and worked for various UG and PG projects till date. He has more than ten years of experience in the field of teaching and research. He has more than 40 publications in reputed peer journal reputed journal Springer,Inderscience,(sci and socopus indexed journal and others) and 3 patent.He has many awards such PM Scholarship etc . He session chair in IEEE International Conference on Advent Trends in Multidisciplinary Research and Innovation (ICATMRI-2020) on December 30, 2020 organized by Pankaj Laddhad Institute of Technology and Management Studies; Buldhana, Maharashtra, India

**Amitava Podder** is working as an Assistant Professor at Computer Science & Engineering Department at Brainware University, West Bengal, India.

**Piyal Roy** is working as an Assistant Professor at Computer Science & Engineering Department at Brainware University, West Bengal, India.

**Brendon Shongwe** is a material and metallurgy researcher and academician.

**Kannadhasan Suriyan** is working as an Associate Professor in the department of Electronics and Communication Engineering in Study World College of Engineering, Tamilnadu, India. He is Completed the Ph.D in the field of Smart Antenna for Anna University in 2022. He is Twelve years of teaching and research experience. He obtained his B.E in ECE from Sethu Institute of Technology, Kariapatti in 2009 and M.E in Communication Systems from Velammal College of Engineering and Technology, Madurai in 2013.

**Blessy Thankachan** is currently working as Associate Professor and Dy. HoD in School of Computer Applications, JECRC University, Jaipur. She has more than a decade of teaching experience. She also have an illustrious experience of research and administrative work too. She has one patent to her credit, and multiple research publications in national and international journals. Dr. Blessy has endued her time in authoring books for University students in Computer Science. She adheres to equal participation in presenting papers in Conferences too. Currently she is guiding around half a dozen research scholars on the topics Software Engineering and Reliability, Big Data, Internet of Things and Machine Learning.

**Burak Tuzun**, PhD, at Plant and Animal Production Department, Technical Sciences Vocational School of Sivas, in Sivas Cumhuriyet University, Sivas, Turkey. His research area includes theoretical calculations, pharmaceutical chemistry, drug design, metal corrosion, spectroscopic analysis of molecules. He has published around 153 articles in Scopus and Web of Science journals. H index is 30 and i-10 index is 80 in Google Scholar. He has published with ELSEVIER, the American Chemical Society, the Royal Society of Chemistry, John Wiley & Sons, CRC Press, and Springer.

# Index

## S



## T



## W

www.ingramcontent.com/pod-product-compliance
Lightning Source LLC
LaVergne TN
LVHW062110300926
849161LV00010B/2418